28th International Geological Congress

Geology of Grand Canyon, Northern Arizona
(with Colorado River Guides)

American Geophysical Union

TITLES IN THE IGC SERIES

Coal and Hydrocarbon Resources of North America *2 volumes*
Coastal and Marine Geology of the United States *1 volume*
Environmental, Engineering, and Urban Geology in the United States
   2 volumes
Geology of Grand Canyon, Northern Arizona (with Colorado River Guides)
   1 volume
Glacial Geology and Geomorphology of North America *2 volumes*
Mesozoic/Cenozoic Vertebrate Paleontology: Classic Localities, Contemporary
Approaches *1 volume*
Metamorphism and Tectonics of Eastern and Central North America
   3 volumes
Mineral Deposits of North America *2 volumes*
Sedimentation and Basin Analysis in Siliciclastic Rock Sequences *3 volumes*
Sedimentation and Stratigraphy of Carbonate Rock Sequences *2 volumes*
Sedimentation and Tectonics of Western North America *3 volumes*
Tectonics of the Scotia Arc, Antarctica *1 volume*
Volcanism and Plutonism of Western North America *2 volumes*

Library of Congress Cataloging-in-Publication Data

Geology of Grand Canyon, northern Arizona (with Colorado River guides).
   p. cm.
   Prepared for the International Geological Congress field trips
T-115 and T-315 during the summer of 1989.
   ISBN 0-87590-642-7
   1. Geology—Arizona—Grand Canyon—Guide-books.   2. Grand Canyon
(Ariz.)—Description and travel—Guide-books.   3. Colorado River
(Colo.–Mexico)—Description and travel—Guidebooks.   I. American
Geological Union.
QE86.G73G47 1989   89-7004
557.91′32—dc20   CIP

Copyright 1989 by the American Geophysical Union, 2000 Florida Avenue, NW, Washington, DC 20009, U.S.A.

Figures, tables, and short excerpts may be reprinted in scientific books and journals if the source is properly cited.

   Authorization to photocopy items for internal or personal use, or the internal or personal use of specific clients, is granted by the American Geophysical Union for libraries and other users registered with the Copyright Clearance Center (CCC) Transactional Reporting Service, provided that the base fee of $1.00 per copy plus $0.10 per page is paid directly to CCC, 21 Congress Street, Salem, MA 10970. 0065-8448/89/$01. + .10.
   This consent does not extend to other kinds of copying, such as copying for creating new collective works or for resale. The reproduction of multiple copies and the use of full articles or the use of extracts, including figures and tables, for commercial purposes requires permission from AGU.

Printed in the United States of America.

DECADE OF NORTH AMERICAN GEOLOGY
GEOLOGIC TIME SCALE

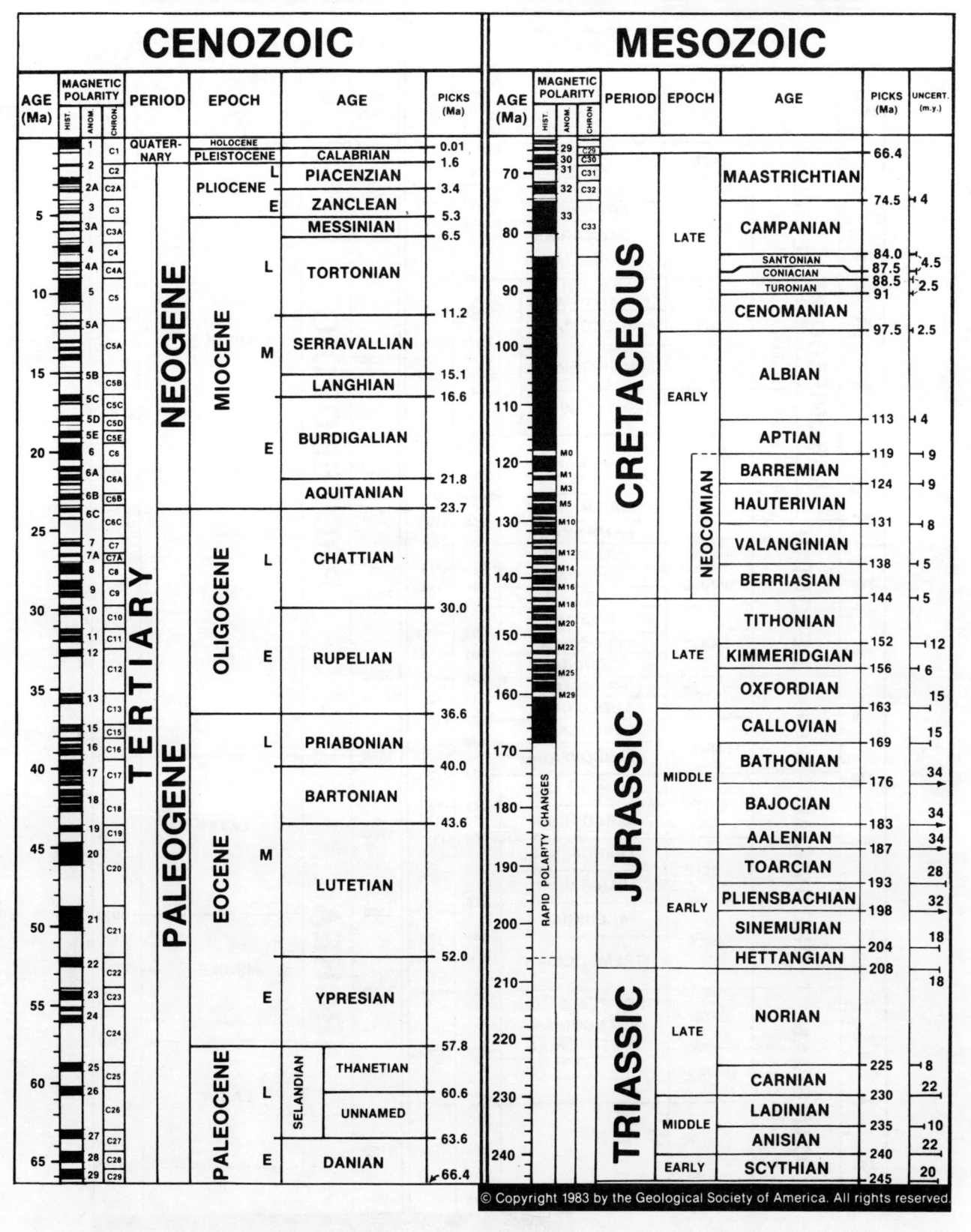

CENOZOIC

AGE (Ma)	MAGNETIC POLARITY (HIST. / ANOM. / CHRON)	PERIOD	EPOCH	AGE	PICKS (Ma)
	1 / C1	QUATERNARY	HOLOCENE		0.01
	2 / C2		PLEISTOCENE	CALABRIAN	1.6
	2A / C2A		PLIOCENE L	PIACENZIAN	3.4
5	3 / C3		E	ZANCLEAN	5.3
	3A / C3A			MESSINIAN	6.5
	4 / C4		L	TORTONIAN	
	4A / C4A	NEOGENE			
10	5 / C5		MIOCENE		11.2
	5A / C5A		M	SERRAVALLIAN	15.1
15	5B / C5B			LANGHIAN	16.6
	5C / C5C				
	5D / C5D		E	BURDIGALIAN	
	5E / C5E				
20	6 / C6				21.8
	6A / C6A			AQUITANIAN	
	6B / C6B				23.7
	6C / C6C				
25	7 / C7		L	CHATTIAN	
	7A / C7A				
	8 / C8				
	9 / C9				30.0
30	10 / C10	TERTIARY	OLIGOCENE		
	11 / C11				
	12 / C12		E	RUPELIAN	
35	13 / C13				
	15 / C15				36.6
	16 / C16		L	PRIABONIAN	
40	17 / C17				40.0
	18 / C18	PALEOGENE		BARTONIAN	
	19 / C19				43.6
45	20 / C20		M	LUTETIAN	
	21 / C21		EOCENE		
50	22 / C22				52.0
55	23 / C23		E	YPRESIAN	
	24 / C24				
					57.8
60	25 / C25		PALEOCENE L	SELANDIAN THANETIAN	
	26 / C26			UNNAMED	60.6
	27 / C27				63.6
65	28 / C28		E	DANIAN	
	29 / C29				66.4

MESOZOIC

AGE (Ma)	MAGNETIC POLARITY (HIST. / ANOM. / CHRON)	PERIOD	EPOCH	AGE	PICKS (Ma)	UNCERT. (m.y.)
70	29 / C29, 30, 31 / C30, 32 / C31, C32		LATE	MAASTRICHTIAN	66.4	
	33 / C33				74.5	4
80				CAMPANIAN		
		CRETACEOUS		SANTONIAN	84.0	4.5
				CONIACIAN	87.5	
90				TURONIAN	88.5	2.5
				CENOMANIAN	91	
					97.5	2.5
100			EARLY	ALBIAN		
110					113	4
				APTIAN		
120	M0, M1, M3, M5				119	9
			NEOCOMIAN	BARREMIAN	124	9
130	M10			HAUTERIVIAN	131	8
	M12, M14			VALANGINIAN		
140	M16, M18				138	5
	M20			BERRIASIAN	144	5
150	M22, M25		LATE	TITHONIAN	152	12
				KIMMERIDGIAN	156	6
160	M29			OXFORDIAN	163	15
				CALLOVIAN	169	15
170		JURASSIC	MIDDLE	BATHONIAN	176	34
180				BAJOCIAN	183	34
				AALENIAN	187	34
190	RAPID POLARITY CHANGES			TOARCIAN	193	28
			EARLY	PLIENSBACHIAN	198	32
200				SINEMURIAN	204	18
210				HETTANGIAN	208	18
220			LATE	NORIAN		
					225	8
230		TRIASSIC		CARNIAN	230	22
			MIDDLE	LADINIAN	235	10
240				ANISIAN	240	22
			EARLY	SCYTHIAN	245	20

© Copyright 1983 by the Geological Society of America. All rights reserved.

DECADE OF NORTH AMERICAN GEOLOGY
GEOLOGIC TIME SCALE

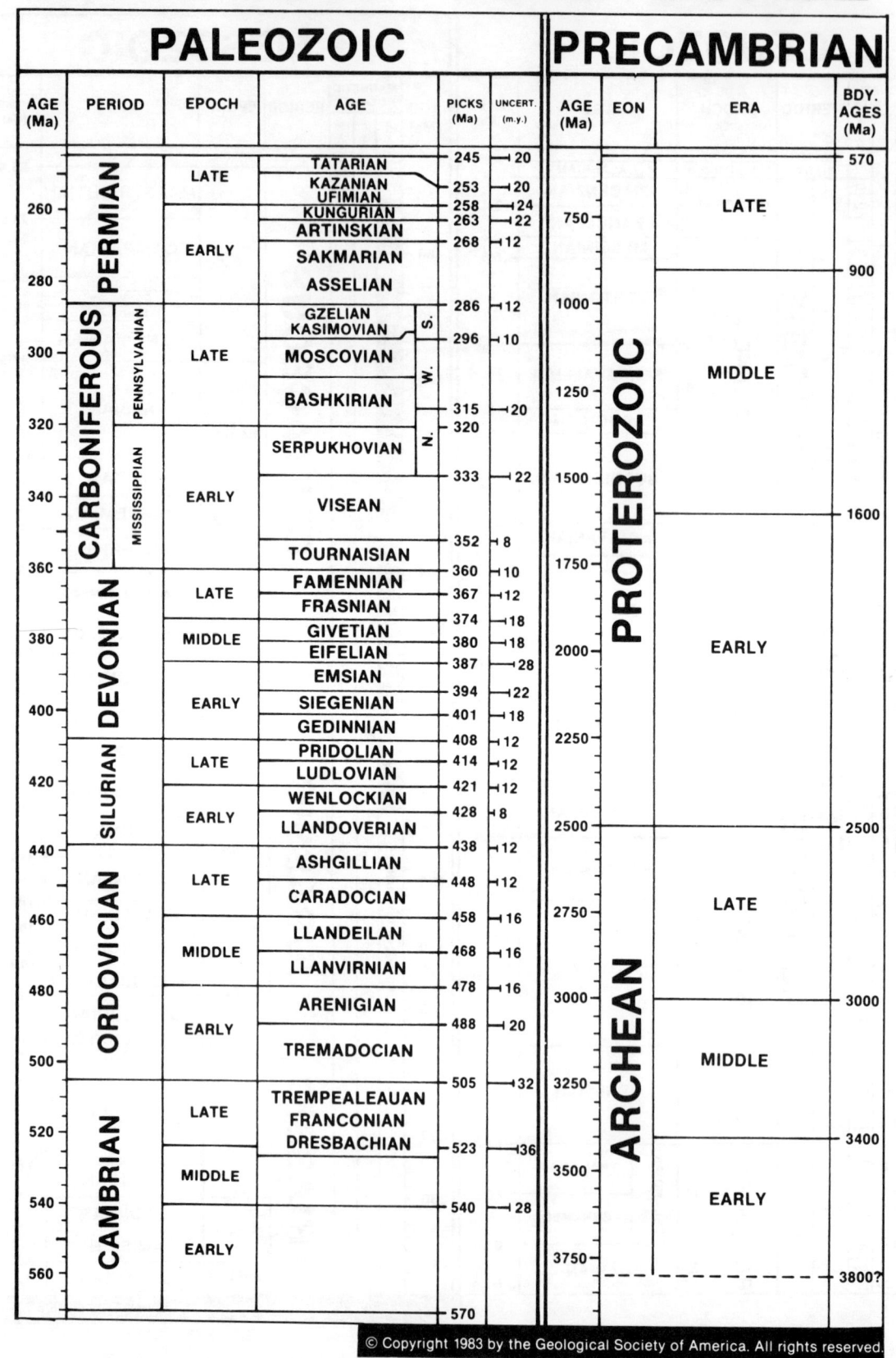

PALEOZOIC

AGE (Ma)	PERIOD	EPOCH	AGE	PICKS (Ma)	UNCERT. (m.y.)
260 — 280 —	PERMIAN	LATE	TATARIAN	245	20
			KAZANIAN	253	20
			UFIMIAN	258	24
		EARLY	KUNGURIAN	263	22
			ARTINSKIAN	268	12
			SAKMARIAN		
			ASSELIAN		
300 — 320 —	CARBONIFEROUS (PENNSYLVANIAN)	LATE	GZELIAN / KASIMOVIAN (S.)	286	12
				296	10
			MOSCOVIAN (W.)		
			BASHKIRIAN	315	20
				320	
340 — 360 —	CARBONIFEROUS (MISSISSIPPIAN)	EARLY	SERPUKHOVIAN (N.)	333	22
			VISEAN	352	8
			TOURNAISIAN	360	10
380 —	DEVONIAN	LATE	FAMENNIAN	367	12
			FRASNIAN	374	18
		MIDDLE	GIVETIAN	380	18
			EIFELIAN	387	28
400 —		EARLY	EMSIAN	394	22
			SIEGENIAN	401	18
			GEDINNIAN	408	12
420 —	SILURIAN	LATE	PRIDOLIAN	414	12
			LUDLOVIAN	421	12
		EARLY	WENLOCKIAN	428	8
			LLANDOVERIAN	438	12
440 — 460 —	ORDOVICIAN	LATE	ASHGILLIAN	448	12
			CARADOCIAN	458	16
		MIDDLE	LLANDEILAN	468	16
			LLANVIRNIAN	478	16
480 — 500 —		EARLY	ARENIGIAN	488	20
			TREMADOCIAN	505	32
520 — 540 — 560 —	CAMBRIAN	LATE	TREMPEALEAUAN / FRANCONIAN / DRESBACHIAN	523	36
		MIDDLE		540	28
		EARLY		570	

PRECAMBRIAN

AGE (Ma)	EON	ERA	BDY. AGES (Ma)
750 —	PROTEROZOIC	LATE	570
1000 —			900
1250 —		MIDDLE	
1500 —			1600
1750 — 2000 — 2250 —		EARLY	
2500 —			2500
2750 —	ARCHEAN	LATE	
3000 — 3250 —		MIDDLE	3000
3500 —			3400
3750 —		EARLY	3800?

© Copyright 1983 by the Geological Society of America. All rights reserved.

Geology of Grand Canyon, Northern Arizona
(with Colorado River Guides)

Lees Ferry to Pierce Ferry, Arizona

Editors:
Donald P. Elston   *George H. Billingsley*   *Richard A. Young*

American Geophysical Union, Washington, D.C.

Copyright 1989 American Geophysical Union

2000 Florida Ave., N.W., Washington, D.C. 20009

ISBN: 0-87590-642-7

Printed in the United States of America

Field Trip T115

June 26–July 5, 1989
June 27–July 6, 1989

Leaders:
W. K. Hamblin   George Billingsley   Peter Huntoon
Stanley Beus   Donald P. Elston

Associate Leaders:
Larry Stevens   Julie Graf   Karen Wenrich
Charles Barnes   Mitchell Reynolds

Field Trip T315

July 20–29, 1989
July 21–30, 1989

Leaders:
W. K. Hamblin   George Billingsley   Peter Huntoon
Mitchell Reynolds   Donald P. Elston

Associate Leaders:
F. G. Poole   Susan W. Kieffer   Richard A. Young
James Mead   John Hendricks

Leaders:

W. K. Hamblin
Department of Geology
Brigham Young University
Provo, UT 84602

George Billingsley
U.S. Geological Survey
2255 N. Gemini Drive
Flagstaff, AZ 86001

Peter Huntoon
Department of Geology
University of Wyoming
Laramie, WY 82071

Stanley Beus
Department of Geology
Northern Arizona University
Flagstaff, AZ 86001

Donald P. Elston
U.S. Geological Survey
2255 N. Gemini Drive
Flagstaff, AZ 86001

Mitchell Reynolds
U.S. Geological Survey
Box 25046, M.S. 905
Denver Federal Center
Denver, CO 80225

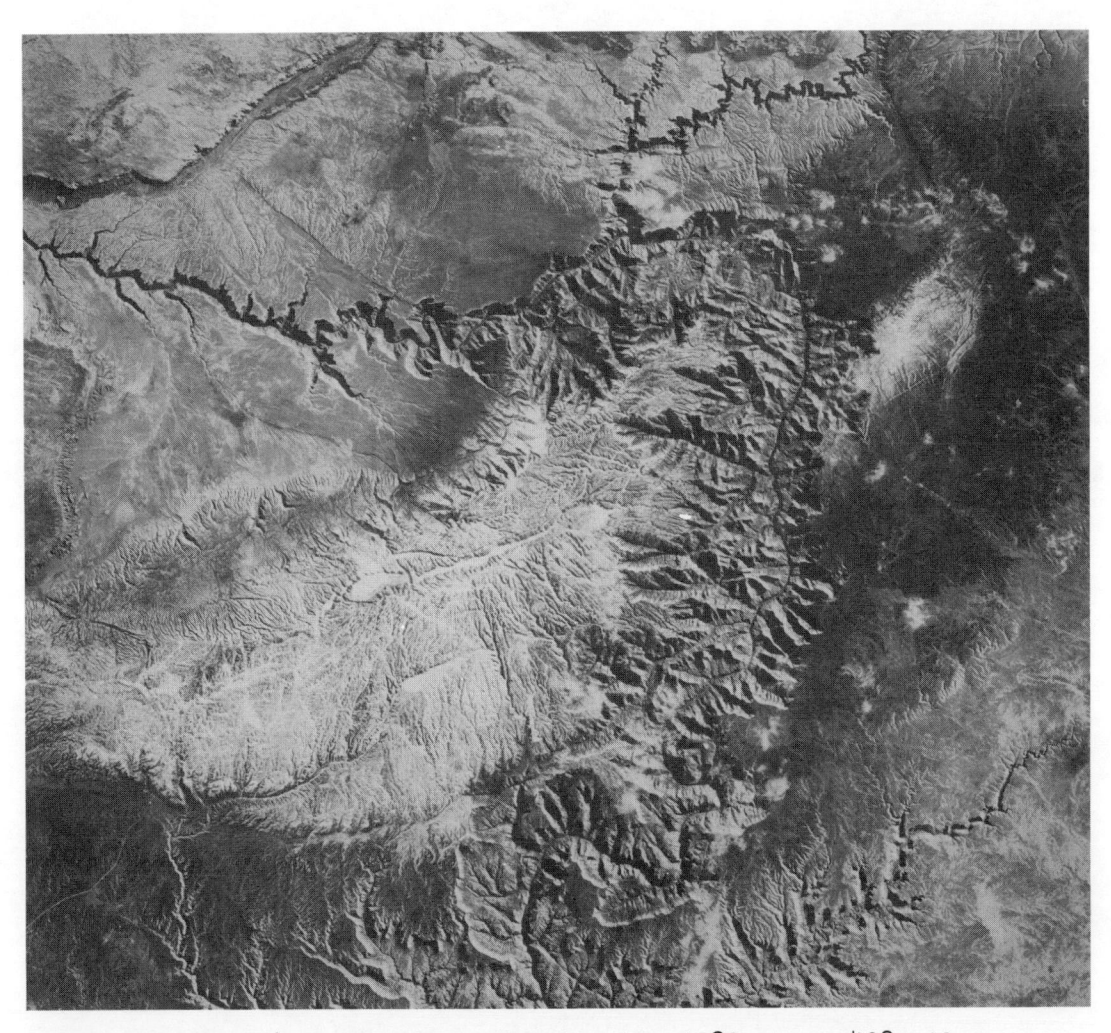

COVER Landsat 4 photo taken January 15, 1983 about 438 miles (705 km) above the eastern Grand Canyon region of northern Arizona. Photo credit: Photo courtesy of U.S. Geological Survey, National Mapping Division, Flagstaff, Arizona.

CONTENTS

CHAPTER   *PAGE*

1: GEOLOGIC LOG OF THE COLORADO RIVER FROM LEES FERRY TO TEMPLE BAR, LAKE MEAD, ARIZONA
George H. Billingsley and Donald P. Elston   1

2: HYDRAULIC LOG OF THE COLORADO RIVER FROM LEES FERRY TO DIAMOND CREEK, ARIZONA
Julia B. Graf, John C. Schmidt, and Susan W. Kieffer   37

3: HYDRAULICS AND SEDIMENT TRANSPORT OF THE COLORADO RIVER
Susan W. Kieffer, Julia B. Graf, and John C. Schmidt   48

4: PHYSIOGRAPHIC FEATURES OF NORTHWESTERN ARIZONA
George H. Billingsley and John D. Hendricks   67

5. MODERN TECTONIC SETTING OF THE GRAND CANYON REGION, ARIZONA
Peter W. Huntoon   72

6. SETTING OF PRECAMBRIAN BASEMENT COMPLEX, GRAND CANYON, ARIZONA
Peter W. Huntoon   74

7. PHANEROZOIC TECTONISM, GRAND CANYON, ARIZONA
Peter W. Huntoon   76

8. EARLY PROTEROZOIC ROCKS OF GRAND CANYON, ARIZONA
Charles W. Barnes   90

9. MIDDLE AND LATE PROTEROZOIC GRAND CANYON SUPERGROUP, ARIZONA
Donald P. Elston   94

10. PETROLOGY AND CHEMISTRY OF IGNEOUS ROCKS OF MIDDLE PROTEROZOIC UNKAR GROUP, GRAND CANYON SUPERGROUP, NORTHERN ARIZONA
John D. Hendricks   106

11. POTENTIAL PETROLEUM SOURCE ROCKS IN THE LATE PROTEROZOIC CHUAR GROUP, GRAND CANYON, ARIZONA
Mitchell W. Reynolds, James G. Palacas, and Donald P. Elston   117

12. PRELIMINARY POLAR PATH FROM PROTEROZOIC AND PALEOZOIC ROCKS OF THE GRAND CANYON REGION, ARIZONA
Donald P. Elston   119

13. PALEOZOIC STRATA OF THE GRAND CANYON, ARIZONA
Stanley S. Beus and George H. Billingsley   122

14. CAMBRIAN STRATIGRAPHIC NOMENCLATURE, GRAND CANYON, ARIZONA - MAPPERS NIGHTMARE
Peter W. Huntoon   128

15. CORRELATIONS AND FACIES CHANGES IN LOWER AND MIDDLE CAMBRIAN TONTO GROUP, GRAND CANYON, ARIZONA
Donald P. Elston   130

16. MESOZOIC STRATA AT LEES FERRY, ARIZONA
   George H. Billingsley   137

17. FISSION-TRACK DATING: AGES FOR CAMBRIAN STRATA, AND LARAMIDE AND POST-MIDDLE EOCENE COOLING EVENTS FROM THE GRAND CANYON, ARIZONA
   Charles W. Naeser, I. R. Duddy, Donald P. Elston, T.A. Dumitru, and P.F. Green   139

18. DEVELOPMENT OF CENOZOIC LANDSCAPE OF CENTRAL AND NORTHERN ARIZONA: CUTTING OF GRAND CANYON
   Donald P. Elston and Richard A. Young   145

19. PALEONTOLOGY, CLAST AGES, AND PALEOMAGNETISM OF UPPER PALEOCENE AND EOCENE GRAVEL AND LIMESTONE DEPOSITS, COLORADO PLATEAU AND TRANSITION ZONE, NORTHERN AND CENTRAL ARIZONA
   Donald P. Elston, Richard A. Young, Edwin H. McKee, and Michael L. Dennis   155

20. PALEOGENE-NEOGENE DEPOSITS OF WESTERN GRAND CANYON, ARIZONA
   Richard A. Young   166

21: PRE-PLEISTOCENE(?) DEPOSITS OF AGGRADATION, LEES FERRY TO WEST-CENTRAL GRAND CANYON, ARIZONA
   Donald P. Elston   174

22. PETROLOGY AND GEOCHEMISTRY OF LATE CENOZOIC BASALT FLOWS, WESTERN GRAND CANYON, ARIZONA
   J. Godfrey Fitton   186

23. PLEISTOCENE VOLCANIC ROCKS OF THE WESTERN GRAND CANYON, ARIZONA
   W. Kenneth Hamblin   190

24. QUATERNARY TERRACES IN MARBLE CANYON AND EASTERN GRAND CANYON, ARIZONA
   Michael N. Machette and John N. Rosholt   205

25. BRECCIA PIPES AND ASSOCIATED MINERALIZATION IN THE GRAND CANYON REGION, NORTHERN ARIZONA
   Karen J. Wenrich and Peter W. Huntoon   212

26. GRAVITY TECTONICS, GRAND CANYON, ARIZONA
   Peter W. Huntoon   219

27. MINING ACTIVITY IN THE GRAND CANYON AREA, ARIZONA
   George H. Billingsley   224

28. BAT CAVE GUANO MINE, WESTERN GRAND CANYON, ARIZONA
   Peter W. Huntoon   228

29. SMALL METEORITE IMPACT IN THE WESTERN GRAND CANYON, ARIZONA
   Peter W. Huntoon   228

REFERENCES   229

PREFACE

The scheduling of the International Geological Congress field trips T-115 and T-315 through the Grand Canyon during the summer of 1989 has provided an unparalleled opportunity not only to prepare detailed river trip logs describing geologic and hydraulic features that may be observed from Lees Ferry to Lake Mead, but also to compile a modern summary of Grand Canyon geology. To persons unfamiliar with details of the geology, it is commonly supposed that no major problems exist because of the superb and extensive exposures. One objective of this volume is to identify and place in perspective some of the salient problems that remain.

Geologic and hydraulic river trip logs (chapters 1 and 2) are designed to be used during a river trip through Marble Canyon and the Grand Canyon. These logs are followed by a review of hydraulic characteristics of the Colorado River (chapter 3). Physiographic, geologic, and structural settings, found in Chapters 4-7, serve as a general review for the geologist and non-geologist alike. Geologic characteristics of the Early Proterozoic crystalline basement, and of stratified and intrusive rocks of the Middle and Late Proterozoic Grand Canyon Supergroup, are summarized in Chapters 8-11. Of particular interest is an interpretation that the Late Proterozoic Chuar Group accumulated mainly in a lacustrine rather than a marine environment of deposition, and that carbonaceous strata of the Chuar Group may have served as a potential source of Precambrian oil. A preliminary, stratigraphically controlled, apparent polar wandering path developed from Proterozoic and Paleozoic rocks of the Grand Canyon and environs is shown in Chapter 12; the polar path and polarity zonation lead to correlations with poles reported from Proterozoic rocks elsewhere in North America, and the character of the polar path may reflect the nature of movement of the North American plate with respect to episodes of tectonism.

The Paleozoic stratigraphy of the Grand Canyon is summarized in Chapter 13. Although one might suppose that problems in correlation of the Paleozoic strata are non-existent because of virtually continuous exposures in the walls of the canyon, such a supposition is dispelled in Chapters 14 and 15. Problems concerning the stratigraphic nomenclature and nature of correlations in the Cambrian Tonto Group are raised in Chapter 14. A revised correlation scheme, proposed in Chapter 15, leads to a somewhat different understanding of the nature and timing of the west-to-east Cambrian marine transgression across the area of the Grand Canyon. Fission track dates from zircon in ash at two horizons in the Cambrian Tonto Group are reported in Chapter 17; the date from one horizon provides an approximate age for the Early-Middle Cambrian boundary.

The stratigraphy of Mesozoic rocks exposed in the area of Lees Ferry, the point of departure for the river trips, is summarized in Chapter 16. The Mesozoic strata remain out of view following departure from Lees Ferry. Although Mesozoic strata were deposited across the Grand Canyon region, most were removed by an episode of late Mesozoic-early Cenozoic erosion related to the Laramide orogeny. This episode of erosion led to an ancient, stripped (and now high-level) plateau surface which forms the rim of the Grand Canyon.

Chapters 17-24 deal with a variety of subjects and problems that bear on Cenozoic development of the landscape of the region. The time of canyon cutting remains the most widely discussed question among travelers who view and pass through the Grand Canyon. Chapter 17 reviews evidence from cooling ages derived from fission track dating on apatite, which indicate plateau uplift occurred in two separate increments: 1) an older event related to Laramide folding and faulting about 65 m.y. ago, and 2) a younger, post-middle Eocene event. The second event would appear to best correspond with the lack of confirmed Oligocene deposits in central and northern Arizona for the interval 37-26 m.y. Evidence bearing on development of the landscape of central and northern Arizona during the Cenozoic, and on restrictions that can be placed on the time of cutting of the Grand Canyon, are summarized in Chapter 18. Evidence for the regional accumulation of of clastic and lacustrine deposits ("Rim gravels") across the stripped plateau surface during Paleocene and Eocene time is summarized in Chapter 19, and these deposits appear to correlate with Paleogene deposits of the high plateaus of Utah to the north. The ancestral Colorado River clearly needed to be established across these Rim gravels prior to establishment of the Colorado River on underlying Paleozoic strata and cutting of the Grand Canyon. In Chapter 20, evidence is reviewed for a southwestern margin of the Colorado Plateau elevated during Late Cretaceous and early Paleocene time, which set the stage for deposition of the Paleocene-Eocene clastic and lacustrine deposits.

A major hiatus encompassing most of Oligocene time exists in the region. The hiatus, for which there is no demonstrated geologic record, presumably reflects an increment of uplift recorded by the post-middle Eocene cooling event recorded by the fission track dates. From regional considerations, the Oligocene may have been the time that the present major topographic elements below the high plateau surface became defined, and possibly was the time of cutting of a major part of the Grand Canyon. Within the Grand Canyon, the potentially oldest Neogene deposits are inferred to be much younger than Oligocene in age. The nature and distribution of a group of undated deposits of aggradation within Marble Canyon and the Grand Canyon are summarized in Chapter 21. Because these may be interpreted to record an episode of aggradation following cutting of the Grand Canyon to essentially its present depth, it is proposed

that these deposits may correlate with late Miocene and Pliocene deposits of aggradation found elsewhere in the region. The petrology of late Cenozoic basaltic flows (~<7 Ma) of the western Grand Canyon region is summarized in Chapter 22. The younger, spectacular Pleistocene lava flows that formed a series of dams in the western Grand Canyon are described in Chapter 23 (from their normal polarity, they could be entirely younger than 720,000 years); speculations are presented on the character and distribution of lakes and deposits that may have been formed behind the temporary lava dams. The distribution and uranium-trend ages of alluvium and soil overlying Quaternary terraces in Marble Canyon and the eastern Grand Canyon are given in Chapter 22; in this chapter, gravel deposits that are proposed to be late Miocene and Pliocene age in Chapter 21 are assigned to the Quaternary. Thus, the the stage is set in Chapters 21 and 24 for further study of the late Cenozoic deposits of aggradation within the Grand Canyon..

Miscellaneous geologic structures and features are described in Chapters 25-29. In Chapter 25, the distribution and origin of breccia pipes developed in Paleozoic rocks of the region are described, and information is presented indicateing a Mesozoic age for the time of mineralization. Gravity structures related to erosion and unloading in the Grand Canyon are described in Chapter 26. Mining activity in the region is summarized in Chapter 27, and in Chapter 28, an account is given of an attempt at mining guano from a cave in the western Grand Canyon. Lastly, a meteorite strike in the western Grand Canyon that resulted in abandonment of an Indian settlement is recounted in Chapter 29.

Acknowledgements

This volume would not have been possible without the support of the U.S. Geological Survey. The cooperation of the many authors and reviewers have made the editors' tasks feasible. The authors and editors particularly wish to thank the U.S. National Park Service, Grand Canyon National Park, for permission to work and sample in the Grand Canyon. The cooperation and willingness to allow geological research in the Park during the past several decades has been greatly appreciated by all of us. We trust that research in the Park can continue in a similar manner in the decades to come. Many important problems in the earth sciences, and in the earth sciences laboratory known as the Grand Canyon, remain to be resolved.

This volume was prepared at the Flagstaff Field Center of the U.S. Geological Survey. The editors particularly appreciate and wish to acknowledge the yeoman support of several individuals. John D. Hendricks prepared the figures in Chapters 1 and 4, and assisted in many other aspects involved with the preparation of the volume. Shirley L. Elston was responsible for compilation of the References section, for assuring accuracy and uniformity in citation, and for proofing much of the volume. Sue Priest, aided by Eileen Haney, was primarily responsible for the paste-up of text and captions on the blue-line masters submitted for photo-reduction and printing; Sue also proofed the paste-up copy. Hugh F. Thomas and his staff prepared the photomechanical reproductions of most of the line figures. Ramona F. Boudreau and Karl A. Zeller photocopied the many photographs. Ramon Sabala provided drafting support for several of the illustrations. Michael L. Dennis' drafting of figures in Chapters 9, 12, 15, 19, and 21 is greatly appreciated.

Marjorie E. MacLachlan, Geologic Names Unit, U.S. Geological Survey, Denver, Colorado, reviewed the manuscripts for conformity in usage of geologic names. Juergen Reinhardt and Penelope Hanshaw, U.S. Geological Survey, Reston, Virginia, arranged for and provided logistical and financial support that allowed this volume to come into being. Janet Evans and Lori Klinzmann, American Geophysical Union, Washington, D.C., compiled the camera-ready copy into this coherent volume. We thank all of the above persons for their efforts and willing support, and we wish to apologize to anyone whom we inadvertently may have overlooked and failed to acknowledge.

D. P. E.
G. H. B.
R. A. Y.
December 1988

CHAPTER 1: GEOLOGIC LOG OF THE COLORADO RIVER FROM LEES FERRY TO TEMPLE BAR, LAKE MEAD, ARIZONA

George H. Billingsley and Donald P. Elston
U.S. Geological Survey, Flagstaff, Arizona

INTRODUCTION

Geological features of interest along the Colorado River from Lees Ferry to Temple Bar on Lake Mead, a distance of 312 miles (502 km), are indicated on this log. The log is meant to be used with one of many commercially available Colorado River map guides. Geological maps of the Grand Canyon, references to which are found at the end of this volume, would usefully supplement this guide. An * preceding the river mile indicates a feature or structure of interest, or the initial appearance of a formation or member.

Colorado River mileage starts at Lees Ferry, Arizona (mile 0). Elevation of the Colorado River at mile 0 is 3,116 feet (949.8 m). Elevation at Lake Mead (high pool level) is 1,157 feet (352.7 m). The drop of 1,959 feet (597 m) in 235 miles (378 km) results in an average gradient of 8.3 feet/mi (2.5 m/km). For a geographic and physiographic overview of the region, see Figures 4.1 and 4.2. Figure 1.1 is a schematic cross section of rocks of the Grand Canyon.

MILE (KILOMETER)

*0 (0)   Boat ramp is on river alluvium that covers the Petrified Forest Member of the Chinle Formation (Upper Triassic). The multicolored shales of this member (fig. 1.2) are seen to the north behind old Fort Lee (established 1876; Mormon river crossing station connecting with communities to the south and route of the "Honeymoon Trail"). The prominent cliff across the river at river level (fig. 1.3), in which a cable for a water gauging station is anchored, is formed by the basal Shinarump Member of the Chinle Formation (chapter 16). Pre-Pleistocene(?) gravel mantles much of the Mesozoic outcrop near the river in the Lees Ferry area (chapter 21). A discussion of the flow and sediment transport at the Lees Ferry gauge is given in Chapter 3.

0.1 (0.16)   Moenkopi Formation (Lower and Middle? Triassic) on left, below gauging station.

0.3 (0.2)   Debris fan of Paria River on right.

0.5 (0.4)   Echo Cliffs Monocline; strata are tilted northeast about 15°. Several joints and fractures in the next mile (1.6 km) are probably related to stresses associated with development of monocline during Laramide deformation, which began about 65 million years ago (chapter 7).

*0.8 (1.3)   Unconformity between the red Lower and Middle(?) Triassic Moenkopi Formation and underlying gray Lower Permian Kaibab Formation (or Limestone) on left (fig. 1.4). Terrace deposits of pre-Pleistocene(?) age on right (chapter 21). (For a summary description of the Paleozoic strata of the Grand Canyon, see chapter 13 and figure 13.1).

0.8 (1.3)   Mouth of Paria River.

1.5 (2.4)   Fault; downstream side is up about 35 feet. (10.7 m) *NOTE: All faults are normal unless otherwise stated.*

*2.1 (3.4)   Contact between Lower Permian Kaibab Formation and underlying Toroweap Formation.

3.4 (5.5)   Entering small graben; strata drops about 20 feet (6.1 m). Kaibab strata are draped across the fault. Toroweap has a three-fold subdivision characterized by a slope-cliff topography with a small recess at the base, commonly seen on the right side of the river. Because of an abrupt facies change farther downstream, the Toroweap forms an unbroken cliff with the Kaibab Formation and Coconino Sandstone mostly on the left side of the river. Cross-bedded sandstone of the lowermost Toroweap can be mistaken for Coconino Sandstone.

4.1 (6.6)   Leaving graben, downstream strata is up about 20 feet (6.1 m). Kaibab strata are draped across fault.

4.3 (6.9)   Navajo Bridge, built in 1928, is 467 feet (142 m) above Colorado River. Permian Coconino Sandstone appears on right just upstream from bridge.

*4.5 (7.3)   Contact of Toroweap with crossbedded Coconino Sandstone.

*4.9 (7.9)   Contact of Coconino Sandstone with underlying (earthy red) Hermit Shale. Best seen on left. Note thin clastic dikes in Hermit filled with Coconino detritus, indicative of an hiatus in deposition.

7.2 (11.6)   Fresh rock fall on left came from canyon rim in the spring of 1970. Note cliff-slope relations between Kaibab and Toroweap Formations here (fig. 1.5) in contrast to sheer cliff formed by these formations seen downstream at mile 50 (km 80) (fig. 1.15).

7.3 (11.8)   Fault; downstream side is down about 2 feet (.6 m).

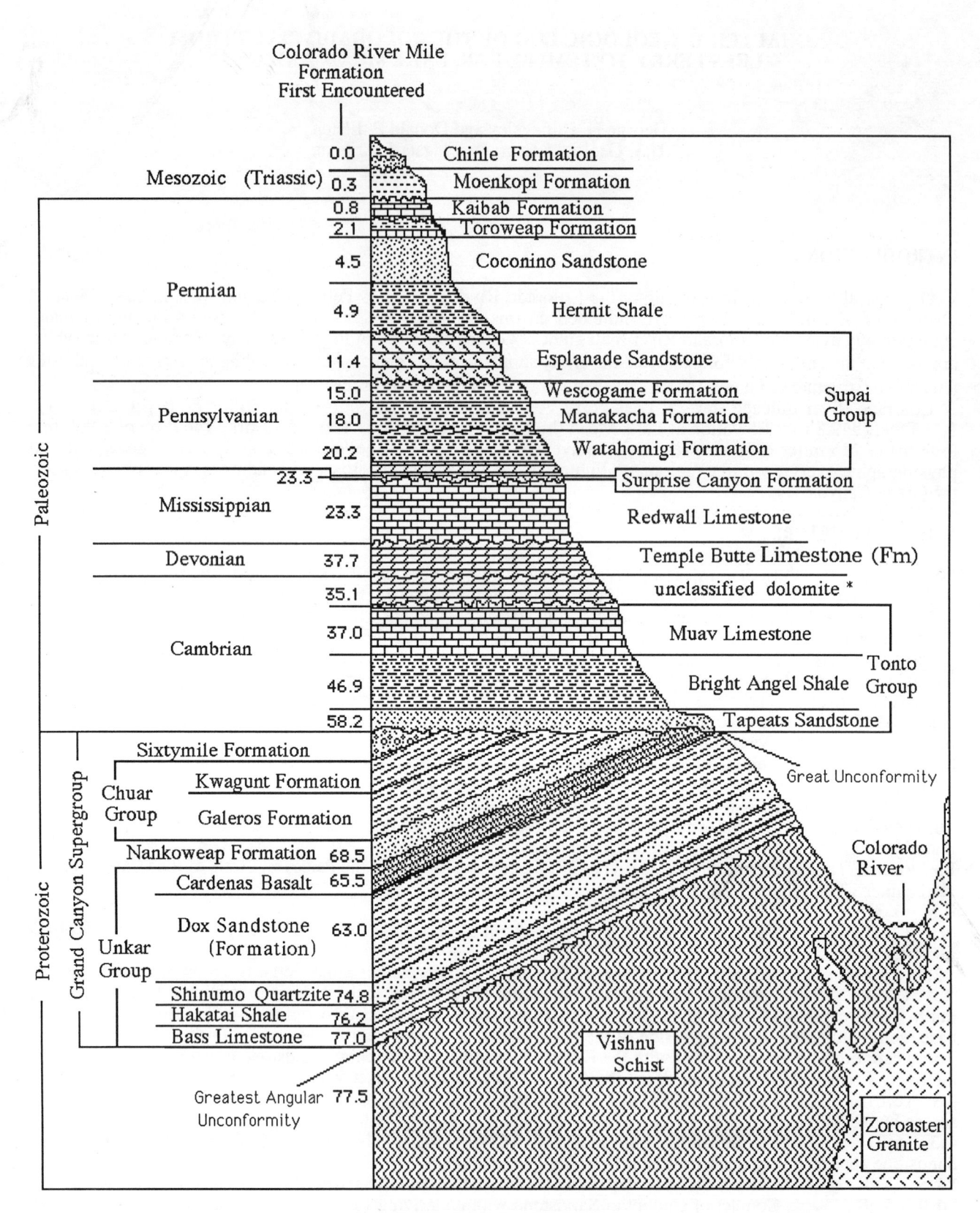

FIGURE 1.1. Schematic cross section of the Grand Canyon. * - Nopah equivalent.

FIGURE 1.2. View upstream from near Gauging Station at Lees Ferry looking north at Triassic and Jurassic strata in the Vermilion Cliffs. Navajo Sandstone (J℞n); Kayenta Formation (℞k); Moenave Formation (℞m); Chinle Formation (℞c).

FIGURE 1.3. View across the Colorado River near the Gauging Station at Lees Ferry. Triassic rocks here dip 14° to the northeast on Echo Cliffs monocline. Contacts dashed where approximately located. Chinle Formation, Petrified Forest Member (℞cp), Shinarump Member (℞cs); Moenkopi Formation (℞m).

FIGURE 1.4. Unconformity between the Triassic Moenkopi Formation (Ṟm) and Permian Kaibab Formation (Pk). Left bank, looking downstream at mile 0.8 (km 1.3).

7.6 (12.2) Fault; downstream side is up about 10 feet (3.1 m).
7.7 (12.4) Above Badger Rapid. The highest Colorado River pre-Pleistocene? gravel deposit lies 750 feet (230 m) above and behind the visible rim of the canyon, on right (chapter 21).
7.8 (12.6) **Badger Creek Rapids**; a 15 feet (4.6 m) drop.
*8.5 (13.7) Begin cemented cones of colluvium and alluvium on Hermit Shale, about 200 feet (60 m) above river on right and left (fig. 1.6); apexes of cones abut zones of carbonate-salt seepage at base of Coconino Sandstone. Cones extend to the south to at least mile 28 (km 45) rising high above river level. The cemented talus is stratified, above which unstratified talus accumulated prior to dissection of the cones and unconsolidated talus by erosion.
10.0 (16.1) **Ten Mile Rock** is a calcareous sandstone block derived from the Toroweap Formation. On both sides of the river for the next mile (1.6 km) are examples of small scale landslide blocks that have rotated backwards towards the parent wall from which they slid due to undercutting of the softer Hermit Shale. Most canyon widening results from this type of mass-wasting.
10.1 (16.3) Fault; downstream side is down about 2 feet (0.6 m).
11.2 (18.0) **Soap Creek Rapids**; a 17 feet (5.2 m) drop.
*11.4 (18.3) Contact of red-brown to grayish red Hermit Shale with underlying cliff-forming, cross-bedded grayish red Esplanade Sandstone (fig. 1.7). The contact is marked by an unconformity, underlain by a zone of bleaching and overlain by a limestone pebble conglomerate; contact is best seen on left side of river on bench.
13.2 (21.2) Well developed low angle cross-bedding in sandstone of the Esplanade, on right. Cross-beds display dominantly south-directed foreset beds.
14.4 (23.2) **Sheer Wall Rapids**; an 8 feet (2.4 m) drop. Tanner Wash enters on left.
*14.8 (23.8) Large fossil reptillian footprints on a slab of Esplanade Sandstone near river level, on left; *Thallasonoides*-like burrows and rare mudcracks occur in some beds. Good cross-beds for photography next 0.2 mile, on left.
*17.0 (27.4) **House Rock Rapids**; a 10 feet (3.1 m) drop. Unconformable contact of Esplanade with underlying Pennsylvanian Wescogame Formation is marked by limestone pebble conglomerate; best seen on left wall adjacent to rapids.
17.5 (28.2) Breccia pipe in Esplanade Sandstone and Wescogome Formation on left. Pipe exhibits some bleaching; easily missed, but this is the first of several such structures to be seen in next 9 miles (14.5 km).
18.0 (29.0) Approximate area of contact of Pennsylvanian Wescogame and Manakacha Formations; difficult to identify; no change in stratigraphy evident (figs. 1.8 and 1.9).
18.3 (29.5) Collapse structure (breccia pipe at depth) in Pennsylvanian Manakacha Formation, on left; structure is inferred to root in Redwall Limestone.

FIGURE 1.5. View looking upstream showing the cliff/slope relationships. Contacts dashed where approximately located. Moenkopi Formation (Trm); Kaibab Formation, Harrisburg Member (Pkh), Fossil Mountain Member (Pkf); Toroweap Formation, Woods Ranch Member (Ptw), Brady Canyon Member (Ptb), Seligman Member (Pts); Coconino Sandstone (Pc); Hermit Shale (Ph).

18.5 (29.8) **Boulder Narrows.** This large block from Toroweap Formation was part of a large prehistoric rock fall. Driftwood on top is from the last pre-Glen Canyon dam flood in 1957.
*18.8 (30.3) Collapse structure (breccia pipe at depth) in Manakacha Formation, on left; 30+ feet (10 m) across; note downwarped beds.
19.0 (30.6) Fault; downstream side is up about 16 feet (5 m).
20.0 (32.2) Small arch in Esplanade Sandstone, near top of cliff on left.
*20.2 (32.5) Unconformity between Pennsylvanian Manakacha and Watahomigi Formations (figs. 1.8 and 1.9); overlain by limestone pebble conglomerate (exposed at mouth of North Canyon).
20.5 (33.0) **North Canyon Rapids**; a 12 feet (3.7 m) drop. Watahomigi Formation well exposed downstream for next half mile (0.8 km).
21.2 (34.1) **21-Mile Rapids**; a 12 feet (3.7 m) drop.
21.8 (35.1) Fault; downstream side is down about 50 feet (15.2 m). Mudcracks in block of Watahomigi Formation on left (tilted eastward and rotated downward on south side of fault).
21.9 (35.2) Fault; downstream side is down about 8 feet (2.4 m). The Kaibab and Toroweap Formations (Permian) on the high rim dip upstream about 10°, opposite to displacements on faults in the area.

FIGURE 1.6. Stratigraphy of the left (east) wall of the Colorado River at mile 9.2 (14.8 km). Contacts dashed where approximately located. Kaibab Formation, Fossil Mountain Member (Pkf); Toroweap Formation, Woods Ranch Member (Ptw), Brady Canyon Member (Ptb), Seligman Member (Pts); Coconino Sandstone (Pc); Hermit Shale (Ph); cemented talus (t).

FIGURE 1.7. Left bank of Colorado River at mile 11.5 (18.5 km) showing contact between Hermit Shale (Ph) and Esplanade Sandstone (Pe). Contacts dashed where approximately located. Toroweap Formation, Woods Ranch Member (Ptw), Brady Canyon Member (Ptb), Seligman Member (Pts); Coconino Sandstone (Pc).

22.7 (36.5) Fault; downstream side is down about 2 feet (0.6 m).

*23.0 (37.0) Contact of slope-forming Watahomigi Formation with underlying light gray Redwall Limestone (Mississippian); contact is mostly covered (figs. 1.8 and 1.9). Four members of Redwall Limestone will be seen in the next 12 miles (19 km). They are, from top to bottom, 1) Horseshoe Mesa Member forming karsted ledges and cliffs; 2) Mooney Falls Member, forming a sheer banded cliff (alternating limestone and dolomite), with minor chert; 3) Thunder Springs Member, forming a cherty banded, in part large scale, wavy bedded cliff, and 4) Whitmore Wash Member, forming a massive cliff (figs. 1.11 and 1.12).

23.1 (37.2) A large collapse structure, partially brecciated, is well exposed on left wall; affects entire section of Supai Group. Best observed by looking upstream before going into 23-Mile Rapids (fig. 1.8). Reentrant and brecciated zone in Coconino Sandstone and Toroweap Formation in high background is not related to breccia pipe..

23.2 (37.3) **23-Mile Rapids**; about a 5 foot (1.5 m) drop.

*23.3 (37.5) The Surprise Canyon Formation, a dark red, in part regolithic siltstone unit of Late Mississippian and Early Pennsylvanian(?) age, is well exposed on right at bottom of 23-Mile Rapids (fig. 1.9). The Surprise Canyon, preserved in channels and valleys, is discontinuous throughout the Grand Canyon.

23.5 (37.8) **23.5-Mile Rapids**; about a 4 foot (1.2 m) drop.

24.2 (38.9) **24-Mile Rapids**; about a 4 foot (1.2 m) drop. Fault; downstream side is down about 7 feet (2.1 m).

(Note: Mile 24 to 25 is about 1.3 mi. (2.9 km) in length due to an error on old maps.)

24.3 (39.1) Large collapse structure with bleached sandstone in Pennsylvanian Manakacha Formation, exposed high on left.

24.5 (39.4) **24.5-Mile Rapids**; a 9 foot (2.7 m) drop. Patch of Surprise Canyon Formation on the right bank.

24.8 (39.9) Fault; downstream side is up about 9 feet (2.7 m).

25.0 (40.2) **25-Mile Rapids**; an 8 foot (2.4 m) drop. A breccia pipe in the Redwall Limestone is seen on right side of the rapids; a cave associated with this collapse is seen in Redwall Limestone just above river on left, below the rapids.

25.5 (41.0) **Cave Spring Rapids**; a 6 foot (1.8 m) drop. Fault; downstream side is up about 22 feet (6.6 m).

26.4 (42.5) Sediment- and breccia-filled karst cave in Redwall Limestone, on left.

FIGURE 1.8. Breccia Pipe at mile 23.1 (37.2 km), left bank (east). View is looking upstream from 23-Mile Rapids. Contacts dashed where approximately located. Hermit Shale (Ph); Esplanade Sandstone (Pe); Wescogame Formation (₽we); Manakacha Formation (₽m); Watahomigi Formation (₽wa); Redwall Limestone (Mr).

FIGURE 1.9. View downstream to slope of Surprise Canyon Formation near bottom of 23-Mile Rapids. Contacts dashed where approximately located. Kaibab Formation (Pk); Toroweap Formation (Pt); Coconino Sandstone (Pc); Hermit Shale (Ph); Esplanade Sandstone (Pe); Wescogame Formation (Pwe); Manakacha Formation (Pm); Watahomigi Formation (Pwa); Surprise Canyon Formation (PMs); Redwall Limestone (Mr).

26.7 (43.0)   **27-Mile Rapids**; a 7 foot (2.1 m) drop. Breccia pipe well exposed at beginning of rapids on right.
26.8 (43.1)   A small rapid occurs here at low water (below 5,000 cfs) because of a rockfall from the right wall in 1975. Rapids formed entirely by rockfalls are rare in Grand Canyon.
29.1 (46.8)   **29-Mile Rapids**; a 7 foot (2.1 m) drop; Shinumo Wash on left.
30.3 (48.8)   **Fence fault**; downstream side is down about 210 feet (64.0 m). Several warm springs emerge from Redwall Limestone on left bank at low water levels, just downstream from fault.
30.5 (49.1)   Fault; downstream side is down about 60 feet (18.3 m). Small springs emerge from Redwall Limestone on right.
31.0 (49.9)   Fault; downstream side is up about 2 feet (.6 m).
31.2 (50.2)   Arch in a cave in the Redwall Limestone on left, about 150 feet (45.7 m) above river. Figure 1.10 shows Paleozoic section as photographed from the arch.
*31.7 (51.0)   **Stantons Cave** on right; archeological site.
*31.9 (51.3)   **Vasey's Paradise**. Flow of springs fluctuates in response to rains and annual snowfall on Kaibab Plateau. **Beware of poison ivy**.
32.8 (52.8)   Test adit (1952) for proposed Marble Canyon dam site, on right (will not be built).
*33.0 (53.1)   **Redwall Cavern**; carved into Thunder Springs Member of Redwall Limestone. Note wavy bedding of this member down stream, perhaps reflecting a pulse of deformation prior to end of accumulation, or a depositional feature, or possible algal mounds.
34.1 (54.9)   Several small springs emerge from the Whitmore Wash Member of the Redwall Limestone for the next 0.25 mi (0.4 km), on right.
*34.8 (56.0)   **Nautiloid Canyon** on left; several fossil orthocone nautiloids may be observed on canyon floor in Whitmore Wash Member of Redwall Limestone, just below contact with overlying Thunder Springs Member. (Wet for best viewing.). Note karst breccia in Whitmore Wash Member opposite Nautiloid Canyon.
*35.1 (56.5)   Disconformable contact between gray Whitmore Wash Member of Redwall Limestone and underlying gray Cambrian "unclassified dolomite" (figs. 1.11 and 1.12). The unclassified dolomite was originally included with the Muav Limestone of the Tonto Group, however, it disconformably overlies the Havasu Member of Muav Limestone (Middle Cambrian). Note pockets of red mudstone beneath the unclassified dolomite here. This dolomite may correlate with dolomite underlying the Upper Cambrian Dunderberg Shale Member of Nopah Formation exposed along U.S. Route I-10 in the Virgin River Gorge of northwestern Arizona.

FIGURE 1.10. View of stratigraphy in Marble Canyon looking southwest to South Canyon from mile 29.5 (47.5 km). Contacts dashed where approximately located. Kaibab Formation (Pk); Toroweap Formation (Pt); Coconino Sandstone (Pc); Hermit Shale (Ph); Esplanade Sandstone (Pe); Wescogame Formation (Pwe); Manakacha Formation (Pm); Watahomigi Formation (Pwa); Surprise Canyon Formation not present here; Redwall Limestone, Horseshoe Mesa Member (Mrh), Mooney Falls Member (Mrm).

*35.3 (56.8) Contact of Upper Cambrian unclassified dolomite with platy, thin bedded limestone of Havasu Member of Muav Limestone (Middle Cambrian); only about 20 feet (6 m) of Upper Cambrian dolomite preserved here (fig. 1.12).
35.7 (57.4) Arch in Redwall Limestone about 150 feet (46 m) above river in small drainage, on right. This area has many solution caverns forming along joints that trend in a northeasterly direction and parallel a large fault 2 miles (3.2 km) to the southeast. These joints, which cross the river at several places downstream, are referred to here as the 36-Mile joint system.
35.9 (57.8) Large cavern on left filled to roof with blocks and debris.
36.0 (57.9) **36-Mile Rapids**; about a 4 foot (1.2 m) drop.
*36.4 (58.6) Lower unit of Havasu Member of Muav Limestone appears.
36.6 (58.9) 36-Mile joint system crosses river here.
37.0 (59.5) View downstream toward first of several Devonian channels; on left bank.
*37.8 (60.8) Channel occupied by Middle and Late Devonian Temple Butte Limestone (or Formation) is located about 60 feet (18 m) above river level on left bank (fig. 1.12). Channel has about 100 feet (30 m) of relief and cutting through Upper Cambrian unclassified dolomite, well into Havasu Member of Muav Limestone. Channel can be distinguished not only by its U-shaped outline, but also by irregular bedding and a purple to white color of the beds that contrast with the homogeneous bedding and gray color of the Redwall Limestone and the unclassified Cambrian dolomite (Nopah equivalent). Several other channels will be seen down river. A sandy, irregularly bedded dolomite, purplish to red brown and exhibiting a bleached top, occupies the channel; two additional units each about 3 feet (1 m) thick and bounded by unconformities, are present at the top of the channel section along this immediate reach of the river. The lower of the two thin units is a white, bleached, aphanitic dolomite, whereas the upper of the two units is a dark gray dolomite that emits a fetid odor when struck by a hammer. These thin units are best examined in the alcove developed in the Redwall Limestone at the top of this first channel. The three-fold subdivision of the Temple Butte is similar to the stratigraphy of Devonian units of central Arizona, 1) the lower and 2) the upper units of the Beckers Butte Member of the Martin Formation, and 3) the fetid dolomite unit of the Jerome Member of the Martin Formation. The two upper units also can be observed from the river at mile 38.2 (km 61.5) on the left.

FIGURE 1.11. Stratigraphic section looking downstream at mile 36.3 (58.4 km). Contacts dashed where approximately located. Kaibab Formation (Pk); Toroweap Formation (Pt); Coconino Sandstone (Pc); Hermit Shale (Ph); Esplanade Sandstone (Pe); Wescogame Formation (IPwe); Manakacha Formation (IPm); Watahomigi Formation (IPwa); Redwall Limestone, Horseshoe Mesa Member (Mrh), Mooney Falls Member (Mrm), Thunder Springs Member (Mrt), Whitmore Wash Member (Mrw); unclassified dolomite (Єu).

38.0 (61.1) 36-Mile joint system crosses river.
38.3 (61.6) Devonian channel on left is tributary to a larger channel just downstream. Note two thin, discontinuous dolomite units of the Temple Butte at top of section; the lower is light gray and the upper is dark gray and irregularly bedded; each is about 1 m thick.
38.4-38.5 (61.8-61.9) Devonian channel diagonally crosses river.
***39.2 (63.1)** Thin bedded limestone of upper part of Gateway Canyon Member of Muav Limestone makes its appearance. The shaly appearance of this member downstream resulted in its assignment to the Bright Angel Shale of the Tonto Group on an early geologic map. This member becomes rather sandy a short distance to the south.
39.3 (63.2) Test adits on left and right mark site of proposed (1946) Marble Canyon Dam (will not be built).
39.8 (64.0) Devonian channel on both sides of river. The 36-Mile joint system crosses river.
40.5 (65.2) 36-Mile joint system crosses river.
40.8 (65.6) Devonian channel trend crosses river. Thin-bedded, green shaly sandstone of Gateway Canyon Member of Muav Limestone on right and left.
41.0 (66.0) **Buck Farm Canyon.** Begin well cemented talus on Gateway Canyon Member of Muav Limestone, on left.

FIGURE 1.12. First view, looking downstream, of erosional unconformity (Devonian channel) cut into Cambrian unclassified dolomite; left bank at mile 37.8 (km 60.8). Contacts dashed where approximately located. Redwall Limestone, Mooney Falls Member (Mrm), Thunder Springs Member (Mrt), Whitmore Wash Member (Mrw); Temple Butte Limestone (Formation) (Dt); unclassified dolomite (€u); Muav Limestone, Havasu Member (€mh), Gateway Canyon Member (€mg).

41.3	(66.5)	Devonian channel crosses river. Good channel for photographs on right (fig. 1.13).
*41.5	(66.8)	**Royal Arches** on right are result of solution and chemical weathering.
43.1	(69.4)	Devonian channels can be seen the next 0.5 mile (0.8 km) on both sides of the river.
43.3	(69.7)	Fault, part of the Eminence Break Graben; downstream side is down about 22 feet (6.6 m).
43.4	(69.8)	Fault; downstream side is down about 20 feet (6.1 m).
43.7	(70.3)	**President Harding Rapids**; a 4 foot (1.2 m) drop. A small fault, part of the Eminence

Break fault system and graben, parallels the river for the next 0.5 mi (0.8 km); left side is down about 40 feet (12.2 m). The east side of the graben is seen high on left above an old rock slide in Redwall Limestone. Displacement across the Eminence Break fault and graben system is about 212 feet (64.6 m) down to the west.

FIGURE 1.13. View downstream at mile 41 (66.0 km) showing outline of a Devonian channel cut into Cambrian rocks, filled with purplish sandy dolomite and limestone of the Temple Butte Limestone. Contacts dashed where approximately located. Redwall Limestone, Mooney Falls Member (Mrm), Thunder Spring Member (Mrt), Whitmore Wash Member (Mrw); Temple Butte Limestone (Dt); unclassified dolomite (€u); Muav Limestone, Havasu Member (€mh), Gateway Canyon Member (€mg).

44.4 (71.4) Large rock fall on left fell during winter of 1975 (fig. 1.14).
*44.5 (71.6) Fault; downstream side is up about 20 feet (6.1 m). Gray, thin-bedded limestone of Kanab Canyon Member of Muav Limestone makes its appearance; a doublet unit, it forms a steep ledge or cliff for next 2+ miles (3.2+ km) to Triple Alcoves. A thin ledge-forming limestone that marks the top of the member here can be traced into sandstone facies south of Triple Alcoves.
44.8 (72.1) Fault; downstream side is up about 10 feet (3.1 m).
46.4 (74.7) Well cemented talus overlying Gateway Canyon and Havasu Members of Muav Limestone, left and right.
*46.5 (74.8) Top of thin, brown, single sandy unit of Peach Springs Member of Muav Limestone; it first appears near river level on right before Triple Alcoves and is overlain by greenish sandy thin-bedded limestone.
46.6 (75.0) **Triple Alcoves** are developed by solution and chemical weathering similar to that of Royal Arches.
46.9 (75.5) Gradational contact between Peach Springs Member of Muav Limestone and underlying green and red sandstone of Bright Angel Shale; mostly concealed (fig. 1.15). Note facies change in the Kanab Canyon Member from limestone to sandstone taking place in a southerly direction; a marker carbonate bed at top of Kanab Canyon passes into dolomitic sandstone, remaining identifiable in a section that becomes mostly sandstone.
47.1 (75.8) **Saddle Canyon**, on right; good outcrops of a Devonian channel can be visited on foot about 0.5 mile (0.8 km) up the canyon.
47.3 (76.1) Devonian channel on left.
47.4 (76.3) Peach Springs Member of Muav Limestone, at river level on left, overlain by doublet section of Kanab Canyon Member (fig. 1.14).
48.0 (77.2) Springs emanating from Muav Limestone, on right.
48.5 (78) Gateway Canyon Member becoming sandy; well cemented talus both sides of river.
48.6-49.5 (78.2-79.6) Peach Springs Member at river level on right; lower part of Kanab Canyon Member becomes sandy (fig. 1.15).
49.0 (78.8) Devonian channel, both sides of river, also small fault, downstream side down about 8 feet (2.4 m).

FIGURE 1.14. Rockfall that occurred during the winter of 1975, on left bank at mile 44.4 (71.4 km). Small fault parallels river (arrows show direction of displacement). Contacts dashed where approximately located. Redwall Limestone, Mooney Falls Member (Mrm), Thunder Springs Member (Mrt), Whitmore Wash Member (Mrw); unclassified dolomite (€u); Muav Limestone, Havasu Member (€mh).

49.7 (80.0)   **Eminence Break fault**; downstream side is up about 100 feet (30 m), exposing uppermost red and green, glauconitic sandstone of Bright Angel Shale, overlain by single brown sandstone unit of Peach Springs Member of Muav Limestone. These relations are best observed on left, below small bend in river (fig. 1.15). From here to mile 59 (km 95) the river is in Bright Angel Shale where five red horizons of glauconitic sandstone can be identified. These may correspond with five dolomite and limestone members of Bright Angel Shale and Muav Limestone identified in the central and western parts of the Grand Canyon. The thickest of the red intervals, in middle part of formation, is thought to correlate with a prominent dolomite unit in the Bright Angel Shale of central Grand Canyon and to be equivalent to the Rampart Cave Member of the Muav Limestone of western Grand Canyon (chapter 15).
51.0 (82.1)   Small springs on left, seen only at very low river level.
51.7 (83.2)   Devonian channel crosses river.
51.8 (83.3)   **Little Nankoweap Creek**; end of Marble Canyon (old boundary between Marble Canyon National Monument and Grand Canyon National Park).

FIGURE 1.15. View downstream of Paleozoic section in lower Marble Canyon at mile 49.8 (80.1km). Contacts dashed where approximately located. Kaibab Formation (Pk); Toroweap Formation (Pt); Coconino Sandstone (Pc); Hermit shale (Ph); Esplanade Sandstone (Pe); Wescogame Formation (₽We); Manakacha Formation (₽m); Watahomigi Formation (₽wa); Surprise Canyon Formation not present; Redwall Limestone (Mr); Temple Butte Limestone not present; unclassified dolomite (Єu); Muav Limestone, Havasu Member (Єmh), Gateway Canyon Member (Єmg), Kanab Canyon Member (Єmk), Peach Springs Member (Єmp); Bright Angel Shale (Єb).

*52.1 (83.8) **Nankoweap Canyon and Nankoweap Rapids**; a drop of 25 feet (7.6 m). Nankoweap Creek-Colorado River gravel deposit on right; overlaps debris cone derived from the east wall of the canyon. The river gravel once occupied course of Colorado River (chapters 21 and 24). Old course of river once flowed along western margin of gravel deposit (right bank), incised into upper part of Bright Angel Shale. The present river course is now incised to the east (left) into the debris cone and fan. Note the imbrication of cobbles derived from the gravel deposit; cobbles of non-local origin are in lower part of the gravel deposit and may correlate with the gravel deposits at Lees Ferry. Access to Late Proterozoic Chuar Group rocks in Nankoweap basin (chapter 9) is gained by means of Nankoweap Creek.
54.8 (88.2) Fault; downstream side is up 2 feet (0.6 m).
56.0 (90.1) **Kwagunt Rapids**; a 7 foot (2.1 m) drop. Kwagunt Canyon on right provides access to Chuar Group rocks.
*56.3 (90.6) Travertine-cemented talus with river gravel on left (fig. 1.16). Travertine is seen in varying amounts on left for the next 5 miles (8 km), deposited by springs and seeps that emerged from strata of the basal Muav Limestone (chapter 21).
56.4 (90.7) Fault; downstream side is down about 15 feet (4.6 m).
57.0 (91.7) Devonian Temple Butte Limestone (or Formation) forms a thin (100 ft; 30 m) but relatively continuous unit that extends westward to Lake Mead; west of Matkatamiba Canyon (mi 148; km 238), it gradually increases in thickness to slightly more than 450 feet (137 m) in the western end of Grand Canyon.
57.3 (92.2) Fault; downstream side is up about 2 feet (0.6 m).
*58.2 (93.7) Gradational contact between green and red Bright Angel Shale and underlying brown Tapeats Sandstone. A full section of the Bright Angel Shale can be seen for next 1.5 miles (2.4 km). The red intervals are not true red beds, but rather are slightly weathered intervals of glauconitic sandstone.
59.6 (95.9) **60-Mile Rapids**; about a 3 foot (0.9 m) drop.
61.0 (98.2) Salt (halite) deposits on left are from seepage from (bleached) marine Tapeats Sandstone. The original red color of the Tapeats Sandstone will be seen at Deer Creek Falls (mi 136; km 219).

FIGURE 1.16. Travertine cemented talus and river gravel (t) mantling slope of Bright Angel Shale (Єb) and transition zone of Tapeats Sandstone (Єt). View is downstream towards left bank near mile 60 (96.5 km).

*61.3 (98.6)   **Junction of Little Colorado and Colorado Rivers.** Note thick deposit above pull-in for boats of well cemented gravel, in part of up-stream origin. The gravel here grades laterally into cemented talus and travertine deposits above the Colorado and Little Colorado Rivers. Extensive travertine and associated talus deposits are found only on the eastern side of the Colorado River south of Kwagunt Creek, and for approximately 7 miles (11 km) up the Little Colorado River along the north side.
*63.0 (101.4)   **Great Unconformity**; separates Paleozoic strata from Proterozoic rocks (chapter 9). Cambrian Tapeats Sandstone here overlies red, upper part of lower member (Escalante Creek Member) of Dox Sandstone (or Formation) (Unkar Group, Middle Proterozoic); representing a nearly 600 million year gap in the stratigraphic record. Salty water seeping from the Tapeats Sandstone is forming encrustations on left.
*64.5 (103.8)   Thrust fault in Dox Sandstone below Tapeats Sandstone, on right.
*65.4 (105.2)   **Palisades fault and monocline**, a splay from north-trending Butte fault, 1 mile (1.6 km) to west. Strata of the lower member of Dox Sandstone (Unkar Group) on the up-stream (north) side of the fault are adjacent to strata of the upper part of the Dox and the overlying Cardenas Basalt, indicative of about 2,800 feet (850 m) of down-to-the-south displacement, which occurred about 800 m.y. ago. However, displacement of the overlying Cambrian Tapeats Sandstone is about 400 feet (120 m) in the opposite sense, or down on the upstream side, which occurred during the Laramide. This served to reduce the displacement of the Proterozoic rocks to about 2,400 feet (730 m) (fig. 1.17). Copper prospects of the late 1800's and early 1900's are located on the Palisades fault, both sides of river. Note cemented gravels overlying cemented talus preserved on bench cut into fault gouge on left side of river. Note, also, sill in lower part of Dox upstream from fault on east side of river; it correlates paleomagnetically with thick diabase sills of central Grand Canyon. Access to Chuar Group and Butte fault to west is gained by means of Chuar Creek, on right.
*65.5 (105.4)   **Lava Canyon Rapids**; a 4 ft (1.2 m) drop. Numerous faults are seen in the Dox Sandstone (Formation) the next 2 miles (3.2 km). Upper member (dark red Ochoa Point Member) and upper middle member (red-orange and dark red Comanche Point Member) of Dox form slopes and benches beneath the Cardenas Basalt, on left (fig. 1.18). Lower part of Cardenas is green spilitic unit; overlying flows are separated by one prominent and several thin beds of sandstone (chapter 10). The basalt has been dated as 1,070±70 Ma by the Rb-Sr isochron method. Cemented gravels virtually at river level on right below rapids are discontinuously preserved for next 0.5 mile (0.2 km); then are seen on left side to Comanche Creek. A major, well cemented talus cone is preserved at the head of Comanche Creek. Debris flows that originated here join gravel deposits preserved within about 200 feet (60 m) of river level.

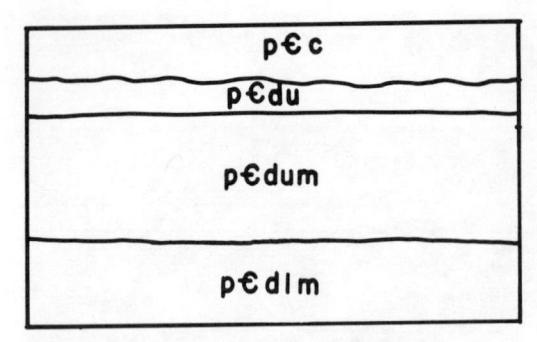

Grand Canyon Supergroup at Palisades Creek before faulting.

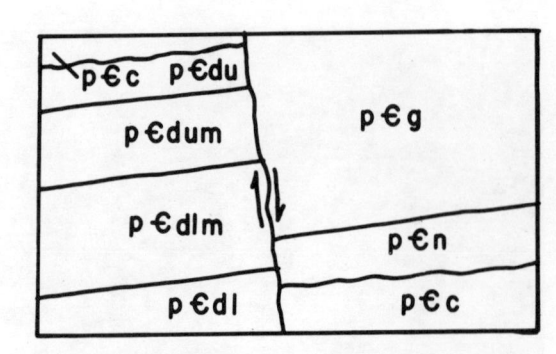

Normal faulting of supergroup about 800 m.y.; about 2,800 ft (850 m) displacement on Palisades fault.

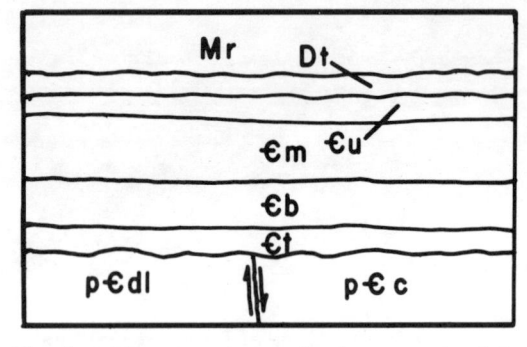

Erosion of supergroup and subsequent burial by Paleozoic rocks.

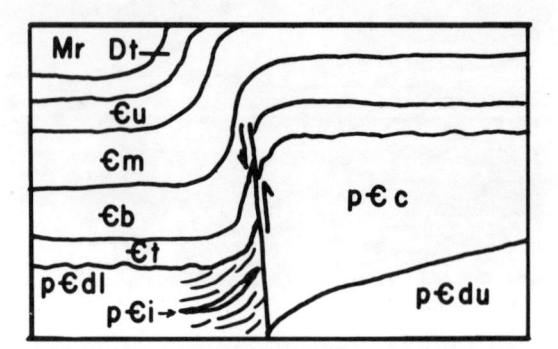

Reverse movement of Palisades fault, about 400 ft (120m) in Laramide time (60-50 m.y.), resulting in monoclinal flexing of upper Paleozoic rocks.

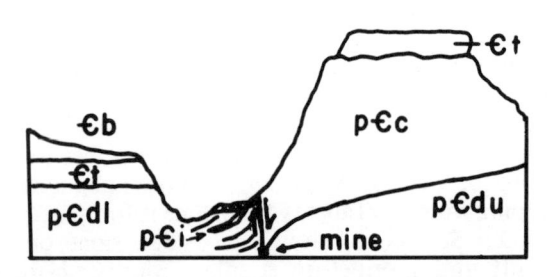

Palisades fault today. Net displacement after reverse faulting, about 2,400 ft (730 m).

FIGURE 1.17. Schematic diagram showing development of the Palisades fault and monocline at mile 65.4 (105.2 km). Paleozoic strata: Redwall Limestone (Mr); Temple Butte Limestone (Formation) (Dt); unclassified dolomite (Nopah equivalent) (€u); Muav Limestone (€m); Bright Angel Shale (€b); Tapeats Sandstone (€t). Proterozoic rocks: Galeros Formation (Chuar Group) (p€g); Nankoweap Foramtion (p€n); Cardenas Basalt (p€c); Dox Sandstone (Formation), upper member (Ochoa Point Member) (p€du), upper middle member (Comanche Point Member) (p€dum), lower middle member (Solomon Temple Member) (p€dlm), lower member (Escalante Creek Member) (p€dl); mafic sill (p€i).

*67.0 (107.8)   The Great Unconformity is well exposed in cliffs beneath Comanche Point on left (fig. 1.18); the Tapeats Sandstone forms a sheer cliff above the Cardenas Basalt, nearly wedging out where it crosses dark red upper member of the Nankoweap Formation high in amphitheater. Dark red ledgy beds of upper member of Dox Sandstone overlie variegated smooth slopes of the upper middle member of Dox, which is of tidal flat and near-shore terrestrial origin. Well cemented gravel near Tanner Creek, on left, correlates with river gravel plastered on cliff of Cardenas Basalt on right side of river above Tanner Canyon Rapids.

FIGURE 1.18. View to east of Comanche Point (skyline) from near Colorado River, mile 67 (107.8 km). Contacts dashed where approximately located. Kaibab Formation (Pk); Toroweap Formation (Pt); Coconino Sandstone (Pc); Hermit Shale (Ph); Supai Group (PIPs); Redwall Limestone (Mr); Temple Butte Limestone (Dt); unclassified dolomite and Muav Limestone, undivided (€um); Bright Angel Shale (€b); Tapeats Sandstone (€t); Great Unconformity (A); Nankoweap Formation (p€n); Cardenas Basalt (p€c); Dox Formation (p€d); terrace gravel of Colorado River (t). Photograph courtesy of H.G. Stephens.

68.5 (110.2) Butte fault and Tanner Rapids; a drop of 20 feet (6.1 m). Butte fault underlies East Kaibab monocline in Paleozoic rocks to the south on the south rim and Coconino Plateau (fig. 1.21). Cardenas Basalt (black) abuts the red sandstone of the middle members of the Dox on both sides of Basalt Graben (fig. 1.19). The Butte fault is on the east (upstream) side of Basalt Graben and Basalt Canyon fault on downstream (west) side. The Cardenas Basalt is overlain by red, ledge-forming sandstone of upper member of Nankoweap Formation in graben. The contact with the Chuar Group is near the middle of the cliff-face above the slope-forming, red-bed interval. Here, dolomite of the Tanner Member of the Galeros Formation rests unconformably on bleached sandstone of the upper member of the Nankoweap Formation. On the upstream (east) side of the graben, dark cliff-forming, ferruginous sandstone of the lower member of the Nankoweap, about 50 feet (15 m) thick,

FIGURE 1.19. View to north of Basalt Graben at Tanner Rapids, mile 70 (112.6 km). Contacts dashed where approximately located. Butte fault (A); Basalt fault (B); Nankoweap Formation (pЄn); lower (ferruginous) member of Nankoweap Formation (pЄnl); Cardenas Basalt (pЄc); Dox Formation (pЄd). Terrace gravels of Colorado River in foreground. Photograph courtesy of H.G. Stephens.

unconformably overlies and locally caps the Cardenas Basalt. The base of the Cardenas Basalt is seen locally near river level adjacent to small faults in the graben; post-Cardenas erosion has removed the physical relief at the top of the Cardenas owing to the faults. Proterozoic displacement on the Butte fault was about 1,700 feet (520 m) down on the downstream (west). However, Laramide displacement of about 600 feet (180 m), down on the upstream (east) side, resulted in the present displacement of about 1,100 feet (335 m).

Displacement across the Butte fault at Tanner Rapids is much less than a few miles to the north (fig. 1.20). On the plateaus north and south of the Grand Canyon, only Laramide displacement is reflected by the flexure of the East Kaibab monocline (fig. 1.21). The post-Laramide Colorado River, whose course was deflected southward by the East Kaibab monocline, was able to flow westward in this area as it rounded the southern extension of the Kaibab Upwarp. This bend in the river is known as the Big Bend area.

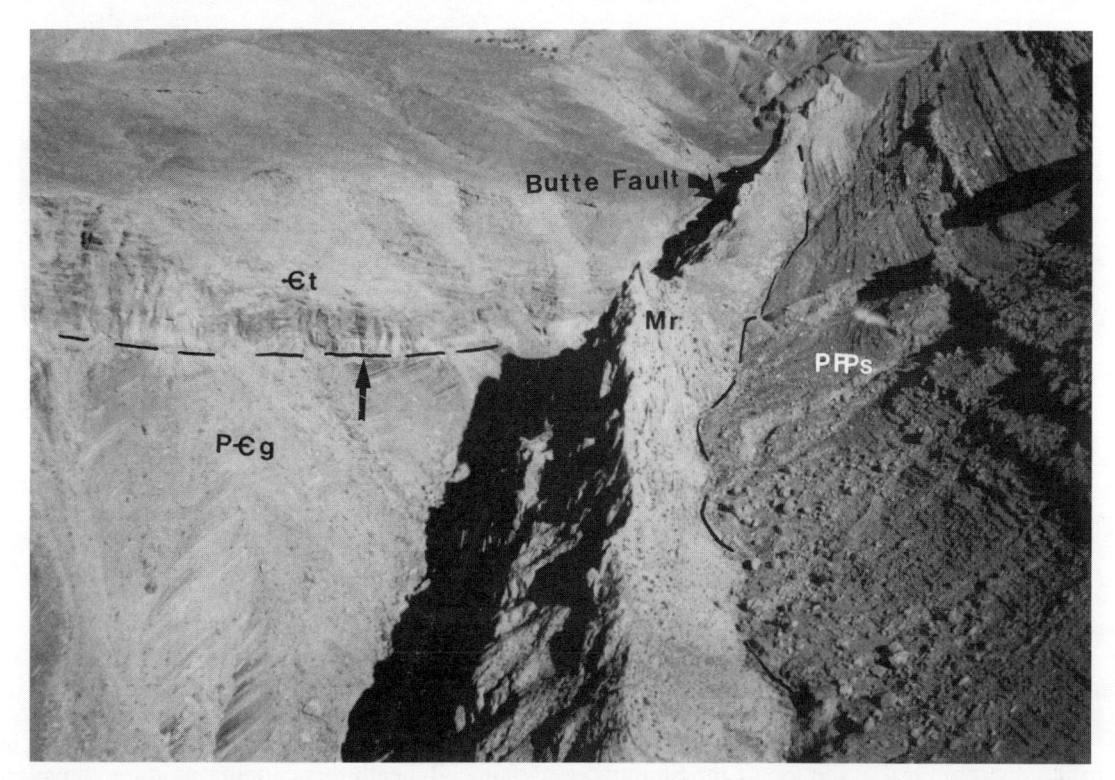

FIGURE 1.20. Butte fault in lower to middle Paleozoic and Late Proterozoic rocks. View is north at north arm of Carbon Creek Canyon (west of mile 63 or 101.4 km). Contacts dashed where approximately located. Supai Group (PPs); Redwall Limestone (Mr); Tapeats Sandstone (€t); Galeros Formation (p€g); Arrow points to Great Unconformity.

*69.0 (111.0)   **Basalt Canyon fault**, west side of Basalt Graben (fig. 1.19); downstream side is up about 430 feet (130 m). Cross back into upper middle member of Dox. No Laramide movement on this fault.
*69.4 (111.7)   Fluvial, well cemented gravels with up-stream components occupies old stream course of Colorado River, indicating an episode of aggradation. Alluvium on gravel near the river is a few tens of thousands of years old (chapter 24).
*69.5 (111.8)   Mouth of Basalt Canyon. Gains access to upper members of Dox Sandstone, Cardenas Basalt, Nankoweap Formation, and Galeros Formation of Chuar Group.
69.7 (112.2)   Several small faults in Dox on right.
70.5 (113.4)   Fault occupied by mafic dike, on left; downstream side is down about 80 feet (24 m).
*70.7 (113.8)   Gravel deposits and small gravel benches resting on middle members of Dox occur along this reach. View of "Hill Top Ruin" to west, on south side of river, is capped by gravels about 425 feet (130 m) above river.
71.1 (114.6)   Mafic dike, very near river on right intruded lower middle member (Solomon Temple Member) of Dox; extension of dike seen at mile 70.5 (km 113.4) correlates paleomagnetically with thick mafic sills of central Grand Canyon.
71.9 (115.7)   A small graben in cliff-forming, Solomon Temple (lower middle) Member of Dox, on right; about 4 feet (1.2 m) of displacement.
72.4 (116.5)   **Unkar Rapids**; a 25 foot (7.6 m) drop. A large debris fan from Unkar Creek is on right, behind which is a bench and a hill capped by the last of gravels of local and upstream origin in the Big Bend area.

FIGURE 1.21. Aerial view of East Kaibab monocline in upper Paleozoic rocks (overlying Butte fault). View looking southeast from near Desert View Tower on the south rim. East-dipping strata are Kaibab Formation (Pk).

72.9   (117.3)   Fault; downstream side is down about 3 feet (1 m).
*73.3 (117.9)   Begin traverse through dark red lower member (Escalante Creek Member) of Dox ; red upper part grades into gray strata at about mile 74 (km 119).
73.6   (118.4)   Graben on right; about 35 feet (11 m) of displacement.
*74.8 (120.4)   Sharp contact between gray sandstone of lower member of Dox and an underlying gray, cliff-forming quartzite bed at the top of the Shinumo Quartzite. Beneath this capping unit are convoluted fluid evulsion structures (fig. 1.22). These characterize most of the upper member of the Shinumo Quartzite and perhaps were the result of earthshocks. Multiple sets of these structures, separated by planar diastems indicative of episodic events, are beautifully displayed in Seventyfive Mile Canyon, above Nevills Rapids.
75.1   (120.8)   Fault; downstream side is up about 3 feet (1 m).
75.2   (121.0)   **Nevills Rapids**; a 15 foot (4.6 m) drop.
*76.2 (122.6)   Contact of Shinumo Quartzite with underlying pale purple, crossbedded sandstone of upper member of Hakatai Shale. Upper member of Hakatai tends to form a recess beneath the Shinumo Quartzite, whereas the middle (red-orange) and lower (dark red) members (seen downstream) form slopes with subordinate ledges (fig. 1.23).
76.4   (122.9)   Fault; downstream side is down about 12 feet (3.7 m).
*76.5 (123.1)   Thick mafic dike on right at top of Hance Rapids, follows a small fault in Hakatai Shale, and thins abruptly where it passes into Shinumo Quartzite; correlative dikes are found in Red Canyon on left. Paleomagnetic directions suggest the dikes correlate with Cardenas Basalt.
76.5   (123.1)   **Hance Rapids**; a 30 foot (9.1 m) drop.
*77.0 (123.9)   Contact of Hakatai Shale with underlying Bass Limestone, Unkar Group, placed at highest laterally persistent sandstone bed. Hance asbestos mines (1880's and 1890's) can be seen high on right about a mile (1.6 km) downstream, as a white slope at the top of a dark sill near the base of the Bass Limestone (fig. 1.23). Mafic sill is not connected with or related to mafic dikes of Red Canyon.
*77.5 (124.7)   Hotauta Conglomerate Member of Bass Limestone. Unconformity between late Early Proterozoic crystalline rocks and overlying Middle Proterozoic strata; unconformity dips northeastward (upstream) and represents an ~425 million year gap in the geologic record (~1,250-~1,675 m.y.). This unconformity is called the Greatest Angular Unconformity; it merges with the Great Unconformity a short distance downstream and about 1,000 feet (305 m) above the river, where the combined unconformities represent a 1,200 million year gap in the geologic record.

FIGURE 1.22. Contorted beds of the Shinumo Quartzite seen in 75-Mile Creek, about mile 75.4 (121.3 km). Hammer for scale. Photograph courtesy of H.G. Stephens.

*77.5 (124.7) **Entering Upper Granite Gorge.** Early Proterozoic metasedimentary and mafic metaigneous rocks (chapter 8) are exposed the next several miles; they consist of mica schist and quartzo-feldspathic schist, and subordinate paragneiss, amphibolite, and calc-silicate rocks, commonly referred to as the Vishnu Schist (fig. 1.24). Foliated and non-foliated granitic plutons and associated pegmatites (fig. 1.25) commonly referred to as Zoroaster Granite. This and other plutons encountered down river variously exhibit foliated and non-foliated textures, implying that intrusion occurred during late stages of deformation. The deformation is commonly considered to be related to the Mazatzal orogeny of central Arizona dated at about 1,650-1,700 Ma). Note that the foliation of the Vishnu Schist is extensively embayed by the plutonic complex.
78.6 (126.5) **Sockdolager Rapids**; a 19 foot (5.8 m) drop.
81.1 (130.5) Vishnu fault; downstream side is down about 40 feet (12.2 m) as measured in overlying Paleozoic rocks.
81.5 (131.1) **Grapevine Rapids**; an 18 foot (5.5 m) drop.
83.6 (134.5) **83-Mile Rapids**; a 7 foot (2.1 m) drop.

FIGURE 1.23. Looking downstream from mile 77 (124 km) near bottom of Hance Rapids. Landslide debris (L) on left; Shinumo Quartzite (pЄs); Hakatai Shale (pЄh); Bass Limestone (pЄb) intruded by dark diabase sill (s); Vishnu Schist (pЄv); Hance asbestos mines at arrow 1; Greatest Angular Unconformity at arrow 2. Contacts dashed where approximately located.

84.6   (136.1)   **Zoroaster Rapids**; a 5 foot (1.5 m) drop.
84.7   (136.3)   Contact between Vishnu Schist (upstream) and foliated Zoroaster Granite.
86.1   (138.5)   Contact between Zoroaster Granite (upstream) and Vishnu Schist.
87.2   (140.3)   Well cemented river gravels at river level (low water), on right.
87.3   (140.5)   Cableway and gaugehouse mark location of the U.S. Geological Survey gauging station, established in 1922 (chapter 3).
87.4   (140.6)   **Kaibab (Trail) suspension bridge**; completed in 1928.
87.5   (140.8)   **Cremation fault and Bright Angel Rapids.** Bright Angel Rapids has about a 7 foot (2.1 m) drop. Bright Angel Creek debris fan on right. Cremation fault crosses river just downstream from Kaibab suspension bridge (fig. 1.26); downstream side is down about 780 feet (238 m). Proterozoic displacement was about 1,000 feet (305 m) down on downstream side; reverse movement during the Laramide offset Paleozoic rocks up on the downstream side about 220 feet (67.1 m), forming the Grandview monocline in upper Paleozoic rocks at the rim of the canyon.
87.8   (141.3)   **Bright Angel (Trail) suspension bridge**; completed in 1967.
88.0   (141.6)   **Tipoff fault**, downstream side is up; Proterozoic displacement was about 400 feet (122 m); Laramide displacement has reversed the displacement about 15 feet (5 m). The structural wedge between the Cremation and Tipoff faults was a graben during the Proterozoic, and became a horst during the Laramide (fig. 1.26).
88.3   (142.1)   **Bright Angel fault**; displacement was about 800 feet (245 m), downstream side down in Proterozoic time. Reversed movement during the Laramide reduced the Proterozoic displacement to about 470 feet (145 m) and displaced Paleozoic strata about 330 feet (100 m) up on downstream side.
88.9   (143.0)   **Pipe Springs Rapids**; about a 7 foot (2 m) drop.
89.1   (143.4)   Contact between Vishnu Schist (upstream) and Zoroaster Granite.
89.2   (143.5)   Fault; downstream side is up about 400 feet (122 m).
89.8   (144.5)   Contact between Zoroaster Granite (upstream) and Vishnu Schist.
90.1   (150.0)   **Horn Creek Rapids**; a 10 foot (3.1 m) drop.
91.3   (146.9)   Contact between Vishnu Schist (upstream) and Trinity Gneiss (chapter 8).
92.5   (148.8)   **Salt Creek Rapids**; a 5 foot (1.5 m) drop.
92.5   (148.8)   Contact between Trinity Gneiss (upstream) and Vishnu Schist.
93.4   (150.3)   **Granite Rapids**; a 17 foot (5.2 m) drop.
94.0   (151.2)   Fault; downstream side is up about 5 feet (1.5 m).

FIGURE 1.24. Near-vertical foliation in Vishnu Schist, about mile 82 (km 132), left bank.

FIGURE 1.25. Pegmatite dikes (light) intrude Vishnu Schist (black); about mile 82 (km 132), left bank. Pegmatite dikes intrude both the Vishnu Schist and Zoroaster Granite.

FIGURE 1.26. Aerial view looking southeast towards the mouth of Bright Angel Creek (A) and lower part of south Kaibab Trail (B) at mile 87.8 (km 141.3). View shows a Proterozoic age graben that is now a Paleozoic horst between the Cremation (arrow C, on left) and Tipoff (arrow D, on right) faults. Kaibab Suspension Bridge (arrow E). Contacts dashed where approximately located. Colorado River terrace (t); Bright Angel Shale (€b); Tapeats Sandstone (€t); Shinumo Quartzite (p€s); Hakatai Shale (p€h); Bass Limestone (p€b); Vishnu Schist and Zoroaster Granite (p€). U/D=sense of Paleozoic displacement on fault; d/u=sense of Proterozoic displacement on fault.

95.0 (152.9) **Hermit Rapids**; a 15 foot (4.6 m) drop. This rapid usually has large standing waves that make it one of the most memorable rides on the river.
*95.3-95.5 (153.3-153.7) Isolated bodies of travertine on Bright Angel Shale and Tapeats Sandstone, 800+ feet (244+ m) above river on left; travertine extends down over Vishnu Schist to about 150 feet (45 m) above river. Source for this travertine were springs emerging from the Muav Limestone high above the river.
96.2 (154.8) Granitic pluton exposed the next 0.3 mile (0.5 km).
96.7 (155.6) **Boucher Rapids**; a 13 foot (4.0 m) drop.
97.7 (157.2) Small Zoroaster Granite pluton exposed the next 0.3 mile (0.5 km).
98.0 (157.7) **Crystal Rapids**; a 17 foot (5.2 m) drop (chapter 3). Crystal Creek debris fan on right. Slate Creek fault crosses river; offset in Paleozoic strata indicate downstream side up about 100 feet (30 m).
98.7 (158.8) Small Zoroaster granitic pluton next 0.4 mile (0.6 km).
99.0 (159.3) Fault; downstream side up about 80 feet (24.4 m) as indicated in Paleozoic strata. Fault underlies Crazy Jug monocline in upper Paleozoic strata.
99.2 (159.6) **Tuna Creek Rapids**; a 10 foot (3.1 m) drop.
99.6 (160.3) **Lower Tuna Rapids**; about a 3 foot (1 m) drop.
100.5 (161.7) **Agate Rapids**; about a 1 foot (0.3 m) drop.
101.2 (162.8) **Sapphire Rapids**; a 7 foot (2.1 m) drop.
102.0 (164.1) **Turquoise Rapids**; a 2 foot (0.6 m) drop.
102.8 (165.4) Contact between Vishnu Schist (upstream) and Zoroaster Granite pluton.
*102.9 (165.6) View downstream (northwest) toward Modnadnock Amphitheater. Tapeats Sandstone thins across a Precambrian hill. Dark resistant ledge-forming beds in overlying green slope of Bright Angel Shale are equivalent to the Rampart Cave, Sanup Plateau, and Spencer Canyon Members of the Muav Limestone to the west (chapter 15). A tan unit at the base of the middle cliff is the Peach Springs Member of the Muav Limestone (a single bed); it is overlain by cliff-forming Kanab Canyon Member (two beds). Above this is the less resistant Gateway Canyon Member (three units forming a slope/ledge/slope), which is in turn overlain by a cliff containing

the Havasu Member of the Muav (two beds), overlain by cliff-forming unclassified dolomite, followed by less resistant Temple Butte Limestone (~30 m thick), and cliff-forming Redwall Limestone at very top.

103.2 (166.1) Arch, on left.

103.9 (167.2) **104-Mile Rapids**; about a 4 foot (1.2 m) drop.

104.8 (168.6) **Ruby Rapids**; an 11 foot (3.4 m) drop. View downstream of tilted cliff-forming Shinumo Quartzite (Unkar Group) unconformably overlain by basal Paleozoic strata (fig. 1.27).

104.9 (168.8) **Serpentine Rapids**; an 11 foot (3.4 m) drop.

*106.5 (171.4) About 1 mile (1.6 km) above Bass Camp; Bass Limestone (Unkar Group) appears beneath Cambrian Tapeats Sandstone on right, then on left. View down river to Vishnu Schist overlain by Bass Limestone that is in turn intruded by diabase sill in middle distance and on left; light colored rocks are thermally altered and bleached Bass Limestone. Above the bleached zone are ledges of the upper Bass Limestone, slope-forming red Hakatai Shale, and cliff-forming Shinumo Quartzite (extending to top of third cliff, characterized by smooth face and capped with uppermost quartzite bed of Shinumo). Above the Shinumo is cliff-forming sandstone of the lower member (Escalante Creek Member) of Dox Sandstone (Formation).

*107.2 (172.5) Full section of diabase sill intruded into Bass Limestone comes into view. Paleomagnetic evidence indicates that the diabase sills were intruded during deposition of the upper middle member (Comanche Point Member) of Dox Sandstone (Formation). The Proterozoic rocks here dip about 12° northeast (fig. 1.27).

107.3 (172.6) West-dipping Wheeler Fold in Bass Limestone; displacement is about 500 feet (152.4 m).down on downstream side.

107.7 (173.6) Contact of Zoroaster Granite (upstream) with Vishnu Schist.

107.7 (173.6) **Bass Rapids**; a 4 foot (1.2 m) drop.

108.5 (174.6) **Shinumo Creek**; enters on right.

108.6 (174.7) **Shinumo Rapids**; an 8 foot (2.4 m) drop. Tapeats Sandstone laps onto Bass Limestone; exposures are complicated by faulting.

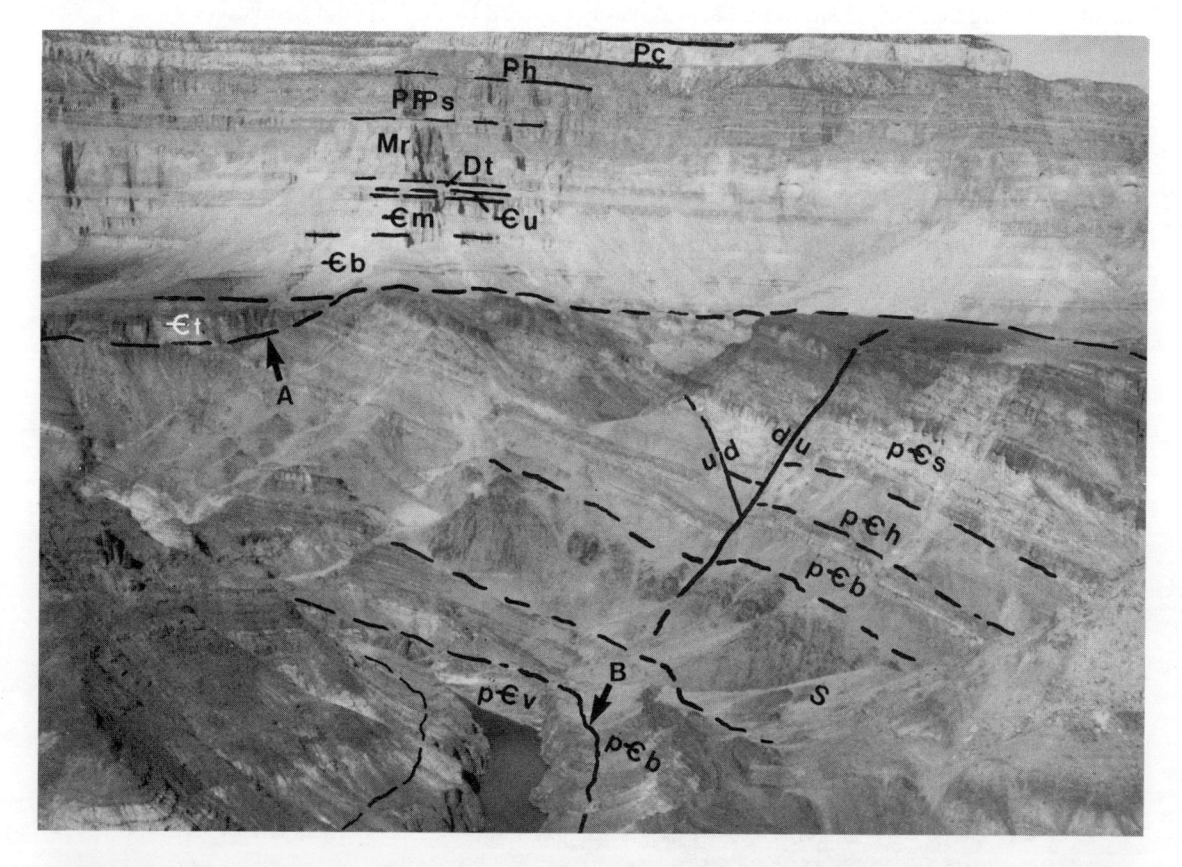

FIGURE 1.27. Aerial view looking northwest (downstream) to Shinumo Creek near mile 107 (km 172), showing tilted Proterozoic Grand Canyon Supergroup dipping about 20° to the northeast. Contacts dashed where approximately located. Coconino Sandstone (Pc); Hermit Shale (Ph); Supai Group (PPs); Redwall Limestone (Mr); Temple Butte Limestone (Dt); unclassified dolomite (Єu); Muav Limestone (Єm); Bright Angel Shale (Єb); Tapeats Sandstone (Єt); Great Unconformity (A); Shinumo Quartzite (pЄs); Hakatai Shale (pЄh); Bass Limestone (pЄb) intruded by diabase sill (S); Vishnu Schist (pЄv); Great Unconformity at Arrow A; Greatest Angular Unconformity at Arrow B

110.7 (178.1) **Hakatai Rapids**; an 8 foot (2.4 m) drop.
112.0 (180.2) **Waltenberg Rapids**; a 15 foot (4.6 m) drop.
112.5 (181.0) Arch, on left.
*114.4 (184.0) Travertine deposits having an apparent source in Kanab Canyon Member of the Muav Limestone appear on left; they are remnants of a formerly continuous wall of travertine that extended downstream the next 3 miles (4.8 km). The two ledge-forming beds in the otherwise slope-forming Bright Angel Shale are equivalent to the Rampart Cave and Sanup Plateau Members of the Muav Limestone to the west (fig. 13.1; chapter 15).
115.5 (185.8) **Monument fault.** On the right the sharply flexed fold in the Tapeats Sandstone becomes a fault in the overlying Redwall Limestone with a displacement of about 20 feet (6.1 m) down on the upstream side. This fold, and the next fold just downstream (fig. 1.28), are part of the Monument fault system.
115.6 (186.0) Monument fault, on right; downstream side is up about 45 feet (13.7 m). Tapeats Sandstone, draped over the fault, dips upstream. Vishnu Schist exposed at river level downstream.
*116.6 (187.6) **Royal Arch Creek (Elves Chasm)**, on left, is cut through a deposit of travertine that extends from the Tapeats Sandstone near river level to the top of the Muav Limestone. View of complete section of Cambrian Tonto Group and overlying strata up and including the Redwall Limestone in amphitheater to the west of Elves Chasm and a curious channel-like feature can be seen at about the level of the Temple Butte Limestone (Formation) in the back of the amphitheater. Note the several sandstone members of the Bright Angel Shale that become limestone beds in western Grand Canyon. Remnants of travertine and talus that once mantled the entire area are present within and north of the amphitheater.
117.7 (189.4) Continuation of Monument fault; downstream side is down about 35 feet (11 m), bringing Tapeats Sandstone to river level. Arch on left. Begin Stephen Aisle, a long stretch of river allowing excellent views of entire Paleozoic section.
118.8 (191.1) Contact of Tapeats Sandstone with Vishnu Schist exposed in next mile (1.6 km). The Vishnu formed an island during the Early Cambrian that was barely covered by the Tapeats Sandstone.
119.0 (191.5) Fault; downstream side is up about 3 feet (0.9 m).
*120.2 (193.4) Begin Conquistador Aisle; stratigraphy of Tonto Group well exposed. Contact of Vishnu Schist (upstream) with Tapeats Sandstone. Excellent exposures of Great Unconformity in Blacktail Canyon on right.
121.5 (195.5) Well cemented gravel at river level, on right; well cemented talus above river on left.
122.0 (196.3) Well cemented gravel, on right.
122.7 (197.4) **Forster Rapids**; a 5 foot (1.5 m) drop.
123.2 (198.2) Gradational contact between Tapeats Sandstone (upstream) and Bright Angel Shale.

FIGURE 1.28. Lower Monument fault (not in view) with folded Tapeats Sandstone dipping upstream at mile 115.6 (km 186.0).

123.2 (198.2) Butchart fault; downstream side is up about 10 feet (3.1).
124.0 (199.5) Gradational contact between the Bright Angel Shale (upstream) and the Tapeats Sandstone. Arch in Supai Group high on left.
124.2 (199.8) Travertine on left.
125.0 (201.1) **Fossil Rapids**; a 15 foot (4.6 m) drop. Fossil Canyon debris fan on left.
126.6 (203.7) Great Unconformity; Tapeats Sandstone on Vishnu Schist (amphibolite).
126.8 (204.0) **127-Mile Rapids**; about a 3 foot (1 m) drop.
*127.4 (205.0) Entering **Middle Granite Gorge**. Salt stalactites hang from Tapeats Sandstone at Great Unconformity, on left.
128.5 (206.8) **128-Mile Rapids**; drops about 6 feet (1.8 m).
129.0 (207.6) **Specter Rapids**; a 4 foot (1.2 m) drop.
129.3 (208.0) Three faults in Proterozoic basement here, but senses and amounts of displacement are unknown. These faults extend into upper Paleozoic rocks in Specter Chasm on left with an overall Paleozoic displacement of about 35 feet (11 m), up on downstream side. A small moncline is associated with these faults and dips upstream.
*130.4 (209.8) **Bedrock Rapids**; a 7 foot (2.1 m) drop. Bedrock Rapids is one of a few rapids caused by a constriction arising from bedrock outcrops. Fault cuts Middle Proterozoic strata of lower Unkar Group; downstream side is down about 50 feet (15 m).
130.6 (210.1) Contact between Vishnu Schist (upstream) and Bass Limestone.
*130.8 (210.5) Diabase sill within Bass Limestone forms a cliff on right for next 3 miles (5 km). The sill and lower Unkar Group strata here dip about 6° northeast (fig. 1.29).
131.6 (211.7) **Deubendorff Rapids**; a 15 foot (4.6 m) drop. Fault cuts Middle Proterozoic rocks; downstream side is up about 150 feet (45 m).
132.1 (212.5) Fault in Middle Proterozoic rocks; downstream side is up about 50 feet (15 m).
132.9 (213.8) Fault in Middle Proterozoic rocks; downstream side is up 50 feet (15 m).
*133.6 (215.0) **Tapeats Creek and Rapids**; a 15 foot (4.6 m) drop. The water in Tapeats Creek (Thunder River) comes from a joint-controlled spring in the Peach Springs Member, Muav Limestone, about 3 miles (5 km) from the Colorado River (fig. 1.30).

FIGURE 1.29. View downstream toward 133-Mile Creek from mile 132.4 (km 213.0), above a diabase sill (S) intruded into the Bass Limestone (pЄb). Contacts dashed where approximately located. Hakatai Shale (pЄh); Great Unconformity (A); Tapeats Sandstone (Єt); Bright Angel Shale (Єb); Muav Limestone (Єm); Surprise Valley Landslide (L).

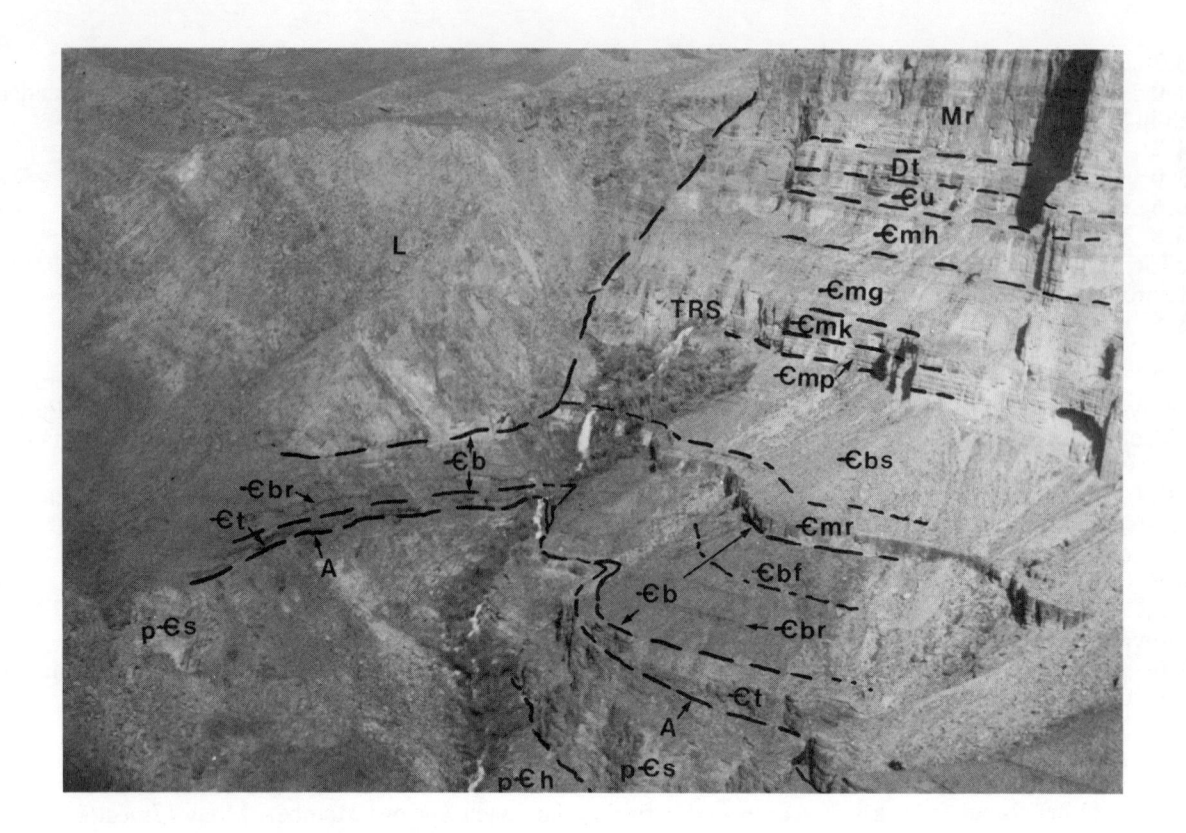

FIGURE 1.30. Aerial view looking west to Thunder River Spring (TRS), which emerges from the Peach Springs Member of the Muav Limestone (Cmp) about 3 miles (4.8 km) from the Colorado River. Part of the Surprise Valley landslide (L) lies just west (left) of the spring, its base resting partly on the Rampart Cave Member of the Muav Limestone (€mr) and partly on the shaly Flour Sack Member of the Bright Angel Shale (€bf). Contacts dashed where approximately located. Hakatai Shale (p€h); Shinumo Quartzite (p€s); Great Unconformity (A); Tapeats Sandstone (€t); Bright Angel Shale (€b), red brown member (€br), Flour Sack Member (€bf); Rampart Cave Member (€mr) of Muav Limestone of western Grand Canyon; upper Bright Angel Shale of eastern Grand Canyon (€bs); Muav Limestone, Peach Springs Member (€mp); Kanab Canyon Member (€mk); Gateway Canyon Member (€mg); Havasu Member (€mh); unclassified dolomite (€u); Temple Butte Limestone (Dt); Redwall Limestone (Mr). For problems in Cambrian nomenclature see Chapters 14 and 15.

134.3 (216.1) Fault in Middle Proterozoic rocks; downstream side is up about 400 feet (122 m). Diabase sill upstream; Hakatai Shale downstream at fault.
134.6 (216.6) Contact of Hakatai Shale (upstream) with Bass Limestone.
*135.0 (217.2) Entering Little Granite Gorge. Bass Limestone overlies Vishnu Schist. Travertine cemented gravels, beginning about 100 feet (30 m) above river, on right, occupy and fill an old Colorado River channel cut through Tapeats Sandstone and into Bass Limestone (fig. 1.31). The filled channel is overlain by the Surprise Valley landslide (chapters 21 and 26). The Colorado River was subsequently established slightly to the south along its present course. The west end of the abandoned channel, very near present river level, joins with the present course of the river at mile 135.8 (218.5 km).
*135.2 (217.5) Narrowest point along Colorado River in Grand Canyon, 76 feet (23 m) wide.
*136.2 (219.2) Deer Creek Falls on right. Part of Surprise Canyon landslide is seen on right as a rubbly slope downstream from the falls. Travertine and travertine cemented gravels are exposed high on left, overlain by landslide that contains dolomite of the Bright Angel Shale (equivalent to the Rampart Cave Member of the Muav Limestone) derived from the right (north) bank; the dolomite is displaced up slope from undisturbed dolomite on the left (south) side of the river. Note, also, that nearly the entire section of Tapeats Sandstone is red in this area. The Tapeats presumably escaped Cenozoic bleaching because of its structurally low position in this area.
*136.6 (219.8) View upstream, a) into west end of old Deer Creek Canyon filled with Surprise Valley landslide on north side of river; b) landslide overlying travertine-cemented gravels on south side of river.
137.0 (220.4) Fault in Paleozoic rocks nearly parallels river; downstream (north) side is up about 33 feet (10 m).

FIGURE 1.31. Abandoned Colorado River channel in the Bass Limestone (p€b) filled with travertine-cemented river gravel (tg). The broken material above the channel fill is part of the Surprise Valley landslide (L), consisting mainly of Redwall Limestone. View is downstream from mile 135 (km 217) to narrowest part of Colorado River. Contacts dashed where approximately located. Vishnu Schist and Zoroaster Granite (p€v); Bass Limestone (p€b); Tapeats Sandstone (€t).

137.2 (220.8) Fault in Paleozoic rocks, nearly parallels river; downstream (north) side is up about 80 feet (25 m). Cambrian Tapeats Sandstone overlies Proterozoic Bass Limestone.
137.6 (221.4) **Doris Rapids**; about a 5 foot (1.5 m) drop
137.7 (221.6) Contact of Bass Limestone with overlying Tapeats Sandstone.
137.9 (221.9) Spring in Tapeats Sandstone; water is seeping down from the Deer Creek area under the Surprise Valley landslide.
138.0 (222.0) "Fishtail island." Tapeats Sandstone and Bright Angel Shale overlap an island of altered Bass Limestone, which in turn rests on Vishnu Schist. Last outcrop of Grand Canyon Supergroup rocks.
138.5 (222.8) Fault in Proterozoic rocks, on left; downstream side is down about 50 feet (15 m). Tapeats Sandstone is at river level. Arch in cemented gravels about 50 feet (15 m) above river, on right. Extensive deposits of cemented gravel and talus are on right (north) side of river.
138.7 (223.2) **Sinyala fault**, on left side of river, covered by gravel on right side; downstream side is down about 12 feet (4 m). The Sinyala fault is small in throw compared to many faults of its length but it has influenced the erosion of more tributary canyons than any other fault in Grand Canyon. All fault displacements from here to Lake Mead are estimated by offset of Paleozoic rocks.
139.0 (223.7) **Fishtail Rapids**; a 10 foot (3 m) drop.
140.4 (225.9) Gradational contact of Tapeats Sandstone with Bright Angel Shale.
141.3 (227.4) Fault; downstream side is down about 4 feet (1 m).
141.8 (228.2) Approximate place where dolomite unit in Bright Angel Shale grades into limestone and becomes assigned to Rampart Cave Member of Muav Limestone, a terminology and convention still currently in use.
*142.0 (228.5) Begin "river (canyon) anticlines" developed in Rampart Cave, Sanup Plateau, Peach Springs, Kanab Canyon, and Gateway Canyon Members of Muav Limestone.
143.6 (231.1) **Kanab Rapids**; about a 12 foot (3.7 m) drop.
*146.0 (234.9) Strongly developed river anticlines. Limestone beds dip variably away from the river, mostly near and at river level, but also, at a few places, substantial vertical distances above the river. The anticlines appear to have resulted from vertical loading stresses of the high canyon walls on saturated clay-bearing strata (chapter 26). The anticlines mostly parallel the Colorado River for the next several miles. A short reach above Upset Rapids, at right angles to the trend of the Sinyala fault, lacks an anticlinal structure.
147.0 (236.5) Arch, high on left
147.8 (237.8) Small spring on the right.

*148.0 (238.1) **Matkatamiba Canyon**; Gateway Canyon Member of Muav Limestone dips upstream as part of the Matkatamiba Syncline. Shelf at entrance to Matkatamiba Canyon is at top of Kanab Canyon Member. The Gateway Canyon Member here is much thicker than reported from sections measured to the west and east.

149.7 (240.9) **Upset Rapids**; a 15 foot (4.6 m) drop.
151.0 (243.0) Small springs and travertine deposits on right next 0.5 mile (0.8 km).
153.3 (246.7) **Sinyala Rapids**; about a 1 foot (0.3 m) drop.
155.4 (250.0) Spring on right.
156.6 (252.0) Supai monocline; dips downstream (southwest).
*156.7 (252.1) **Havasu Canyon and Havasu Rapids**; about a 2 foot (0.5 m) drop.
159.1 (256.0) **159-Mile dikes.** Basalt dikes on both sides of the river extend through Supai Group; they appear to be associated with Pleistocene (780,000 yrs.; unpublished date) cinder cones on Esplanade Sandstone on both sides of river.
164.4 (264.5) **164-Mile Rapids**; about a 3 foot (1 m) drop.
164.7 (265.0) Arch, high on left.
166.5 (267.9) **National Rapids**; about a 2 foot (0.5 m) drop.
168.0 (270.3) **Fern Glen Rapids**; about a 3 foot (1 m) drop. The best view of the Surprise Canyon Formation since Marble Canyon can be seen as a dark red channel fill at top of Redwall Limestone cliff; look downstream, high on right.
171.2 (275.5) **Mohawk-Stairway fault**; downstream side is up about 90 feet (27 m). Strata dip to the east (upstream) about 3°. This fault controls the drainage of Mohawk and Stairway Canyons. Peach Springs and Kanab Canyon Members of Muav Limestone on upstream side of fault; Bright Angel Shale on downstream side.
171.4 (275.8) **Gateway Rapids**; about a 3 foot (1 m) drop. Mohawk Canyon debris fan on left.
173.2 (278.7) Travertine deposits high on the left next 1 mile (1.6 km).
175.0 (281.6) "Red Slide;" a large landslide and talus deposit, on right.
175.4 (282.2) Travertine deposits high on left.
176.5 (284.0) Several travertine deposits high on the left.
177.0 (284.8) Basalt flow remnant in alcove high on left; source unknown.
177.2 (285.1) Pleistocene basalt flows (chapters 22 and 23) on left bank. These are remnants of flows that erupted downstream.
177.5 (285.6) Travertine deposits on both sides of river.
*178.0 (286.4) **Vulcan's Forge** (Vulcan's Anvil); part of a Pleistocene dike and sill system seen on right bank of river and wall of canyon.
*178.2 (286.7) Pleistocene basalt flows that came from volcanos mainly on north rim filled Toroweap Canyon and flowed down the course of the Colorado River for a distance of at least 74 miles (119 km). See Chapter 23.
178.8 (287.7) Contact of Bright Angel Shale (green) on Tapeats Sandstone (brown).
*179.0 (288.0) **Toroweap fault**; downstream side is down about 580 feet (176.8 m). The basalt flows are displaced about 15 feet (5 m), indicating some Pleistocene movement. Tapeats Sandstone, upstream; Rampart Cave Member of Muav Limestone, downstream.
*179.0 (288.0) **Lava Falls Rapids**; a 13 foot (4 m) drop. Note high terrace deposits on left bank which are remnants of debris flows derived from Prospect Canyon. Prospect Canyon drainage was blocked by basalt flows emplaced at and near its mouth that were locally derived from feeder dikes and vents or from flows down Toroweap Canyon on the north. The drainage above the blockage at the mouth of Prospect Canyon is presently filled with alluvium to the level of the highest basalt flows. The fill is currently being excavated along the Toroweap fault in Prospect Canyon. Toroweap Canyon, opposite Prospect Canyon, is filled with basalt and interbedded alluvium and remains unexcavated because multiple basalt flows farther up canyon have interrupted the drainage.
179.5 (288.8) Warm springs and travertine on left. These springs are the warmest in Grand Canyon (80 °F; 26.6 °C).
*179.8 (289.3) Bench on left below Lower Lava Rapid is on top of Peach Springs Member of Muav Limestone; thin bed of tuff marks contact with Kanab Canyon Member, and provided zircon for a fission track age of 535 Ma (chapter 17).
180.7 (290.7) Fault; downstream side is down about 40 feet (12 m).
181.3 (291.7) Spring on left; seen only at very low water level..
181.4 (291.9) Contact between Rampart Cave Member of Muav Limestone, and underlying Bright Angel Shale.
183.4 (295.1) Several small faults seen on left but do not cross the river for next few miles.
183.5 (295.3) Spring on left.
184.8 (297.3) Lava cascade on right.
185.2 (298.0) **185-Mile Rapids**; about a 2 foot (0.5 m) drop.
185.3 (298.2) Fault; downstream side is down about 10 feet (3 m).

*187.0 (300.9)   Lava-filled channel of old Whitmore Canyon, on right. Whitmore Trail traverses the lava fill. Base of narrow channel is about 30 ft (10 m) above river.
187.5 (301.7)   Fault; downstream side is down about 180 feet (55 m).
188.0 (302.5)   Gradational contact between Bright Angel Shale and Tapeats Sandstone. Present (post-basalt) Whitmore Canyon and debris fan on right.
189.5 (304.9)   Fault; downstream side is up about 70 feet (21 m). Tapeats Sandstone upstream; Vishnu Schist and Zoroaster Granite, downstream. Great Unconformity on left.
189.6 (305.1)   Large landslide exposed high on left.
189.8 (305.4)   Travertine on left.
190.0 (305.7)   Upper part of Tapeats Sandstone on granite, left and right.
*190.7 (306.8)   **Hurricane fault zone**; downstream side is down about 1,300 feet (400 m). Here, landslide deposits cover trace of fault on right; terrace deposits and basalt flows cover fault on left. Zoroaster Granite, upstream; Muav Limestone, downstream. The Hurricane fault is encountered again from mile 220 to 224 (km 354-360).
192.1 (309.1)   Fault; downstream side is down about 30 feet (10 m). Two faults, paralleling the river with their right sides (west) up about 40 feet (12 m) and 60 feet (18 m), intersect the fault that crosses river. These faults are seen ahead on left.
192.5 (309.7)   Fault paralleling river is seen downstream on left; downstream side is up about 100 feet (30 m). Thick river terrace deposits and interbedded basalt flows on right.
193.1 (310.7)   Fault; downstream side is down about 50 feet (15 m).
193.7 (311.7)   Four small faults on right; upstream side up on all, about 15 feet (5 m) each.
195.0 (313.8)   Fault; downstream side is up 10 feet (3 m).
196.0 (315.4)   Fault; downstream side is up 40 feet (12 m).
196.3 (315.9)   Lone Mountain monocline; dips upstream (east) about 15°.
196.3 (315.9)   Frogy fault; downstream side is down about 400 feet (120 m).
196.7 (316.5)   Fault; downstream side is up about 10 feet (3 m). Several late Miocene (~6 Ma) basalt dikes cross the river, following small faults. Arch in river conglomerate on right spans a dike and fault.
197.4 (317.6)   Approximate position of north-south-trending axis of Lone Mountain anticline.
198.0 (318.6)   Entering Parashant graben. Five faults; overall displacement is 100 feet (30 m) down on downstream side.
198.5 (319.4)   Fault; downstream side is down, about 200 feet (60 m)
199.3 (320.7)   Fault; downstream side is up about 180 feet (55 m).
199.8 (321.5)   Leaving Parashant graben. Fault; downstream side is up about 100 feet (30.5 m). Rampart Cave Member, Muav Limestone, upstream; Bright Angel Shale, downstream, at river level.
201.0 (323.4)   Fault follows river; left bank is down about 10 feet (3 m).
203.0 (326.6)   Fault; downstream side is down about 30 feet (9 m).
204.3 (328.7)   Fault; downstream side is down about 60 feet (18 m).
205.2 (330.2)   **205-Mile Rapids**; a 13 foot (4 m) drop.
205.3 (330.3)   Spring on left bank in back eddy of 205-Mile Rapids; seen only at very low water level.
205.5 (330.6)   River cuts through landslide debris.
*205.7 (331.0)   Tapeats Sandstone at river level on right is the upper part of the massive sandstone unit of the Tapeats (figs. 15.2 and 15.3), closely overlain by the *Olenellus* horizon. This unit of the Tapeats of Figure 15.3 is correlated with (the upper part of) the Prospect Mountain Quartzite of Whitney Ridge, Nevada, on the basis of stratigraphic position and the presence of *Olenellus*. This horizon is overlain by a slope-forming interval containing shale and sandstone of Tapeats lithology, overlain by cliff-forming sandstone of Tapeats lithology called the "red-brown sandstone" (figs. 1.32 and 15.3); transitional beds in the slope above this red cliff grade into overlying Bright Angel Shale. In current terminology, the lower slope and overlying red-brown sandstone are assigned to the Bright Angel Shale. A prominent green bed beneath the cliff is a tuff that provided zircon for a fission track age of 563 Ma assigned to the Tapeats (chapter 17).
205.8 (331.1)   Large landslide on left; toe of slide at river level is still active. Several large landslides of probable late Tertiary age are seen in the next several miles.
207.6 (334.0)   Fault; downstream side is up about 250-300 feet (75-90 m). Tapeats Sandstone, upstream; Vishnu Schist, downstream.
208.0 (334.7)   **Entering Granite Park.** Granite Park fault; parallels river on left in Proterozoic crystalline rocks; left (east) side of fault is up about 200 feet (60 m) or more.
208.8 (336.0)   **209-Mile Rapids**; about an 8 foot (2.4 m) drop.
209.0 (336.3)   Fault; downstream side is down about 400 feet (120 m). Vishnu Schist, upstream; Bright Angel Shale, downstream.
209.5 (337.1)   Fault; downstream side is up about 130 feet (40 m), on left. A large landslide covers this fault on right.
209.9 (337.7)   Fault; downstream side is up about 170 feet (52 m).
210.2 (338.2)   Red brown sandstone member of Bright Angel Shale

211.0 (339.5) Contact of shale and sandstone that has a Tapeats lithology (assigned to Bright Angel Shale) with underlying Tapeats Sandstone.
212.0 (341.1) Tapeats Sandstone thins and wedges out across a Proterozoic "island" of amphibolite.
212.3 (341.6) End of "island."
***212.9 (342.6)** **Pumpkin Spring**; a warm, carbonated mineral spring, on left. Graben on right, having about 20 feet (6 m) of displacement; does not reach the river.
213.4 (343.3) Fault on right does not reach the river; downstream side is down about 40 feet (12 m).
214.5 (345.1) Two small grabens exposed high on right wall do not reach river level.
215.0 (345.9) Two small faults, forming a horst, cross the river, exposing Zoroaster Granite for a short distance; displacement is about 60 feet (18 m).
215.6 (346.9) **Three Springs Canyon**, on left. Greatest amount of displacement, about 2,400 feet (730 m), on the Hurricane fault in the Grand Canyon occurs about 1.5 miles (2.4 km) to east, upstream in Three Springs Canyon.
215.8 (347.2) **Begin Lower Granite Gorge**. Fault; downstream side is up about 50 feet (15 m). Tapeats Sandstone, upstream; Zoroaster Granite, downstream.
216.0 (347.5) Arch at top of Redwall Limestone, high on right.
217.3 (349.6) **217-Mile Rapids**; a 16 foot (4.9 m) drop.
219.0 (352.4) Fault; downstream side is down about 15 feet (5 m).
219.2 (352.7) Fault; downstream side is down about 40 feet (12 m).
219.3 (352.9) Three Springs fault, a scissors fault, follows river for .0.5 mi (0.8 km); left side is up about 15 feet (5 m).
220.4 (354.6) **Granite Spring Rapids**; about 8 foot (2.4 m) drop.
221.0 (355.6) Three Springs fault; left side is now down, about 20 feet (6 m). This fault follows the river for the next 3 miles (4.8 km); displacement increases to about 200 feet (60 m) at mile 224.0 (km 360.4).
222.8 (358.5) Hurricane fault, on left, parallels but does not cross Colorado River; displacement is about 1,550 feet (470 m), right (west) side down.
223.5 (359.6) Begin Diamond Creek tonalite pluton, a Zoroaster-type of granite.
225.3 (362.5) Small remnant of basalt flow about 100 feet (30 m) above river, on right.
***225.7 (363.2)** **Diamond Creek Rapids**; a 25 foot (7.6 m) drop. Elevation at top of rapids is 1,343 feet (410 m). Take-out point for many river trips.
227.0 (365.2) End of Diamond Creek pluton; beginning of 229-Mile gneiss. Deepest part of Lower Granite Gorge, 1,115 feet (340 m) deep.
228.2 (367.2) Fault; downstream side is down about 10 feet (3 m).
229.2 (368.8) **Travertine Canyon** on left. Travertine deposits in western Grand Canyon have accumulated from springs in and below Rampart Cave Member, Muav Limestone (fig. 1.32).
229.6 (369.4) Fault; downstream side is up about 30 feet (9 m).
230.0 (370.1) End of 229-Mile gneiss; begin Travertine Falls pluton.
***230.5 (370.9)** **Travertine Falls** on left.
230.9 (371.5) **231-Mile Rapids**; about an 8 foot (2.4 m) drop. End Travertine Falls pluton; begin Vishnu Schist.
232.3 (373.8) **232-Mile Rapids**; about a 5 foot (1.5 m) drop. Beginning of highly contorted migmatite gneiss of the 232-Mile pluton (Zoroaster Granite), which continues for next 4.4 miles (7 km); it is the largest continuous exposure of migmatite in Grand Canyon.
233.5 (375.7) **234-Mile Rapids**; about a 6 foot (1.8 m) drop.
233.6 (375.9) Fault; downstream side is down about 10 feet (3 m).
233.8 (376.2) Fault; downstream side is up about 15 feet (5 m).
***235.2 (378.4)** **Bridge Canyon Rapids**; about an 8 foot (2.5 m) drop. A fault is located about half way through the rapid; downstream side is up about 8 feet (2.5 m). A large bridge formed in conglomerate spans a small ravine about 0.5 mile (0.8 km) up canyon; hence the name Bridge Canyon.
***236.0 (379.7)** **Gneiss Rapids**; last rapid, now a riffle, in Grand Canyon; **begin Lake Mead**. Fault; downstream side is down about 5 feet (1.5 m).
236.7 (380.9) End 232-Mile pluton; begin 237-Mile pluton (Zoroaster Granite).
237.0 (381.3) End 237-Mile pluton; begin Vishnu Schist with Zoroaster Granitic plutons and mafic dikes.
237.5 (382.1) **Bridge Canyon dam site**; elevation 1,220 feet (372 m). The proposed dam would have been nearly 700 feet (210 m) high, backing water as far as 81 miles (130 km) to Havasu Canyon. The dam will probably not be built because of environmental concerns.
238.5 (383.7) End Vishnu Schist; begin Separation granite pluton.
239.6 (385.5) Separation fault; downstream side is up about 20 feet (6 m). Separation Canyon, on both sides of river, is a spectacular example of a fault-controlled canyon.
240.0 (386.2) End Separation granite pluton; begin Vishnu Schist.
242.0 (389.4) Fault; downstream side is up about 5 feet (1.5 m).
242.3 (389.9) End of Vishnu Schist; begin Spencer granitic pluton.

FIGURE 1.32. View of Paleozoic section northwest of the mouth of Separation Canyon, north side of Colorado River at mile 240 (386 km). Redwall Limestone (Mr), Horseshoe Mesa Member (h), Mooney Falls Member (m), Thunder Springs Member (t), Whitmore Wash Member (w); Temple Butte Limestone (Dt); unclassified dolomite (€u); Muav Limestone (€m), Havasu Member (h), Gateway Canyon Member (g), Kanab Canyon Member (k), Peach Springs Member (p), unnamed shale (s), Spencer Canyon Member (sp), Sanup Plateau Member (sa), Rampart Cave Member (r); Bright Angel Shale (€b), Flour Sack Member (f), Meriwitica Member (m), Tincanebits Member (t), unnamed shale (s), red brown sandstone member (rb); Tapeats Sandstone (€t).

242.9 (390.8) Fault; downstream side is up about 20 feet (6 m).
243.3 (391.5) On left, small remnant of Pleistocene basalt that originated from Toroweap area (mi 180; km 290), Lower Granite Gorge here is about 420 feet (130 m) deep.
245.2 (394.5) End of Spencer granitic pluton; begin Vishnu Schist. Two prominent buttes in Redwall Limestone on southern skyline ahead are Spencer Towers.
*246.0 (395.8) Remnant of Pleistocene basalt on right. Before being covered by Lake Mead in 1936, this basalt flow marked the beginning of Lava Cliff Rapids, considered then the largest and roughest rapid in the Grand Canyon.
246.1 (396.0) Fault; downstream side is up about 5 feet (1.5 m). End Vishnu Schist; begin Surprise-Quartermaster granite pluton.
247.0 (397.4) Fault; downstream side is down about 10 feet (3 m).
249.0 (400.6) A small remnant of Pleistocene basalt in protected re-entrant, on left.
249.2 (401.0) Fault; downstream side is down about 20 feet (6 m).
252.2 (405.8) **Meriwitica fault**; parallels canyon on left. Displacement here is about 450 feet (135 m), up on left. The fault plane dips to the left (west) about 83°, making this a reverse fault. In overlying upper Paleozoic rocks, the reverse movement has produced the east-dipping Meriwitica monocline. The lake parallels the fault for the next 3.3 miles (5 km). The Meriwitica fault is a scissors fault that has a hinge point about 6 miles (10 km) to the north. Beyond this, the fault is a normal fault, east side down, with a maximum displacement of about 175 feet (50 m).
*254.2 (409.0) Last remnant of Pleistocene basalt flow, on right.
255.4 (410.9) Leave Meriwitica fault; downstream side is up about 185 feet (55 m).
256.8 (413.2) Fault; downstream side is up about 100 feet (30 m).
257.1 (413.7) Fault; downstream side is up about 10 feet (3 m).
257.7 (414.6) Fault; downstream side is down about 40 feet (12 m).

259.0 (416.7) A large deposit of travertine on left bank resulted from several now inactive springs.
259.5 (417.5) Two faults, downstream side up, have a combined displacement of about 30 feet (9 m). Excellent view of Great Unconformity on right. Near the river, water from a large spring in Quartermaster Canyon (left) travels under and through travertine that has filled the canyon to a depth of about 400 feet (120 m). A travertine dam, the result of springs about 1 mille (1.6 km) up-canyon, has blocked the transport of stream gravel; surface water and gravel now overflow the dam on its west side, resulting in a drainage that now falls back into the original canyon over a cliff of Tapeats Sandstone. Similar dams occur upstream in Meriwhitica and Havasu Canyons.
261.0 (419.9) End Surprise-Quartermaster granitic pluton; begin Tapeats Sandstone.
266.5 (428.8) Several faults cross the lake the next 2.5 miles (4 km), forming a series of horsts and grabens. Large deposits of travertine have accumulated on right bank for next several miles. The toes of several landslide masses on left have been partially re-activated since 1936 because of the lake water.
272.9 (439.1) Spring on left.
274.0 (440.9) Springs on left.
274.2 (441.2) Travertine deposits on right next 1 mile (1.6 km).
***274.3 (441.3)** **Columbine Falls (Emery Falls)**, on left.
***275.5 (443.3)** **Rampart Cave fault**; downstream side is up about 600 feet (180 m). Rampart Cave is 0.5 mi (0.8 km) to the west at the top of the Rampart Cave Member of the Muav Limestone. The cave contains Pleistocene giant-sloth dung deposits, 11,000 to more than 40,000 years old.
***277.6 (446.7)** **Muddy Creek Formation** (late Miocene) accumulated against Cambrian to Mississippian strata along the **Grand Wash Cliffs** to an average elevation of about 3,600 feet (1,100 m), approximately 2,400 feet (730 m) above lake level. The base of the Muddy Creek lies at at an unkown depth below this part of Lake Mead. Muddy Creek sediments occupy older tributary drainages cut into the Grand Wash Cliffs, including an older tributary canyon that the present Colorado river occupies seen on the right and higher up on the walls of the Grand Wash Cliffs behind us. See Chapter 18.
***278.0-278.5 (447.3-448.1)** **Grand Wash fault**, approximate location; marks boundary between Colorado Plateau and Basin and Range geologic provinces. The Grand Wash fault is buried by deposits of Upper Miocene Muddy Creek Formation. Displacement on fault may be as much as 10,000 feet (3,050 m); downstream

FIGURE 1.33. Aerial view looking northeast along Wheeler fault (A) and Paleozoic strata dipping to the southeast. Grand Wash Cliffs (B) in background. Contacts dashed where approximately located. Muddy Creek Formation (Tmc); Kaibab and Toroweap Formations (Pkt); Hermit Shale (Ph); Esplanade Sandstone and Pakoon Limestone (Pep); Callville Limestone (℔c); Redwall Limestone (Mr); Temple Butte Limestone (Dt); basalt of Grand Wash (gw).

FIGURE 1.34. God's Pocket, Lake Mead, mile 283 (455 km). View is north towards tilted Permian strata of Wheeler Ridge. Contacts dashed where approximately located. Toroweap Formation (Pt); Hermit Shale (Ph); Esplanade Sandstone (Pe); Pakoon Limestone (Pk).

side is down. Minor displacement on the Grand Wash fault occurred during early deposition of the Muddy Creek. About 2,600 feet (800 m) of Muddy Creek material has been removed by erosion at this locality.
*281 (453)   **Wheeler Ridge**, north-trending ridge on left. Tilted strata on left are of Pennsylvanian age; ahead, are tilted strata of Permian age overlain by Muddy Creek Formation (fig. 1.33).
282 (454)   Kaibab and Toroweap Formations and Hermit Shale, ahead in Gods Pocket (fig. 1.34). Coconino Sandstone was not deposited here.
283 (455)   Hermit Shale overlies pink and white Esplanade Sandstone (fig. 1.34). The Esplanade intertongues with a thick sequence of gray limestone (Permian Pakoon Limestone). Red shale and sandstone of the basal Esplanade underlies the Pakoon Limestone. The Esplanade and Pakoon overlie the Pennsylvanian Callville Limestone. The name Callville, applied to Pennsylvanian strata in the Basin and Range Province, replaces the Grand Canyon names of Wescogame, Manakacha, and Watahomigi Formations of the Supai Group.
284 (456)   Callville Limestone overlies Mississippian Redwall Limestone.
285 (458)   Light gray Redwall Limestone overlies dark gray Devonian Temple Butte Limestone. Approximate location of **Wheeler fault**; downstream side is down about 4,500 feet (1,370 m). The basalt of Grand Wash on right, ~3.8 Ma, rests on gravels of a tributary to the Colorado River cut into the Muddy Creek Formation (fig. 1.35).
286 - 287 (460 - 462) **Arizona - Nevada State line** is now at about the middle of Lake Mead for the next 26 mles (42 km) to Temple Bar. Arizona on left; Nevada on right. Paleozoic sequence dips from 25° to 65°.
*288 (463)   **Entering Iceberg Canyon** (fig. 1.35). Redwall and Temple Butte Limestones on left (Iceberg Ridge); Callville Limestone, on right (Azure Ridge), dips east about 65° to 75°. Iceberg fault follows Iceberg Canyon, mostly covered by Lake Mead, but it can be seen in Driftwood Cove to rear (north). The fault is an oblique slip fault. In Driftwood Cove, the Hermit-Toroweap contact rests against the lowermost Redwall Limestone on the fault, indicating an approximate vertical displacement of about 2,750 feet (840 m); the right (west) side is down when looking downstream (or south) from the cove.
291 (468)   Devil's Cove on right. Cambrian through Pennsylvanian rocks overlying the Proterozoic crystalline basement are faulted at several places. The Great Unconformity is tilted eastward about 65°. Highly faulted Paleozoic strata are partially covered with deposits of the Muddy Creek Formation, on left.
*294 (473)   Sandy Point on left. Recent sand dunes overlie Sandy Point basalt (~3.8 Ma; equivalent to basalt of Grand Wash), which rests on gravel of Colorado River.

FIGURE 1.35. Aerial view looking north at Iceberg Canyon; Azure Ridge on left (in Nevada); Iceberg Ridge on right (in Arizona). Arizona/Nevada State line and Iceberg fault are located about the middle of Iceberg Canyon. D/U indicates relative movement on faults. Contacts dashed where approximately located. Driftwood Cove (dc); Devil's Cove (dco); basalt of Grand Wash (gw); Muddy Creek Formation (Tmc); Kaibab Formation (Pk); Toroweap Formation (Pt); Hermit Shale (Ph); Esplanade Sandstone and Pakoon Limestone (Pek); Callville Limestone (₽c); Redwall Limestone (Mr); Temple Butte Limestone (Dt); Cambrian undivided (Єu); Muav Limestone (Єm); Bright Angel Shale (Єb).

300 (484) Fine-grained pink sediments of Muddy Creek Formation (on left) have been gradually sliding into Lake Mead since 1936 when Lake Mead was formed by Boulder Dam. The Muddy Creek sediments overlie granite on both sides of the lake. Paleozoic rocks were eroded before deposition of the Muddy Creek. In this area, the Muddy Creek sediments are displaced nearly 1,000 feet (305 m), west (right) side down, by the Wheeler fault.

301 (485) Granite Cove on left; Hiller Mountains on right. Proterozoic basement rocks similar to those seen in Grand Canyon are on both sides of the lake, at places overlain by deposits of Muddy Creek Formation.

302 (486) Entering Virgin Canyon. Proterozoic basement rocks are exposed for the next 5 mles (8 km).

307 (494) Gateway Cove on left; Temple Mesa on right. Temple Mesa is composed of alluvial fan and lake sediments of the Muddy Creek Formation, capped by a basalt flow that may be correlative with the 5.8 Ma Fortification Basalt Member of the Muddy Creek Formation near Boulder Dam to the west. The Muddy Creek overlies Proterozoic granite and schist.

312 (503) Temple Bar, end of trip; located on reworked sediments of Muddy Creek Formation.

CHAPTER 2: HYDRAULIC LOG OF THE COLORADO RIVER FROM LEES FERRY TO DIAMOND CREEK, ARIZONA

Julia B. Graf[1], John C. Schmidt[1,2], and Susan W. Kieffer[3]

INTRODUCTION

The following log describes the major hydraulic characteristics of sections of the main river channel (reaches) and selected rapids, recirculation zones, and sand deposits from Lees Ferry to Diamond Creek, a distance of 225 mi (362 km) (fig. 2.1). The log is meant to be used in conjunction with Chapter 1 of this book as well as the "Guide to the Colorado River in the Grand Canyon" (Stevens, 1983). The log draws from references cited in Chapter 3 and few references are repeated here.

Recirculation zones, composed of one or more eddies, typically form above and below rapids. Sand is selectively deposited in these zones because sediment-transport capacity is less than in the main channel. These sand deposits form the major campsites along the river. Within recirculation zones, the loci of sediment deposition are at the two stagnation points at either end of the zones--the point where flow is separated from the channel banks and the point where downstream flow is again attached to the channel banks. The former deposits of sand are termed separation bars; they typically mantle debris fans. The latter deposits are termed reattachment bars. Outside of recirculation zones, sand is deposited in narrow strips which line the channel banks. These channel-margin deposits may in some cases have been deposited in small eddies. See Figure 3.6 for a schematic diagram illustrating these features.

The elevation of the surface of the Colorado River at Lees Ferry (mi 0, km 0) is about 3,116 feet (949.8 m). The elevation of Lake Mead (mi 235; 376 km) at a high level is 1,157 feet (352.7 m). The drop of 1,959 feet (597.1 m) over this distance results in an average gradient of 8.3 ft/mi (2.5 m/km). The gradient of the river surface through the rapids is much steeper than the average river gradient; in fact, most of the drop along the river occurs through the rapids (see fig. 3.2). However, the drop in water surface elevation in the rapids depends on discharge, and the gradient can vary by as much as a factor of two. Accurate changes in elevation with discharge have been measured at ten different rapids (fig. 2.1), and results can be found on U.S. Geological Survey Miscellaneous Investigations Maps I-1897 A-J. Elevation changes through the rapids given in this log are characteristic of low flows (less than 10,000 ft3/s, 283 m3/s).

The river can be divided into 11 reaches--stretches of a few tens of kilometers--within which the hydraulic and sedimentary characteristics are relatively uniform and distinct from adjacent reaches. General characteristics of each reach are described below, and specific individual hydraulic features of interest within the reach are also mentioned.

REACH 1. MILE 0 TO 11.4 (0 TO 18.3 KM), UPPER MARBLE CANYON

The river channel is moderately wide and shallow with an average flow width of 280 feet (85.3 m) and a mean depth of 23.9 feet (7.3 m) at a discharge of 24,000 ft3/s (680 m3/s). Average slope through this section is 0.00099, and about 42 percent of the river bed is covered by boulders and bedrock. This section averages 1 campsite per 2.5 miles. A reattachment bar in this reach is shown in Figure 2.2. Most of this bar is covered by a recirculation zone at moderate discharge and could not be used as a campsite. However, the high-elevation parts of the bar at the downstream end of the bar are occasionally used for camping. No separation bar exists here because the talus slope at the separation point is too high and steep for deposition to occur. At debris fans in this reach, however, campsites are typically separation bars, as shown in Figure 2.3.

0.0 (0.0) Cableway for discharge measurements at U.S. Geological Survey gaging station "Colorado River at Lees Ferry". A ferry was operated in this area from 1873 until 1928, and Lees Ferry is named for John D. Lee who operated the ferry from 1873 to 1877 (Rusho and Crampton, 1981).

Changes in bed elevation and sediment transport at this gaging station are described in Chapter 3. This gaging station and the gaging station on the Paria River, 1.1 mi (1.8 km) upstream from its mouth, were established in 1923 to document the amount of water flowing from the upper Colorado River basin to the lower Colorado River basin. The annual average amount of water passing the gage at Lees Ferry and passing the gage on the Paria River must total 8.23 million acre feet (1.02×10^{10} m3) to fulfill the Colorado River Compact of 1922.

[1]U.S. Geological Survey, Tucson, Arizona
[2]Now at Geology and Geography Departments, Middlebury College, Middlebury, Vermont
[3]U.S. Geological Survey, Flagstaff, Arizona

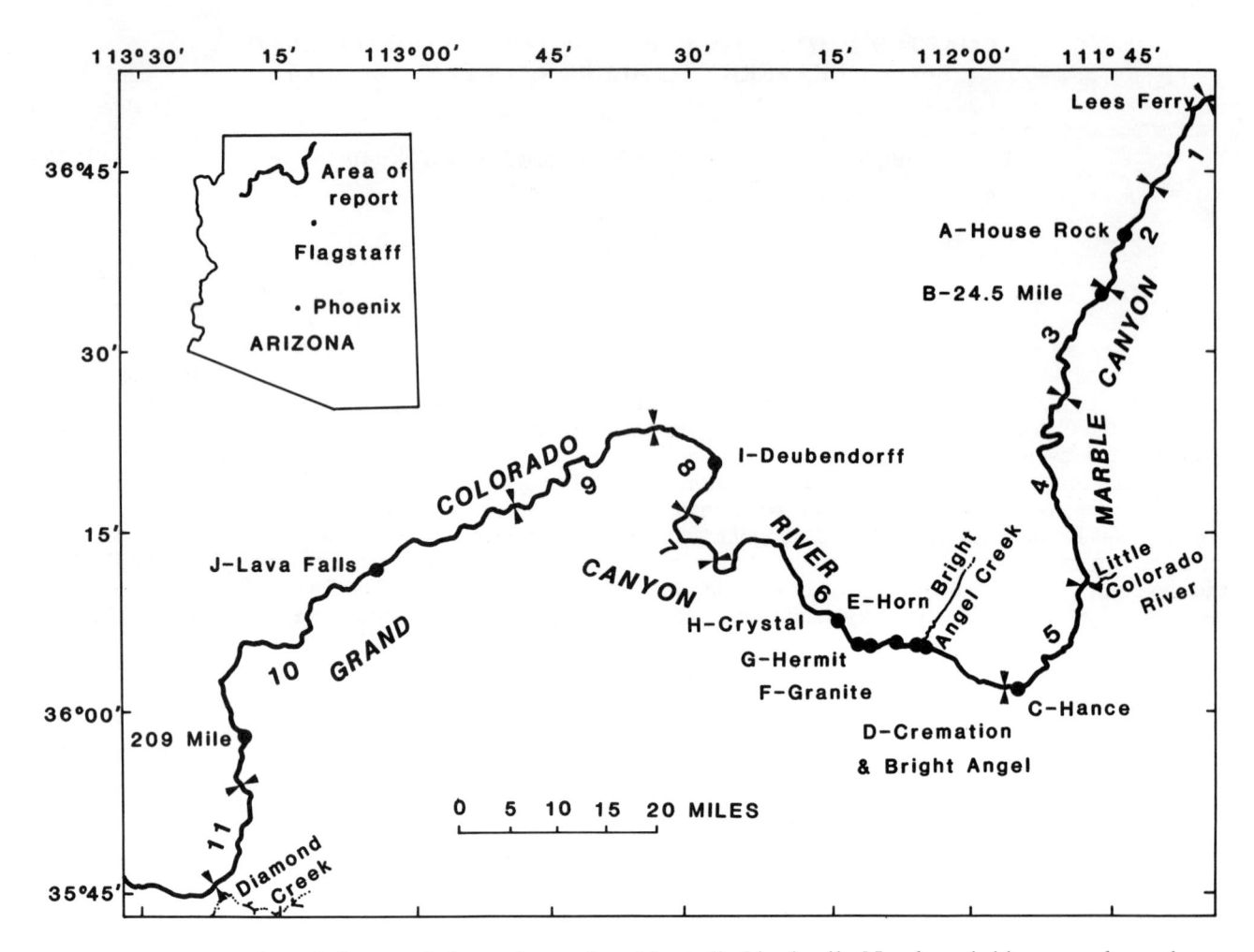

FIGURE 2.1. Location of characteristic reaches and rapids studied in detail. Numbers 1-11 are reach numbers: (1) Upper Marble Canyon; (2) Supai Gorge; (3) Redwall Gorge; (4) Lower Marble Canyon; (5) Furnace Flats; (6) Upper Granite Gorge; (7) The Aisles; (8) Middle Granite Gorge; (9) Muav Gorge; (10) Lower Canyon; (11) Upper Part of Lower Granite Gorge. Capital letters A through I refer to the individual map in the series of U.S. Geological Survey Miscellaneous Investigations Maps I-1897.

0.8 (1.3)   Mouth of the Paria River. The Paria River drains about 1,410 mi2 (3,652 km2) in northern Arizona and Utah and is the second largest tributary drainage basin of the Colorado River between Glen Canyon Dam (15 mi or 24 km upstream from river mile 0) and Lake Mead. The Paria River carries very high suspended sediment loads and is a major source of sand, delivering an average of 3.6 megatons of suspended sediment per year to the Colorado River.
1.4 (2.3)   "Compact point" The arbitrary (and invisible--don't look for it!) point dividing the drainage basin of the Colorado River into the "Upper Basin" and "Lower Basin"; this point is referenced for the division of water allocations between the two parts of the basin.
7.8 (12.6)   Badger Creek Rapids, a 15-foot (4.6-m) drop (see fig. 2.2). Separation bars located just downstream from the rapid mantle the alluvial fans at the mouths of Badger Creek (right bank) and Jackass Creek (left bank) (fig. 2.3). Opposing tributary drainages and their debris fans, such as this, are commonly associated with fault-controlled drainages. Reattachment bars, more susceptible to erosion, have been largely stripped from the recirculation zones at this site since completion of Glen Canyon Dam in 1963.

REACH 2. MILES 11.4 TO 22.6 (18.3 TO 36.4 KM), SUPAI GORGE

The river channel is narrower and steeper than Upper Marble Canyon. The average width of Supai Gorge is 210 feet (64.0 m) and mean depth is 29.6 feet (9.0 m) at a discharge of 24,000 ft3/s (680 m3/s). Average slope for this reach is 0.0014, and about 81 percent of the river bed is covered with boulders and bedrock. This section averages 1 campsite per 1.1 miles; campsites typically are on separation bars.
17.0 (27.4)   House Rock Rapids, a 10-foot (3.1-m) drop (fig. 2.4). This will be the best stop in the early part of the trip for a detailed look at the hydraulics of a rapid and zone of constriction and recirculation; however,

FIGURE 2.2. A recirculation zone above Cathedral Wash in the Upper Marble Canyon reach showing a reattachment bar. Flow is from left to right. Discharge is less than 10,000 ft3/s (283 m3/s).

FIGURE 2.3. Badger Rapids as seen from the plateau rim on the left bank. Flow is from right to left.

FIGURE 2.4. House Rock Rapids. Note Rider Canyon debris fan forming constriction. Discharge about 7,000 ft3/s (200 m3/s).

most other rapids exhibit similar features. The rapid occurs in the channel constriction caused by the debris fan from Rider Canyon. Figure 3.5 illustrates the hydraulic features of this rapid at 5,000 ft3/s (141 m3/s) and at 30,000 ft3/s (850 m3/s).

18.1 (29.1) **18-Mile Wash.** A debris flow of flash flood origin, on left, occurred here in the summer of 1987, covering a popular camping beach on the separation bar with about 3 feet (1 m) of debris. Several small, steep tributaries in this area flowed at the same time and those flows significantly changed the alluvial fans and associated sand bars.

REACH 3. MILES 22.6 TO 36.0 (36.4 TO 57.9 KM), REDWALL GORGE

The river channel is relatively narrow, having an average width of 220 feet (67.1 m) and mean depth of 24.4 feet (7.4 m) at a discharge of 24,000 ft3/s (680 m3/s). This reach contains many rapids popularly known as "The Roaring Twenties." One of these, 24.5-Mile Rapids, has been studied in detail (figs. 2.1 and 2.5). A fascinating hydraulic feature here is the shift of the tailwaves from one side of the river to the other as discharge changes (fig. 2.5). Average slope through this section is 0.0015, and about 72 percent of the river bed is covered with boulders and bedrock. This reach has 1 campsite per 1.1 mi; campsites typically are on separation bars.

26.8 (43.1) A small rapid occurs here at low water (below 5,000 ft3/s; 141 m3/s) because of a rockfall from the right wall in 1975. Rapids formed entirely by rockfalls are rare in the Grand Canyon.

31.9 (51.3) **Vasey's Paradise.** In addition to Vasey's Paradise Spring and many smaller springs, four large springs [Blue Spring on the Little Colorado River (mi 61.3; 98.6 km), Havasu Spring on Havasu Creek (mi 156.7; 256.0 km), Thunder Spring on Tapeats Creek (mi 133.6; 215.0 km), and Roaring Springs on Bright Angel Creek (mi 87.5; 140.8 km)], contribute water to the Colorado River between Lees Ferry and Lake Mead. The discharge of these springs is poorly documented, but flow in Vasey's Paradise has ranged from 0.3 to 10 ft3/s (0.01 to 0.3 m3/s) in five measurements.

REACH 4. MILES 35.9 TO 61.5 (57.9 TO 99.0 KM), LOWER MARBLE CANYON

The river channel is wide and shallow; it has an average flow width of 350 feet (106.7 m) and a mean depth of 18.3 feet (5.6 m) at a discharge of 24,000 ft3/s (680 m3/s). Average slope across this section is 0.00101, and only about 36 percent of the river bed is covered by boulders and bedrock. This reach has 1 campsite per 0.4 mi, a

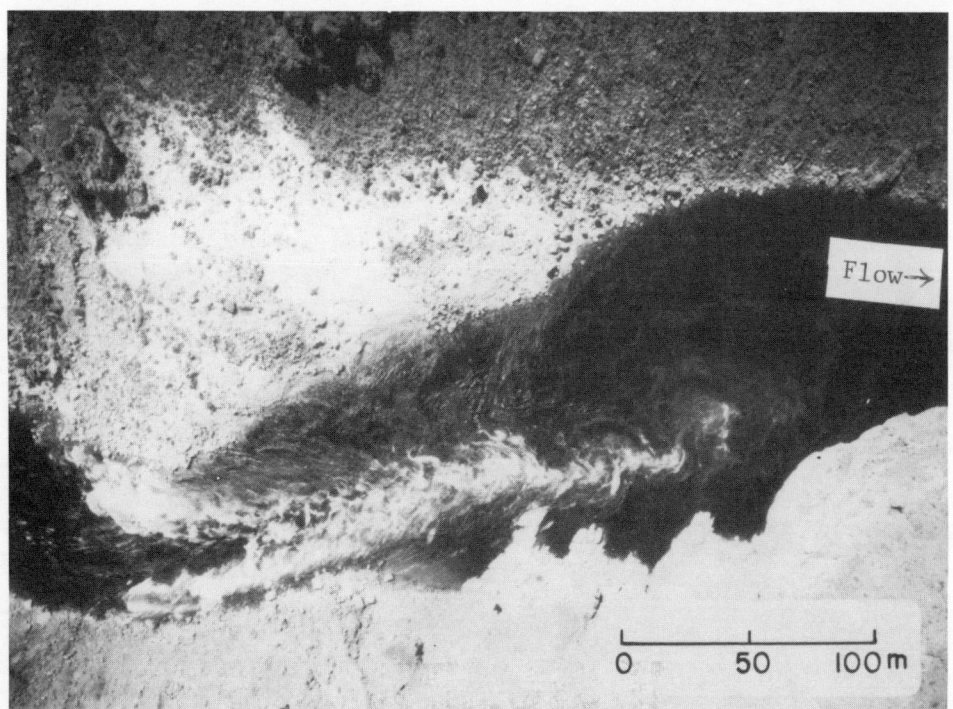

FIGURE 2.5. Tailwaves at 24.5-Mile Rapids at discharges of 5,000 ft3/s (140 m3/s) (top) and at 30,000 ft3/s (840 m3/s) (bottom). Note the dramatic change in the orientation of the tail jet. Photographs by U.S. Bureau of Reclamation, 1984 (top), and National Park Service, 1986 (bottom).

factor of three greater than the first three reaches. This reach contains many large recirculation zones which contain sand deposits, and both separation and reattachment bars serve as campsites. Large parts of reattachment bars are exposed at moderate to low flows [less than about 20,000 ft3/s (566 m3/s)]. Areas that are channels for upstream-flowing current at higher flows become quiet backwaters at these lower flows. These low velocity zones become warmer than main channel flow, and serve as important nursery areas for native fish, which prefer warm water (16-17 °C) to the cold water of the post-dam era (11-12 °C) (Pre-dam river temperatures ranged from nearly freezing in winter to a high of (26.7 °C) in July and August.)

44.2 (71.1) **Eminence Break Camp.** Large recirculation zone and sand bars on the left bank below the riffle (fig. 2.6) are typical of areas below constrictions in wider reaches of the canyon. The large backwater, or quiet pool, shown on the shoreward side of the reattachment bar is typical of these recirculation zones. Characteristics of flow and sediment at this location are described in Chapter 3, and in Figure 3.7. A large rockfall occurred during the winter of 1975 at the lower end of the beach on the left (see fig. 1.12).

61.0 (98.2) Cableway marks location of U.S. Geological Survey gaging station "Colorado River above the Little Colorado River near Desert View". The station was established in 1983 as a part of the Glen Canyon Environmental Studies; flow and sediment transport data were obtained for about six months in 1983 and for a 4-month period in 1985-86.

61.3 (98.6) Junction of the Little Colorado and Colorado Rivers. The Little Colorado River drainage basin is more than 24,000 mi2 (62,000 km2) in area and is the largest tributary to the Colorado River between Glen Canyon Dam and Lake Mead. It is also a major source of sediment, delivering an average of about 10 megatons of suspended sediment per year. When not in flood stage, the water of the Little Colorado River is derived from Blue Spring and several smaller springs about 13 mi (21 km) above the confluence. The flow in Blue Spring is very constant, ranging from 217 to 232 ft3/s (6.1 to 6.6 m3/s) in 13 measurements made over a 17-year period. The spring is the natural ground-water discharge point for the Black Mesa hydrologic basin (about 28,000 mi2; 45,000 km2). The water has a high dissolved solids concentration (2,000-4,000 parts per million) with high concentrations of chloride, sodium, calcium, and bicarbonate. (Johnson and Sanderson, 1968) The water of Blue Spring is warmer than the Colorado River water (20 °C), and this is believed to be the reason that a once common and now endangered native fish (humpback chub) successfully spawns only in the mouth of the Little Colorado River.

REACH 5. MILES 61.5 TO 77.4 (99.0 TO 124.5 KM), FURNACE FLATS

The river channel is very wide and shallow (fig. 2.7), with an average width of 390 feet (118.9 m) and a mean depth of 14.7 feet (4.5 m) at a discharge of 24,000 ft3/s (680 m3/s). Average channel slope is 0.0021, and about

FIGURE 2.6. Reattachment bar and backwater at Eminence Break Camp, October 12, 1985; discharge about 3,000 ft3/s (84 m3/s). Flow is from lower right to top center; backwater is the ponded area at the lower left of the photo.

FIGURE 2.7. Sand deposits which resemble point-bar deposits above Unkar Rapids, Furnace Flats reach. View is downstream.

30 percent of the river bed is covered by boulders and bedrock. This section has 1 campsite per 0.4 mi, and channel margin deposits typically serve as campsites. Large gravel bars are common, and sand deposits along the banks have a configuration similar to point bars more typical of alluvial rivers.
76.5 (123.1) Hance Rapids, a 30-foot (9.1-m) drop. Hance Rapids is one of the rockiest rapids on the river, formed by debris flows from Red Canyon on the left and boulders of Shinumo Quartzite from the cliffs above the Hakatai Shale on the right. Figure 3.1 can be used to identify characteristic features of the rapids.

REACH 6. MILES 77.4 TO 117.8 (124.5 TO 189.5 KM), UPPER GRANITE GORGE

The river channel is narrow and deep, with an average width of 190 feet (57.9 m) and a mean depth of 27.1 feet (8.3 m) at a discharge of 24,000 ft3/s (680 m3/s). Average channel slope is 0.0023, and about 62 percent of the river bed is covered by boulders and bedrock. This section has 1 campsite per 1.7 mi. Separation and channel-margin deposits typically serve as campsites.
81.2 (130.7) Vishnu camp. Campsite on the left bank (fig. 2.8) is one of the largest in a reach of small, scarce sand bars typical of these narrow parts of the canyon. This sand deposit has characteristics of a channel margin deposit, but the deposit was probably formed by the same processes that form reattachment bars. This camp (also called "Grapevine camp") and "Cremation camp" at mile 86.0 (138.4 km) (it is not at the mouth of Cremation Canyon) are the only significant sand deposits between Hance Rapids and Bright Angel Creek, a distance of about 10 miles (16.1 km).
87.3 (140.5) Two cableways and gage houses mark the location of the U.S. Geological Survey gaging station called "Colorado River near Grand Canyon," established in 1922. Data are sent from the upper gage, and the gage at Lees Ferry, via satellite to USGS offices in Tucson, Arizona, where they are stored on computer. The apparatus visible on top of the gage house on the left bank is part of the instrumentation that makes this transfer possible. This gaging station has the longest record of water-quality information of any gaging station in the southwestern United States. A discussion of flow and sediment transport at this gage and the effect of changes in Bright Angel Rapid, which controls the stage-discharge relation at the gage, is given in Chapter 3.
90.1 (150.0) Horn Creek Rapids, a drop of 10 feet (3.1 m). Whereas most major rapids in the Grand Canyon are formed by constrictions caused by debris fans at tributary mouths, Horn Creek Rapids exists not only because of the debris from Horn Creek, but also because the canyon is narrowed by bedrock walls and large boulders of the talus slope on the right bank.

FIGURE 2.8. Sand deposit at Vishnu (Grapevine) camp in Upper Granite Gorge reach; discharge about 45,000 ft3/s (1260 m3/s).

93.4 (150.3) Granite Rapids, a drop of 17 feet (5.2 m). The features of this rapid are labelled in detail in Figure 3.1.

95.0 (152.9) Hermit Rapids, a drop of 15 feet (4.6 m). At discharges exceeding about 20,000 ft3/s, the rapid has beautiful standing waves that make it one of the most memorable "rides" on the river. This rapid best mimics the textbook features seen in converging-diverging laboratory flumes. Figure 2.9 shows the rather calm conditions at this rapid at 5,000 ft3/s and the waves that develop in it at higher discharges.

98.0 (157.7) Crystal Rapids, a drop of 17 feet (5.2 m). Crystal Creek debris fan on the right; Slate Creek enters from the left in the middle of the rapid. See discussion of the history of Crystal Rapids in Chapter 3. Figure 2.10 is a mosaic of the river taken from the terrace downstream of Crystal Creek when the discharge was 92,000 ft3/s (2,600 m3/s) in 1983.

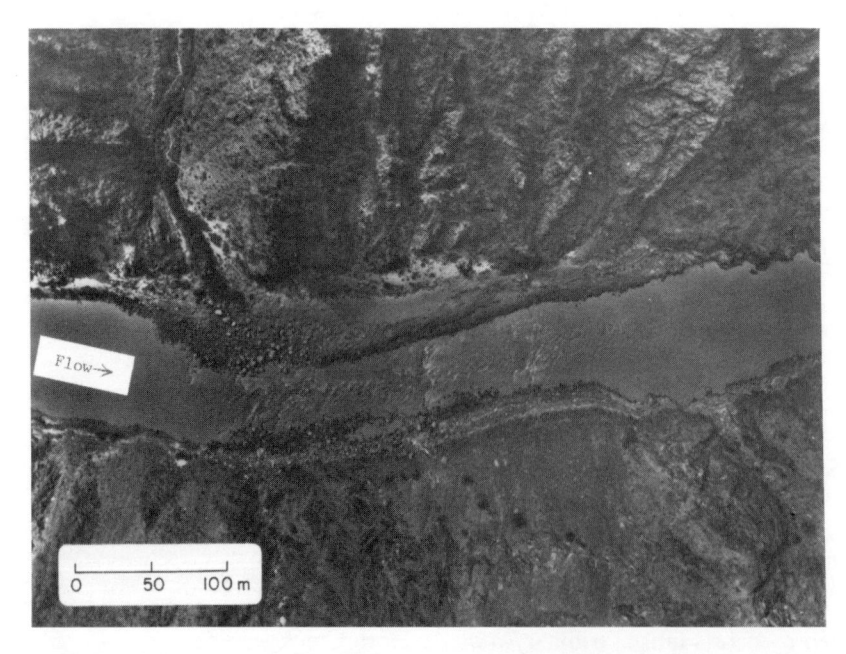

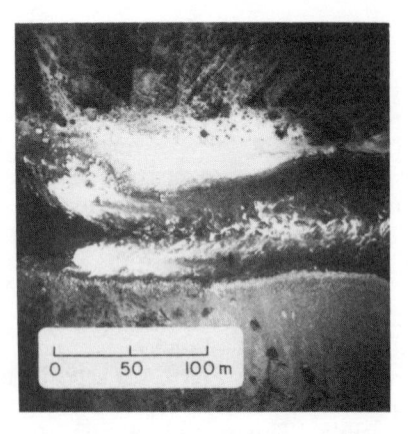

FIGURE 2.9. Tongue and lateral waves at Hermit Rapids at 5,000 ft3/s (140 m3/s) (left) and 30,000 ft3/s (840 m3/s) (right). Note the differences in tongue lengths, angle of lateral waves to shore, and nonbreaking rollers in the tongue. Top photo by U.S. Bureau of Reclamation, 1984. Bottom photo by National Park Service, 1986.

FIGURE 2.10. Mosaic of Crystal Rapids at 92,000 ft3/s (2,600 m3/s) taken from the terrace west of Crystal Creek. Note that the entire debris fan is covered with water.

REACH 7. MILES 117.8 TO 125.6 (189.5 TO 202.0 KM), THE AISLES

The river channel is relatively narrow, with an average width of 230 feet (70.1 m) and a mean depth of 20.9 feet (6.4 m) at a discharge of 24,000 ft3/s (680 m3/s). Average channel slope is 0.0017; about 48 percent of the river bed is covered by boulders and bedrock. This section has 1 campsite per 0.25 mi, and all types of deposits serve as campsites.

120.2 (193.4) Begin **Conquistador Aisle.** Note the large recirculation zone below **Blacktail Rapids** (fig. 2.11).

REACH 8. MILES 125.6 TO 139.9 (202.0 TO 225.1 KM), MIDDLE GRANITE GORGE

The river channel is relatively narrow, with an average width of 219 feet (64.0 m) and a mean depth of 25.6 feet (7.8 m) at a discharge of 24,000 ft3/s (680 m3/s). Average channel slope is 0.0020; 68 percent of the river bed is covered by boulders and bedrock. This reach has one campsite per 0.43 mi, and channel margin deposits serve as campsites.

130.4 (209.8) **Bedrock Rapids**; a 7-foot (2.1-m) drop. Bedrock Rapids is one of a few rapids caused by a constriction of the river caused by bedrock outcrops, such as the protruding mass of granite seen at water levels below 60,000 ft3/s (1,704 m3/s). Other rapids of this type seen in low water flows include Lower Tuna Rapids at mile 99.6 (160.3 km), a riffle at mile 113 (181.8 km), and 231-Mile Rapids at mile 230.9 (371.5 km).

131.6 (211.7) **Deubendorff Rapids**, a 15-foot (4.6-m) drop. A fault cuts Middle Proterozoic rocks; downstream side is up about 150 feet (45.7 m). Note that two tributary drainages, Galloway and Stone Canyons, intersect the Colorado River here, one above and one near the bottom of the rapid. At the present time, most of the debris in Deubendorff Rapids has come from Galloway Canyon. However, the erosional setback of the river into the wall across from Stone Canyon suggests that, in the past, there was a larger debris fan at the mouth of Stone Canyon that forced the river toward the left bank. See Chapter 3 for details. Figure 2.12 shows the configuration of this rapid at low water.

FIGURE 2.11. Typical view of The Aisles reach. Photo looking downstream showing large recirculation zone on the right bank below Blacktail Canyon.

FIGURE 2.12. Deubendorff Rapids at about 5,000 ft3/s (140 m3/s), showing rocky area at the head of the rapid. Photograph by U.S. Bureau of Reclamation, 1984.

REACH 9. MILES 139.9 TO 159.9 (225.1 TO 257.3 KM), MUAV GORGE

The river channel is narrow and deep, with an average width of 180 feet (54.8 m) and a mean depth of 22.0 feet (6.7 m) at a discharge of 24,000 ft3/s (680 m3/s). Average channel slope is 0.0012; 78 percent of the river bed is covered by boulders and bedrock. This section has 1 campsite per 0.91 mi. Because of the less resistant nature of the Muav Limestone, debris fans are small and have low elevations. The low debris fans in this reach make separation sand bars more susceptible to erosion than in reaches with higher debris fans, such as Upper Granite Gorge, because they are exposed to downstream flow as the flow overtops the debris fans.

REACH 10. MILES 159.9 TO 215.8 (257.3 TO 347.2 KM), LOWER CANYON

The river channel is wide and shallow (fig. 2.13), with an average width of 310 feet (94.5 m) and a mean depth of 19.3 feet (5.9 m) at a discharge of 24,000 ft3/s (680 m3/s). Average channel slope is 0.0013; only 32 percent of the river bed is covered by boulders and bedrock. This reach has one campsite per 0.42 mi.
166.0 (267.1)   Cableway marks location of the U.S. Geological Survey gaging station "Colorado River at National Canyon near Supai". Flow and sediment transport data were collected here in 1983 and 1985-86.
166.5 (267.9)   **National Rapids**, about a 2-foot (0.6-m) drop. Recirculation zone below rapid on left is especially interesting because it is long, narrow, and very turbulent. The reattachment bar here is unusual in that it is not attached to the bank at many discharges and forms an island (compare with the more usual configuration illustrated in fig. 3.6b).

FIGURE 2.13. Separation bar below National Canyon in the Lower Canyon reach. View looking downstream.

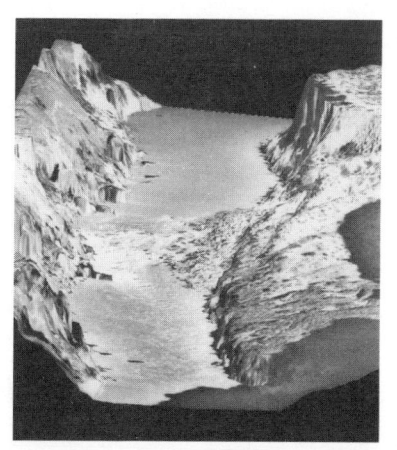

FIGURE 2.14. Lava Falls Rapids. Digitized images, vertically exaggerated. (Top) stereographic pair. Middle: view from cliff on right (north) bank looking south into Prospect Canyon showing set of old, high terraces. Bottom; view from left (south) bank at Prospect Creek looking toward the north wall and lava cliffs.

179.0 (288.0) Lava Falls Rapids, a 13-foot (4.0-m) drop (fig. 2.14). Note the high terraces on left bank, remnants of old debris flows derived from Prospect Canyon.

REACH 11. MILES 215.8 TO 225.7 9347.2-363.2 KM), UPPER PART OF LOWER GRANITE GORGE

The river channel is narrow and deep, with an average width of 240 feet (73.2 m) and a mean depth of 29.6 feet (9.0 m) at a discharge of 24,000 ft3/s (680 m3/s). Average channel slope is 0.0016; about 58 percent of the river bed is covered by boulders and bedrock. This reach has one campsite per 0.43 mi.

212.9 (342.6)   Pumpkin Springs. Travertine deposits from a small spring can been seen at low flow on the left bank. The flow in this reach is fast and turbulent, and rocks along the banks have been polished and sculpted into potholes. Just off the spring itself, the river bottom dives to reach a depth of about 90 feet (27 m). What appears to be a recirculation zone on the left bank below the spring is in fact downstream-directed flow under some flow conditions, but trenching of the bar demonstrated that it formed entirely by upstream flow. Sand deposits here may be channel margin deposits, or may have been formed in a recirculation zone at very high flows.

225.3 (362.5)   Cableway marks location of the U.S. Geological Survey gaging station "Colorado river above Diamond Creek near Peach Springs", at which flow and sediment transport data were collected here in 1983 and 1985-86. Elevation here is 1,343 feet (409.3 m).

CHAPTER 3: HYDRAULICS AND SEDIMENT TRANSPORT OF THE COLORADO RIVER

Susan W. Kieffer[1], Julia B. Graf[2], and John C. Schmidt[2,3]

INTRODUCTION

For much of its length, the Colorado River in the Grand Canyon is confined by bedrock, large blocks of talus, or by alluvial debris too coarse to be transported except under high flows. Rivers located in bedrock gorges, like the Colorado River, differ in a fundamental way from alluvial rivers: the immovable material of the bed and banks causes flow depth, flow velocity, and bed shear stress to increase more rapidly with discharge than in alluvial rivers (Baker, 1984). The flow conditions caused by the bedrock conditions also cause distinctive hydraulic phenomena. Examples are sudden transitions between subcritical and supercritical flow, hydraulic jumps, large-scale turbulent structures, and strong flow-separation phenomena with associated unique deposits. The Colorado River is a good place to study these phenomena because in the Grand Canyon the processes that produce the phenomena occur over a broad range of discharges. On many other bedrock rivers, these features exist only at high discharges when field work is difficult.

The long history of hydraulic and sediment monitoring at two U.S. Geological Survey gaging stations also makes the Colorado River in the Grand Canyon a good place to study hydraulic and sediment transport phenomena. The gages are officially referred to as the gage at the Colorado River at Lees Ferry, Arizona, located at river mile 0.0 (referred to as the Lees Ferry gage in this report) and as the gage at the Colorado River near Grand Canyon, located at river mile 87 (referred to as the gage near Phantom Ranch in this report).

Unique features of the flow and associated erosional and depositional effects have been described by Eddy (1929); Maxson and Campbell (1935); McKee (1938a); Matthes (1947); Leopold (1969); Dolan, Howard and Gallenson (1974); Howard and Dolan (1976); Laursen and others (1976); Dolan and others (1978); Turner and Karpiscak (1980); Howard and Dolan (1981); Brian and Thomas (1984); Beus and others (1985); Kieffer (1985); Burkham (1986); Schmidt (1986); Webb (1987); Wilson (1986); Marley, (1987); Randle and Pemberton (1987); Schmidt (1987); Webb and others (1987); Schmidt and Graf (1988); Kieffer (1987a,b and 1988); Webb and others (1988); and Julia Graf and Durl Burkham (written communication, 1988). The studies postdating 1985 were part of the Glen Canyon Environmental Studies project conducted by the U.S. Bureau of Reclamation, in collaboration with the U.S. National Park Service, U.S. Geological Survey, U.S. Fish and Wildlife Service, Arizona Department of Game and Fish, and university and private researchers. These studies were conducted between 1983 and 1987 to provide technical data to aid in management of Glen Canyon Dam to have minimal impact on the environment along the river corridor (National Research Council, 1987; U.S. Bureau of Reclamation, 1988).

Hydraulic characteristics vary along the river; rock type and tributary location are the major controls on channel characteristics (Dolan and others, 1978; Howard and Dolan, 1981). Average channel width in reaches of resistant rock may be as narrow as 55 m, whereas in reaches of less resistant rock it may be as wide as 120 m. The average channel gradient is also variable, with the steepest reaches (Upper and Middle Granite Gorges) having the greatest transport capability. The size distribution of sediment stored in the channel is related to average channel conditions: the rockiest bottom conditions (in many places, bedrock) are found in the narrow reaches.

Tributary canyons, which are spaced along the river at intervals commonly determined by regional jointing and fault structures (Dolan and others, 1978) are the source of major debris flows which constrict the river channel laterally and form underwater dams or weirs (fig. 3.1 a, b, c). These debris dams cause the "pool-and-rapid" sequences of the Colorado River, because each debris dam causes an upstream backwater (the "pool") and a section of fast shallow water across the debris dam (the "rapid"). The river can be thought of as a series of linked segments of main channel characteristics, pools, and rapids. Most of the total drop in elevation of the river occurs in short steep drops through the rapids (fig. 3.2) (Leopold, 1969). Sand storage occurs primarily in main channel pools and in zones of recirculating flow associated with constrictions caused by the debris dams. Where the channel is constricted, changes in flow and sediment supply are accommodated by changes in sediment storage on the channel bed.

In this chapter, mixed units are used: metric units are used where possible and reasonable, but because of the long history of expression of discharge in cubic feet per second (cfs), we present discharge data in metric units (cubic meters per second, cms) with the English equivalent in parentheses. Finally, the reader should be aware that there are inconsistencies in the English language in the use of the words "rapid" and "rapids": On U.S. Geological Survey maps, the formal names of the rapids are given with the plural form, for example,

---

[1] U.S. Geological Survey, Flagstaff, Arizona
[2] U.S. Geological Survey, Tucson, Arizona
[3] Now at Geology and Geography Departments, Middlebury College, Middlebury, Vermont

Crystal Rapids, whereas in some guidebooks to the river (for example, Stevens, 1983) the formal names omit the "s" on the word Rapid. The ambiguous use of singular and plural forms leads to confusion in discussion of hydraulics and sediment transport if one wishes to compare and contrast the properties of one rapid with other rapids. Therefore, in this paper, proper names of rapids are given with the plural form as shown on the U.S. Geological Survey maps, but text discussion often uses the word rapid to indicate a single stretch of "white water", and the word "rapids" as the plural form.

HISTORY OF FLOW REGULATION AND MAIN-CHANNEL SEDIMENT TRANSPORT

To understand the modern sedimentary record along the river, it is necessary to understand the history of flow and sediment transport (summaries of this history can be found in River and Dam Management, 1987, and in Glen Canyon Environmental Studies Final Report, 1988). The Colorado River and its tributaries drain much of the southwestern United States, ultimately emptying into the Gulf of California in Mexico. Only the Mississippi

FIGURE 3.1 a. Aerial photo of Granite Rapids at a discharge of 141 cms (5,000 cfs), showing the geomorphic features common to many rapids. Photograph by U.S. Bureau of Reclamation, 1984. Regional faults from Dolan et al. (1978).

FIGURE 3.1 b. Aerial photographs of Granite Rapids at 850 cms (30,000 cfs) showing typical wave structures in rapids. Photographs by U.S. National Park Service, June 12, 1986. (b) is at the same scale as (a).

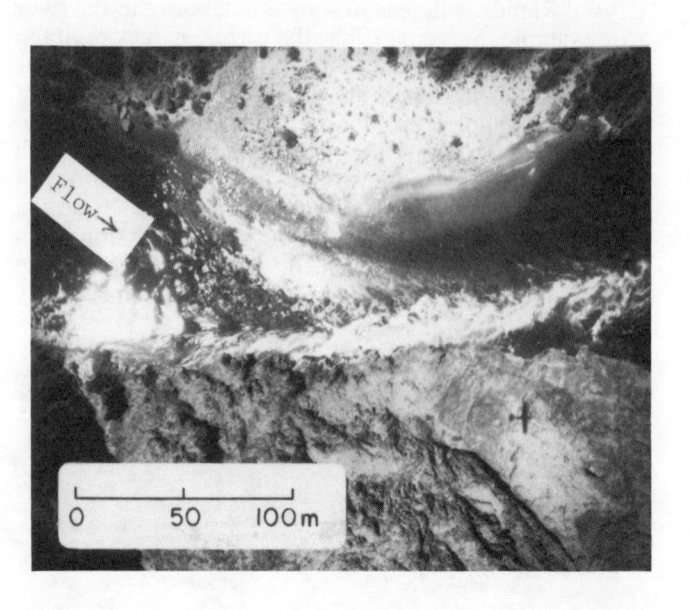

FIGURE 3.1 c. Part of 3.1 b enlarged so that individual hydraulic features can be identified.

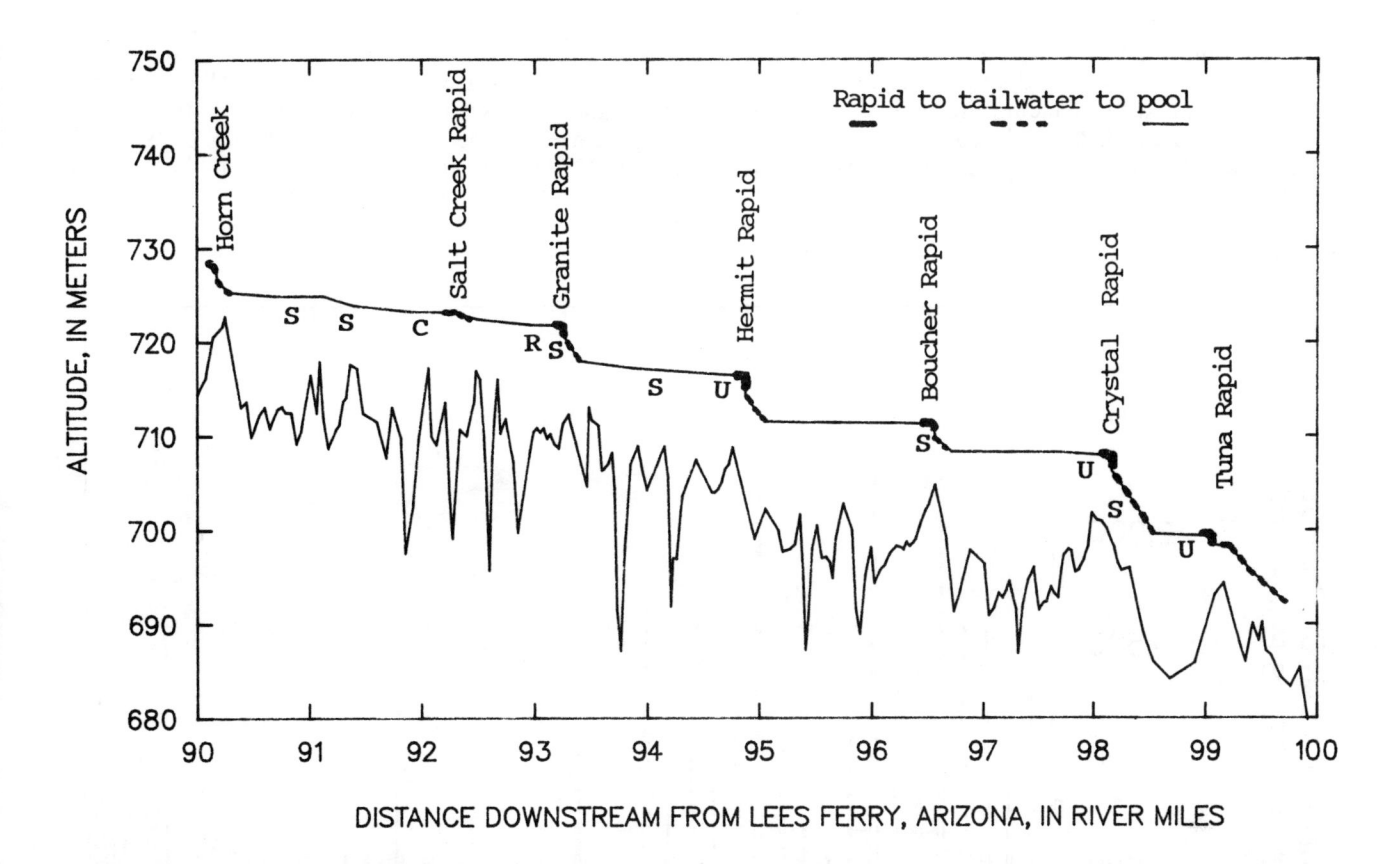

FIGURE 3.2. Profile of elevation drop along a reach of the river from mile 90 (144 km) (Horn Creek Rapids) to mile 100 (160 km) (Tuna Rapids) (personal communication, Timothy Randle, 1987). The solid line shows the bottom profile; the dotted line shows the water surface profile at about 700 cms (25,000 cfs). Note how the water surface profile shows that the river consists of a series of main channel segments of gentle slope; pools in which the water slope is nearly horizontal; and rapids, in which the gradient is steepest. Rapids, pools (backwaters), and channel reaches are indicated. Sand deposits and camping beaches are shown by the symbols present in this section are indicated: S-separation bar; R-reattachment bar; C-channel margin deposit; UP-upper-pool deposit; PB-point bar. Data on beaches from Brian and Thomas (1984); bar type classification from Schmidt and Graf (1988).

River exceeds the Colorado in length within the United States. The potential of the Colorado River waters being diverted for use in irrigation and settlement was recognized more than a century ago, and proposals to use the river for hydroelectric power are nearly a century old. In 1905, severe floods on the Colorado River caused extensive damage in the Imperial Valley of California, and political pressure arose for the construction of storage dams on the river.

In 1922 a formal agreement, the Colorado River Compact, allocated water between the upper-basin states (Colorado, Utah, Wyoming, New Mexico) and lower-basin states (Arizona, California, Nevada). This resulted in the construction of Hoover Dam, which formed Lake Mead, dedicated by President Franklin D. Roosevelt on September 30, 1935. Subsequent political pressures throughout the 1940's and 1950's resulted in development of more dams on the Colorado River system, so that it is now sometimes referred to as "the world's most regulated river" (National Research Council, 1987, p. 18).

The goals of development and regulation of the river expanded rapidly from simple irrigation and flood control: current usages also include hydroelectric power generation; municipal and industrial water supplies as well as irrigation; artificial environments for trout fishing; and recreation on both the river and the impounded lakes. The Bureau of Reclamation coordinates a complex set of mandated usages with many agencies and interested parties, and must do so based on the forecasts of available runoff, available storage, and requirements or demand for water--all under applicable laws.

The Colorado River Storage Project Act of 1956 authorized the construction, operation, and maintainance of 4 storage projects and 11 irrigation projects. One of these was Glen Canyon Dam, which is located 15.5 miles (25 km) upstream from Lees Ferry. Glen Canyon Dam was completed in 1963, and Lake Powell Reservoir then began filling. The release of water into the Grand Canyon is now controlled by the penstocks to eight generators, capable of releasing 890 cms (31,500 cfs), two river outlets that can release up to 425 cms (15,000 cfs), and two concrete-lined spillways with a combined design release capacity of 5,890 cms (208,000 cfs).

The emplacement of the dam has altered the river downstream by stabilizing the flow and the temperature of the water, and by reducing the sediment loads (fig. 3.3). Prior to 1963, the Colorado River in the Grand Canyon was a relatively unregulated river. We will refer to this as the period of <u>unregulated flow</u>. During the gaged years from 1921 to 1962, the average daily discharge at the gage station near Phantom Ranch was 478 cms (16,900 cfs). The mean annual flood was 2180 cms (77,500 cfs); if the large flood of 1884 is included, this value increases to 2320 cms (82,700 cfs). The 10-year recurrence flood between 1921 and 1962 was 3,500 cms (123,000 cfs); if the 1884 flood is included in this calculation, the value increases to 3,710 cms (131,000 cfs).

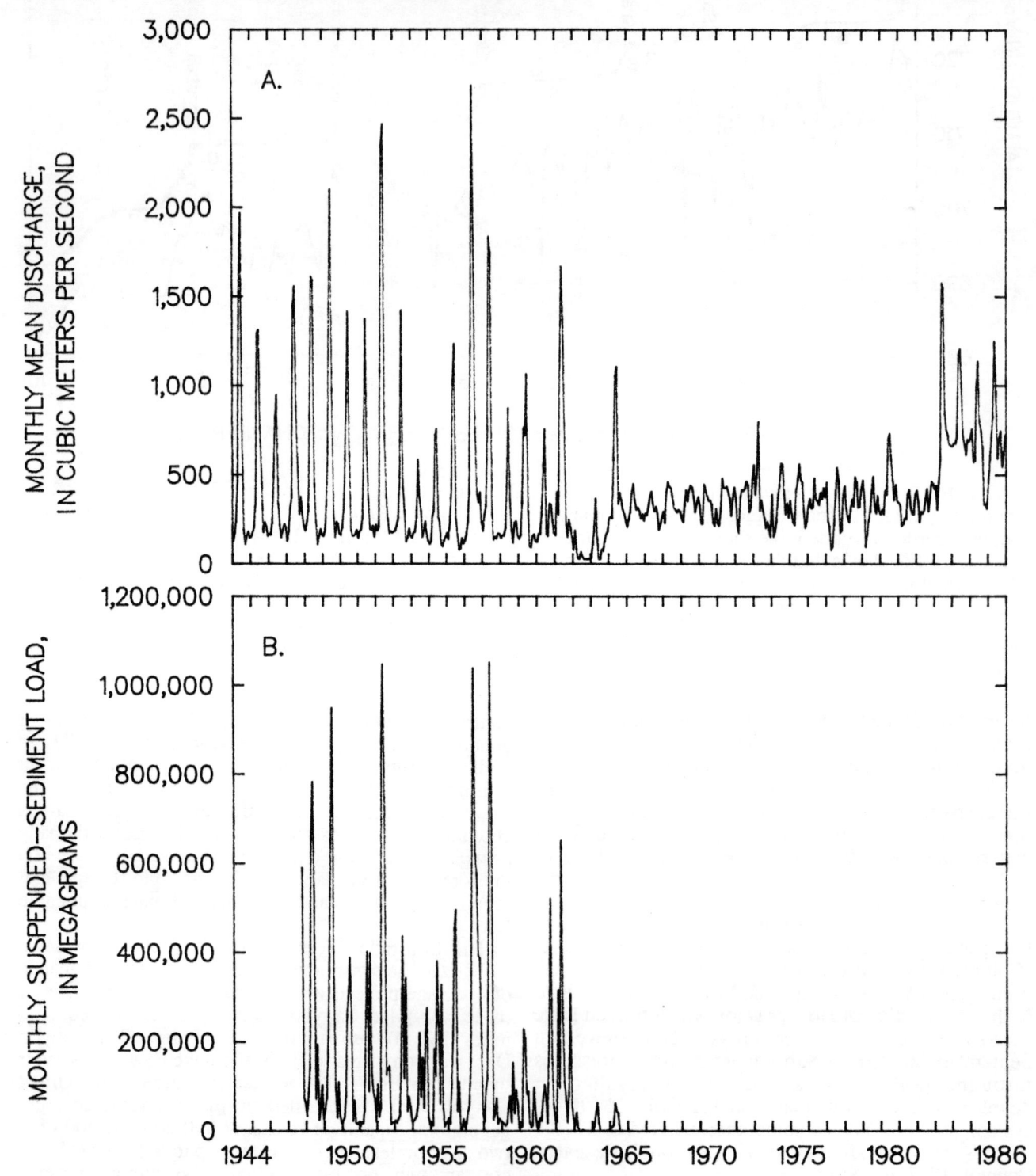

FIGURE 3.3 a,b. Monthly mean discharge from 1943-1986 (a) and monthly suspended sediment load (1947-1986) (b) at the U.S. Geological Survey gaging station, Colorado River at Lees Ferry.

During unregulated flow, discharges annually ranged from a winter low of the order of 100 cms (a few thousand cfs) to a peak of that was typically around 700 cms (25,000 cfs) in early summer as a result of snowmelt runoff and regional rainfall (fig. 3.3a). From year to year, the magnitude of the high flows varied significantly. Thunderstorms in July and August produced smaller flow peaks. Before flow regulation, the sediment load was large--annual suspended-sediment load past the Phantom Ranch gage was 288 megagrams/day (318 tons/day). Although most sediment was carried by the spring floods, thunderstorms in the summer over tributary watersheds produced flows with very high suspended sediment concentrations (up to a mean daily concentration of 780,000 milligrams/liter in the Paria River); the Colorado was thus also heavily laden with sediment at times other than the major spring runoff time (fig. 3.3b).

Each year, pools in the main channel of the Colorado River filled with sediment during periods of low flow and were scoured during spring floods (fig. 3.4a) (Howard and Dolan, 1981, fig. 7, p. 279). These scour-and-fill cycles represent the process by which sediment was moved through the canyon. Sediment accumulated in the channel during low flow because velocities in the pools during these periods (usually the late summer through early winter seasons) were not sufficient to move the debris delivered into the river channel by the tributary flows. When the Colorado was at high flow, the bed scoured and sediment was flushed downstream. Bed elevation changes at gaging stations show that as discharge and velocity increased with passage of the spring snowmelt floods, the bed began to scour when the mean velocity reached a threshold value of about 1.5-1.7 m/s (these values are based on detailed analyses at the two gaging stations when the bed was at a relatively high elevation). Scour continued as long as the mean velocity continued to increase. Sediment was deposited on the bed when velocity dropped below that required for scour if sand and gravel were available. At Lees Ferry the bed scoured and filled up to 6 m during the year (fig. 3.4a). Maximum scour was only about 2.5 m at the gage near Phantom Ranch, and the bed did not refill every year as did the bed at Lees Ferry (compare figs. 3.4a and 3.4b) (Burkham, 1986).

At both gaging stations, the ability of flow to transport sand is several orders of magnitude less when the bed is at a low elevation than when the bed is at a relatively high elevation. At low bed elevation, channel cross-sectional area is greater than at high bed elevation for the same discharge; therefore mean section velocity is less at low bed elevation. The relation between sand concentration and discharge at the two gage stations is poor -- much of the variation in sand concentration cannot be accounted for by variations in discharge and bed elevation. Prior to construction of the dam, transport of sand, especially at the gage near Phantom Ranch, appears to have been controlled to a large extent by the supply of sand within the main channel and from tributaries. Sand supply was apparently limited, and much of the time less sand was available for transport than the flow was capable of moving. Also, much of the sediment contributed by tributaries (sand as well as silt and clay) was transported through the channel even at relatively low discharges.

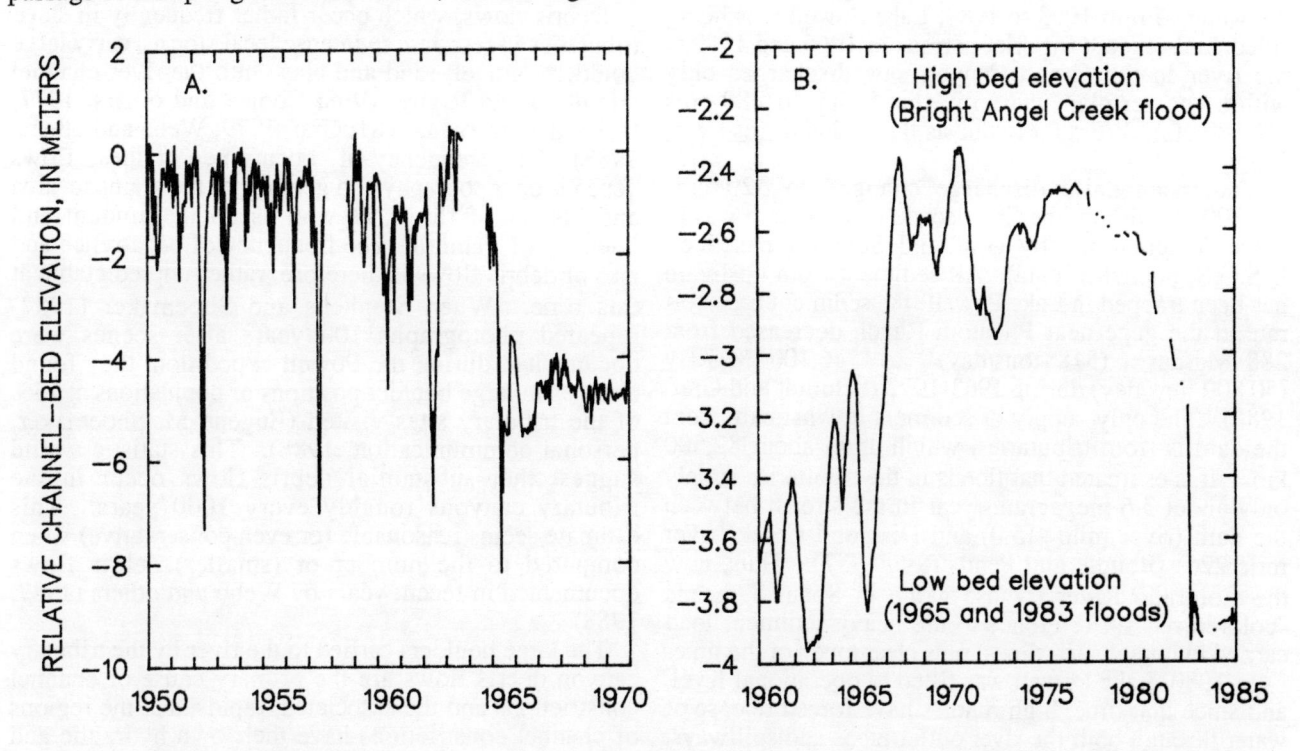

FIGURE 3.4 a,b. Bed elevation at (a) the U.S. Geological Survey gaging station, Colorado River at Lees Ferry (river mile 0), 1950-1970, and (b) at the U.S. Geological Survey gaging station, Colorado River near Grand Canyon (river mile 87.2, 139 km), 1960-1985.

Although we do not know how representative the two gage stations are of the rest of the channel, we assume that narrow, high-velocity reaches behave in much the same way as the reach surrounding the gage near Phantom Ranch, and that wider, lower-velocity reaches behave more like that at Lees Ferry. Sand deposits along the channel margins are major habitats of flora and fauna, and are extensively used for recreational camping. The sand deposits have always been fewer and smaller in narrow reaches than in wide reaches. This suggests that the behavior of the bed reflects the sand-storage characteristics of the channel as a whole-- that in reaches where the capacity to transport sand is greater than supply much of the time (such as near Phantom Ranch), little sand is stored on either the bed or banks of the channel whereas in reaches where capacity to transport sand is less than supply much of the time (such as at Lees Ferry), larger quantities of sand are stored in both the bed and banks of the channel.

Flow regulation at Glen Canyon Dam drastically changed the discharge regime and sediment transport. The Dam, which was begun in 1956 and completed in 1963, controls flow from a drainage area of about 2.8×10^5 km² (108,000 mi²). The stored water forms Lake Powell Reservoir, which, at peak capacity, holds 3.3×10^{10} m³ (27 million acre-feet or 8.8×10^{12} gallons). The dam was built to ensure delivery of water to Mexico (mandated by a treaty between the U.S. and Mexico), and to ensure agreed-upon partitioning of water between the upper- and lower-basin states. Simultaneously, it has been mandated that dam operations, including the level to which the lake is filled, maximize the power generation that can be obtained by storage and release of the water. From 1963 to 1980, Lake Powell was being filled, and--except for brief periods in 1965 and 1980,-- the river in the Grand Canyon saw discharges only within power-plant capacity, that is, up to 880 cms (31,500 cfs). We refer to this as the period of <u>reservoir filling</u>.

The mean daily discharge changed to 320 cms (11,300 cfs) during the first decade that the dam was in place (calculations by Graf and Schmidt based on U.S.G.S. published data). All sediment from upstream has been trapped in Lake Powell: the sediment transport rate at the gage near Phantom Ranch decreased from 288 Mg/day (318 tons/day) to 46,100 Mg/day (50,800 tons/day) during 1963-1972 (Schmidt and Graf, 1988). The only supply of sediment downstream from the dam is from tributaries, which drain about 82,500 km². It is estimated that floods in the tributaries supply only about 3.6 megagrams/year in the stretch between the dam (river mile -15.0) and Diamond Creek (river mile 225) (Randle and Pemberton, 1987). Thus, now the Colorado River (whose name in Spanish means "colored red" in reference to the heavy sediment load carried in unregulated times) runs clear much of the time.

In 1980, Lake Powell was filled to operational level, and since that time, high runoffs have forced release of water through both the river outlet tubes and spillways. The current (1987) estimate by the Bureau of Reclamation is that operations will require that in approximately 1/4 of future years water will need to be discharged through either the river outlet works or spillway. In other words, now that Lake Powell is filled to operational capacity there will be more frequent "floods" (flows exceeding power-plant capacity) on the Colorado River than there were during the reservoir-filling years. We refer to this as the <u>current operational period</u>. If all eight generators of the power plant are running, the generators plus the river outlet tubes can discharge about 1,415 cms (50,000 cfs). The spillways were designed to discharge 2,830 cms (100,000 cfs) each, but severe erosion occurred in the spillways in 1983 when they were used to raise the total discharge to 2,600 cms (92,600 cfs). (The maximum instantaneous peak discharge during the 1983 flood was 2,720 cms (96,200 cfs); however this was of very short duration (a few hours at most. The maximum daily flow was 92,200 cfs, and we use 92,000 cfs as the peak discharge during this time for convenience in this article). Even though the spillways were redesigned and repaired, their use in the future will be discouraged.

Large tributaries (especially the Paria and Little Colorado River which together drain more than 72,000 km²) respond to regional precipitation and carry very large suspended sediment loads. Sediment loads vary by an order of magnitude from year to year, and may have longer term variations. Floodplain studies (Hereford, 1984a, 1986) have revealed long-term variations in sediment storage in these tributaries, sediment loads measured in the last 40 years may show less sediment input to the Colorado River than at other times in history. A decrease in both the amount of sediment stored on the bed of the river and of sediment transported past gage stations since about 1940 has been documented.

Debris flows, which occur rather frequently in steep tributaries in response to intense local storms, carry large boulders, gravel, sand and clay into the river channel (Hamblin and Rigby, 1968; Cooley and others, 1977; Howard and Dolan, 1981; Graf, 1979; Webb and others, 1988). The frequency of occurrence of these flows depends on topography and geology of the drainage area and history of flows, as well as on the amount and intensity of rainfall. The frequency of occurrence and size of debris flows is therefore rather unpredictable at this time. When Stephens and Shoemaker (1987) repeated photographs 100 years after scenes were documented during the Powell expedition, they found changes in large boulder positions or populations at 15% of the tributary sites visited (Eugene M. Shoemaker, personal communication, 1988). This statistic would suggest that substantial debris flows occur in the tributary canyons roughly every 1000 years. This estimate seems reasonable (or even conservative) when compared to the number of (smaller) debris flows documented in recent years by Webb and others (1987, 1988).

The large boulders carried to the river by the tributary canyon debris flows are the primary cause of channel constrictions and the associated rapids, and the regions of channel constrictions have their own hydraulic and sediment patterns that are distinctively different from the main channel flow and sediment transport and storage patterns.

HYDRAULIC AND GEOMORPHIC FEATURES OF RAPIDS

Hydraulic and geomorphic features found at most rapids are identified on the air photos of Granite Rapids shown as Figures 3.1a, b, c. A detailed hydraulic map and cross sections of House Rock Rapids are provided in Figure 3.5. Most camping spots and many lunch spots are located at settings generically similar to those specifically illustrated and discussed in this guide, so the reader may find it useful and or entertaining to try to find the features discussed in the photos and in the text at different rapids as time permits. The major rapids on the river that are particularly good for examination of hydraulic and geomorphic features have been mapped in detail by Kieffer (1988). Maps of these rapids are available as U.S. Geological Survey Miscellaneous Investigations Maps 1897 A-J and in the following list, the alphabetic prefix indicates the proper map designation in the series: (A) House Rock, (B) 24.5-mile, (C) Hance, (D) Cremation-Bright Angel (a small rapid, but important monitoring site); (E) Horn, (F) Granite, (G) Hermit, (H) Crystal, (I) Deubendorff, and (J) Lava Falls. (See figure 4.2 for location of these rapids.)

The hydraulic features shown in Figure 3.1 are given names which arise from a combination of river-runners descriptive terms and formal hydraulics nomenclature; river-guides are familiar with these words and can explain them to passengers.

The shoreline at a rapid can vary dramatically as discharge changes. For example, Figure 3.5a shows that at House Rock Rapids much of the debris fan is exposed at 140 cms (5,000 cfs) (a typical low discharge from Glen Canyon Dam), whereas Figure 3.5b shows that water spills across most of the fan at 850 cms (30,000 cfs) (a typical high discharge through the dam). The drop in water-surface elevation through this rapid is greatest at low discharges (fig. 3.5c) and decreases with increasing discharge (this shallowing of gradient can be seen by comparing the shoreline above and below the rapid in figs. 3.5a and 3.5b). This "shallowing" of the gradient through rapids with increasing discharge is typical of all rapids studied up to discharges of 1,275 cms (45,000 cfs) (Schmidt and Graf, 1988), and is probably typical of the older rapids at higher discharges. It may not be true at recently formed rapids, like Crystal Rapids, where nonequilibrium conditions between the hydraulics and geomorphology exist (see following discussion).

Velocities and depths change by approximately an order of magnitude through a rapid (velocities along different streamlines are shown in fig. 3.5d). These dramatic changes produce sudden transitions between the two possible flow regimes: subcritical and supercritical. These regimes differ in the relation of the average flow velocity to the critical velocity, $c = (gd)^{1/2}$, where g is the acceleration of gravity and d is the average flow depth. For example, if the flow is 1 m deep, the critical velocity is 3.1 m/s; if the flow is 10 m deep, the critical velocity is 9.9 m/s. If the flow velocity is less than the critical velocity, the flow is subcritical; if the flow velocity is greater than the critical velocity, the flow is supercritical. In subcritical flow, the role of gravity forces is more pronounced than the role of inertial forces: the velocity is "low" (the words "tranquil" and "streaming" are used to describe subcritical flow in some hydraulics literature). In supercritical flow, on the other hand, inertial forces are dominant. The flow has a "high" velocity (the words "rapid", "shooting", or "torrential" are used).

The response of a flow field to an obstacle in the channel or to changes in channel shape and alignment depends on whether the flow is subcritical or supercritical. Changes in channel configuration cause subcritical flow to accelerate or decelerate smoothly, and there are no standing waves. On the other hand, in supercritical flow, changes in channel configuration cause standing wave patterns, and the flow adjusts to the channel geometry changes primarily in passing through these waves. In the rapids of the Colorado River, one generally sees a transition from subcritical conditions in the upstream pool, to supercritical conditions within the rapid, to subcritical conditions in the downstream runout and main channel.

In a backwater above a rapid, the velocity is very low--on the order of, or less than, 1 m/s or less (depending on discharge) (fig. 3.5e, cross-section a). Upstream velocities actually develop in recirculating zones above the debris fan, allowing the formation of small beaches (e.g., near the boat shown in fig. 3.5a). Water accelerates as it flows across the upstream side of the underwater debris dam (fig. 3.5e, cross section b). At approximately this position or slightly downstream, subaerial exposures of the debris fan constrict the river (fig. 3.5e, cross section c). The combination of lateral and vertical constriction causes the river to flow fast and shallow, and supercritical flow conditions result. The highest velocities are obtained in and slightly downstream from the most tightly constricted part of the channel (fig. 3.5e, cross section d). Velocities exceeding 7 m/s have been measured at discharges of only 140 cms (5,000 cfs); during the 1983 flood of 2,600 cms (92,000 cfs), kayaks were measured travelling at 9 m/s in Crystal Rapids. Standing waves develop in the supercritical regions of the rapid. Their presence, height, and structure depend on discharge. The largest wave reported on the river was at Crystal Rapids during 1983--5-6 m, but waves of 3 m in height occur at several other rapids.

Downstream from the constriction, the channel width typically exceeds the upstream main-channel width for a distance of the order of 100 m (figs. 3.5, 3.6, 3.7). Most of the water leaves the rapid in a narrow jet that shoots into the downstream reach (tailwater) at a velocity of about 4-5 m/s (fig. 3.1; fig. 3.5, cross-section e). Because of the momentum of the jet, and the distance required for substantial mixing and expansion, the jet--narrowed by the constriction--does not immediately spread out to fill the entire downstream channel. Zones of recirculating current (commonly called "eddies," but see more precise definition in next paragraph) exist adjacent to the expanding jet.

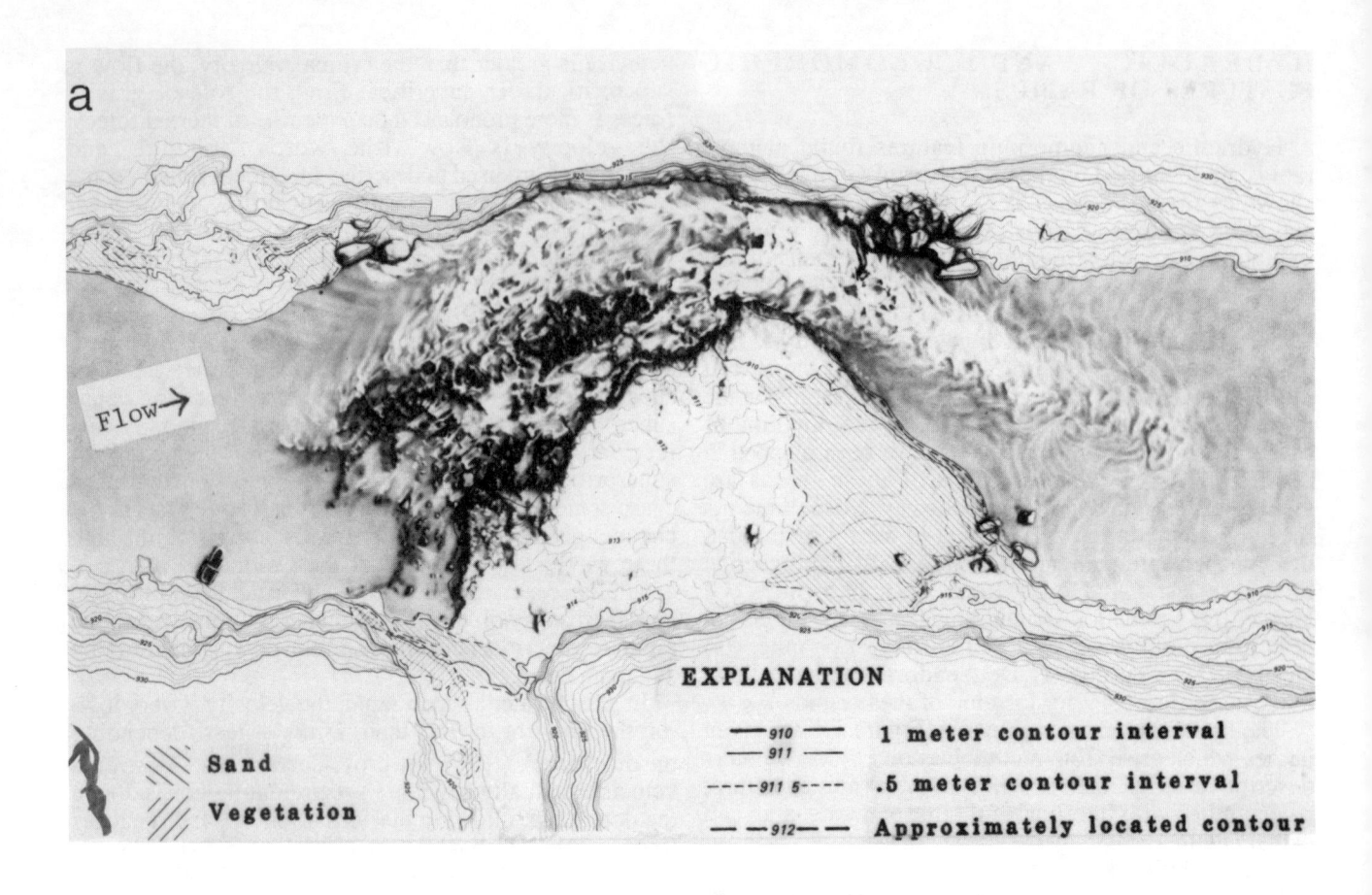

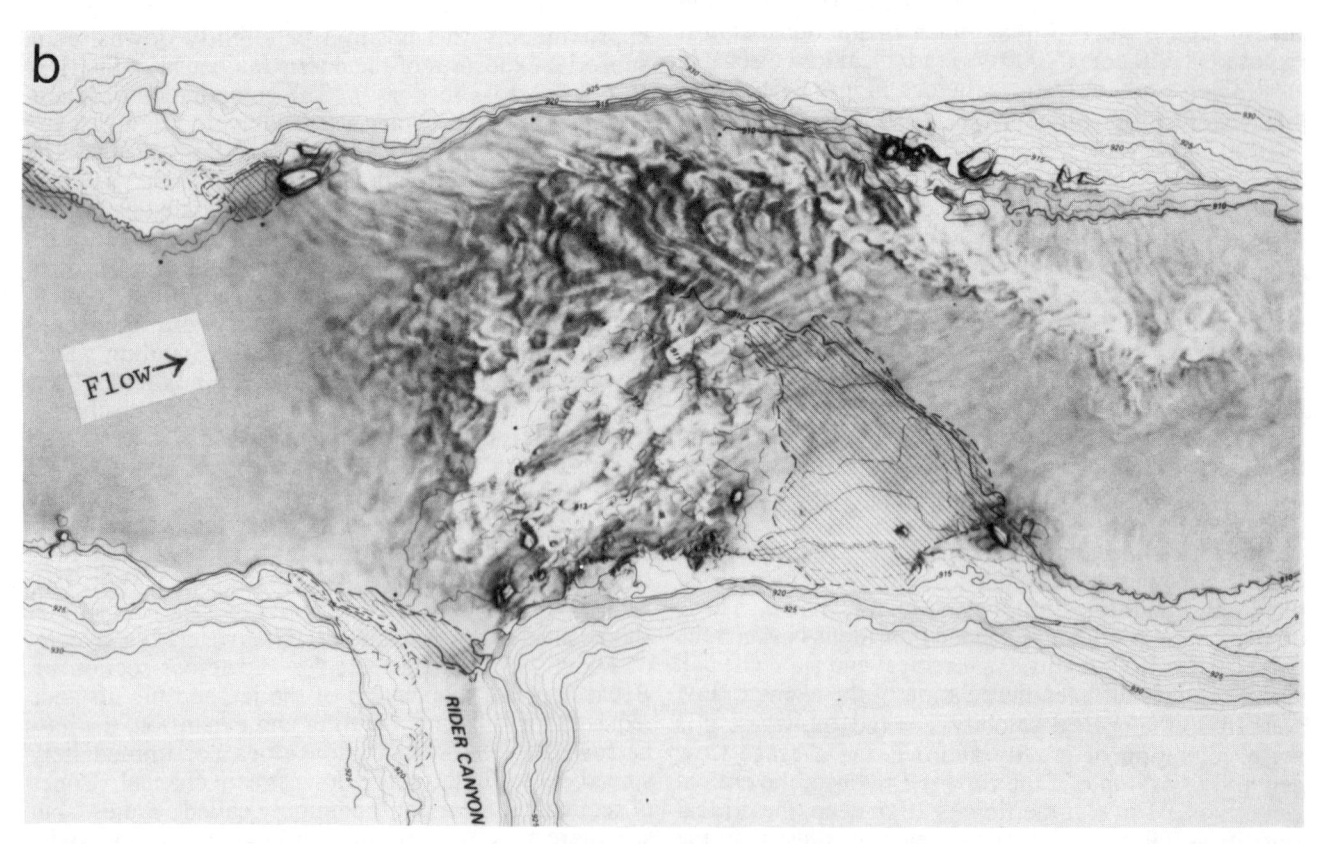

FIGURE 3.5 a,b. Hydraulic maps of House Rock Rapids. (a) Topographic contours and standing waves at 141 cms (5,000 cfs). (b) The same at 850 cms (30,000 cfs).

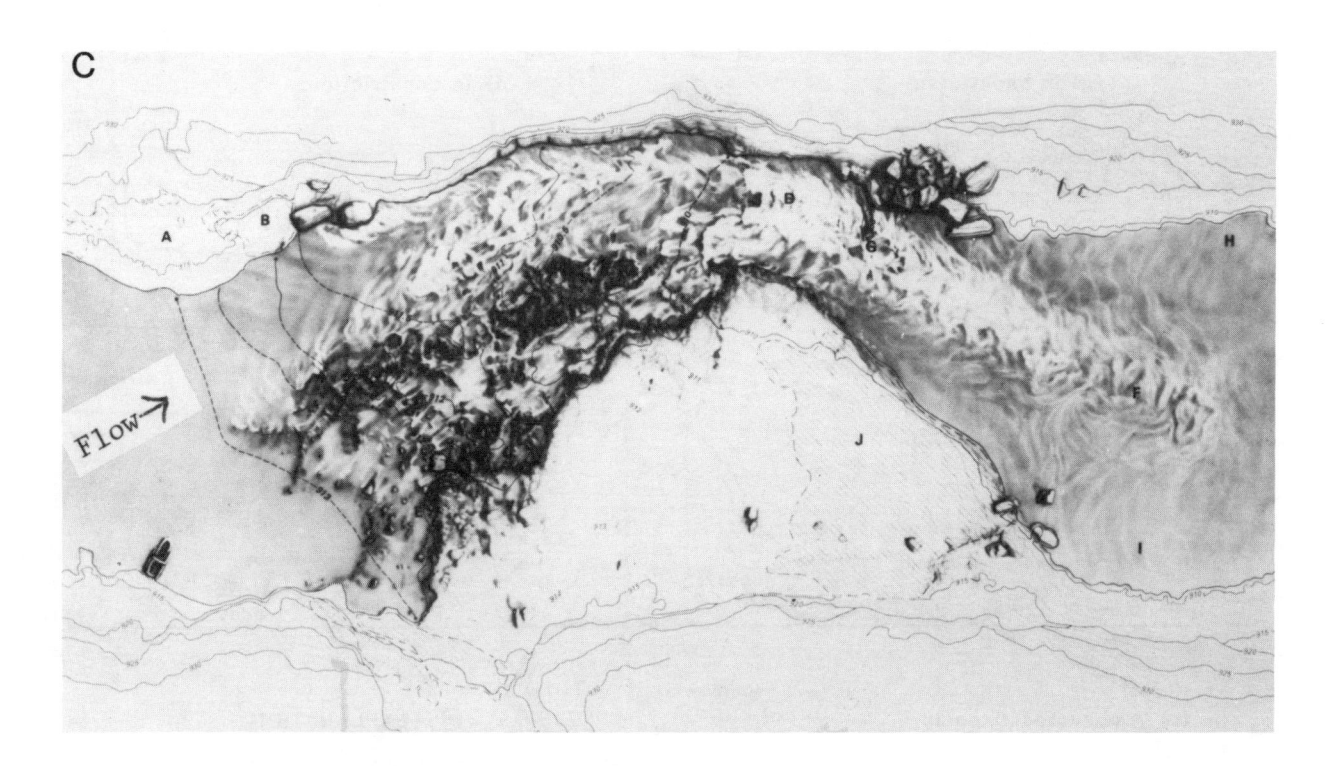

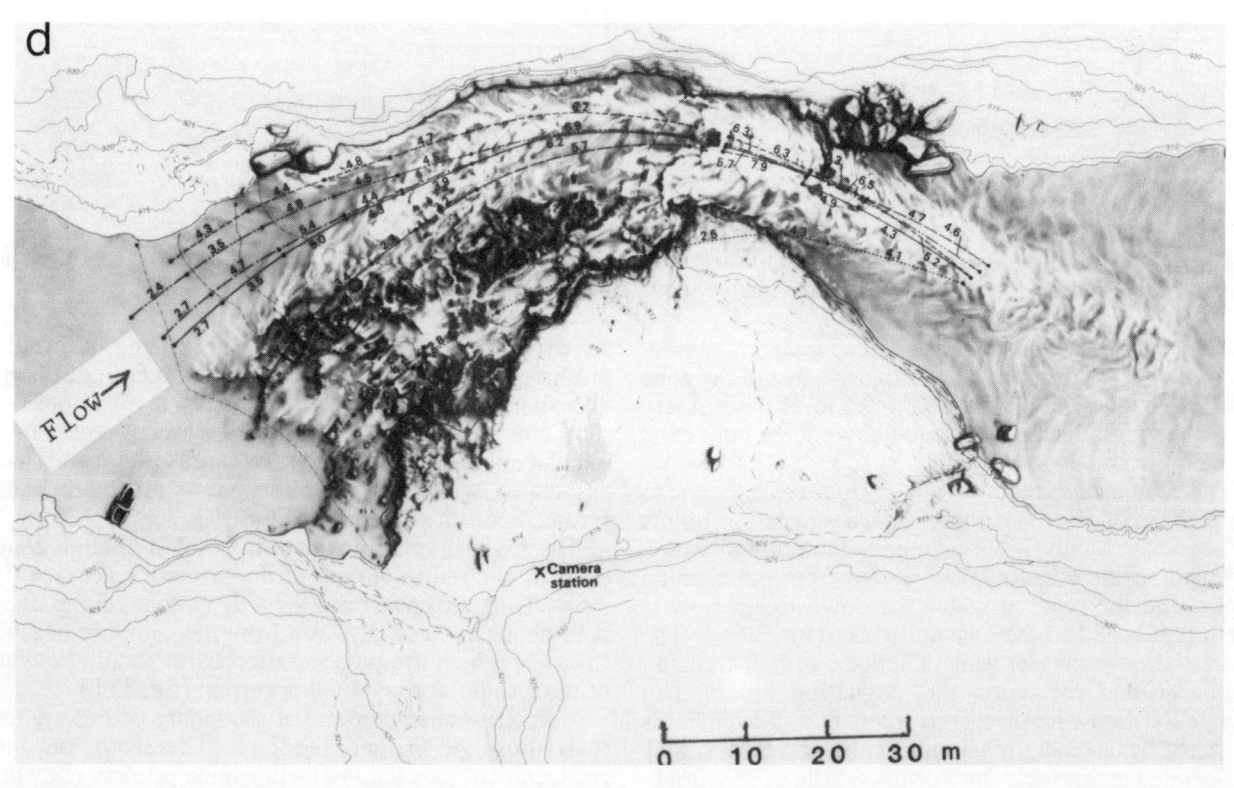

FIGURE 3.5 c,d. (c) Water-surface contours at 141 cms (5,000 cfs). (d) Velocities and streamlines at 140 cms (5,000 cfs). Flow direction is from left. Scale is 1:2000. Contour intervals indicated with solid lines are 1 m (dashed lines = 0.5 m). Diagonal pattern slanting left indicates sand; diagonal pattern slanting right indicates vegetation. The large sand deposit mantling the downstream part of the debris fan is a separation bar deposit.. In (d) numerals indicate velocities along the streamlines between the adjacent dots; velocities are in m/s. Trajectories of the floats were determined from movies taken from the camera station indicated. The boat showing is a standard commercial motor raft that is 10 m (33') in length. <u>THE BOAT IS SHOWN ONLY FOR SCALE; THESE MAPS ARE FOR ILLUSTRATION OF HYDRAULIC FEATURES ONLY, AND ARE NOT INTENDED FOR NAVIGATIONAL PURPOSES</u>.

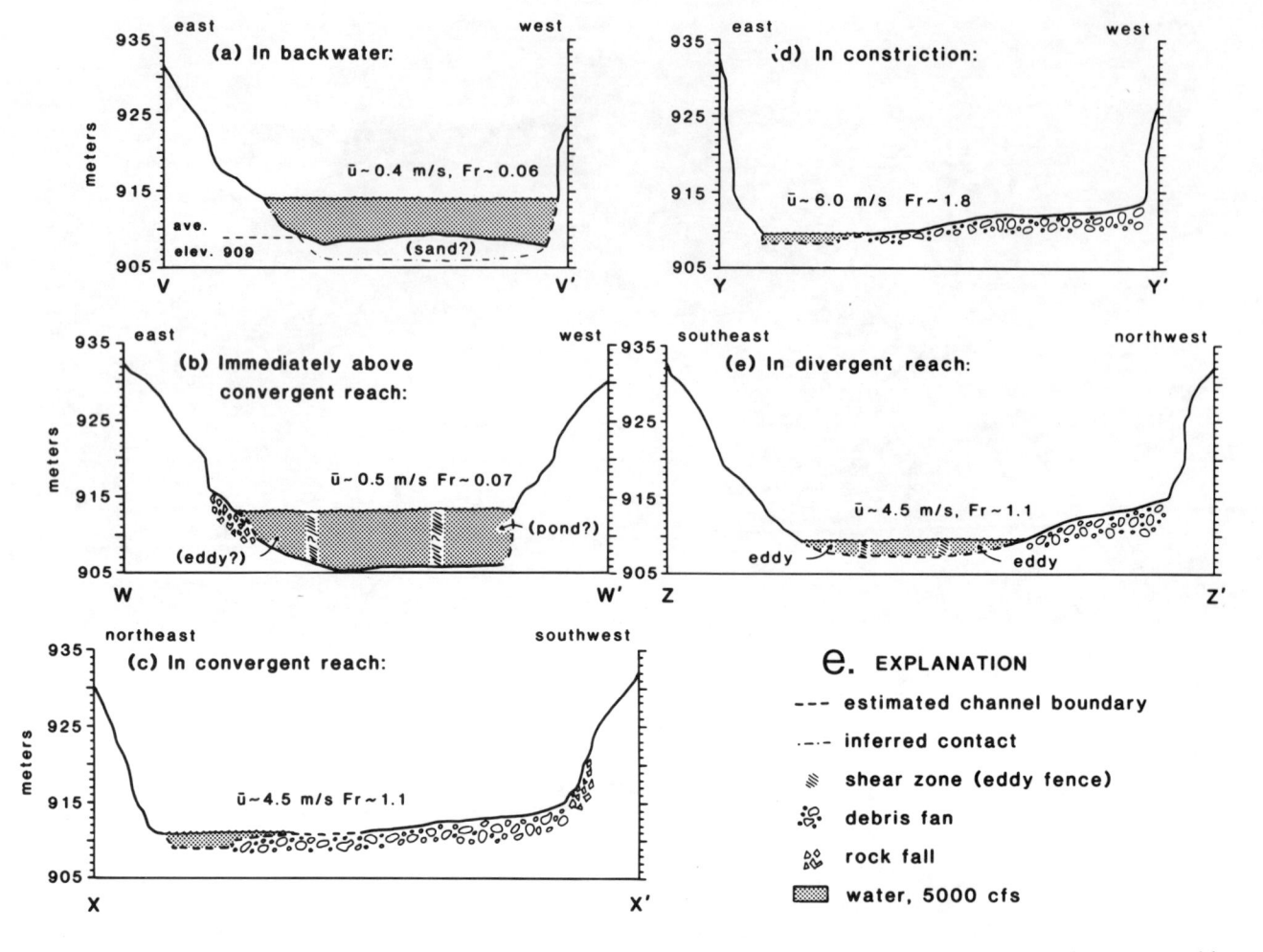

FIGURE 3.5 e. Five cross sections through different reaches of House Rock Rapids, based on the topographic map shown in part (a). Location is approximately described on each cross section.

Flow patterns within the zones of recirculating-current are remarkably consistent at all rapids. Most of the zone is filled by one eddy, which we refer to as the primary eddy. At high discharges, secondary eddies may exist upstream from the primary eddy.

The existence of recirculating-current zones is due to the process of flow separation. Two stagnation points are associated with flow separation--the point at which downstream flow becomes separated from the channel banks, and the point at which the downstream flow is again reattached to the channel banks (fig. 3.6a). The recirculation zones extend in length with increasing discharge and therefore, the stagnation points may change location with discharge unless their movement is restricted by channel irregularities or by gravel bars.

Despite the high velocities characteristic of a bedrock-gorge river such as the Colorado, stagnation points are areas of relatively low velocity and, therefore, of very low sediment-transport capacity. Sand bars develop in the stagnation regions. There are two types of sand bars corresponding to the two stagnation points--separation bars located near the separation points and reattachment bars located near the reattachment points. Because stagnation points can shift in location as discharges change, the shape and characteristics of sand bars is not simply related to the flow field at any particular discharge, but is determined by the changes in discharge and stagnation points (see figs. 3.7c, d, e) Note that sand bars associated with the reattachment points form beneath the primary eddy, whereas the sand bars associated with the separation points are more likely formed beneath secondary eddies (fig. 3.6a).

The highest velocities within the recirculation zones occur in the return current of the primary eddy (see fig. 3.6). This current excavates a deep channel. Bathers who chose to wade offshore from the camping beaches (most of which are on separation bars) should be wary of the channel formed by this current (fig. 3.6b).

Velocities characteristic of secondary eddies are less than those of primary eddies. Therefore, because reattachment bars tend to form from a primary eddy and separation bars tend to be formed beneath secondary eddies near shore (see fig. 3.6a), the typical mean grain size of separation bars is finer (very fine to fine sand) than that of reattachment bars (fine to medium sand).

Characteristic sedimentary structures of these sand bars are dune-foreset cross-lamination, and various climbing-ripple structures. Along the river margin, deposits of white, well-sorted sand, at places overgrown by tamarisk trees less than 5 years old--were deposited in

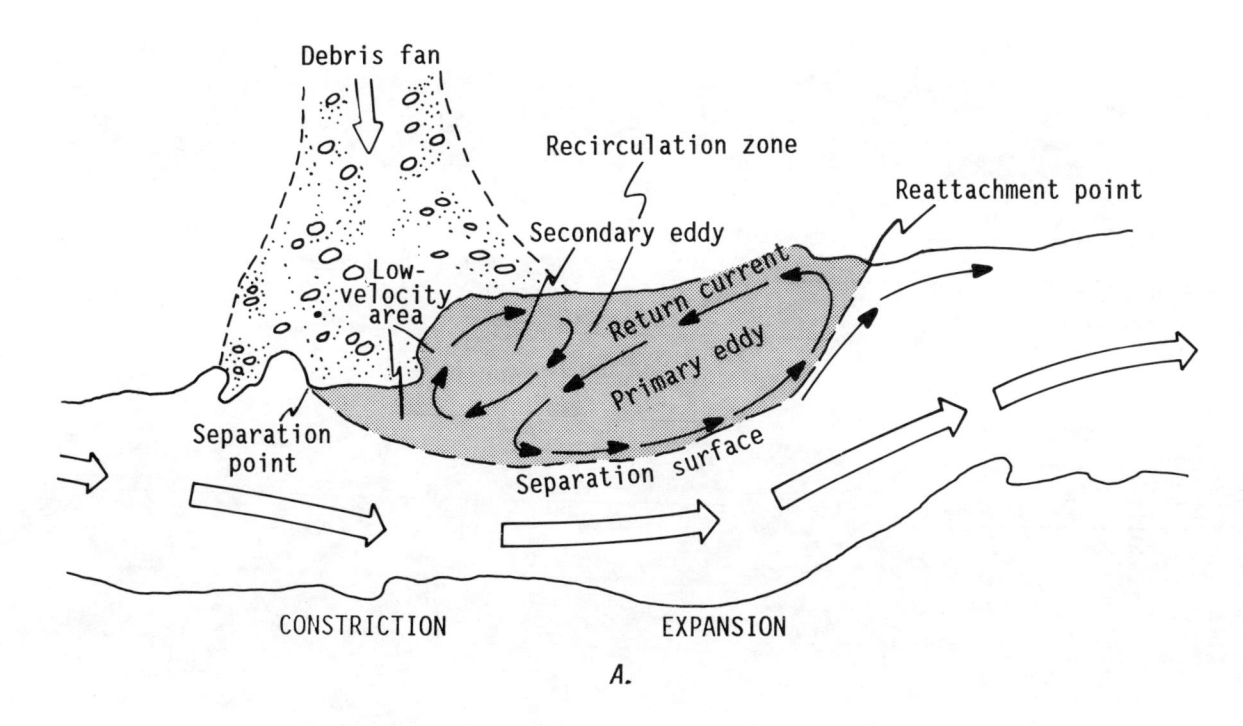

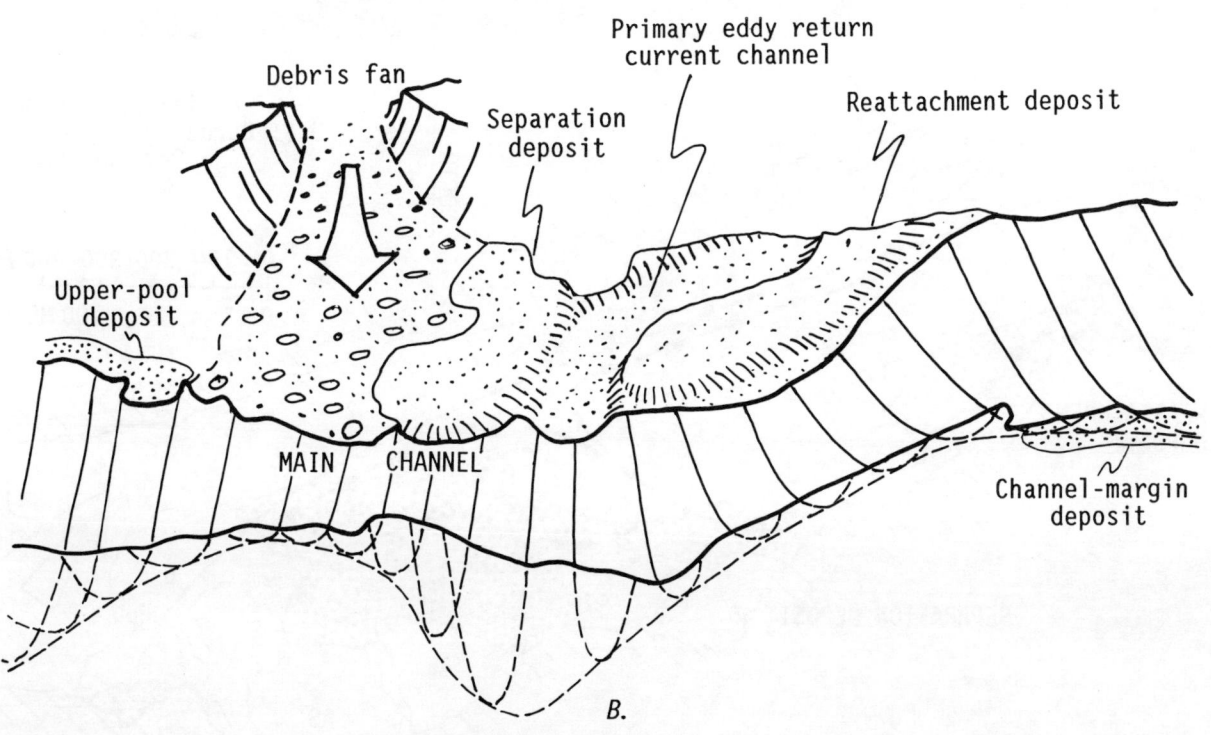

FIGURE 3.6 a,b. Flow patterns and configuration of bed deposits in a typical recirculation zone. (a) Flow patterns in the recirculation zone downstream from a debris fan. The large open arrows indicate the main current direction; smaller dark arrows indicate flow directions within the recirculating zone. The separation and reattachment points, primary and secondary eddies, and return current channels are discussed in the text. The separation surface between the primary eddy and the main current is indicated schematically; this separation surface seems to correspond directly with the outer break in slope of the edge of the bar crest (fig. 3.6b, and 2.7 c, d, e. (b) Three-dimensional view of the configuration of bed deposits in a recirculating zone and their relation to flow characteristics shown in part (a). Note the small channel-margin deposit indicated downstream in the main current where the channel has returned to full width. The solid and dashed lines indicate schematically the elevation of the bed of the river in the vicinity of a constriction caused by a debris fan.

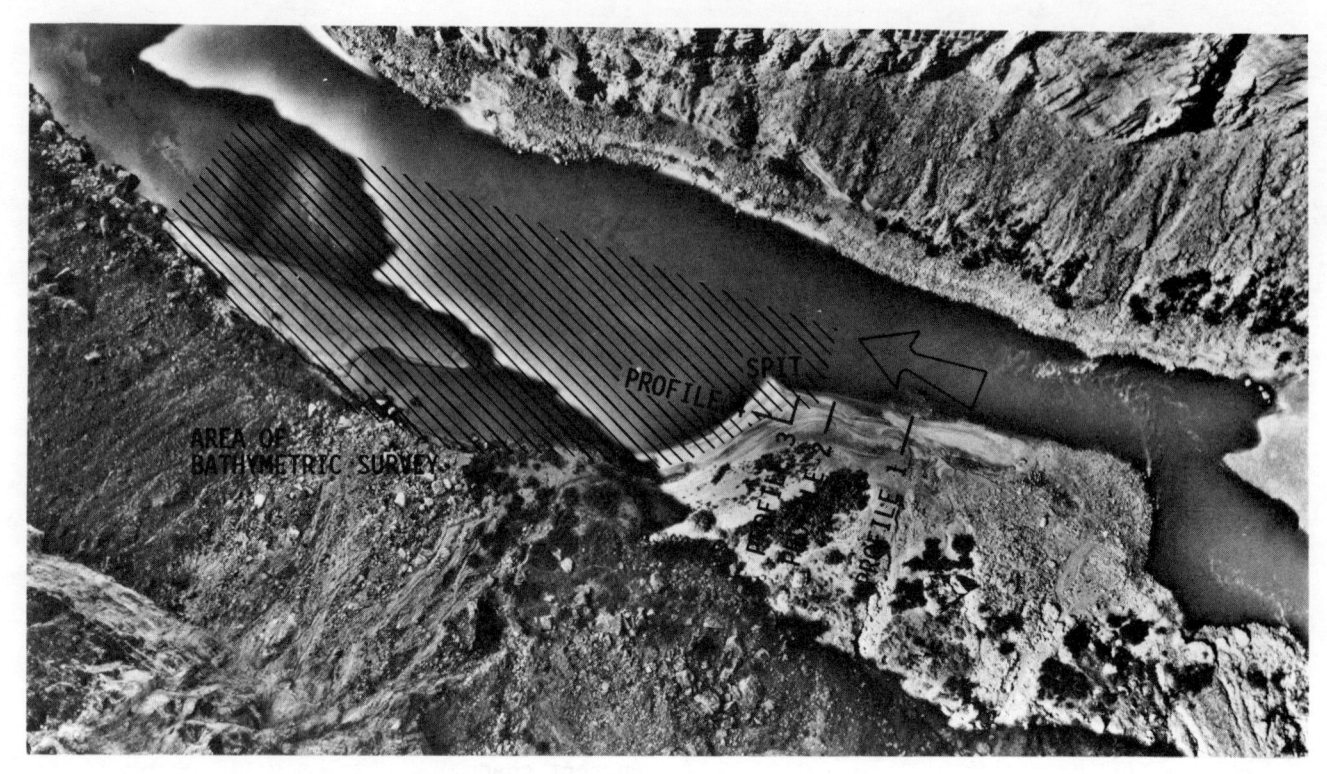

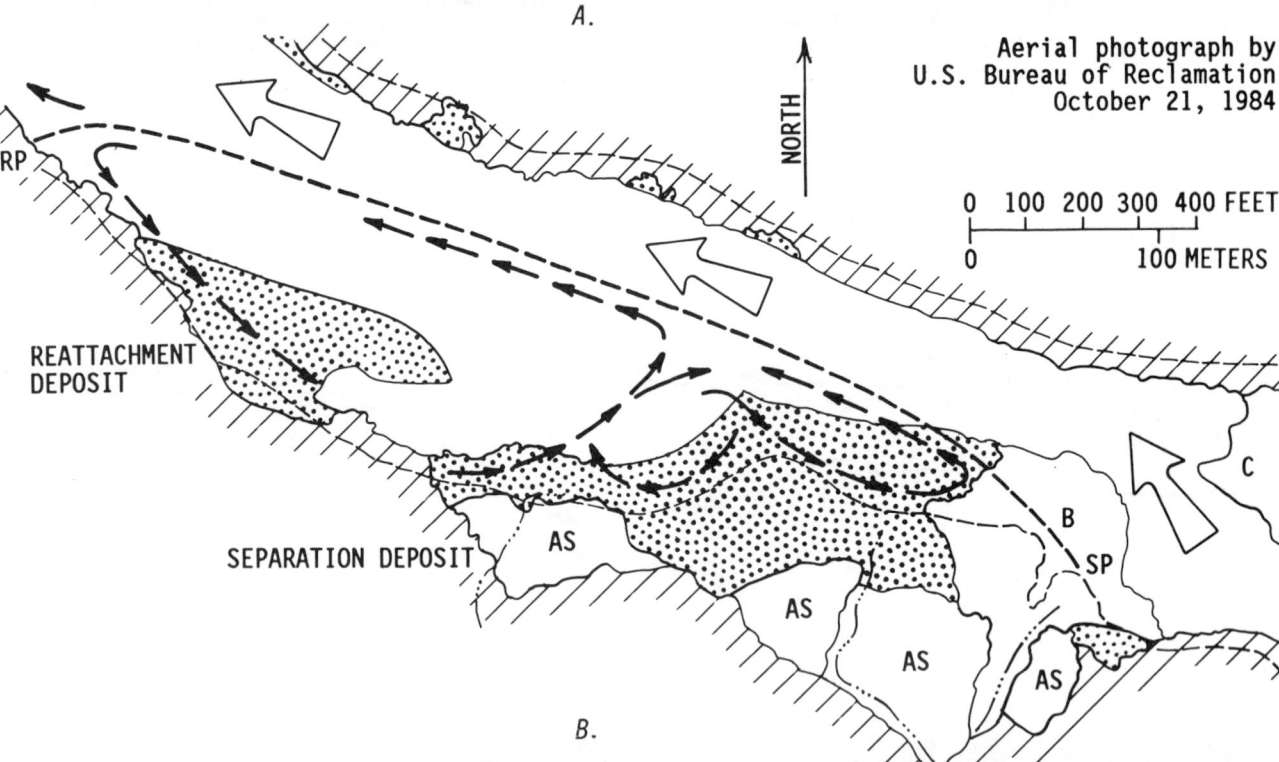

FIGURE 3.7 a,b. (a) Aerial photo of surficial geology and hydraulic features at Eminence Break Camp (river mile 44.2; 70.7 km). The discharge in this photo is 140 cms (5,000 cfs). Photo taken October 21, 1981. Cross-hatching indicates the area of bathymetric maps given in (c), (d), and (e). (b) Surface features mapped from photo in (a): hydraulic features superimposed were measured on May 25, 1985, at a discharge of 11600 cms (41,000 cfs). The stippled area indicates river-deposited very-fine to medium sand. AS indicates aeolian sand deposits; B is tributary debris fan deposits; C is cobbles and gravel; SP is the separation point; and RP is the reattachment point. The arrows show flow directions in the primary and secondary eddies; the dashed line indicates the separation surface between the recirculation zone and the main current, whose flow direction is indicated by the large open arrows. The light dashed line indicates the edge of the water at 1,160 cms (41,000 cfs).

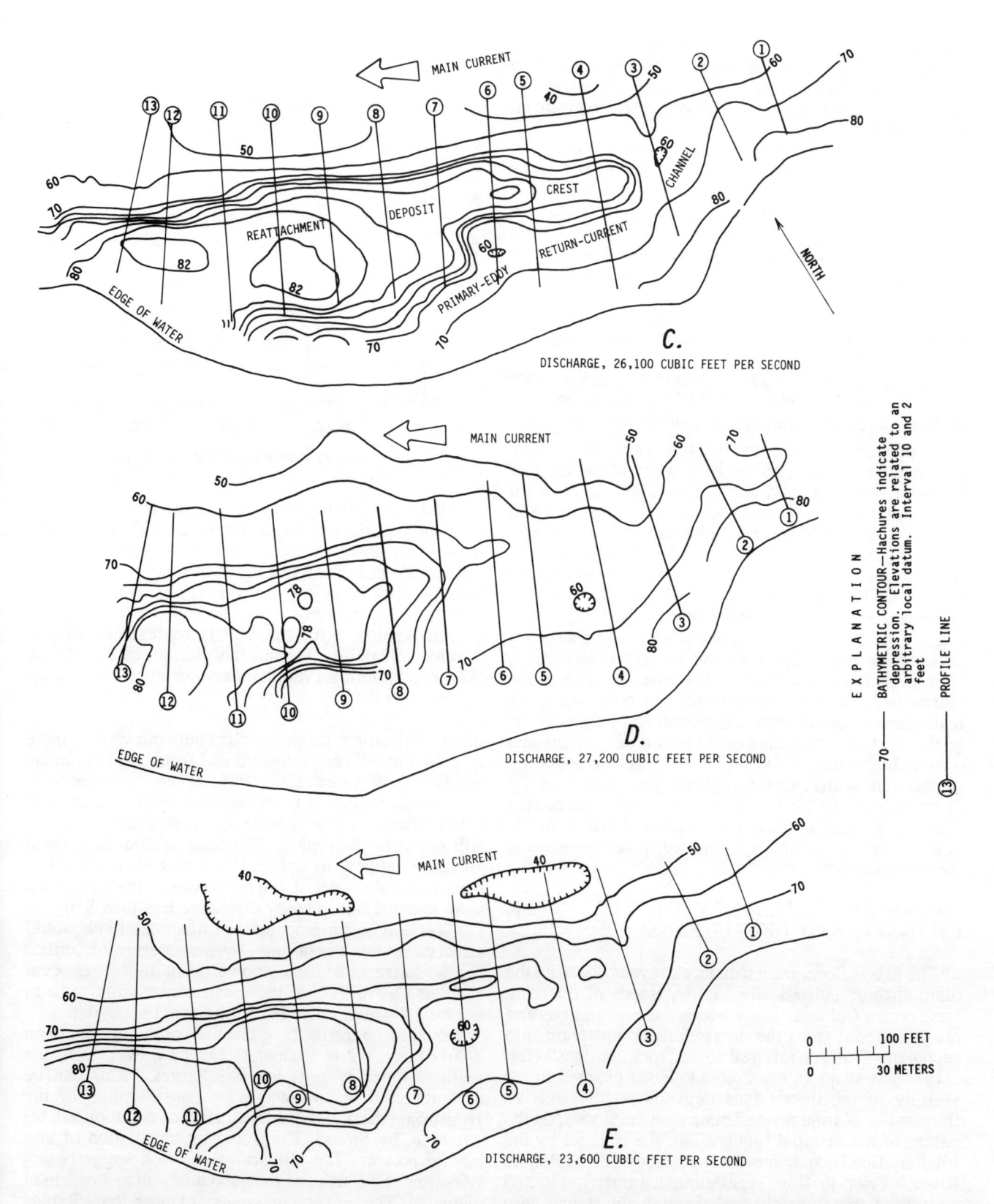

FIGURE 3.7 c,d,e. Bathymetric contours within the recirculation zone shown in (b) on April 26, 1985, September 2, 1985, and January 16, 1986, respectively. Note that the flow direction in these photos is not quite aligned with the flow direction of the same region shown in Figure 3.6, and that the scale is somewhat different. Note that the discharges at which these three maps were made are within the narrow range of 670 cms-770 cms (23,600 cfs-27,200 cfs). See text for discussion of the differences in shapes of the deposits.

or since 1983. Higher deposits of tan silty sand are associated with larger floods that occurred prior to completion of the dam. These latter deposits are particularly prominent in the first four miles downstream from Lees Ferry.

Classification of sand bars throughout the Grand Canyon, and analysis of changes in these bars since completion of Glen Canyon Dam, show that separation bars may be more stable under changing flow conditions than reattachment bars because they are formed from secondary eddies whose velocity is less dramatically influenced by discharge changes. Separation bars are also located downcurrent (in the sense of local flow direction) from a nearby sediment source--the reattachment bar (note the relative current directions in fig. 3.6; see also fig. 3.7b).

Bathymetric maps of a recirculation zone at Eminence Break Camp (river mile 44.2 (70.7 km)) made on three different occasions illustrate one of the ways in which sand deposits respond to different flows (fig. 3.7). Releases from Glen Canyon Dam were unusually high between the April and September 1985 surveys for approximately two months--between 708 cms and 850 cms (25,000 and 30,000 cfs); discharges were between 1130 and 1415 cms (40,000 and 50,000 cfs) for two months of this period. During that time, the reattachment deposit at Eminence Break Camp was significantly degraded, and the separation deposit slightly aggraded (figs. 3.7c and d). Dam releases fluctuated strongly between the September 1985 and January 1986 surveys (figs. 3.7d and 2.7e). During that time, the crest of the reattachment deposit was eroded. At other sites, the reattachment deposit crest was sharpened by aggradation of the crest and deepening of the return current channel during high flows. During the strongly fluctuating flows, the reattachment deposit was flattened by degradation of the crest and filling in of the return current channel. Because of the limited amount of data available at this time, we cannot say which of these responses is the more common.

GEOMORPHIC HISTORY OF THE RIVER CHANNEL AND DEBRIS FANS

The debris flows from tributary canyons constrict the main channel episodically. Later, floods of differing sizes on the Colorado River widen the constrictions and move material from the debris fans downstream into secondary features referred to as "rock gardens" (fig. 3.1a). The shape of the Colorado River channel in the vicinity of the debris fans depends on the relative frequencies of tributary and main-channel floods, on the nature of the material brought into the channel by the tributary floods, and on the competence of the Colorado River. Prior to flow regulation, natural floods had contoured the river channel through the debris into remarkably uniform shapes, for which the ratio of narrowest channel width to average channel width (the "constriction", herein also called the "shape parameter") was about 0.5 (fig. 3.8).

In the first decade after 1963 (the dam closure), 27% of the tributary fans had built outward because of

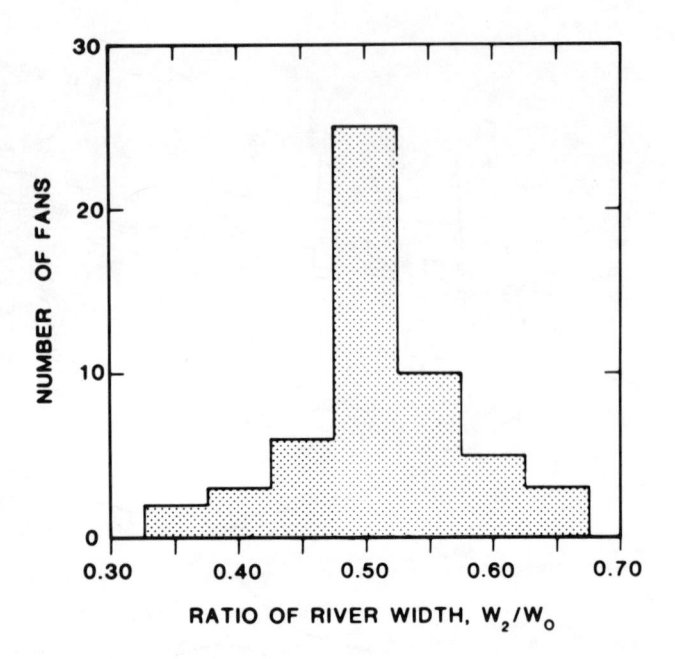

FIGURE 3.8. Histogram showing the ratio of surface river width at the narrowest part of a rapid to the width of the river upstream from the constriction (the shape parameter) Data for 59 of the largest debris fans in the 400-km stretch below Lees Ferry. Note that the histogram is strongly peaked near the value 0.50. Crystal Rapids, prior to 1983 is represented by the extreme left point. During the discharges 1,770 cms (92,600 cfs) during 1984, its shape parameter moved up toward 0.40-0.45.

tributary flooding; ten percent had built outward by more than 14.8 m (49 feet) (Howard and Dolan, 1979). In the additional 12 years until 1986, severe changes have occurred in enough of the tributary canyons to lead us to believe that, on the time scale of decades, major changes will occur in the rapids. We have arbitrarily defined "severe changes" as debris flows that have emplaced boulders on the order of 1 m diameter into that part of the main channel occupied by discharge less than 850 cms (30,000 cfs) because changes of this magnitude would affect campsites and rafting. Events observed (described below) suggest that if only power-plant discharges occur at Glen Canyon Dam, the mean constriction ratio in Figure 3.8 would move from 0.5 toward about 0.3.

Boulders on the order of the size commonly found on debris fans (1-2 m diameter) can be moved from the main channel even at low discharges. This can be demonstrated in two ways: by consideration of the Hjulström criterion (Hjulström, 1935 as extrapolated, for example, by Strand, 1986) and by calculation of unit stream power. The Hjulström criterion suggests that velocities of 7-9 m/s can move boulders up to 1 or 2 m in diameter. The unit stream power at House Rock Rapids at 7,000 cfs is 3300 Newtons per meter per second, sufficient to allow movement of such boulders. Therefore, the velocities of 6-8 m/s (20-27 ft/s) measured at many rapids even at low discharges (200 cms, or 7,000 cfs, is the lowest discharge at which velocity measurements have been made, see fig. 3.5d) are adequate to move the boulders that comprise most debris fans.

An example of the capability of the river to contour its own channel has been provided by the history of Crystal Rapids. Before 1966, there was only a small debris fan at the mouth of Crystal Creek (fig. 3.9a). In 1966, a debris flow down Crystal Creek substantially enlarged this fan, probably temporarily damming the river (fig. 3.9b). The storm that caused this debris flow in Crystal Creek also caused a large flow in Bright Angel Creek that deposited debris at the mouth of the creek--a hydraulic control that controls the stage-discharge relation at the gaging station just upstream. The increased depth and decreased velocities in the pool upstream allowed sand and gravel to be accumulated all the way upstream to the gage station (about 1 km), raising the bed elevation at the gage station (fig. 4b).

Crystal Rapids became a rapid equalled only by Lava Falls in difficulty of raft navigation because of the severe constriction of the river and the large boulders in the main channel. From 1966 to 1983, the debris fan was subjected only to discharges typical of the power plant: about 880 cms (31,000 cfs). These discharges were sufficient to cut the channel to about 1/4 of the upstream width, i.e., to a constriction ratio of 0.25, but the channel remained much more tightly constricted than it is at other places where unregulated floods had cut the channel in pre-dam times (in the histogram of fig. 3.8, Crystal Rapids is represented by the extreme left data bar).

The high discharges of 1983 caused velocities in the channel at Crystal Rapids to rise to values where further widening and downcutting of the channel could occur. Many boulders were carried downstream into the rock garden; the channel enlarged laterally by about 12 m; if readers walk down the north bank of the debris fan at Crystal Rapids, they will notice a steep cut-bank at approximately the narrowest section of the rapid. This cut-bank was formed by the lateral erosion in 1983. Similar cut banks can be observed at other rapids where

FIGURE 3.9. A debris flow in December 1966 greatly increased the constriction of the channel at Crystal Creek (river mile 98.1; 157 km). Left photo by A.E. Turner, Bureau of Reclamation, March 1963, discharge 141 cms-170 cms (5,000 to 6,000 cfs). Right photo by Mel Davis, Bureau of Reclamation, March 1967, discharge 450 cms (16,000 cfs).

fresh debris flows--even small ones--have been eroded by the river, e.g., at the head of Granite Rapids.

The main channel at Crystal Rapids also deepened by headward erosion. Similar deepening of the channel through the Bright Angel Creek debris fan probably caused the scour measured at the Phantom Ranch gage during this time (fig. 3.4). The constriction at Crystal Rapids at the time of this writing (1988) is about 0.40. Because of the headward erosion, much of the drop through Crystal Rapids occurs over a very short plunge at the top of the rapids. At discharges in the range of about 850 cms (30,000 cfs) to 1415 cms (50,000 cfs), spectacular hydraulic jumps at the bottom of this plunge pose a major hazard to navigation (Marley, 1987).

The hydraulic and geomorphic changes in Crystal Rapids in 1983 cause us to inquire what level flood might be required to produce the wider channel widths seen at most rapids (the constriction of 0.50). Extrapolation of the calculations for Crystal Rapids suggests that a flood of about 400,000 cfs could account for such widening (Kieffer, 1985). This is not an unreasonable flood for the Colorado River prior to emplacement of Glen Canyon Dam: a flood of about 8,500 cms (300,000 cfs) occurred in 1884, and a flood of 6,230 cms (220,000 cfs) in 1921.

At the present time, the Colorado River generally flows at discharges much less than the discharges that occurred during annual floods in unregulated times. If the river becomes severely constricted by debris flows, even normal operational levels of Glen Canyon Dam are sufficient to clear the channel to approximately 1/4 of the unconstricted width. Maximum dam releases could bring the channel morphology to a configuration approaching that of the older debris fans that have seen the largest natural floods. The behavior of Crystal Rapids suggests that a discharge of 2,600 cms (92,000 cfs) could probably bring the constriction ratio at most rapids to about 0.4.

THE EFFECT OF DEBRIS FLOWS ON THE GEOMETRY, HYDRAULICS, AND SEDIMENT TRANSPORT IN THE COLORADO RIVER CHANNEL

A schematic diagram of the evolution of the debris-fan river-channel morphology that may be applied to most of the debris fans in the Grand Canyon is shown in Figure 3.10. The sequence shown in Figure 3.10 represents but one cycle in recurring episodes in which debris fans are enlarged by floods or debris flows in tributary canyons and then modified by floods of varying sizes in the Colorado River. Both types of events probably having time scales on the order of 10^2-10^4 years (Shoemaker and Stephens, 1969; Stephens and Shoemaker, 1987; Webb and others, 1987, 1988).

The beginning of the sequence (fig. 3.10a) is arbitrarily chosen as a time when the channel is relatively unconstricted (e.g., fig. 3.10a might be compared with the configuration of Crystal Rapids, fig. 3.9a). The river is suddenly disrupted and ponded by catastrophic debris-fan enlargement (fig. 3.10b), forming a temporary "lake" behind the debris dam. The surface of the debris fan is temporarily a "waterfall" in this model. As the ponded water overtops the debris dam, it erodes a channel, generally in the distal end of the debris fan (fig. 3.10c). This is the beginning of evolution of a "rapid" from a "waterfall". Breaching of earth dams comparable to these debris dams generally occurs within hours to days in settings like the confined gorge of the Grand Canyon, if the discharges are fairly high as they tend to be when the meteorologic conditions for formation of debris flows are favorable (Schuster, 1986).

Unless the debris dam is massively breached by the first breakthrough of the ponded water (i.e., more than half washed away), the constriction of the main river is initially severe (as it was at Crystal Rapids from 1966 to 1983). Floods of differing sizes and frequency erode the channel to progressively greater widths, as shown in Figures 3.10c, d, and e. Small floods enlarge the constriction somewhat, but the constriction tends to remain so tight that supercritical flow with large standing waves occurs. Moderate floods (fig. 3.10d) enlarge the channel further, and may widen the channel so that at lower discharges the flow is only weakly supercritical or even subcritical (this condition occurs now at Crystal Rapids when discharges are below about 15,000 cfs). At the same time that lateral widening occurs during these floods, headward erosion of the channel also occurs. Thus, the local gradient within the rapid is changing, and new waves can arise as the channel geometry changes. Prediction of wave behavior and safety of navigation is particularly difficult in a newly formed rapid like this because of the constantly shifting morphology of the channel. Large floods widen the channel even further.

Because of the movement of material out of the constricted part of the channel, a rapid evolves into two parts: the original debris deposit, and a reworked debris deposit below it, the so-called rock garden. Because the flow velocities in a rapid change with discharge and change as the channel changes shape, the distance to which rocks can be transported also changes continually, so the rock garden is a constantly changing feature of a rapid. Rock gardens can extend on the order of 1 km below the original debris deposit.

Debris flows in tributaries also affect sand transport and storage characteristics in the vicinity of the rapids. In the case of Crystal Rapids, the 1966 tributary flood caused the flow to deepen and the velocity in to be decreased in the main channel for nearly 1.5 km upstream. At Bright Angel Creek, the 1966 debris flow affected the hydraulics and sand storage characteristics at the Phantom Ranch gage (0.6) km upstream from the mouth of the Creek. The deeper pool and lower velocity at a given discharge create conditions which favor deposition of sand and gravel in the pool. Under these conditions, sand transport rates through the pool and below the rapid would be lower than pre-debris flow rates. Sand transport rates would increase with time: bed elevation in the pool increase because of deposition and elevation at the rapid decreases because of erosion. These two effects decrease the depth of flow in the pool over time, and increase its velocity. Therefore, the sand

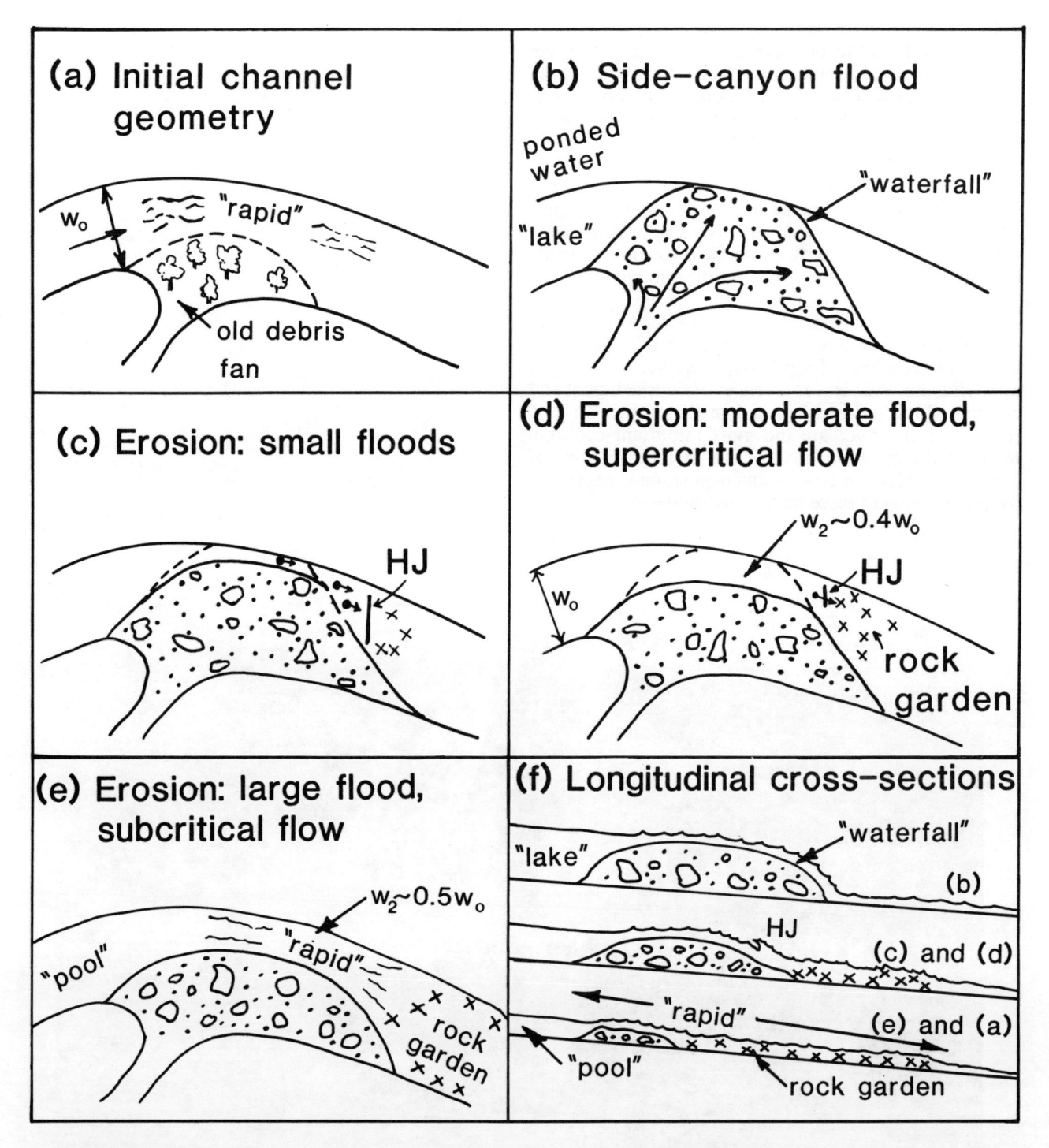

FIGURE 3.10. Schematic illustration of the emplacement and modification of debris fans and channel morphology, the formation and evolution of rapids, and the formation of rock gardens. (a)-(e) Map views. (f) Cross sections corresponding to (a) - (e). See text for details.

transport rates increase with time after the debris flow emplacement.

Recirculation zones associated with debris fans may be greatly changed by new debris flows. Changes would depend on the geometry of the channel and the geometry of the newly emplaced material. Recirculation zones may be increased in size by extension and steepening of the debris fan, or decreased in size by deposition of material too coarse to be moved by regulated flow.

Sand deposits in the recirculating zone in reaches affected by the hydraulic changes may be lost because of erosion by the debris flow itself, because of deposition of coarse debris, because of increased water surface elevation in the pool upstream of the rapid, or because of decreased bed elevation and sand transport rates in the pool. If floods under regulated flow are insufficient to return the rapid to its pre-debris flow constriction, then bed elevation and sand transport rates in the pool upstream should be higher than before the debris flow. This may favor the deposition of sand in the recirculating zone above the rapid, and favor the erosion of sand from the recirculating zone below the rapid.

In summary, it should be remembered that the shape of the channel and the hydraulic and sediment transport characteristics of the river at any instant of geologic time are strongly influenced by debris flow (or flood) events from tributary canyons. The equilibrium shape of the channel that results from these events and the contouring of the deposits by floods along the Colorado River may well preserve a record of the largest discharge in the shape of the channel. Each flood, as well as eroding and removing material, leaves a veneer of new sediment and cobbles along the beaches and in the eddies. These regions of the river are the most important ecologic niches and recreational places, and an understanding of the balance between erosion and deposition is necessary for preservation of these fragile environments.

View north towards lava-filled Whitmore Canyon (mi 187.8; km 302.2; A), and lava in tributary canyon (B). New Whitmore Canyon at C (mi 188; km 302.5). See chapter 23 for discussion.

CHAPTER 4: PHYSIOGRAPHIC FEATURES OF NORTHWESTERN ARIZONA

George H. Billingsley and John D. Hendricks
U.S. Geological Survey, Flagstaff, Arizona

Three physiographic provinces are recognized in Arizona; the Colorado Plateau, Arizona Transition Zone, and Basin and Range (fig. 4.1). These provinces strike northwest across the state with the northeastern one-third in the Colorado Plateau, the southwestern half in the Basin and Range, and an intervening central mountainous region called the Arizona Transition Zone (Peirce, 1984). Part of northwestern Arizona lies within the Colorado Plateau and part within the Basin and Range Provinces. Peirce (1984) extended the Arizona Transition Zone into the western Grand Canyon region in northwestern Arizona, but we prefer a restricted definition of the Arizona Transition Zone which limits its extent to central Arizona (fig. 4.1).

For many years, only two provinces were formally recognized in Arizona; the Colorado Plateau and Basin and Range (Fenneman, 1946). The boundary between the two provinces was never clearly defined, often vague, and placed at different locations based on a variety of geologic or physiographic reasons. Only along the Grand Wash Cliffs in northwestern Arizona was the boundary clearly defined by the trace of the Grand Wash fault. Geologists working in central Arizona recognized that the mountainous terrane in that part of the state was physiographically and structurally distinct from either the Colorado Plateau or the Basin and Range, and indeed did incorporate some aspects of both provinces. They argued that the terrane was transitional between the two provinces with a Cenozoic geologic history that has resulted in a confused geomorphic and physiographic topography.

One approach to the province boundary problem was to designate a new geologic province. With this in mind, Peirce (1984), reviewed the history of the boundary problems in central Arizona and proposed a new province, the Arizona Transition Zone, originally and informally referred to as the transition zone (Wilson and Moore, 1959, p. 90). Peirce (1984, p. 11) suggested "a word like transition has the advantage of being noncommital with regard to a particular topographic style."

A distinctive geomorphic feature, the erosional scarp of the Permian strata known as the Mogollon Escarpment, was designated by Peirce (1984) as the boundary between the Colorado Plateau and Arizona Transition Zone. This scarp trends from near Seligman, Arizona, diagonally to the southeast across central Arizona, and ends an undefined distance into New Mexico. This southeastern segment of the scarp does correspond in general to a logical, well-defined physiographic and geomorphic boundary between the Plateau and Transition Zone. The Mogollon scarp also extends northwestward from Seligman, crossing the western Grand Canyon until it intersects the Grand Wash Cliffs (fig. 4.1). However, the northwestern segment of the Permian scarp does not define a distinct physiographic boundary; in fact, regions southwest of the scarp, such as the Aubrey Valley, Juniper Mountains, and the Hualapai Plateau are physiographically more similar to and continuous with the Colorado Plateau. For this reason, the Aubrey Valley, Juniper Mountains, and the Hualapai Plateau northwest of Seligman, as defined by Peirce (1984), are not considered to be part of the Arizona Transition Zone in this report. Instead, we suggest a boundary that better fits the physiography and geology of this region. The boundary extends southwestward from Seligman, Arizona, and follows the basal outcrops of Paleozoic rocks (covered in many places by Tertiary volcanics) as an inferred line that nearly parallels Interstate Highway 40 to the Grand Wash Cliffs, just north of Interstate 40 (fig. 4.1).

The southern boundary between the Arizona Transition Zone and the Basin and Range is somewhat more vague and not easily defined. It is marked by the interface of extensive bedrock exposures with low-elevation, alluvial desert basins or valleys and a sharp contrast in structure and deformation (Peirce, 1985) (fig. 4.1).

The physiography of the Grand Canyon lies within the southwestern part of the Colorado Plateau province; it is a region of gently folded Paleozoic and Mesozoic strata that have an average regional dip of about one degree to the northeast. Several north-south–trending monoclines disrupt the strata of the Plateau with a strong eastward dip producing local topographic breaks. The Grand Wash fault marks the boundary between the Basin and Range and Colorado Plateau at the mouth of Grand Canyon.

Local physiographic areas or subprovinces surrounding the Grand Canyon and within the Colorado Plateau were described by early geologists on the basis of similar structure, climate, and a unified geomorphic history. Near the Grand Canyon, each subprovince has its own geomorphic or physiographic characteristics with good local structural or topographical boundaries that are well defined. However, away from the Canyon, some of the boundaries become vague and ill-defined due to deterioration of their characteristics as originally described. In Figure 4.2, these vague boundaries are indicated by long dashed lines and are placed where the regional terrane is most typical of the original physiographic description.

Figure 4.2 shows the numerous place-names and cultural features referred to in this guidebook. In addition, Figure 4.2 also includes the Colorado River mileage below Lees Ferry in five-mile increments and political subdivisions. Figures 4.1 and 4.2 will help orient the user of this guidebook as to the location of features on your journey through Grand Canyon.

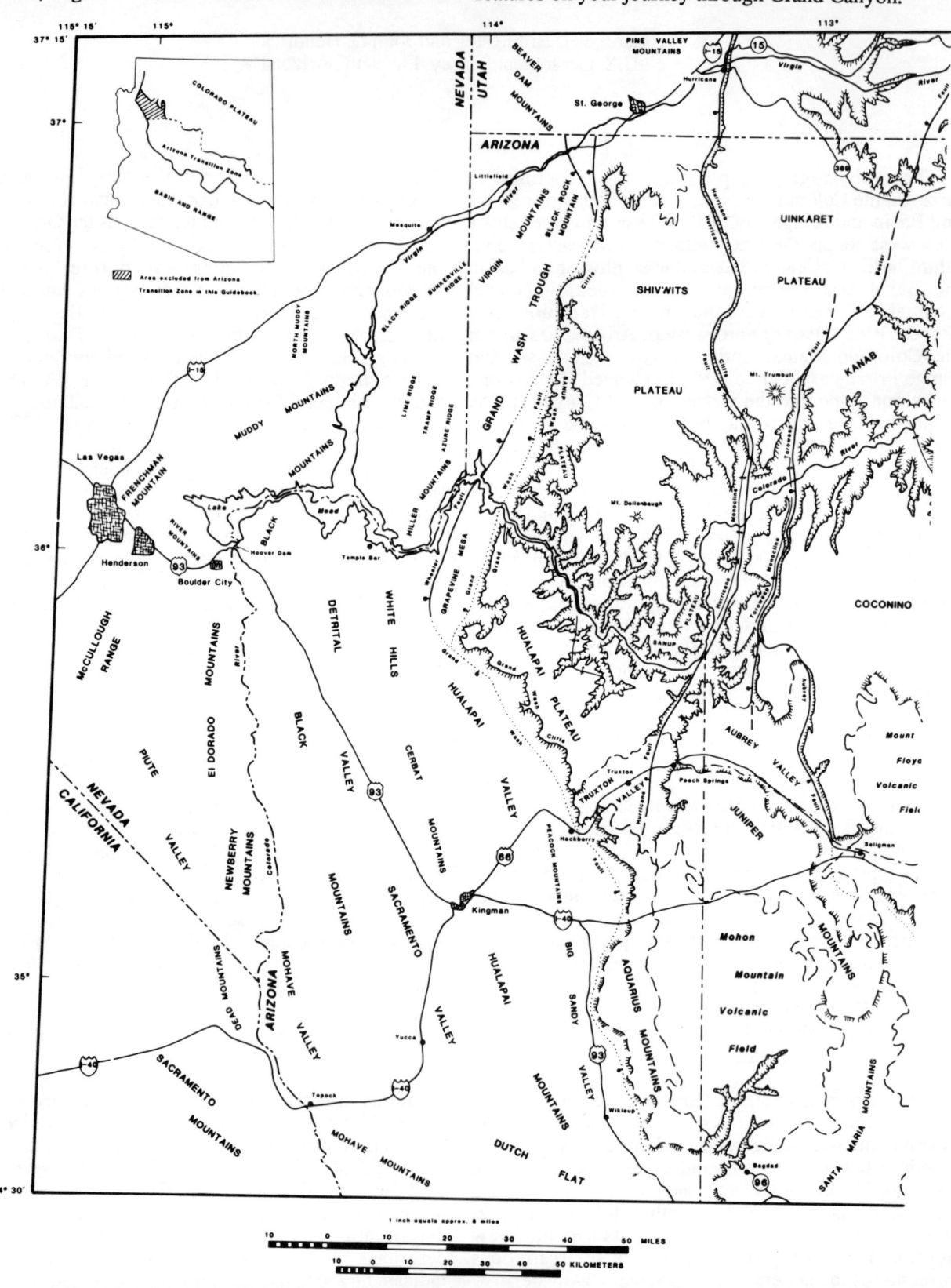

FIGURE 4.1. Map showing physiographic and cultural features of the Grand Canyon region.

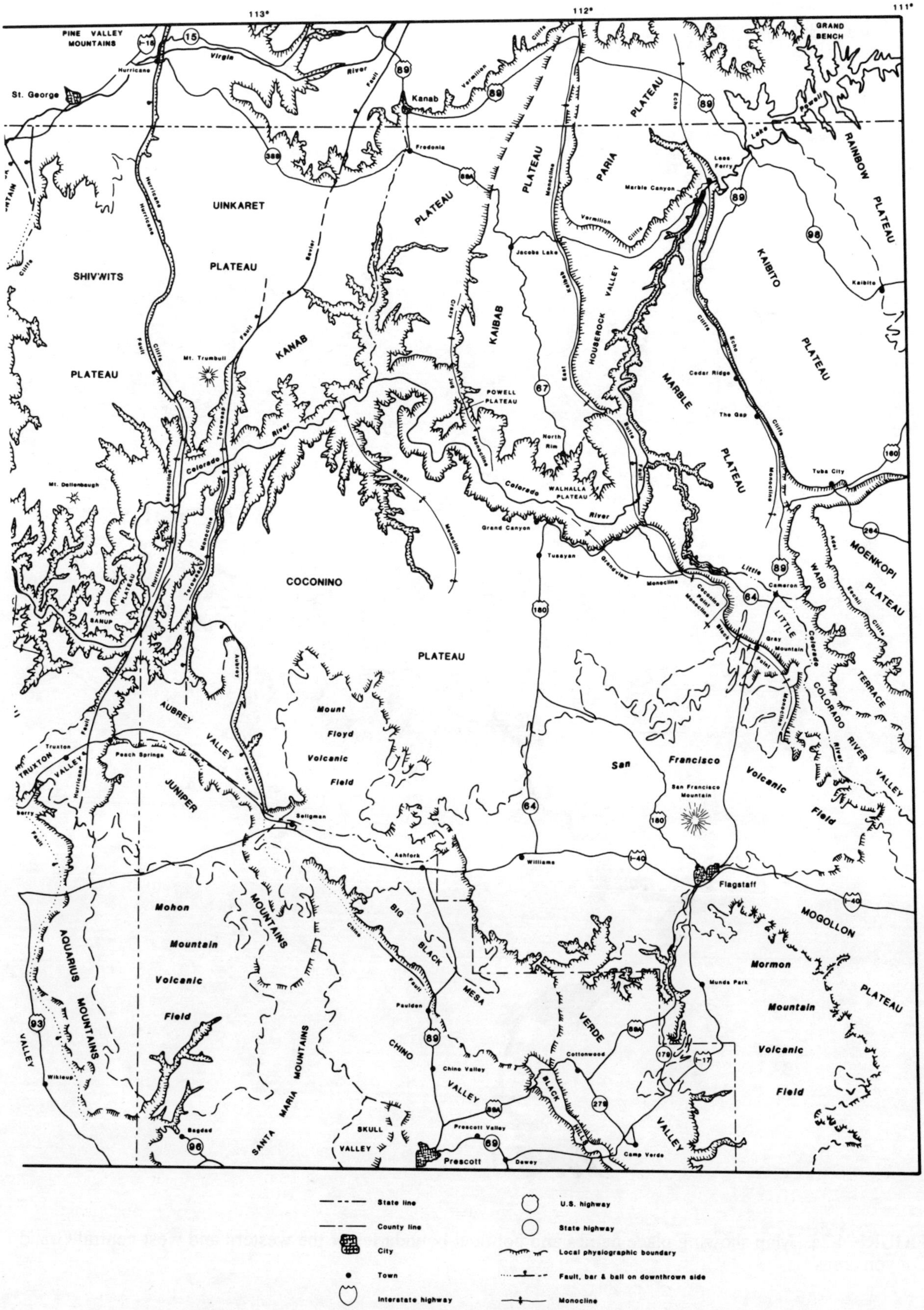

113°
112°
111°
PINE VALLEY MOUNTAINS
St. George
Hurricane
Virgin
River
Kanab
Fredonia
Vermilion
Cliffs
GRAND BENCH
Lake Powell
Lees Ferry
Marble Canyon
PARIA PLATEAU
RAINBOW PLATEAU
KAIBITO PLATEAU
Kaibito
UINKARET PLATEAU
SHIVWITS PLATEAU
KANAB PLATEAU
KAIBAB PLATEAU
Jacobs Lake
HOUSEROCK VALLEY
MARBLE PLATEAU
Cedar Ridge
The Gap
Echo
Cliffs
Tuba City
Mt. Trumbull
POWELL PLATEAU
North Rim
WALHALLA PLATEAU
Colorado
River
Grand Canyon
Tusayan
Mt. Dellenbaugh
SANUP PLATEAU
Monocline
Toroweap
Hurricane
Fault
MOENKOPI PLATEAU
WARD TERRACE
LITTLE COLORADO RIVER VALLEY
Little
Cameron
Gray Mountain
COCONINO PLATEAU
Aubrey
AUBREY VALLEY
Truxton
TRUXTON VALLEY
Peach Springs
JUNIPER MOUNTAINS
Mount Floyd Volcanic Field
San Francisco Volcanic Field
San Francisco Mountain
Seligman
Ashfork
Williams
Flagstaff
MOGOLLON PLATEAU
Mormon Mountain Volcanic Field
Munds Park
Mohon Mountain Volcanic Field
AQUARIUS MOUNTAINS
BIG BLACK MESA
Chino
Paulden
CHINO VALLEY
Chino Valley
VERDE VALLEY
Cottonwood
BLACK HILLS
Camp Verde
SANTA MARIA MOUNTAINS
SKULL VALLEY
Prescott Valley
Prescott
Dewey
Bagdad
Wikieup
State line
County line
City
Town
Interstate highway
U.S. highway
State highway
Local physiographic boundary
Fault, bar & ball on downthrown side
Monocline

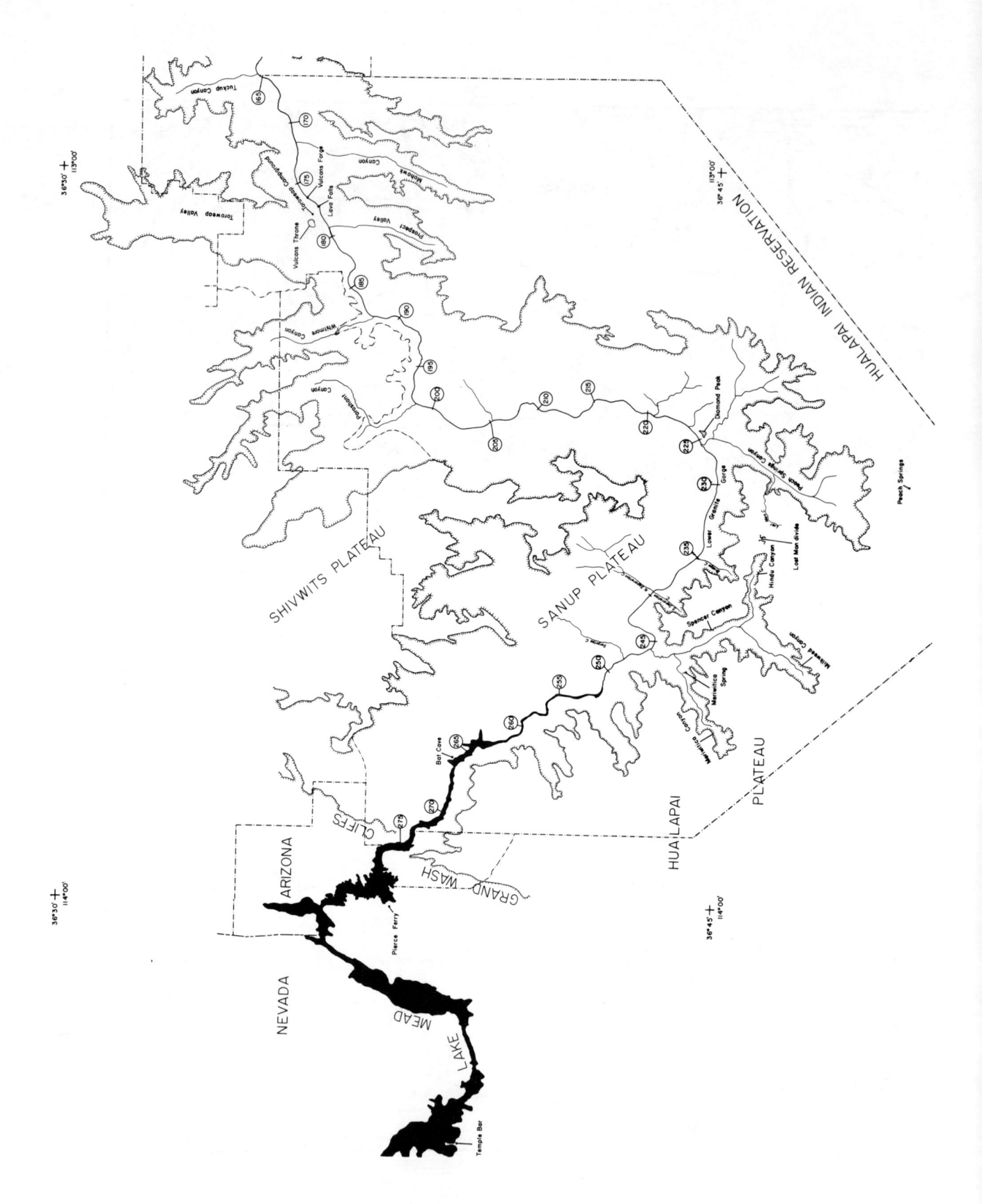

FIGURE 4.2a. Map showing place names and political boundaries of the western and west-central Grand Canyon areas.

FIGURE 4.2b. Map showing place names and political boundaries of the central and eastern Grand Canyon, and Marble Canyon areas.

CHAPTER 5: MODERN TECTONIC SETTING OF THE GRAND CANYON REGION, ARIZONA

Peter W. Huntoon
Department of Geology and Geophysics,
University of Wyoming, Laramie, Wyoming

INTRODUCTION

The Grand Canyon is located in the southwestern corner of the Colorado Plateau (fig. 5.1), a geologic province underlain by thick continental crust that became slightly separated from the continental craton during the Cenozoic Era along bounding zones of intense deformation. These zones include the extensional Rio Grande rift on the east, the compressional Uinta uplift on the north, and the drastically extended Basin and Range Province on the south and west. The Paleozoic rocks in the Grand Canyon are about a mile (1.6 km) thick, and they rest on a 19 to 25 mile- (30 to 40 km-) thick rigid crust (Smith, 1978) comprised of an ancient (1700+ Ma) basement complex. The lower Paleozoic rocks in the Grand Canyon region accumulated in the equatorial region of the earth as platform sediments on the then northwestern part of a growing continent. Subsequently, the continent was rafted through plate tectonic processes first southward and then northward some 2,000 miles (3,220 km), and simultaneously rotated a few tens of degrees counterclockwise (Elston and Bressler, 1977, 1984; Elston, 1988b; chapter 12).

Between the beginning of Cambrian and the end of Cretaceous time, an interval of roughly 480 million years, net subsidence of the Precambrian surface in the Grand Canyon region amounted to between 1.5 to 2 miles (2.4 to 3.2 km). During the last 70 million years, a period dominated by uplift, the Precambrian surface has risen in pulses a total of approximately 2 miles (3.2 km).

MODERN GEOPHYSICAL SETTING AND TECTONISM

Although the Grand Canyon is situated in the southwestern part of the Colorado Plateau physiographic province, the present geophysical properties of the crust under the Canyon appear to be transitional in character between those of the stable interior of the plateau and the extensional Basin and Range Province to the west. For example, the thickness of the crust is 25 miles (40 km) or more under most of the plateau, including the eastern Grand Canyon, the crust tapers to the west to about 19 miles (31 km) in thickness near the Grand Wash fault (Keller and others, 1976; Smith, 1978). The crust is only 15 miles (24 km) thick in southern Nevada and western Utah. Heat flow through the plateau interior is typically 1.5 heat flow units (1 heat flow unit = 10^{-6} calorie/centimeter-second), but from east to west across the Grand Canyon it increases to 2 units (Blackwell, 1978). Increased heat flow in the western Grand Canyon appears to be substantiated by a warming of ground water discharged from springs west of Kanab Canyon. Temperatures of waters from springs in the lower Paleozoic section in the eastern part of the Grand Canyon range from 50-73 °F (10-22 °C) whereas those sampled from the same rocks to the west range from 64-86 °F (17-30 °C) (Loughlin and Huntoon, 1983, Table 1).

The intermountain seismic belt (Smith and Sbar, 1974) coincides with the Basin and Range - Colorado Plateau crustal transition zone. The seismic belt is characterized by high seismicity (Sbar and others, 1972) and by tectonic extension (Doser and Smith, 1982). Seismicity associated with it extends as far eastward in the Grand Canyon as the West Kaibab fault zone (Smith and Sbar, 1974, p. 1206; Anderson and Huntoon, 1979, fig. 1; Woodward-Clyde Consultants, 1982, fig. 6-2). Fault plane solutions for earthquakes in the intermountain seismic belt in southern Utah reveal that east-west extension is occurring and that focal depths are shallow, mostly at depths of less than 10 miles (16 km) (Smith and Sbar, 1974). The faults have steep dips, and motion on them tends to be near-vertical (Sbar and others, 1972, p. 26). Smith and Sbar (1974, p. 1210) report that an earthquake on November 11, 1971, near the Hurricane fault zone north of Cedar City, Utah, produced three north-trending fractures in alluvium. The longest fracture exhibited horizontal east-west extension and could be traced for half a mile (0.8 km). Focal depths during the earthquake were from near-surface to slightly more than a mile (1.6 km) deep.

Citing a combination of refraction, reflection, low resistivity, and high Pn velocity data, Keller and others (1975) postulate the presence of a mantle upwarp occupying a band at least 50 miles (80 km) wide under the transition zone and extending approximately 30 miles (50 km) eastward under the Colorado Plateau. The eastern limit of the upwarp boundary appears to correlate with a lateral change in crustal magnetization reported by Shuey and others (1973).

The coincidence between the postulated mantle upwarp and Smith and Sbar's (1974) intermountain seismic belt led Keller and others (1975, p. 1097) to speculate that the presence of the upper crustal low-velocity layer may be related to the mechanism of Cenozoic faulting and seismicity. Specifically, the presence of a low velocity layer east of the Wasatch front could provide an explanation for the presence of

Cenozoic block faulting east of the province boundary. The shallow seismicity characteristic of the intermountain seismic belt also extends east of the Wasatch front and roughly coincides with the eastern limits of Cenozoic normal faulting. Late Cenozoic basaltic volcanism also characterizes much of this same region (Best and Brimhall, 1970), which is compatible with the timing of the extensional tectonism that has occurred here.

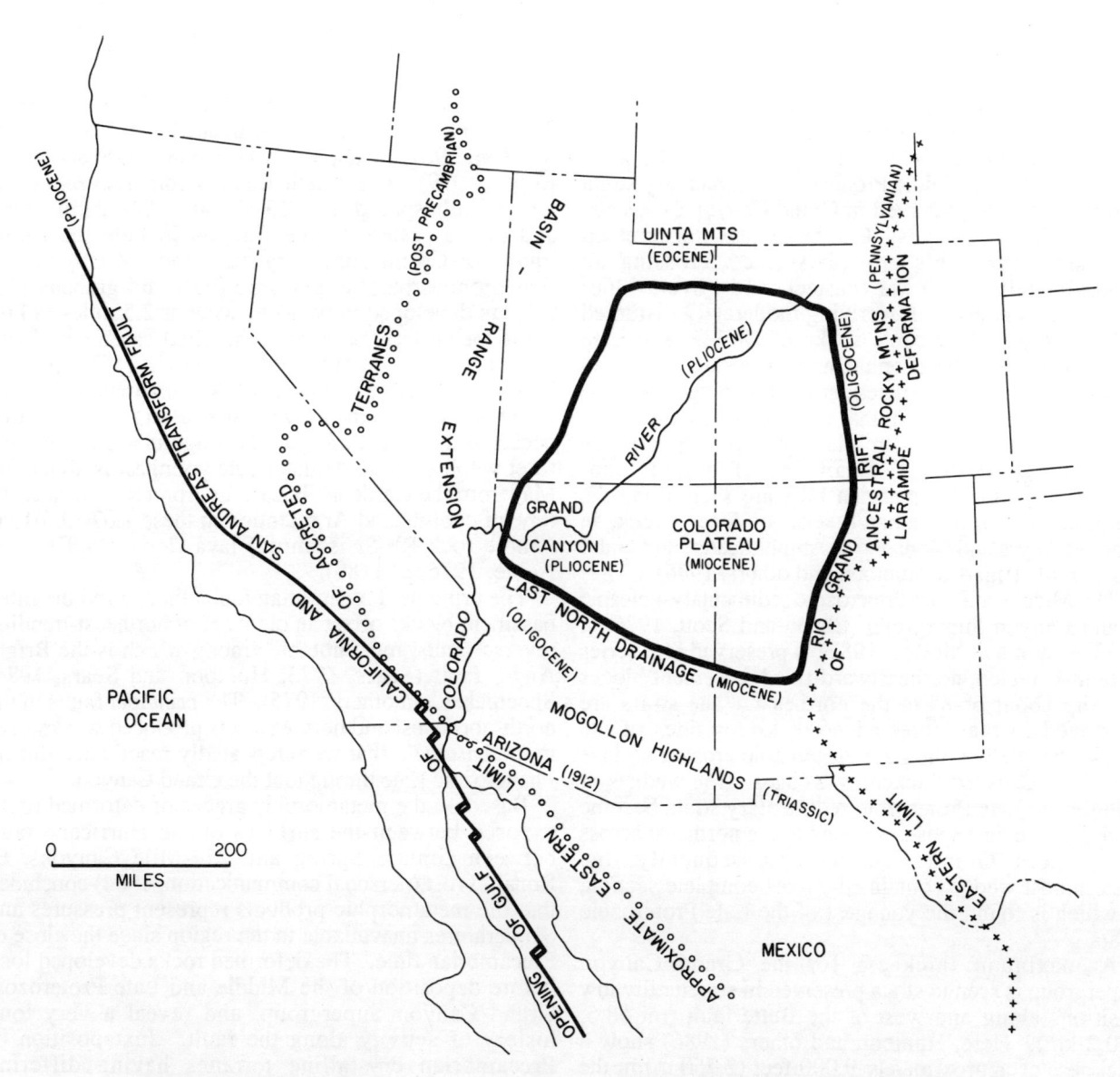

FIGURE 5.1. Selected tectonic elements in the western North American cordillera and the timing of their inception.

CHAPTER 6: SETTING OF THE PRECAMBRIAN BASEMENT COMPLEX, GRAND CANYON, ARIZONA

Peter W. Huntoon
Department of Geology and Geophysics,
University of Wyoming, Laramie, Wyoming

Erosion during Late Proterozoic time beginning about 800 Ma ago served to expose progressively older rocks of the Precambrian basement complex toward the west. Consequently, no Middle Proterozoic sedimentary strata or mafic sills are preserved in Grand Canyon exposures west of Fishtail Canyon. Cambrian seas transgressed eastward across this erosion surface, crossing an approximately 1,700 Ma (Pasteels and Silver, 1966) basement complex and overlying moderately disturbed sedimentary and volcanic rocks of Middle and Late Proterozoic age. Relief on the erosion surface consisted of scattered, largely isolated, small but rugged hills as high as 1,200 feet (400 m) that were buried by Cambrian strata. The highest Precambrian hills were covered by the Muav Limestone. Examples of buried topographically high Precambrian hills are seen along the Colorado River immediately east of Deer Creek, in Modred Abyss and Monadnock Amphitheater, and under Isis Temple (fig. 4.2; Huntoon and others, 1986).

The Middle and Late Proterozoic sedimentary-volcanic Grand Canyon Supergroup (Elston and Scott, 1976, p. 1763; Elston and McKee, 1982) is preserved in a series of fault-bounded, northeastward-tilted, basement blocks that dip about 6°-8° to the northeast. The strata are preserved in the depressed northeastern sides of the blocks, beveled by the Late Proterozoic erosion surface into northeastward thickening wedges. The wedges of Middle and Late Proterozoic sedimentary strata become thicker and their exposures widen to the northeast across the eastern Grand Canyon. Consequently, the easternmost wedge contains the most complete section, in which is found the youngest of the Late Proterozoic strata.

A maximum thickness for the Grand Canyon Supergroup is seen in strata preserved in structurally low positions along and west of the Butte fault (mi 68.5; 110.2 km). Here, Huntoon and others (1986) show a thickness of approximately 9,000 feet (2,750 m) in the vicinity of Malgosa Canyon (mi 58; 93 km). Elston and Scott (1976, p. 1770) reported a thickness between 12,000-14,000 feet (3,650 m and 4,270 m) based on sections measured elsewhere, and comparable thicknesses were since reported by Elston and McKee (1982) and Elston (1988a). Recent measurements of the thickness of the Chuar Group have been incorporated with the thicknesses of underlying units of the Grand Canyon Supergroup in Table 9.1.

The principal faults that define the boundaries of the tilted basement blocks in the eastern Grand Canyon are north- and north-northeast-trending Precambrian normal faults. These have extensive records of recurrent Precambrian displacement (Walcott, 1890; Noble, 1914; Maxson, 1961; Sears, 1973; Huntoon and Sears, 1975; Elston, 1979). The principal faults comprise zones that are typically spaced from 15-20 miles (24-32 km) apart, and in the eastern Grand Canyon include the Butte, Phantom-Cremation, Crystal, and Muav faults. Synchronous parallel parasitic folds and grabens (fig. 6.1) are developed in bands as wide as 2.5 miles (4 km) along the eastern margins of the tilted basement blocks (Huntoon and others, 1986; Elston, 1979, p. 15). Elston and McKee (1982, p. 689) estimated that the latest Precambrian movements along these faults occurred at about 823 Ma. This age is a mean of five reset whole-rock K-Ar dates (the youngest is about 790 Ma) from the Cardenas Basalt, interpreted to reflect the time of cooling and Ar retention in these 1,070 ± 70 Ma (whole rock Rb-Sr isochron) lava flows (McKee and Noble, 1976, p. 1189).

The principal Precambrian faults that bound the tilted basement blocks offset an older set of northeast-trending reverse faults, most notable among which is the Bright Angel fault (Sears, 1973; Huntoon and Sears, 1975; Shoemaker and others, 1975). The principal faults in the north-northwest and northeast sets produced weaknesses in the basement that were repeatedly reactivated during Phanerozoic time throughout the Grand Canyon.

Based on the metamorphic grades of deformed rocks exposed between the surfaces of the Hurricane fault between Granite Spring and 224-Mile Canyons, B. Ronald Frost (personal communication, 1984) concluded that the metamorphic products represent pressures and temperatures unavailable in the region since the close of Precambrian time. The deformed rocks developed long before deposition of the Middle and Late Proterozoic Grand Canyon Supergroup, and reveal a very long history of activity along the fault. Juxtaposition of Precambrian crystalline terranes having differing fracture-foliation fabrics and lithologies are also common along the Hurricane fault in the Grand Canyon exposures.

Huntoon (1981, p. 130) observed the juxtaposition of discordant foliation trends in Proterozoic metamorphic outcrops across northeast-trending, segments of the Meriwitica fault, concluding that this was caused by a substantial but unknown magnitude of Precambrian offset. Shoemaker and others (1975, p. 40) concluded there is a good possibility that considerable right lateral strike slip displacement occurred along northeast-trending faults prior to deposition of the Middle and Late

Proterozoic Grand Canyon Supergroup.

FIGURE 6.1. Middle Proterozoic Shinumo Quartzite in hanging wall (left) against Early Proterozoic Vishnu Schist (right) along a normal basement fault that is part of a graben in Phantom Canyon, Grand Canyon, Arizona. Cambrian Tapeats Sandstone caps bench to left. View is toward the southeast.

CHAPTER 7: PHANEROZOIC TECTONISM, GRAND CANYON, ARIZONA

Peter W. Huntoon
Department of Geology and Geophysics,
University of Wyoming, Laramie, Wyoming

PALEOZOIC AND MESOZOIC SETTING

The Grand Canyon region occupied a position on a western sloping continental shelf during the Paleozoic Era. The large interval of Paleozoic time spanning 325 million years was characterized by gradual net subsidence and net accumulation of sediments that thicken westward from 3,500 to 5,000 feet (1,067 to 1,525 m) across the Grand Canyon (Peterson and Smith, 1986). The top of the sedimentary pile fluctuated within several hundred feet of sea level with roughly half this period spent below the ocean surface.

Mesozoic time in the Grand Canyon region was characterized by low but generally emergent conditions in which continental sedimentation prevailed. Large inland seas mostly lay to the north and northeast but transgressed southwestward over northwestern Arizona for geologically brief periods (Molenaar, 1983, fig. 7). Mesozoic marine rocks in the Grand Canyon region include parts of the Lower and Middle(?) Triassic Moenkopi Formation, the Middle Jurassic Carmel Formation, and the Upper Cretaceous Mancos Shale. Although the Mesozoic section is almost completely eroded within the immediate vicinity of the Grand Canyon, McKee and others (1967, p. 49) estimated that upwards of 4,000 feet (1,220 m) of Mesozoic sediments were deposited in the region based on thicknesses and paleotrends in Mesozoic outcrops to the east and north.

CAMBRIAN TO LATE TRIASSIC DEFORMATION

What was to become the Colorado Plateau formed part of the North American craton during Paleozoic and Mesozoic time. Although buffeted by continental scale tectonic events, the Grand Canyon region remained relatively insulated by distance to the point that virtually nothing happened in the form of discrete offsets along faults in the underlying basement or local volcanism. The forces operating in the Grand Canyon region were sufficiently attenuated that deformation was manifested only as broad scale gentle warping of the crust and variable rates of subsidence or uplift.

The longest hiatus in the Paleozoic stratigraphic record includes Ordovician to Middle Devonian time, an interval of approximately 120 million years, yet the contact between the underlying unclassified dolomites of Late Cambrian age and the late Middle and Late Devonian Temple Butte Formation is only a disconformity throughout the Grand Canyon. Similarly, the erosion surface at the top of the Devonian rocks in the Grand Canyon, which presumably was a direct result of orogenic uplift associated with the Antler orogeny, is disconformably overlain by the Mississippian Redwall Limestone.

The most spectacularly preserved erosional event in the Grand Canyon Paleozoic section is the Late Mississippian emergence of the Mississippian Redwall Limestone. A series of westward draining valleys incised as much as 400 feet (120 m) into the limestone occur in the western Grand Canyon (Billingsley and Beus, 1985, p. 27). The surrounding limestone surface underwent karstification and the landscape took on the appearance of the modern Yucatan Peninsula. The Mississippian-Pennsylvanian(?) Surprise Canyon Formation (Billingsley and Beus, 1985) fills the valleys and caves, yet it and strata of the Supai Group lie disconformably on virtually all of the older rocks.

To date, the only offsets along faults identified as dating from the Paleozoic Era in the Grand Canyon are revealed by minor angular unconformities at the top of the Mississippian Redwall Limestone (Billingsley and McKee, 1982, p. 145). This deformation appears to have occurred prior to deposition of the Chesterian-Morrowian(?) Surprise Canyon Formation based on the fact that the Surprise Canyon Formation occupies channels that were eroded into the Redwall surface, which presumably was uplifted coincident with deformation. McKee and Gutschick (1969, p. 79-80) observed that at least 150 feet (46 m) of the Redwall section, including the entire Horseshoe Mesa Member, was truncated by erosion across the crest of a minor anticline at a site along the Tanner Trail (mi 68.5; 110.2 km). The erosion surface is overlain unconformably by the Watahomigi Formation. They tentatively attributed the folding to displacement along the underlying Precambrian Butte fault. Huntoon and Sears (1975, p. 470) observed a similar truncation of the upper 30 feet (9 m) of the Horseshoe Mesa Member of the Redwall Limestone in a band 1/4 mile (0.4 km) wide across the Bright Angel fault. They concluded that the truncation and an anomalous 10° westward dip were caused by reverse motion along the underlying Bright Angel fault resulting from reactivation of that Precambrian structure. Such deformation would have resulted from horizontal, generally east-west compression in the context of present compass directions.

LATE TRIASSIC TO LATE CRETACEOUS UPLIFT

The major plate tectonic reorganization associated with the opening of the Gulf of Mexico and the Atlantic Ocean between Late Triassic and Late Jurassic time was accompanied in the southern cordillera by landward subduction of Pacific oceanic crust, and inboard development of the Mesozoic magmatic arc (Dickinson, 1981, p. 120). The effects of these events were profoundly felt in the Grand Canyon region through general emergence of the Colorado Plateau region (fig. 7.1), strong uplift of the contiguous Mogollon Highlands to the southwest, and uplift and eastward thrusting in the Sevier Highlands to the west (Woodward-Clyde Consultants, 1982). The uplift of the Mogollon Highlands caused regression of the Moenkopi seas and was accompanied by the initiation of voluminous arc volcanism in the Highlands or arc terranes to the south of the Highlands dating from about 230 million years ago (Hayes and Drewes, 1978; Stewart and others, 1986).

The first of the Late Triassic units was the voluminous Chinle Formation. In the southwestern Colorado Plateau region the Chinle was comprised in part of volcanic detritus originating from the Mogollon Highlands and air fall ash probably originating from the northwest (Stewart and others, 1972).

Late Triassic to Eocene paleoslopes remained northeasterly off the Mogollon Highlands across the region (Young, 1982). Debris eroded from the Mogollon Highlands or arc terranes to the south and

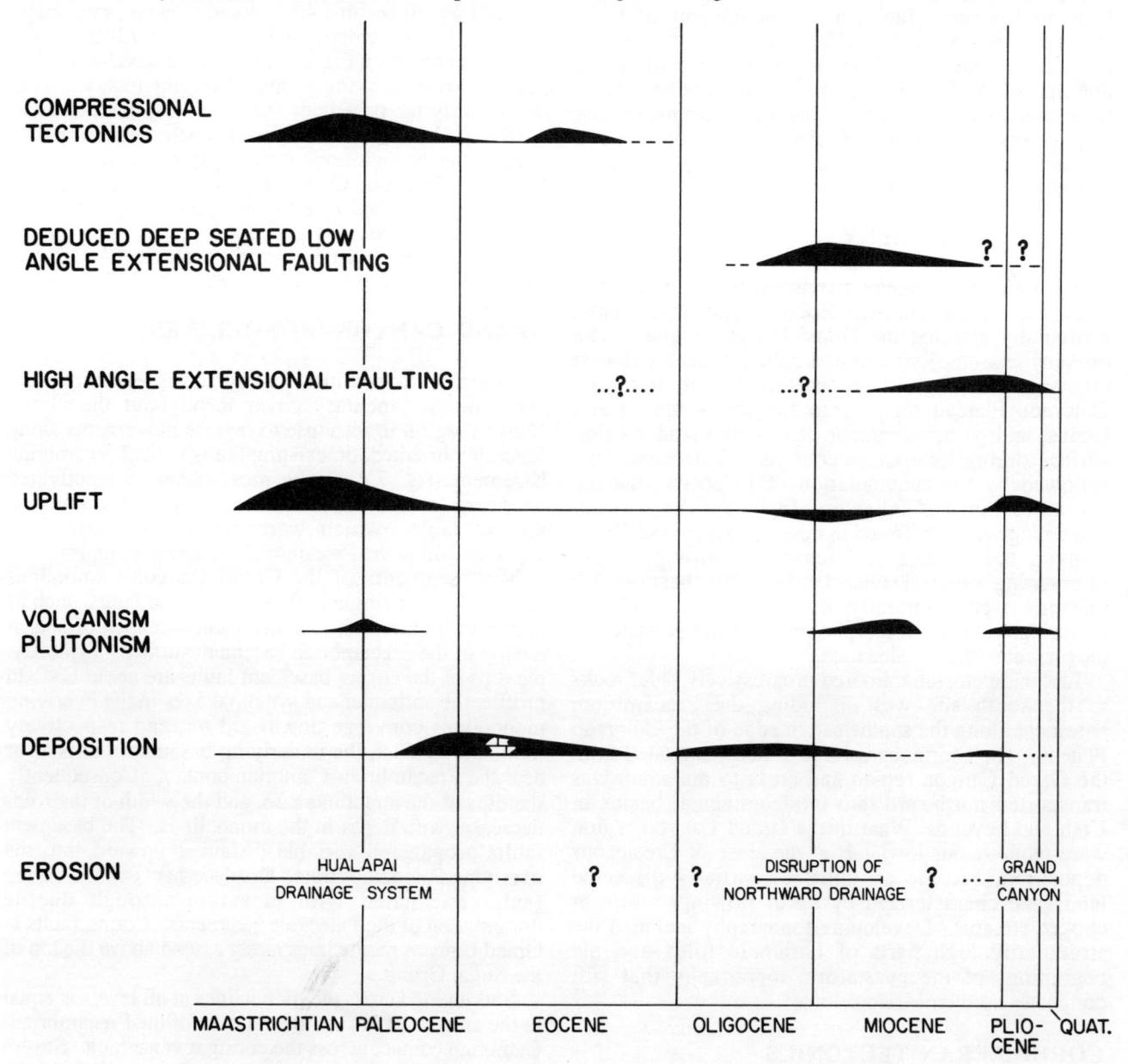

FIGURE 7.1. Temporal relationships between tectonism and erosion in Laramide and post-Laramide time in the Grand Canyon region, Arizona.

west included clasts derived from the arc volcanics, Paleozoic and Mesozoic sedimentary cover, and Precambrian basement. These rocks constituted a primary source of sediment for many extensive Mesozoic units that were deposited in the Grand Canyon region and to the northeast (Stewart and others, 1986). Erosional unroofing of the Mogollon Highlands appears to have exposed Paleozoic and Precambrian terranes as early as Late Triassic time based on clasts contained in the Shinarump Member of the Chinle Formation (Stewart and others, 1972). The last marine transgression occurred in Late Cretaceous time encroaching from the northeast, and resulted in the deposition of the Dakota Sandstone followed by the Mancos Shale (Fisher and others, 1960).

Neither discrete offsets along reactivated Grand Canyon basement faults nor development of local structural basins and uplifts during pre-Laramide, Mesozoic time have been documented within the confines of the Grand Canyon owing to the paucity of Mesozoic strata in the area. Mesozoic tectonism in the region was at least of sufficient magnitude to have influenced fluvial facies and isopachous patterns (Peterson, 1984, 1986).

LARAMIDE OROGENY

The Laramide orogeny, herein defined as embracing latest Cretaceous through Eocene structural events, profoundly affected the Grand Canyon region. The orogeny caused 1) widespread uplift, 2) east-northeast crustal shortening, 3) compartmentalization of the Colorado Plateau region into subsidiary uplifts and basins, and 4) development of a widespread erosion surface during Late Cretaceous-early Paleocene time followed by the accumulation of Paleocene-Eocene arkosic sediments (chapters 18 and 19). Crustal shortening was manifested in development of the Grand Canyon monoclines and regional warping of the intervening crust (Davis, 1978). The basin–uplift margins were generally defined by monoclines, imposing a structural morphology that has persisted to the present with little alteration.

Laramide erosion unroofed progressively older rocks to the south and west including the Precambrian basement along the southwestern edge of the Colorado Plateau. The enormous volume of detritus eroded from the Grand Canyon region and areas to the south was transported northward into intracontinental basins in Utah and beyond. What in the Grand Canyon region were monotonous lowlands at the close of Cretaceous deposition, became an uplifted, partially dissected landscape characterized by north flowing, sediment choked streams. Developing topography included the structurally high parts of Laramide folds and the beginnings of the cuestaform topography that still dominates northern Arizona today.

CORDILLERAN TECTONICS

The currently favored plate tectonic explanation for Laramide orogenesis was a flattening of the angle of subduction of the Pacific oceanic plate under North America resulting from rapid rates of subduction (fig. 7.2). Dickinson (1981, p. 125) summarized this model as follows. 1) The belt of magmatism moved inland as the locus of melting near the top of the subducted slab shifted away from the zone of subduction. 2) Magma generation waned as slab descent became subhorizontal because the slab no longer penetrated as deeply into the asthenosphere. 3) Shallower descent of the slab increased the degree of shear and area of interaction between the descending slab and the overriding cratonic crust. As rapid subduction took place during Laramide time, the subducted hot, buoyant oceanic plate appears to have underplated North America as far eastward as the Great Plains, hence contributing to the uplift of the western United States (Damon, 1979).

What was to become the Colorado Plateau was caught in the eastward compressing Laramide cordillera. The result, as shown on Figure 7.3, was the development of generally north striking, eastward verging monoclines as the underlying basement failed in response to east-northeast, horizontal compressive stresses from Late Cretaceous through Eocene time. Laramide monoclinal folding in the Grand Canyon region was accompanied by gentle regional warping of the intervening structural blocks, a process that produced uplifts such as the Kaibab Plateau and downwarps such as the Cataract Creek basin.

GRAND CANYON MONOCLINES

Laramide monoclines formed in the Paleozoic and Mesozoic sedimentary cover throughout the Grand Canyon region in response to reverse movements along favorably oriented, preexisting faults in the Precambrian basement (fig. 7.4). The most common reactivated basement faults were steep, west-dipping Precambrian normal faults, which were pre-existing structural elements within the Precambrian basement complex.

Most segments of the Grand Canyon monoclines developed over single high-angle reverse faults, such as shown on Figure 7.5, creating abrupt stepped offsets at the top of the Precambrian basement surface. Typically, the dips of the coring basement faults are about 60°. In profile, the anticlinal and synclinal axes in the overlying monoclines converge downward on, and respectively terminate against, the underlying basement faults at or near the Precambrian-Cambrian contact. Consequently the dips of the strata increase, and the width of the folds decrease, with depth in the monoclines. The basement faults propagated variable distances upward into the overlying Paleozoic strata. Displacements on the coring faults attenuated with elevation through ductile deformation of the Paleozoic sediments. Coring faults in Grand Canyon monoclines rarely extend above the top of the Supai Group.

Shortening across the monoclines at all levels is equal to the amount of horizontal override of the Precambrian-Cambrian contact across the coring reverse fault. Severe crowding such as shown on Figure 7.6 develops in the basal Paleozoic rocks in the vicinity of the synclinal axes during deformation. The basal Paleozoic rocks in the

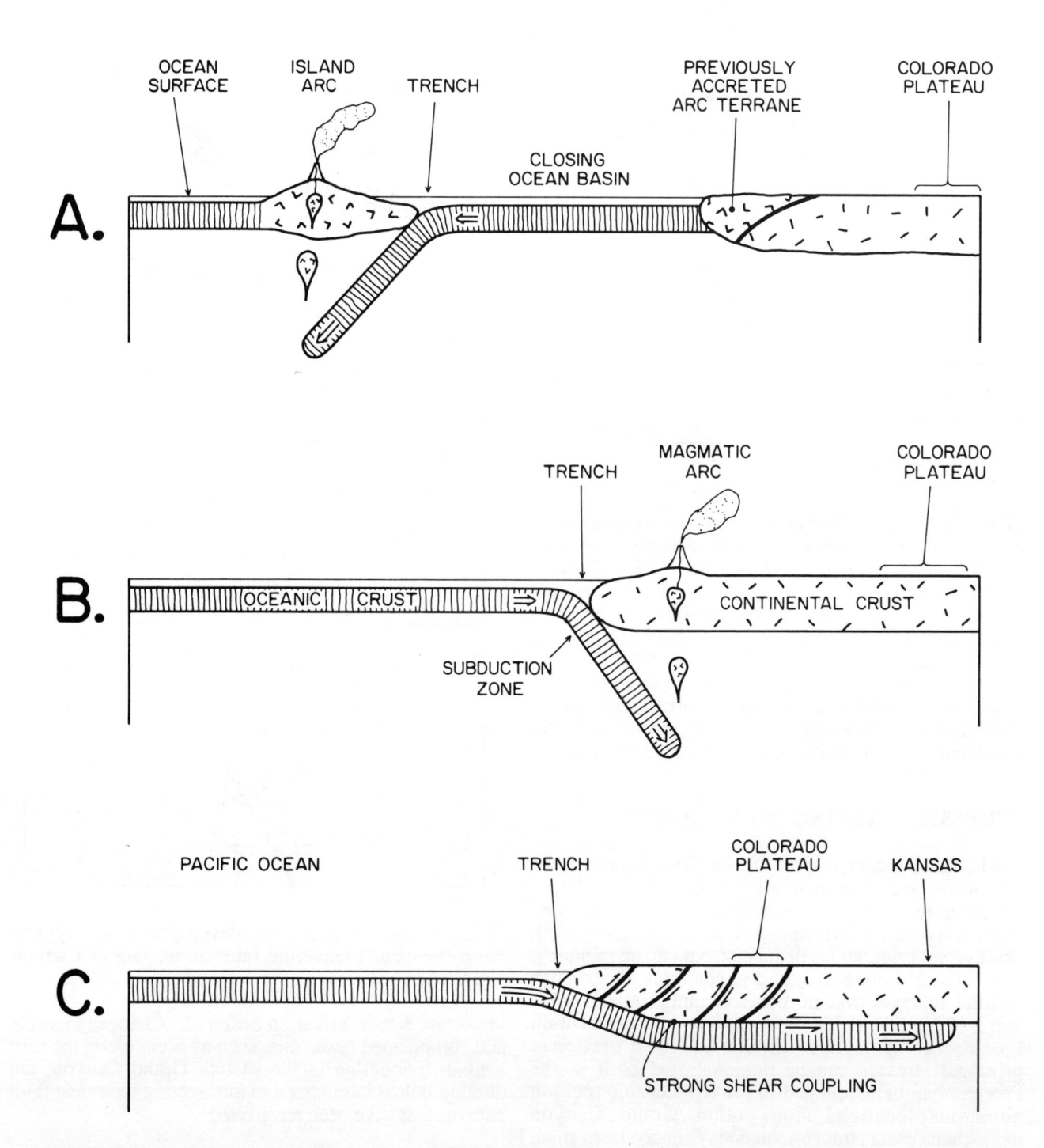

FIGURE 7.2. Convergent margin orogens along western North America. A. Intra-oceanic arc-trench orogen active periodically in post-Precambrian through Late Triassic time. Notice that the ocean basin closes allowing island arc to accrete to continent, then another subduction zone and its island arc can form offshore and likewise eventually accrete to continent. B. Slow landward subduction causing development of magmatic arc inboard on continent above steeply descending slab active from Late Triassic to Late Cretaceous time. C. Rapid subduction resulting in shallow slab descent, and slab underplating of continent to produce buoyant uplift and strong shear coupling with eastward telescoping of continental crust during Laramide time. Vertical scales are greatly exaggerated.

syncline commonly are riven with low angle conjugate thrusts that mechanically thickened the strata in the syncline and operated to move material out of the syncline away from the fold. Basal Cambrian beds in the downthrown limbs are commonly overturned and highly attenuated in thickness adjacent to the coring fault. As shown on Figure 7.7, the thicknesses of the Cambrian through Pennsylvanian strata lying between the anticlinal and synclinal axis in the East Kaibab monocline are attenuated between 30 and 60 percent; the

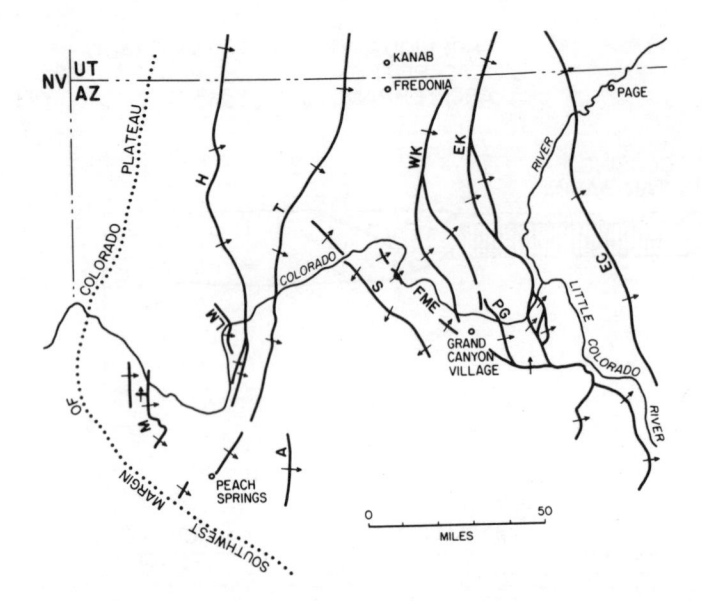

FIGURE 7.3. Locations of the Laramide monoclines in the Grand Canyon region, Arizona. From west to east: M - Meriwhitica, LM - Lone Mountain, H - Hurricane, T - Toroweap, A - Aubrey, S - Supai, FME - Fossil-Monument-Ermita, WK - West Kaibab, PG - Phantom-Grandview, EK - East Kaibab, EC - Echo Cliffs.

Redwall Limestone in many outcrops has undergone low grade metamorphism as a result. This contrasts with comparatively gentle dips of less than 15° and virtually no attenuation at the level of the Permian strata as shown on Figure 7.8.

STRESSES CAUSING MONOCLINES

The stress regime responsible for development of the monoclines was one involving generally east-northeast, horizontally oriented maximum principal stresses, and vertical minimum principal stresses. Reches (1978), using stress orientations deduced from calcite twinning, minor faults, kink bands, and minor folds, determined that the average strike of the maximum principal stress was N67°E in the Paleozoic rocks along segments of the East Kaibab monocline. Similar east-west maximum principal stresses can be demonstrated both in the Precambrian basement and in lower Paleozoic rocks at numerous locations along other Grand Canyon monoclines using stress orientations deduced from small scale low angle conjugate thrust faults and slickenside striations on the thrust surfaces (Huntoon, 1981).

SINUOSITY AND BRANCHING OF MONOCLINES

The Laramide monoclines are sinuous and branch (fig. 7.3). The variable trends of segments and branching character of the monoclines, as well as the presence or lack of folding at specific locations, have been attributed to selective reactivation of favorably oriented basement faults (Noble, 1914; Huntoon and Sears, 1975; Huntoon, 1981; Huntoon and others, 1986). Favorable dips appear to be of paramount

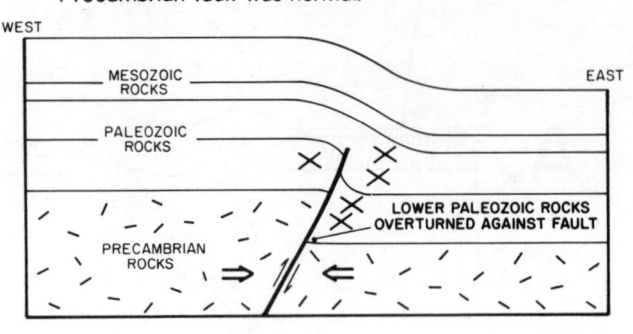

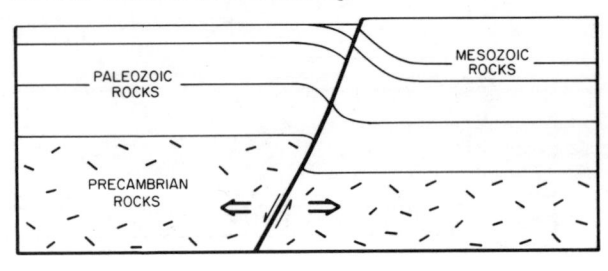

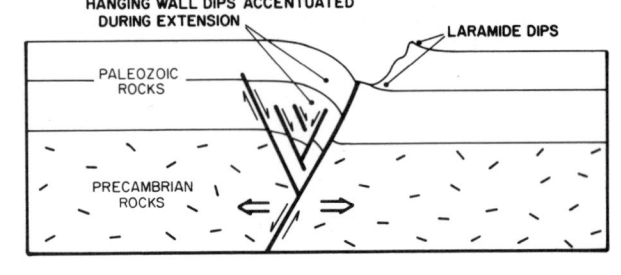

FIGURE 7.4. Stages in the development of a typical north-trending monocline-fault zone, Grand Canyon region, Arizona.

importance as reactivation occurred. Changes in strike and complicated branching, such as occur along the East Kaibab monocline in the eastern Grand Canyon, are directly linked in outcrops to intersecting basement fault patterns that have been reactivated.

AGE OF MONOCLINES

The most convincing evidence that the Grand Canyon monoclines developed during Laramide time is the fact that erosion accompanying uplift in the region resulted in beveling of the monoclines in the western Grand Canyon. Unconformities in the southern high plateaus of Utah north of the Grand Canyon, and elsewhere in the Rocky Mountain region, establish a Maastrichtian age for the initiation of Laramide deformation (Kelley, 1955; Dickinson and others, 1988). The Cretaceous-early Paleocene Laramide erosion surface in the western Grand Canyon region was buried by upper Paleocene-lower Eocene deposits (Young, 1982; chapters 18, 19,

FIGURE 7.5. Middle Proterozoic Cardenas Basalt (left) displaced down in normal fashion against older Middle Proterozoic Dox Sandstone (right) along west-dipping basement fault underlying the East Kaibab monocline, as viewed northward from Tanner Rapids, Grand Canyon, Arizona. In Laramide time, the fault was reactivated with approximately 600 feet (183 m) of reverse (left up) offset to cause monoclinal folding of overlying Paleozoic and younger rocks at this location.

and 20). These lower Tertiary sediments may be slightly involved in the last folding events in the western Grand Canyon.

EOCENE TO MIOCENE STABILITY

Rapid rates of subduction of oceanic crust in trenches along the west coast and attendant shallow angles of descent of the Pacific oceanic plate under the North American plate respectively slowed and steepened about 45 m.y. ago, in Eocene time (Dickinson, 1981, p. 127). The result of the steepened slab descent was significant relaxation of east-west compressive stress across the region deformed during Laramide contraction. Deformation ceased in the Grand Canyon region.

Tectonic quiescence at shallow crustal levels prevailed in the Grand Canyon region from about late Eocene through early Miocene time, longer than in the Basin-Range to the west and south. The southwestern part of the Colorado Plateau remained undifferentiated from the Mogollon Highlands to the south, and together they constituted an uplift that was tilted slightly more steeply toward the north than at present. Rapid early Late Cretaceous-early Paleocene regional erosion that accompanied Laramide uplift and deformation waned, leaving an erosion surface that beveled successively older rocks southward toward the Mogollon Highlands. Drainage was toward the north in incised canyons that originated in the Mogollon Highlands and terminated in basins on the Colorado Plateau (Nations and others, 1985).

LARAMIDE EROSION SURFACE

Remnants of the Laramide drainage system are best preserved on the southwest corner of the Colorado Plateau as hanging valleys west of the Toroweap monocline and south of the Colorado River (Young, 1966, 1970). As shown on Figure 7.9, progressively younger and more deeply incised channels in the drainage system are superimposed along the Hurricane fault zone. The deepest of these is incised more than 1,600 feet (488 m) below the Laramide surface on the top of the Redwall Limestone. The youngest channel is now preserved as Peach Springs Canyon and was reorganized in pre-Miocene time from a meandering pattern into a straight canyon that exited the region over what is now the north rim of the Grand Canyon at some location between the Shivwits and Kaibab Plateaus (fig. 4.1).

Young and Brennan (1974, p. 88) and Young (1982, p. 32) state that regional northward dips in the vicinity of the Hualapai Plateau during the period of maximum incision had to be between 0.5° and 1° greater than at present in order for the streams to clear the present north rim of the Grand Canyon between the Shivwits and Kaibab Plateaus.

LOWER TERTIARY ROCKS

The floors of the Laramide paleocanyons along the Hurricane fault zone and the oldest sediments preserved in them date from Late Cretaceous through middle or late Eocene time (Young, 1982; chapter 20). Northward flow is revealed by channel slope, imbricated pebbles, and increasing Precambrian clast concentrations toward the south (Young, 1979, 1982; Young and Huntoon, 1987).

At Long Point in Cataract Basin south of the Grand Canyon, Young (1982, p. 37) and Young and Hartman (1984) documented that lacustrine limestones within arkosic sediments resting on the Moenkopi Formation are of early Eocene (or older) age based on the occurrence of viviparid gastropods. These lacustrine

FIGURE 7.6. Mechanical crowding within the Cambrian Tapeats Sandstone at the synclinal hinge of the East Kaibab monocline, south wall of Chuar Canyon, Grand Canyon, Arizona. Notice almost vertical dips at this level in the fold. Arrow points to man for scale.

FIGURE 7.7. East Kaibab monocline at Chuar Butte, Grand Canyon, Arizona. Ductile thinning of Supai Group and Redwall Limestone in center is obvious. Note increasing steepness of dips with depth. View is toward the north, Kwagunt Butte to left.

FIGURE 7.8. East Kaibab monocline north of the Grand Canyon, Arizona. Dips in the upper Paleozoic section are moderate at this stratigraphic level. Plateau surfaces are comprised of the Permian Kaibab Formation. View is toward the north, Cocks Comb in center.

FIGURE 7.9. Deeply incised Laramide paleochannels in Peach Springs Canyon area, Arizona. The two meander loops (small arrows) are older and less deeply incised than the Peach Springs paleocanyon (large arrows), which is colinear with the Hurricane fault. Offset lower Tertiary rocks along the floor of Peach Springs Canyon reveal that late Cenozoic normal faulting postdated incision of both channels. Arrows show northward paleocurrent directions. U.S. Geological Survey photo.

limestones and associated fluvial deposits indicate that the drainage on the Laramide erosion surface was locally ponded toward the close of Laramide time Young (1982, p. 37) attributes the ponding to late Laramide uplift of the Kaibab upwarp accompanied by development of the Cataract structural basin to its west (chapter 17)

Tertiary arkosic conglomerates gradually filled the earliest Laramide canyons in the vicinity of Peach Springs Canyon and on the Hualapai Plateau, and spread over the adjacent divides (fig. 4.1). The conglomerate lenses in the arkoses are characterized by abundant Precambrian crystalline clasts and Cretaceous-Paleocene volcanic clasts. The original upper surface, where preserved, exhibits a deeply weathered profile and a dark red color (Koons, 1948; Young, 1966, p. 24-25). Rocks equivalent to these arkoses, called the Music Mountain Conglomerates by Young (1966, p. 24; 1970), are overlain by late Oligocene volcanics in the Aquarius Mountains near the southern margin of the Colorado Plateau (Young and McKee, 1978, p. 1749; Young, 1979, p. 41).

The onset of late Oligocene-Miocene arc volcanism to the southwest of the Colorado Plateau was heralded by increased amounts of volcanic and pyroclastic fragments in the upper part of the Oligocene Buck and Doe Conglomerate (Young, 1966, p. 73a; 1970; chapter 20), which was deposited in aggrading paleocanyons on the Hualapai Plateau. This unit is overlain by younger lavas and the 18.6 Ma Peach Springs Tuff (Young and Brennan, 1974, p. 84). The volcanic activity was a harbinger of extensional tectonics that would strongly affect the region and would structurally differentiate the Colorado Plateau from the Basin and Range Province to the west. Northward, through-flowing drainage onto the Colorado Plateau within the Transition Zone of central Arizona had already been interrupted by Oligocene time.

EARLY TERTIARY TECTONISM

During late Oligocene time, the Pacific-Farallon spreading center and Farallon-American trench obliquely converged upon each other along the continental margin near the U.S.-Mexico border. The result was annihilation of the spreading center and trench at the point of contact, and the mechanical substitution of right lateral transform faulting along the contact zone. These events caused the cessation of subduction, eventual termination of arc magmatism, and widespread extensional tectonism in the region inboard of the San Andreas transform fault system (Dickenson, 1981). The result was wholesale extension within the Basin-Range to the west and south of the Grand Canyon region during late Oligocene through middle Miocene time (Hamilton, 1982, p. 21).

The major phase of Basin and Range extension in the vicinity of the Whipple Mountains along the lower Colorado River southwest of the Colorado Plateau commenced in late Oligocene time, considerably earlier than the first surface manifestations of normal faulting on the Colorado Plateau. Detachment faulting in the Whipple Mountains is characterized by east-northeast dipping, low angle, normal faults in which the upper plate glided downslope to the northeast on an unextended lower plate (Davis and others, 1980). Although presently unproven, similar extension also may have been occurring during this period at middle and lower crustal levels under the southwestern part of the Colorado Plateau (Lucchitta and Suneson, 1982).

A speculative tectonic model accommodating late Oligocene-late Miocene extension at deep crustal levels under the Colorado Plateau and at shallower levels in the Basin and Range involves a gently northeast dipping "Wernicke-style" crustal penetrating normal fault (Wernicke, 1981), as illustrated on Figure 7.10D. The relative sense of motion along the fault surface would

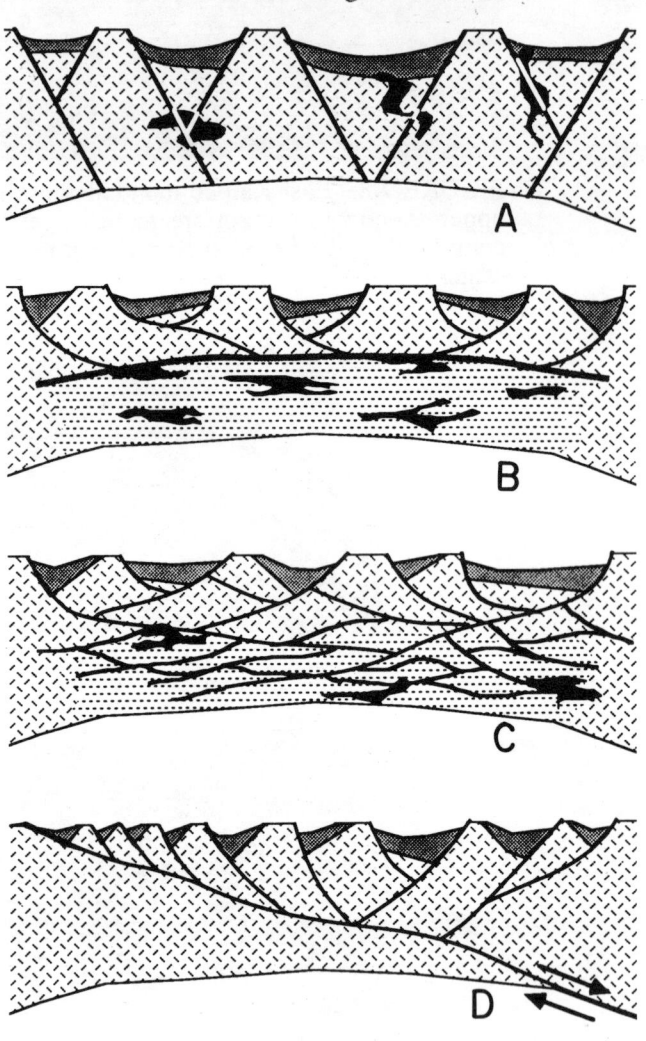

FIGURE 7.10. Summary of models used to explain extension in the Basin and Range Province southwest of the Colorado Plateau, Arizona (from Allmendinger and others, 1987, fig. 7). A. Classic horst-graben model; B. subhorizontal decoupling zone model; C. shear zone bounded lense model; D. crustal penetrating shear zone model. A late Oligocene-early Miocene crustal penetrating shear could have produced down-to-the-southwest rotation and crustal thinning of the western Colorado Plateau.

have been Colorado Plateau down to the northeast. Northwest striking blocks calving off the thin trailing edge of the upper plate could account for tectonic erosion of the southwestern edge of the Colorado Plateau.

Extension along the postulated fault is inferred by Huntoon (1988) to have been underway in late Oligocene time and allowed for 1) progressive northeastward tectonic erosion of the Colorado Plateau, 2) structural differentiation of the Colorado Plateau from the Basin and Range in Miocene time, 3) thinning of the crust under the western part of the Colorado Plateau from about 25 to 19 miles (40 to 30.6 km) (Keller and others, 1975), and 4) nearly 1° of down to the southwest rotational subsidence of the southwestern edge of the Colorado Plateau (Young and Brennan, 1974, p. 84). The latter would easily accommodate sufficient down-to-the-south rotation of the Laramide erosion surface on the Colorado Plateau to force 1) abandonment of Laramide paleocanyons that could no longer clear the north rim of the Grand Canyon and 2) permit establishment of a west-flowing Colorado River at least by late Miocene(?)-Pliocene time). Such subsidence can be used to explain why lower Tertiary rocks in the southern high plateaus of Utah now lie 2,000 feet or more above some of their source areas in the Grand Canyon region.

The relative motion of the Colorado Plateau was down to the northeast on the postulated fault. However, because such motion also corresponded to opening of the Rio Grande rift, absolute regional motions had to include 1) a slight clockwise rotation of the Plateau and 2) southwestward extrusion of the lower plate from under the Colorado Plateau. This model implies that the Colorado Plateau was rotating into extensional space created within the Basin and Range Province, wherein the lower plate was moving into, and westward in concert with, the Basin and Range at a rate slightly greater than the upper plate, which was carrying the Colorado Plateau. Thus, the entire Basin and Range Province was translating to the southwest away from the Colorado Plateau at all crustal levels.

MIOCENE TO HOLOCENE EXTENSION

By late Oligocene time, the western part of the Colorado Plateau was probably undergoing the first significant shallow east-west crustal extension to affect the region since late Precambrian time. The extensional tectonism defined the modern western boundary of the Plateau itself (Rowley and others, 1978), and the Grand Canyon region within it was experiencing east-west extension that caused down-to-the-west normal faulting. This tectonic stress regime has persisted in the region to the present with a probable peaking of activity in Pliocene time. The principal Miocene and younger faults, coupled with the Laramide monoclines, served to outline the plateaus and basins within the Colorado Plateau as we know them today.

NORMAL FAULTING

The strain resulting from late Cenozoic east-west extension in the Grand Canyon region was mostly accommodated by normal displacements along faults in north-south trending zones. The first failures, such as along the Hurricane fault zone (fig. 7.11), reactivated the principal Precambrian basement faults in the zone. In most instances, as shown on Figure 7.12, the principal normal faults severed the Laramide monoclines, displacing them down to the west opposite to Laramide downfolding to the east.

Fault displacements along individual faults and fault densities increase with depth in the Paleozoic section within major fault zones such as the Hurricane, Toroweap, and Eminence zones. This observation reveals that the subsidiary faults propagated upward from shallow levels within the brittle lower Paleozoic carbonate section, and in some cases from the Precambrian basement. Displacements attenuate with elevation in the Paleozoic section through ductile deformation within the clastic Pennsylvanian and Permian sediments.

EXTENSIONAL SAG

Infolding of the Paleozoic rocks toward faults within the downdropped hanging wall blocks is common along many large displacement Grand Canyon faults. Dips of normal faults in the Paleozoic section are commonly greater than 70° as compared to dips of the order of 60° along the underlying reactivated basement faults (Huntoon and Sears, 1975, p. 470). In some locations the dips of fault surfaces tend to shallow gradually with depth (Hamblin, 1965, p. 1155). In either case, as displacement proceeded along the less steeply dipping deeper parts of a fault, space tended to develop between the fault surfaces in the steeply dipping upper parts of the fault. The mechanical response to the creation of the space was compensating volume infolding of the upper part of the hanging wall block toward the fault surface.

EXTENSIONAL BASINS

The area encompassed within the Toroweap and Hurricane fault zones along the Colorado River is characterized by modest late Cenozoic structural subsidence resulting from east-west crustal extension across the zones. Extension across numerous northwest-trending faults in the Hurricane fault zone in the area centered on Parashant Canyon has progressed sufficiently to create a 10 mile (16 km) diameter structural basin having approximately 600 feet (180 m) of structural closure (Huntoon, 1977a, fig. 6).

AGE OF FAULTING

Definitive timing for the inception of late Cenozoic faulting has eluded researchers in the Grand Canyon. Tertiary rocks that could be used to bracket the onset of faulting are missing in the eastern Grand Canyon, and those in the western Grand Canyon (such as shown on Figure 7.13), are faulted and have so far provided only maximum ages. It is likely that the first faults to become active in the region were the Grand Wash and Hurricane

FIGURE 7.11. View toward the northeast along the main strand of the late Cenozoic Hurricane fault from the top of Diamond Peak, Grand Canyon, Arizona. Arrows point to the Cambrian Tapeats Sandstone on respective sides of the fault. The east (right) dipping Laramide Hurricane monocline is not developed in this reach.

FIGURE 7.12. Two down-to-the-west late Cenozoic normal faults sever the east (left) dipping Lone Mountain monocline south of Parashant Canyon (foreground). Faulted surface is the top of the Permian Esplanade Sandstone.

FIGURE 7.13. Lower Tertiary Robbers Roost Gravel offset down-to-the east by late Cenozoic normal faulting (arrows) at the head of Diamond Canyon, Grand Canyon, Arizona. View is toward the north.

faults (western Grand Canyon; fig. 4.1). This activity probably dates from middle Miocene time but may have begun as far back as early Miocene or possibly late Oligocene time.

Initial faulting along the Grand Wash fault appears to have been synchronous with the latter part of the major phase of Basin and Range extension to the southwest, which Hamilton (1982, p. 21) assigned to late Oligocene through late Miocene time. Young and Brennan (1974, p. 86-87) observed that eruption of the 18-19 Ma Peach Springs Tuff predated most of the offset along the southwestern part of the Grand Wash fault. The tuff is tilted on modern ranges that are partially buried by younger basin fill. Therefore, major movement occurred along the Grand Wash fault after deposition of the Peach Springs Tuff, but prior to deposition of the upper part of the Muddy Creek Formation (Lucchitta, 1972, p. 1942). These relationships bracket the major offsets across the Grand Wash and associated faults between middle Miocene and early Pliocene time.

RECURRENT LATE CENOZOIC FAULTING

The most spectacularly exposed and complete records of late Cenozoic faulting in the western Grand Canyon region occur along the Hurricane and Toroweap faults. The clarity and quality of this record is made possible by deep cross sections, and the presence of successive late Cenozoic lava flows, each of which exhibits an offset proportional to its age. In addition, Quaternary cinder cones and alluvium, such as shown on Figure 7.14, record the latest offsets. The resolution of the number and timing of late Cenozoic displacements is constrained only by the number of upper Cenozoic rocks deposited across the two faults.

Highlights from outcrops near Whitmore Wash along the Hurricane fault serve to illustrate the character of the record present in that single area within the Grand Canyon. The Paleozoic rocks are offset approximately 1,000 feet (300 m) whereas younger Pleistocene basalt flows which fill an old channel of Whitmore Wash to a depth of about 600 feet (180 m) are displaced only 75 feet (23 m). A slightly eroded 50-foot (15-m) scarp (fig. 7.15), which extends 8 miles (13 km) north of the Colorado River in Whitmore Wash, offsets a Stage IV [Hamblin, 1970] Pleistocene basalt flow that cascaded into the canyon from the east. The overlying alluvium is offset about 15 feet (4.5 m). The offsets of the basalts and alluvium were Quaternary events (Anderson and Huntoon, 1979), but probably not Holocene as originally claimed by Huntoon (1977b, p. 1620). In this one location the variable offsets associated with four distinct units allows resolution of four faulting events. Certainly there were other faulting events during this same interval of time that have eluded detection owing to lack of strata of intermediate age.

DISRUPTION OF DRAINAGE

Tectonic differentiation of the Colorado Plateau from

FIGURE 7.14. West downfaulted Quaternary basalt (right of cone) and alluvium along the Toroweap fault in Prospect Valley south of the Colorado River, Grand Canyon, Arizona. Notice that the offset of the basalt is greater than that of the alluvium. View is toward the southeast.

FIGURE 7.15. West downfault scarps (diagonal lineations across center) in Pleistocene(?) alluvium along the Hurricane fault south of the mouth of Whitmore Canyon (left), Grand Canyon, Arizona. View is toward the northeast.

the extending Basin and Range to the south and west had progressed sufficiently to sever the northward flow of streams across the plateau margin in the western Grand Canyon region by late Early Miocene time. Similar severing of the northward flowing streams took place in early(?) Oligocene time in the region south and east of the eastern Grand Canyon (Peirce and others, 1979). Two other processes operated to further disrupt these streams on the Colorado Plateau itself, 1) partial burial by Miocene volcanics, and 2) down to the south subsidence of the plateau surface inferred by Young and Brennan (1974) of probable late Oligocene to late Miocene age.

The modern Colorado River became integrated from east to west across the region following the close of the late Miocene-Pliocene Muddy Creek deposition in the Grand Wash trough (Lucchitta, 1972). The river incised into the formerly northward sloping Laramide erosion surface and carved the Grand Canyon to within 50 feet (15 m) of its present depth by early Pleistocene time (McKee and others, 1968, p. 135)

The Mogollon Rim is defined as the south-facing erosional escarpment comprised of the northward retreating Permian stratigraphic section (Elston, 1978, 1984a; Peirce and others, 1979; Mayer, 1979; fig. 4.1). The Rim is useless for defining the southern boundary of the Colorado Plateau in the Grand Canyon region because its trend through the western Grand Canyon predates both Oligocene(?) and Miocene tectonism and drainage reorganization that outlined the plateau south of the Grand Canyon. Rather the rim in the area dates from Laramide erosion and thus is independent of the Miocene plateau margin. In fact, the rim lies north of the Colorado River in the area west of Diamond Canyon, little retreated from its Laramide position just north of Hindu Canyon (Young and Brennan, 1974, p. 87).

CHAPTER 8: EARLY PROTEROZOIC ROCKS OF GRAND CANYON, ARIZONA

Charles W. Barnes
Department of Geology, Northern Arizona University, Flagstaff, Arizona

INTRODUCTION

From approximately 77 miles (124 km) downstream from Lees Ferry to just upstream from the Grand Wash Cliffs, Early Proterozoic crystalline rocks are intermittently exposed along a total distance of nearly 184 miles (300 km). The exposures generally consist of nearly vertical cliffs carved by the Colorado River; occasional faults cutting across the trend of the river periodically both raise and lower these old rocks out of view. For most of the lower two-thirds of the river trip, exposures are continuous over distances of tens of miles (kilometers).

These crystalline rocks form the basement for both the strata in the Grand Canyon Supergroup (chapter 9) and the overlying Phanerozoic strata (chapters 13-16). Each of these three major rock units is separated by unconformities whose time gaps range from 0.2 to 1.2 Ga. The rocks comprise both plutonic and metamorphic units whose boundaries range from highly diffused to knife edge along both plutonic and structural contacts. Metamorphic rocks include a wide range of generally mafic schists and gneisses, as well as occasional thin calc-silicate units and felsic migmatitic masses. Protoliths include both volcanic and sedimentary units with an aggregate thickness estimated at 7.5 miles (12 km). The igneous rocks include at least twenty separate plutons emplaced over a time span of at least a hundred million years before, during, and after the regional metamorphic events. The texture in these plutons ranges from strongly foliated to nonfoliated, with compositions ranging from granite to granodiorite to tonalite and diorite. Other intrusives include dikes and sills of fine- to medium-grained granite, as well as granodiorite, aplite, and pegmatite, commonly intimately intermixed with one another. Pegmatitic and aplitic dikes and sills average 20 percent (Babcock, in press) of the total volume of Early Proterozoic rocks in the Grand Canyon.

Nomenclature and Previous Work

The Early Proterozoic rocks within the Grand Canyon were first described by John Wesley Powell (1874). Walcott (1894) named the basement complex the Vishnu terrane, after exposures near the Vishnu Temple, a butte named in 1880 by C.E. Dutton after Vishnu, one of the Hindu trinity of gods. Noble (1914) changed this designation to the Vishnu Schist. Noble and Hunter (1916) delineated granitic and hornblende gneisses, amphibolites, mica schists, quartz diorite and gabbroic intrusions, and alkalic pegmatites. They suggested that some of the gneisses might have been basement for the deposition of the protoliths for the mica schists, with intrusion of the mafic intrusives accompanying the later phases of the regional metamorphism of the Vishnu.

The stratigraphy of the Vishnu protoliths is essentially unknown, and will remain so until structural studies unravel the complex relations among several foliations, compositional banding, and relict bedding. Campbell and Maxson (1938) attempted to subdivide the Vishnu into an older sedimentary unit (the Vishnu Series) and a younger volcanic unit (the Brahma Series) on the assumption that the Brahma was the upper unit in the trough of a large isoclinal syncline whose axial plane foliation was vertical and parallel to the trend of Bright Angel Canyon. Later mapping (Ragan and Sheridan, 1970) challenged that earlier interpretation and demonstrated that the metavolcanic and metasedimentary units are interlayered. Campbell and Maxon's division of the basement complex was erroneous; the term Brahma is no longer used.

Campbell and Maxson (1938) named the granitic body exposed between Zoroaster and Cremation Canyons the Zoroaster Granite and noted that it intruded the Vishnu Schist; the type area was named after Zoroaster, the founder of the national religion of what is now Iran. A series of studies by Babcock and other workers (1974, 1979); Boyce (1972); Lingley (1973, 1976); Walen (1973); Clark (1976, 1978, 1979); and Brown and others (1979) have subdivided the Early Proterozoic rocks into four major units: the Vishnu Metamorphic Complex, Trinity Gneiss, Elves Chasm Gneiss, and the Zoroaster Plutonic Complex. While usage of some of these names violates the principles of the North American Code of Stratigraphic Nomenclature, they are unfortunately in common published usage and are therefore retained for this paper. The names Vishnu Schist and Zoroaster Granite are used in the river log (chapter 1).

Geochronology

Limited field data (Babcock, in press) suggest that the Elves Chasm and Trinity Gneisses may have formed the basement for the supracrustal Vishnu protoliths, an idea first suggested by Noble and Hunter (1916). In the absence of radiometric dating, the field relations are consistent with, but do not prove, that these gneissic units formed the basement for Vishnu protoliths.

Rocks assigned to the Vishnu Metamorphic Complex exhibit at least two distinct stages of regional metamorphism, the earlier one in the greenschist facies

(Babcock, in press) and the later and more extensive one of the Barrovian type in the almandine-amphibolite facies (Boyce, 1972). In limited areas, temperatures reached 650-700 °C at pressures of 3-4 kilobars (Babcock, in press), forming migmatites by partial melting processes. In addition, Vishnu rocks display thermal metamorphism about some plutons and areas of retrograde metamorphism.

Strongly foliated Vishnu rocks exhibit a minimum of three stages of deformation; detailed structural analysis allows two additional events to be recognized (Boyce, 1972; Lingley, 1973). The plutonic and high grade metamorphic rocks of the Zoroaster Complex display boundary and textural relations consistent with their formation prior to, during, and after the main stage regional metamorphism of the Vishnu units. Thus, the limited geochronologic work is beset by the usual difficulties in interpreting which event (or events) have been dated.

The oldest dated rock units are the Ruby Creek pluton within the Zoroaster Plutonic Complex, at about mile 104 (km 167). These plutons appear to have been intruded prior to or during the main stage of regional Vishnu metamorphism and yield Rb-Sr dates ranging up to approximately 1,780 Ma (Babcock, in press). These same plutons have initial-strontium ratios of 0.7012 to 0.7030, suggesting a mantle source (Babcock, in press).

Uranium-lead and rubidium-strontium dating of Vishnu rocks (Pasteels and Silver, 1966; Livingston and others, 1974) is consistent with assigning an age of 1,720-1,710 Ma to the earliest Vishnu metamorphism (Babcock, in press). A highly discordant U/Pb zircon age of 1,705±15 Ma is considered by Pasteels and Silver (1966) as the primary crystallization event for the granitic gneiss in the Zoroaster, a syntectonic body with respect to the early Vishnu metamorphism. The main stage regional metamorphism of the Vishnu protoliths probably occurred between 1,680-1,650 m.y. (Babcock, in press), a time limited by a U/Pb date (Pasteels and Silver, 1966) of 1,675±15 Ma for an unfoliated pegmatite crosscutting migmatized Vishnu schist. Other similar dates include a slightly discordant U/Pb date of 1,675±15 Ma (estimated correction) on monazite from both granitoid and quartzofeldspathic parts of migmatites near Phantom Ranch (mi 88; km 142) and a Rb-Sr date (Livingston and others, 1974) of 1,650 Ma for both paramphibolites from the Vishnu and the late syntectonic to post-tectonic Phantom pluton.

Slightly younger preliminary Rb-Sr dates from anorogenic Surprise Canyon pluton (Babcock, in press) range from 1,630-1,580 Ma, while still younger potassium-argon ages, ranging from 1,550-1,240 Ma were reported by Aldrich and others, (1957), Giletti and Damon (1961), and Damon (1968). These dates may record primary ages for post-tectonic pegmatites and granitic intrusives (Babcock, in press); these young dates may also record nonsystematic error due to argon loss during late stage thermal events.

Unconformities

The surface of nonconformity between the Early Proterozoic crystalline basement and the overlying units of the Grand Canyon Supergroup is exposed over distances of tens of miles (kilometers) and exhibits less than 50 feet (15 m) of relief (Sharp, 1940). In places the upper few yards (meters) of the basement are weathered and a residual, generally arkosic conglomerate and breccia are preserved. Once separated out as the Hotauta Conglomerate, the unit is now regarded as a basal member of the Bass Limestone (Dalton, 1972). Both the residuum and the weathering suggest subaerial weathering of the basement and reworking of the regolith prior to deposition of the Unkar Group. The time interval represented by this surface of nonconformity approximates 450 m.y. (Elston and McKee, 1982).

The sub-Paleozoic unconformity is a widespread surface marking the Precambrian - Paleozoic contact in the Grand Canyon region. Powell (1876) and Walcott (1883) referred to this boundary as the "Great Unconformity." This surface truncates both the structures of the crystalline basement and the tilted strata of the Grand Canyon Supergroup. Where that surface cuts the uppermost Chuar Group, the missing time interval may be as little as 200 m.y. (Elston and McKee, 1982). Where that surface places Paleozoic rocks directly on crystalline basement, the missing time interval approximates 1.1 Ga. Over 95 percent of the surface, relief is less than 150 feet (45 m), but monadnocks of crystalline rocks or resistant units within the Grand Canyon Supergroup may locally produce a maximum relief of 800 feet (240 m) (Wheeler and Kerr, 1936).

PETROLOGY, STRUCTURES, AND TECTONIC SETTING

The following section describes the petrology, structures, and inferred tectonic setting of each of the four major Early Proterozoic units. While it is tempting to attempt regional correlations, the regional extent and bounding geometries of these units is simply unknown. The closest exposed Early Proterozoic rocks are those assigned to the Yavapai Series and other rocks in central Arizona (Karlstrom and others, 1987), where radiometric dates and lithologies both exhibit some overlap. The reader interested in Early Proterozoic provinces in Arizona is referred to Karlstrom and others (1987).

Trinity Gneiss and Elves Chasm Gneiss

Two areas of feldspathic gneiss form mappable units distinguishable from both the Vishnu Metamorphic Complex and the Zoroaster Plutonic Complex along the Colorado River. The Trinity Gneiss is first encountered in an outcrop band trending north-northeasterly from Salt Creek to Trinity Creek from (mi 91.5-92.9; km 147.2-149.5). The Trinity Gneiss is a well-foliated, medium- to coarse-grained quartzofeldspathic gneiss whose foliation dips very steeply northwesterly and southeasterly. Compositional banding is well developed, as are cataclastic textures (Babcock, in press). Interspersed within the gneissic units are layers of quartzite, biotite schist, and a few small calc-silicate lenses. Babcock suggests that the protolith was tuffaceous with flow units of intermediate composition interbedded with sediments that were deposited between eruptions. The age of the Trinity Gneiss is unknown, although field relations suggest it was emplaced prior to the main stage metamorphism of Vishnu protoliths.

The Elves Chasm Gneiss is exposed only along the river from Waltenberg Rapids downsteam to where its exposure is terminated by faulting associated with the Monument Fold (mi 112.4-118.7; km 181-191) downstream to the Conquistador Aisle area. The upstream contact with the Vishnu Metamorphic Complex is an area of cordierite-anthophyllite gneisses that may represent a paleosol (Babcock, in press); if this interpretation is accepted, the Elves Chasm Gneiss formed the basement for deposition of the Vishnu protoliths. Similar rocks in the Bagdad area, about 80 miles (130 km) south of western Grand Canyon, are metamorphosed chloritized footwall rocks to volcanogenic massive sulfide deposits (Conway, 1986). The Elves Chasm Gneiss is predominantly a tonalitic gneiss with small segregations of granodiorite to granodioritic gneiss with minor calc-silicate pods and areas of cataclasis. Rotated xenoliths and zones of intrusive breccia suggest that the Elves Chasm is an orthogneiss. The time relation between the Elves Chasm and Trinity Gneisses is unknown. Together they may have formed the earliest Proterozic craton prior to deposition of Vishnu protoliths (Babcock, in press).

Vishnu Metamorphic Complex

The metasedimentary and metavolcanic rocks assigned to the Vishnu Metamorphic Complex originated as marine sediments consisting primarily of quartz-rich sand, silt, and clay with lesser amounts of feldspar, mica, and iron oxides. Carbonate lenses make up less than one percent of the total Vishnu rock volume (Babcock, in press). Interspersed with those sediments were basaltic to andesitic lava flows, feeder dikes, and sporadic ash layers.

An early greenschist facies metamorphism (about 1,720-1,710 Ma) formed slates and phyllites from the sediments and greenstones and greenschists from the volcanics. In areas where this earliest metamorphism is well preserved, relict structures such as graded bedding and cross-bedding survive. Excellent examples of these relict structures are exposed from mile 96 to 97 (km 154-156) (Babcock, in press) near the mouth of Boucher Canyon (fig. 4.1).

The later (1,680-1,650 Ma) main stage amphibolite facies metamorphism formed micaceous and quartzose schists and gneisses containing garnet, staurolite, and sillimanite from the original slates and phyllites and converted the greenschists and greenstones into amphibolites and hornblende schists.

The dominant structure throughout the Vishnu Metamorphic Complex is a pervasive nearly vertical foliation striking north-northeasterly (fig. 1.24). This foliation is an axial planar foliation to a series of generally upright isoclinal folds whose amplitudes range to 50 feet (15 m) and whose fold axes plunge moderately northeasterly. This foliation also exhibits strong lineation of rod-like minerals whose plunge is generally vertical. The axial planar foliation (S2) clearly transposes compositional layering (S1). Intrusions of pegmatitic bodies several yards (meters) wide form tabular bodies parallel to the foliation; boundinage structures in the pegmatite are common and record a minimum of 20 percent shortening perpendicular to the foliation (Boyce, 1972); the boudins are commonly tablettes-de-chocolat (Wegmann, 1932) structures with double axes of extension plunging in every orientation within the plane of foliation, or simpler sausage-link boudins whose axis of extension is in the plane of foliation.

Some quartzofeldspathic veins and dikes have been quasi-flexurally folded about fold axes coplanar with S2 but whose plunges may be at any angle. These folds originated in an environment of strong horizontally-directed northwest-southeast compression (Boyce, 1972). A second set of folds (F2) flexes S2 into concentric and cylindrical folds about steeply plunging axes whose plunge is perpendicular to S2. A third set of folds (F3) forms open flexures in the axes of F2 folds, with the axial plunge ranging from horizontal to gently northwesterly (Boyce, 1972).

The largest scale structure yet mapped in these Early Proterozoic rocks is an F2 fold. S2 is deformed into a domelike structure that stretches from Zoroaster Creek to Cremation Canyon (mi 85-86; km 137-138) a distance of approximately 1 mile (0.6 km) (Lingley, 1973). No doubt, still larger scale structures remain to be recognized.

The metasediments and metavolcanics that now form the Vishnu Metamorphic Complex have chemical compositions centering about graywacke and basalt. Assuming isochemical metamorphism, such compositions are similar to those typical of modern magmatic-arc systems.

Zoroaster Plutonic Complex

The twenty identifiable plutons, and innumerable dikes and sills of granite, granodiorite, granitic pegmatite and aplite that form the Zoroaster Plutonic Complex record a period of intrusion and ultrametamorphism spanning several hundred million years (from about 1,780 to 1,240(?) m.y. with emplacement sequences prior to, during, and after the two recorded sequences of metamorphism of the Vishnu protoliths. As an example of the earliest events, the Diamond Creek, 245-Mile, 99-Mile, and Zoroaster Grey plutons are pervasively foliated, suggesting pre- to syn-metamorphic formation (Babcock, in press).

Potentially somewhat younger plutons, which include the Surprise Canyon pluton (mi 248; km 399), have <u>preliminary</u> rubidium-strontium ages that range from 1,630 to 1,580 Ma. The Surprise Canyon and other plutons consist of medium-grained apparently anorogenic granites with high initial-strontium ratios (>0.7100), which suggests they formed by partial melting of crustal rock. Babcock (in press) has suggested that such plutons may be related to the 1,500-

1,400 Ma transcontinental anorogenic belt of Silver and others (1977). However, the slightly foliated Phantom Creek pluton in the Bright Angel Creek area of the eastern Grand Canyon, which might be intermediate between the highly foliated and apparently unfoliated granites, has an age that falls within the main episode of deformation and plutonism as determined from U-Pb dating of zircon (Pasteels and Silver, 1966). Although some plutons exhibit no apparent foliation in their cores, they do exhibit foliated margins, indicating they were emplaced prior to the end of the main episode of deformation. The apparent lack of foliation in the cores of some plutons thus does not necessarily imply a distinctly younger age for their time of intrusion and a relation to the anorogenic granites of central Arizona. Potentially all of the plutonic intrusions in the Grand Canyon may have been emplaced in the interval from about 1700-1650 m.y. as determined from the dating of zircons.

The contact zones of the plutons are largely migmatitic, with the fabric ranging from directionless to strongly foliated or lineated (Babcock, in press). This group of plutons has an initial-strontium ratio of about 0.7040, suggesting a less radiogenic (more mantle-like) source than the older Ruby Creek pluton of highly foliated rocks, whose initial-strontium ratios range from 0.7012 to 0.7030 (Babcock, in press). Broadly the Zoroaster Plutonic Complex thus exhibits a very general trend toward increasing alkali content and increasingly directionless fabrics through time.

The dikes and sills, especially the pegmatitic and aplitic variety, form an average of 20 percent of the volume of the entire Early Proterozoic rocks, and record abundant post-tectonic intrusion of aqueous-phase pegmatitic and aplitic melts. Pegmatitic minerals include tourmaline, beryl, apatite, and feldspar crystals up to 8 inches (20 cm) long (Babcock, in press).

The trend toward increasingly alkalic and equigranular bodies may reflect development of a thicker, more mature continental crust through time. Neodymium-isotope evolution paths (Bennett and DePaolo, 1987) suggest that magmatic rocks were derived from materials that separated from the mantle in the Grand Canyon region about 1.8 to 2.0 billion years ago. These materials progressively formed highly concordant, then increasingly discordant, crystalline bodies as a part of an early Vishnu higher pressure-lower temperature metamorphic event followed by a lower pressure-higher temperature main stage event. This then was presumably followed by intrusions that were perhaps related to the transcontinental anorogenic granite belt. The last intrusive events may have been a series of tabular pegmatitic and aplitic intrusions.

At least 200 m.y. of erosion followed, creating a gently rolling surface of subaerial erosion. This surface was eventually covered by marine waters marking a transgression from the west (Dalton, 1972) as the Bass Limestone was deposited. Only pockets of basal arkosic conglomerate in the Bass of the eastern Grand Canyon mark the residuum of the underlying Early Proterozoic rocks that form the basement complex for the entire Grand Canyon.

CHAPTER 9: MIDDLE AND LATE PROTEROZOIC GRAND CANYON SUPERGROUP, ARIZONA

Donald P. Elston,
U.S. Geological Survey, Flagstaff, Arizona

GENERAL FEATURES

The Grand Canyon Supergroup consists of the Unkar Group (~2 km thick) and Nankoweap Formation (~100 m) of Middle Proterozoic age, and the Chuar Group (~1.6 km) and Sixtymile Formation (61 m) of Late Proterozoic age (figs. 9.1 and 9.2; table 9.1). These strata accumulated, with one major and a few minor breaks in sedimentation, in the interval ~1250 m.y. to about 825 m.y. or less. The Grand Canyon Supergroup is known only from exposures in the eastern and central parts of the Grand Canyon. The most complete section is preserved in the eastern Grand Canyon where the strata contain minor mafic intrusions and appear nearly as unmodified by burial metamorphism as the overlying Paleozoic strata. Strata of the lower half of the Unkar Group are preserved in the central Grand Canyon where they commonly display the effects of thermal metamorphism caused by the intrusion of thick mafic sills. A general northeasterly tilting of about 10° allows the Proterozoic sequence to be readily distinguished from the nearly horizontal Paleozoic strata that form the continuous walls of the Grand Canyon.

Strata of the Unkar Group, Nankoweap Formation, and Sixtymile Formation are dominantly red beds. Igneous rocks, which are restricted to the Unkar Group, consist of mafic sills that were intruded at the time of deposition of the upper middle member of the Dox Sandstone (fig. 12.1), and a mafic volcanic unit (Cardenas Basalt) that marks the top of the group. These igneous units are described in detail in Chapter 10. The Unkar Group accumulated very near sea level; four cyclic episodes of deposition in shallow water marine and then near-shore subaerial environments are recognized. Above this, the lower member of the Nankoweap Formation appears to have accumulated subaerially. Following a long episode of erosion, the upper member of the Nankoweap appears to have accumulated mainly in a shallow marine environment. Strata of the Chuar Group consist dominantly of gray shale and subordinate carbonate and red beds. Accumulation of most of the Chuar may have taken place in a lacustrine rather than a marine environment (Reynolds and Elston, 1986), as has been commonly supposed. The Sixtymile Formation at the top of the Grand Canyon Supergroup accumulated in a fluvial and locally a lacustrine environment during and subsequent to an episode of uplift and faulting that brought deposition of the Chuar Group to an end.

The Paleozoic section, about 1,650 m thick, accumulated during a 300-m.y. interval from about 570 to 270 m.y. In contrast, the Grand Canyon Supergroup, ranging from about 3,500 to 4,000 meters in thickness, accumulated during an approximately 425 m.y. interval of time from ~1250 to ~<825 m.y. From paleomagnetic correlations (fig. 12.1), the Grand Canyon Supergroup appears to be one of the most complete and long ranging records of Middle and early Late Proterozoic age preserved on the North American continent.

STRATIGRAPHIC AND STRUCTURAL FRAMEWORK

Strata of the Grand Canyon Supergroup are cut by widely spaced, throughgoing north- and northeast-trending faults, and by a relatively large number of structurally subordinate northwest-trending faults. Several unconformities are present that mark times when structural adjustments occurred on the faults, presumably in response to regional if not continentwide tectonic events. Unconformities are found at the top of the Hakatai Shale, at the top of the Unkar Group (Cardenas Basalt), and within and at the top of the Nankoweap Formation. Multiple unconformities are present in the lower and middle members of the Sixtymile Formation reflecting recurring increments of movement on the adjacent Butte fault. Appreciable regional tilting occurred as a consequence of this youngest of the Proterozoic structural adjustments, called the Grand Canyon-Mackenzie Mountains disturbance (Elston and McKee, 1982). This disturbance was followed by the accumulation of an unknown thickness of continental deposits of the upper member of the Sixtymile Formation, most of which were removed by erosion before deposition of Paleozoic strata.

NOMENCLATURE

The reader is referred to Spamer (1983) for an extensive annotated bibliography of Grand Canyon geology and paleontology. Spamer has summarized the history of development of geologic knowledge of the canyon, and also has provided a summary of the stratigraphic nomenclature. Space is devoted here to the stratigraphic nomenclature because inconsistencies have arisen during the past decade and a half, and some unnecessary revisions to the well-established formation names for the Unkar Group have been proposed. The descriptive names established by Noble (1914) for formations of the Unkar Group (limestone, shale sandstone, quartzite) have been retained in this report because they clearly and quickly convey to the reader the

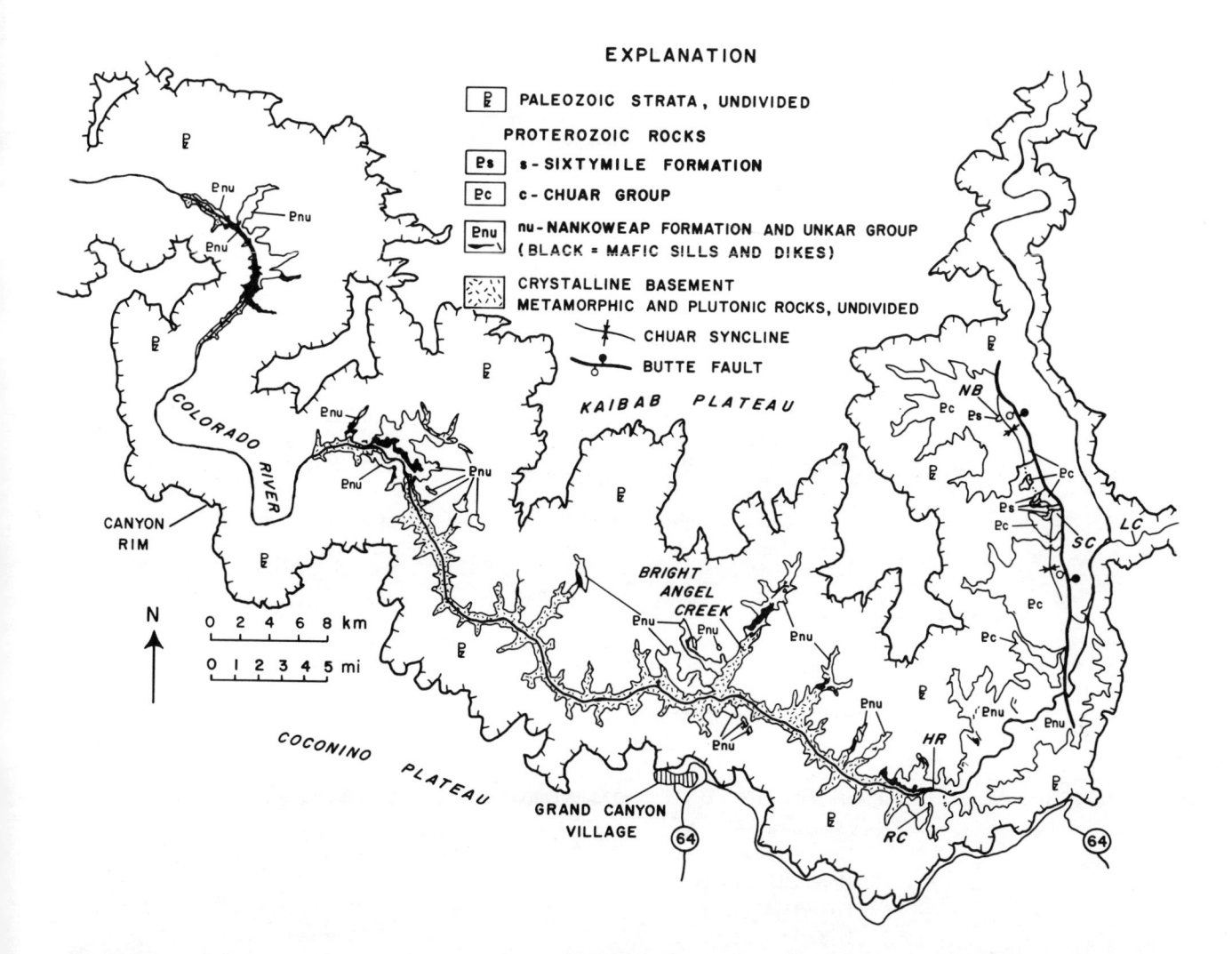

FIGURE 9.1. Generalized geologic map of eastern and central Grand Canyon showing distribution of units of the Middle and Late Proterozoic Grand Canyon Supergroup overlying the Early Proterozoic crystalline basement. The north-south-trending eastern margin of the Chuar Group in the eastern Grand Canyon marks the trace of the Butte fault (bar with ball indicating downthrown side; open ball, Proterozoic displacement; closed ball, Phanerozoic displacement). Adapted from Huntoon and others, 1976 (ed. 1986). (LC, Little Colorado River; NB, Nankoweap Butte; SC, Sixtymile Canyon; HR, Hance Rapids; RC, Red Canyon).

character of the dominant lithology of the formation. In addition, informal position names (lower, middle, upper) are used here for members of formations of the Unkar Group, Nankoweap Formation, and Sixtymile Formation (fig. 9.2; table 9.1) because they serve to readily identify the stratigraphic positions of the members to workers unfamiliar with the geographic distribution of features and their geologic relationships to the members.

The earliest geologic studies in the Grand Canyon were by Powell (1876), whose first expedition was in 1869. In 1882-83, Charles D. Walcott conducted pioneer work in the eastern Grand Canyon. His studies identified all of the units that were formally subdivided by later workers and that are now included in the Grand Canyon Supergroup. For a fascinating account of Walcott's work, see Yochelson (1979). Walcott named the Chuar (1883) and the Unkar (1894). He gave the Chuar group status, but he referred to both of these units in his 1890, 1894, and 1895 reports as terranes. Walcott's 1894 report contains a geologic map of the Proterozoic section in the eastern Grand Canyon that remains fundamentally unchanged to this day. In his 1894 report, Walcott redesignated the Grand Canyon Group of Powell (1876) as the Grand Canyon Series, consisting of two terranes of stratified rocks (the Unkar and Chuar) overlying a crystalline basement that he called the Vishnu terrane.

Strata of the Unkar terrane were subdivided into formations by Levi Noble (1914), who carried out detailed mapping and stratigraphic studies in and near the Shinumo Quadrangle in the central Grand Canyon. Noble used the terms "Unkar Group" and "Chuar Group" at this time, but the Chuar had not yet been subdivided into formations. However, the section of the Unkar Group studied by Noble is incomplete because in the area of the central Grand Canyon pre-Paleozoic erosion had removed strata above the level of the middle

TABLE 9.1. Summary of Middle and Late Proterozoic Grand Canyon Supergroup, northern Arizona[1], showing inferred depositional environments.

	Thickness (meters)		Depositional environment
Cambrian			
Tonto Group	180-395		
	"Great unconformity"		
Late Proterozoic			
Grand Canyon Supergroup	3,585 (+)		
Sixtymile Formation	59-64 (+)		
Upper member	12		Fluvial
	Unconformity		
Middle member	25		Lacustrine?
	Unconformity		
Lower member	22-27		Subaerial
Chuar Group	1,676[2]		
Kwagunt Formation	632		
Walcott Member	281		Lacustrine
Awatubi Member	301		Tidal flat?/lacustrine
Carbon Butte Member	50		Subaerial
Galeros Formation	1,044		
Duppa Member	104		Lacustrine
Carbon Canyon Member	350		Lacustrine/subaerial
Jupiter Member	434[3]		Lacustrine
Tanner Member	156		Marine/lacustrine
	Unconformity		
Middle Proterozoic			
Nankoweap Formation	113-250 (+?)		
Upper member	100 (+)		Marine, subaerial?
	Major unconformity		
Lower (ferruginous) member	13 (+)		Subaerial
	Unconformity		
Unkar Group	1,775		
Cardenas Basalt	224-450 (300[4])		Tidal flat?/subaerial
Dox Sandstone	920		
Upper member	93	Cycle 4	Subaerial?
Upper middle member	167		Tidal flat
Lower middle member	270	Cycle 3	Subaerial, fluvial
Lower member	390		Marine
Shinumo Quartzite	350		
Upper member	118		Tidal flat
Upper middle member	80	Cycle 2	Near shore
Lower middle member	130		marine and subaerial
Lower member	0.15-187.5 (18[4])		Subaerial, tidal flat?
	Unconformity		
Hakatai Shale	125		
Upper member	40		Marine (deltaic)
Middle member	22	Cycle 1	Subaerial
Lower member	62		Marine
Bass Limestone	80		Marine
Hotauta Conglomerate Mbr.	0-2		Marine
	"Greatest unconformity"		
Early Proterozoic			
Vishnu Schist			

[1] Tonto Group from McKee and Resser (1945); Grand Canyon Supergroup from Elston And Scott (1976), Elston (1979), and this report; Sixtymile Formation from Elston (1979) and Elston and McKee (1982); Chuar Group from Ford and Breed (1972a, 1973); Nankoweap Formation and Unkar Group from Elston and Scott (1976); Unkar Group from Beus and others (1974), and unpublished sections measured by D. P. Elston.

[2] Thicknesses for Chuar Group modified from Reynolds and Elston (1986; M. Reynolds, written communication, 1988).

[3] Thickness in part calculated from structure section.

[4] Nominal thickness.

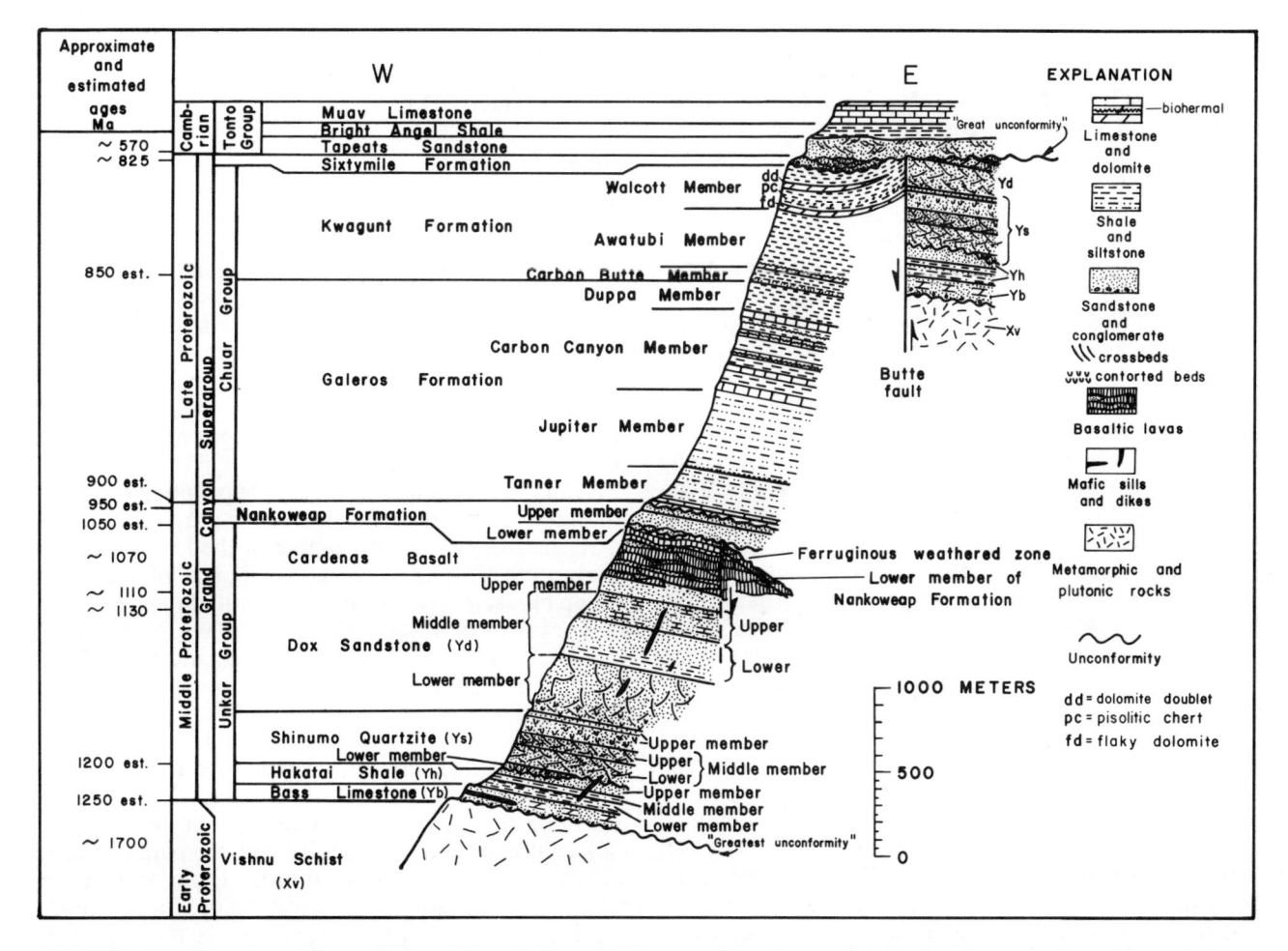

FIGURE 9.2. Stratigraphic outline of Grand Canyon Supergroup showing structural and stratigraphic relations between Proterozoic and basal Cambrian strata, eastern Grand Canyon, Arizona. Stratigraphic offset on Butte fault is portrayed as it existed during early Paleozoic time. Approximate and estimated ages are in part from radiometric ages derived from igneous rocks of the Grand Canyon, and in part from paleomagnetic correlations with other Middle Proterozoic successions dated by U-Pb (zircon) and Rb-Sr isochron methods on igneous rocks (fig. 12.1).

part of the Dox Sandstone. The missing strata, preserved in the eastern Grand Canyon, had been included by Walcott in his Unkar terrane, as were overlying red bed strata later called Nankoweap by Van Gundy (1934, 1951). It was not until recently that Ford and others (1972) formally named the basaltic flows in the eastern Grand Canyon the Cardenas Lavas, employing a name that had been used without documentation by Keyes (1938); (see discussion of Keyes' publications in Spamer, 1983). Because lava is hot fluid and the term basalt applies to solidified rock, the Cardenas Lavas have been formally redesignated the Cardenas Basalt (Elston, 1988a).

The stratigraphy of the Unkar Group has been summarized by Beus and others (1974), whose stratigraphic framework is virtually identical with the framework shown in Figure 9.2 and Table 9.1. However, at about this time and since then, a number of reports have appeared in which modifications to Noble's nomenclature have been proposed. Some of the proposed modifications are given in Spamer (1983), implying that their use has been generally accepted, whereas other proposed modifications are not shown by Spamer. The Bass Limestone (Noble, 1914) was called the Bass Formation by Dalton (1972), but later, Dalton and Rawson (1974) called it the Bass Limestone. Somehow, the name Bass Formation was attributed to Noble by Spamer (1983, p. 67). The original name, Bass Limestone, is favored here because the Bass is the only formation in the Unkar Group that contains appreciable beds of carbonate, even though such beds are not dominant in the easternmost exposures. Daneker (1974) referred to the Shinumo as the Shinumo Quartzite (of Noble, 1914), but he later called this formation the Shinumo Sandstone (Daneker, 1975). Spamer has retained the term "Quartzite" in his compilation. Much of the Shinumo is a sedimentary quartzite, being firmly to extremely well cemented by silica. No change in nomenclature has been proposed for the Hakatai Shale.

The Dox Sandstone (Noble, 1914) has been redesignated the Dox Formation (Stevenson, 1973; Stevenson and Beus, 1982), and formal names derived from local topographic features and drainages were proposed for its four members (in ascending order, the

Escalante Creek, Solomon Temple, Comanche Point, and Ochoa Point Members). These names are listed by Spamer (1983, p. 67), but are not used in this report for reasons given in the initial paragraph of this section.

Van Gundy (1934, 1951) recognized that a thin sequence of red beds unconformably overlying basaltic flows of the Unkar Group, and unconformably overlain by strata of the Chuar Group, belonged to neither the Unkar nor the Chuar. Although called the Nankoweap Group by Van Gundy, the unit later was reduced to formational rank by Maxson (1967) on his geologic map because the Nankoweap had not been subdivided into formations. Elston and Scott (1976) corroborated the conclusions of Van Gundy, and also recognized a locally preserved ferruginous unit (lower member and ferruginous weathered zone) unconformably overlying the Cardenas Basalt, and in turn unconformably overlain by the main (upper) member of the Nankoweap. A major hiatus separates the lower and upper members. The Grand Canyon Series also was redesignated as the Grand Canyon Supergroup at this time (Elston and Scott, 1976), following current practice that allows a supergroup to be established if it consists of some combination of groups and formations totaling at least three.

Stratigraphic studies and mapping by Ford and Breed (1972a, 1972b, 1973) led to the establishment of three formations in the Chuar Group. The lower two formations, the Galeros and Kwagunt, were subdivided into several formal members (fig. 9.2 and table 9.1), which were adopted in Elston (1979). However, the Sixtymile Formation at the top of the Proterozoic section was removed from the Chuar Group and established as a separate unit for reasons given in Elston (1979) and Elston and McKee (1982), resulting in the present fourfold subdivision for the Grand Canyon Supergroup.

Great and Greatest Angular Unconformities

Two major unconformities bound the Grand Canyon Supergroup. The lower separates stratified Proterozoic rocks from the underlying crystalline basement, and the upper separates Proterozoic strata from overlying Paleozoic strata (fig. 9.2). The lower unconformity was called the "greatest angular unconformity" by Noble (1914, p. 31), whereas both Walcott (1894, p. 506, and 1895, p. 317) and Noble (1914, p. 31, and Plate IX) referred to the upper unconformity as the "great unconformity." Where Grand Canyon Supergroup strata have been removed by pre-Paleozoic erosion, Paleozoic strata directly overlie the crystalline basement and the two unconformities merge into a single, truly great unconformity that marks an ~1,200 m.y. loss of record. We now know that the upper unconformity marks an ~230 m.y. hiatus (between ~800 m.y. and 570 m.y.), whereas the lower unconformity marks an ~425 m.y. loss of record (between ~1,675 m.y. and ~1,250 m.y.). With respect to the latter, the age of the crystalline basement in the inner gorge of the Grand Canyon has been reported by Pasteels and Silver (1965), and the age of the Bass Limestone at the base of the Grand Canyon Supergroup is estimated on the basis of paleomagnetic pole position (fig. 12.1).

STRATIGRAPHY

Unkar Group

General characteristics. Strata of the Unkar Group consist dominantly of fine to coarse grained red beds that accumulated in marine, nearshore marine, tidal flat, and near-shore continental environments. Conspicuous carbonate beds that locally contain stromatolites are found in the Bass Limestone; clastic deposits also predominate at places. Rare stromatolite-bearing carbonate beds, a few centimeters to a few decimeters thick, also are found locally in the upper middle member of the Dox Sandstone.

With the exception of the gray capping quartzite of the Shinumo Quartzite, and overlying gray strata of the lower half of the lower member of the Dox Sandstone, marine clastic and carbonate deposits of the Unkar Group characteristically are dark to light purple in color. In contrast, beds considered to have accumulated subaerially are red-brown to brick red and red orange in color. The tidal flat, upper middle member of the Dox Sandstone is variably red brown and purple in color; the red brown hematite is associated with dessication features whereas the purple hematite locally can be associated with stromatolitic horizons. Disseminated, reduction spots (non-red "freckles") that locally exhibit dark central nucleii characterize both the purple (marine) and red brown (subaerial) red beds. From detailed field relations (for example, diversely colored mud chips in dessication cracks in the upper middle member of the Dox Sandstone), the different hematite pigmentations can be concluded to be primary and to have formed during the course of accumulation of the sediments. The color of the hematite thus has aided in the identification and discrimination of four depositional cycles in the Unkar Group that record accumulation in alternating shallow water marine and subaerial near-shore environments. The proposed cycles are outlined on Table 9.1.

Bass Limestone. The Bass consists of interbedded sandstone and silty sandstone, prominent interbeds of conglomerate and dolomite (some stromatolite-bearing), and subordinate interbeds of argillite. The formation becomes generally finer grained toward the top. Carbonate strata are prominent at places, particularly in the central Grand Canyon. Clastic rocks predominate in the east, in the area of Hance Rapids. The upper boundary is arbitrarily drawn at the top of the highest laterally continuous sandstone, and the stratigraphically highest carbonate beds commonly lie some distance below this contact. A discontinous basal conglomerate, called the Hotauta Conglomerate by Noble (1914), lithologically similar to lenses of conglomeratic sandstone in the overlying Bass, is now treated as a member of the Bass (Dalton, 1972; Beus and others, 1974).

Hakatai Shale. The Hakatai Shale is a fine-grained, slope-forming formation that is readily subdivided into three members. The lower member consists of purple to reddish-purple mudstone, interbedded sandy siltstone, and rare thin beds of sandstone. These strata accumulated under quiet water conditions, apparently during the waning stages of marine deposition that marked accumulation of the Bass. Beds of the lower member of the Hakatai grade upward into fine-grained strata of the middle member, with a boundary drawn at or very near a place of change in color.

The middle member of the Hakatai Shale is the most distinctively colored red-bed unit in the Grand Canyon. It consists of mudstone, siltstone, and subordinate sandy siltstone that exhibit a striking reddish-orange color. Small to large (up to 10 cm and more across), non-red, spherical to spheroidal reduction spots (or "freckles") are common. Many contain dark-gray to greenish-gray central nucleii, which in turn contain very dark gray to black central cores of possibly organic (stromatolitic?) origin. The greenish gray central nucleii appear identical to reduction spots observed in the lower middle and upper members of the Dox Sandstone, in strata that are considered to have accumulated in subaerial environments on the basis of their sedimentary structures. The middle member of the Hakatai accumulated either in a shallow, near-shore marine environment (Reed, 1974), or alternatively, in mudflats adjacent to the sea (preferred here, on the basis of the color of the hematite pigment).

Widely divergent paleomagnetic directions and poles have been obtained from strata of the lower and upper members of the Hakatai Shale (fig. 12.1), but no well defined paleomagnetic direction has been obtained from the intervening red-orange middle member. In view of this, the distinctive red-orange pigmentation is inferred to have been developed as a consequence of subaerial exposure of the thin middle member over a prolonged interval of time. From demagnetization analysis, the alteration presumably led to the degradation of detrital specularite (black hematite) and an enhancement of the pervasive red-orange pigmentation. The large difference between paleomagnetic poles for the lower and upper members of the Hakatai Shale (fig. 12.1) suggest that accumulation and alteration of the middle member occurred over a time span at least several million years, perhaps even more than 10 m.y. in length.

The upper member of the Hakatai Shale consists of pale purple or lavender, fine- to coarse-grained, cross-bedded sandstone of probable marine deltaic origin (Reed, 1974). The unit forms ledges and locally sheer cliffs where protected by strata of the overlying Shinumo Quartzite. From a detailed structural study along the Bright Angel monocline, Sears (1973) has shown that deposition of the upper member of the Hakatai Shale in the eastern Grand Canyon ended as a consequence of minor uplift accompanying high-angle reverse faulting, southeast side up, on the northeast-trending Bright Angel fault. Sears (1973) reported similar movements on other faults in the eastern Grand Canyon area arising from a general northwest-directed compression.

Shinumo Quartzite. Four members are recognized here, one less than reported by Daneker (1975). They are: 1) a lower member consisting of purplish arkosic conglomeratic sandstone; 2) a lower middle member composed of purple cross-bedded quartzite (which Daneker, 1975, subdivided into two units); 3) a "rusty-red" upper middle member, which except for color is lithologically similar to quartzite of unit 2; and 4) an upper, red brown and locally purplish sandstone member that exhibits highly disrupted bedding, and which is capped by very well cemented gray quartzite. Much of the upper member beneath the gray quartzite cap has been subjected to bleaching and the removal of hematite pigment. The lower member is possibly of fluvial and tidal-flat origin, whereas the lower middle and upper middle members variably are of nearshore marine, tidal flat, and supratidal-flat origin (Daneker, 1975). The upper member was considered by Daneker (1975) to have accumulated as part of a mouth-bar system in the lower deltaic plain and delta-front facies of a tidal complex.

Along Bright Angel Creek, the lower (conglomeratic sandstone) member of the Shinumo Quartzite accumulated on the northwestern, downthrown side of the northeast-trending Bright Angel fault, overlapping a faulted northwest-facing monoclinal fold (Sears, 1973). The lower member wedges out on the upthrown side of the fault to the east, and its near-feather edge is seen in exposures along the Colorado River farther to the southeast (chapter 1, mile 76.2). Multiple sets of highly contorted beds disrupted by fluid evulsion are present locally in the upper part of the lower member of the Shinumo Quartzite; in addition, such structures are abundant across all but the capping quartzite unit of the upper member of the Shinumo Quartzite.

Crisp, well-defined reduction spots ("freckling") characterize the undeformed cross-bedded lower middle (purple quartzite) member, suggesting that the member has remained unaltered by the passage of ground water or other solutions since the time of its accumulation and lithification. The lower middle member passes upward into "rusty-red" quartzite of the upper middle member through a transition interval several meters thick, and the strata in this interval variously are purple and red-brown. The "freckled" purple beds in this interval contain a paleomagnetic direction identical to underlying beds of purple quartzite that gave rise to the paleomagnetic pole for the Shinumo Quartzite (fig. 12.1). In contrast, the "non-freckled" rusty-red beds contain scattered paleomagnetic directions reflecting secondary directions of magnetization related either to present field overprinting or to other undefined secondary components of magnetization. The differences in color between the members do not appear to be related to differences in lithology or to differing environments of deposition. Some parts of the rusty-red member are extremely well cemented by silica whereas other parts are poorly cemented, suggesting that the movement of ground or connate waters was responsible for the irregular silicification as well as for an inferred secondary hematization. The rusty-red member thus is concluded to owe its color and variable cementation to an episode of alteration.

The upper member of the Shinumo is an

orthoquartzite that is characterized by abundant, complexly contorted, gnarly bedded, fluid evulsion structures found across all but the capping quartzite unit (fig. 1.22). The structures occur in sets that are separated and truncated by several widely spaced, originally horizontal or near-horizontal planar diastems. Near the top of some individual deformed sets of beds, observed locally in cross section in Seventyfive Mile Canyon (fig. 9.1; chapter 1, mile 75), vertical dewatering structures are seen to be truncated by the planar diastems. The field relations indicate that soft-sediment deformation of saturated sets of beds occurred as a consequence of a series of dewatering episodes which served to disrupt and deform an originally horizontal stratification within the sets of beds. The vertical dewatering structures truncated by apparently horizontal diastems suggest that deformation did not occur as a consequence of appreciable downslope movement arising from slurry-slumping, although such an origin has been attributed to the structures (Daneker, 1975; Stevenson and Beus, 1982). The dewatering structures, occurring in the lower and upper members of the Shinumo Quartzite, are present in all exposures in the eastern and central Grand Canyon. Because these structures are an ubiquitous regional feature of the stratigraphy, a series of earthshocks rather than gravity sliding down local depositional slopes would appear to provide a defensible explanation for their origin. In Bright Angel Creek, the Shinumo Quartzite accumulated over a fault and fold system whose activity marked the onset of deposition of the Shinumo; recurrent episodes of movement on this and other throughgoing faults in the region may well be recorded by the dewatering structures in the Shinumo Quartzite.

Enclaves of purple sandstone containing disseminated reduction spots are preserved near the base of the upper member, where they are enclosed by red-brown pigmented sandstone that lacks freckles. (The relations are most conveniently observed at the mouth of Seventyfive Mile Canyon.) The red-brown color is the dominant color of the lower one-third of the upper member, but relict red-brown pigmented sandstone also is observed at places across the middle and upper parts of the upper member. The upper member thus appears to once have been colored by red-brown pigmented hematite. Bleaching of the upper member and removal of this pigment perhaps is related to a Late Cretaceous or Cenozoic mobilization of connate or ground waters accompanying and following uplift of the Colorado Plateau (chapter 18).

The enclaves of purple "freckled" sandstone near the base of the upper member are considered to be relicts of an original, generally purple color, similar to purple freckled sandstone of the lower middle member underlying the rusty-red upper middle member. Hematite from the purple sandstone perhaps was mobilized and re-precipitated to form the secondary, red-brown pigment. Redistribution of hematite perhaps occurred near the end of deposition of the upper member, perhaps as part of an event that also provided the abundant silica that cements the capping quartzite unit. If the fluid evulsion structures were tectonically generated across the time of deposition of the upper member, a genetic relationship also might have existed between an interrelated episode of secondary hematization of the gnarly beds and silicification of the capping quartzite at the end of deposition of the Shinumo Quartzite.

Two discrete, but anomalous, anti-parallel paleomagnetic directions were obtained from the two hematite phases in the upper member. An anomalous reversed polarity direction was obtained from purple enclaves near the base, whereas the enclosing red-brown pigmented sandstone provided an anomalous normal polarity direction. Neither of the paleomagnetic poles calculated from these directions are shown on Figure 12.1.

The capping gray quartzite, commonly of the order of 10-15 m thick, shows no evidence of hematization. A sharp diastem separates the quartzite from much less well cemented gray sandstone (containing some greenish gray argillaceous beds) of the overlying lower member of the Dox. Although a distinct hiatus appears indicated, the disconformable(?) contact may represent only a relatively brief loss of record. The reason that the environment no longer seemed to be amenable for the oxidation of iron to produce hematite in the uppermost Shinumo Quartzite and lower member of the Dox Sandstone is not known, but the cause does not appear attributable to a change in lithology. Some widespread mechanism for reducing the amount of available oxygen would seem to be required.

Dox Sandstone. The Dox Sandstone, more than 900 m thick, is the thickest formation of the Unkar Group. Four distinctive members accumulated in alternating shallow-marine to tidal-flat and continental environments of deposition (table 9.1). The thicknesses of the four members become progressively less in an upward direction (approximately 387 m, 263 m, 161 m, and 93 m). The stratigraphy of the Dox has been described by Stevenson and Beus (1982), who also have discussed its sedimentology and environments of deposition. The lower member (Escalante Creek Member of Stevenson and Beus, 1982), a nearshore marine and supratidal(?) deposit, is gray in its lower half. Pale purple interbeds mark a gradual transition to dark purple, marine or supratidal beds of the upper 194 meters. Sandstone and silty sandstone dominate in the lower member, although some argillaceous beds are present.. Two intervals of contorted bedding, reflecting the stratigraphically highest fluid evulsion structures in the Unkar Group, are present within 30 m of the base of the lower member (Stevenson and Beus, 1982, fig. 2). These would appear to represent the last of the series of earthshocks that began with deposition of the Shinumo Quartzite.

The lower middle member (Solomon Temple Member of Stevenson and Beus, 1982) of the Dox Sandstone consists of interbedded red-brown mudstone, siltstone, silty sandstone, and sandstone. The lower one-third is comparatively fine grained, consisting of interbedded mudstone and sandstone. The upper two-thirds consists dominantly of fine- to medium-grained sandstone containing subordinate siltstone and rare interbeds of claystone. The slope-forming lower part, which accumulated in a mud flat probably close to the sea, contains channellike, ledge-forming interbeds of sandstone. Overlying strata form cliffs, ledges, and subordinate steep slopes, with channel-like festoon cross beds indicative of fluvial conditions. Reduction spots are common in the member. A dark nucleus and its enclosing bleached halo, analysed spectroscopically, was found to contain a two-fold enrichment in organic carbon with respect to enclosing red bed material (written communication, Richard L. Reynolds, 1979).

Deposition of the upper middle member (Comanche Point Member of Stevenson and Beus, 1982) of the Dox Sandstone marked a return to marginal-marine and tidal flat conditions. The deposits consist mainly of interbedded fine grained, slope-forming, argillaceous sandstone and sandy argillite, and subordinate claystone. Variegated colors abound. Intimately associated, the colors variously are purplish and red-brown, and appear to reflect accumulation under alternately very shallow-water marine and subaerial conditions. Salt casts are common, as are ripple marks and dessication cracks. In detail, red-brown chips derived from mudcrack horizons can be observed within mudcracks that cut underlying purple-colored beds of inferred tidal flat origin. A few, thin, widely spaced pale-green beds associated with beds of inferred tidal flat origin very locally contain thin beds of stromatolitic dolomite.

The upper member (Ochoa Point Member of Stevenson and Beus, 1982) of the Dox Sandstone may record sedimentation under upper tidal flat conditions. Bedding types, salt casts, and ripple marks suggest deposition in alternating sand and mudflats marginal to the sea.

Sills and dikes. Intrusive and extrusive igneous activity recorded in the Grand Canyon Supergroup is restricted to rocks of the Unkar Group (fig. 9.2; chapter 10). Although eruption of the Cardenas Basalt was the final event in accumulation of deposits of the Unkar Group, important intrusive igneous activity preceded eruption of the Cardenas.

From paleomagnetic directions and poles (fig. 12.1), intrusion of thick mafic sills in the lower part of the Unkar Group in the area of the central Grand Canyon (fig. 9.1; chapter 10) appears to have occurred during deposition of the lower part of the upper middle member of the Dox Sandstone. A dike and a thin sill that intrude the lower and lower middle members of the Dox in the eastern Grand Canyon (chapter 1, miles 65.4 and 71.1) also contain an "upper middle" Dox paleomagnetic direction. The paleomagnetic directions and poles from these sills and from one dike are distinct from the Cardenas paleomagnetic direction and pole. Additionally, a thin sill intruding the Bass Limestone immediately below Hance Rapids (chapter 1, mile 77) has a paleomagnetic direction that may correlate with deposition of the upper part of the lower middle member of the Dox Sandstone (pole not shown on figure 12.1), which would indicate that intrusion of this thin sill slightly pre-dated intrusion of the thick sills of the central part of the Grand Canyon.

Dikes exposed at the head of Hance Rapids and in nearby Red Canyon (fig. 9.1; chapter 1, mile 76.5) cannot be connected in the outcrop with the thin sill below the rapids. The dikes have a Cardenas paleomagnetic direction and pole (not shown on figure 12.1). The correspondence suggests that the Red Canyon dikes were emplaced at the time of eruption of the Cardenas Basalt, and that they could have been feeder dikes for the volcanic complex.

Cardenas Basalt. This formation, designated the Cardenas Lavas by Ford and others (1972) and now called the Cardenas Basalt (Elston, 1988a), is more than 300 m thick. Descriptions of the Cardenas Basalt have been given by Hendricks and Lucchitta (1974) and Lucchitta and Hendricks (1974, 1983), and also are given in Chapter 10.

Eruption of the Cardenas records an abrupt outpouring of potassium-rich mafic lavas (basalt and basaltic andesite) that brought deposition of the Dox Sandstone to an end. A single flow, a few meters thick and precursor to the main eruption, is found locally in the eastern Grand Canyon a few meters below the top of the Dox (Elston and Scott, 1973, 1976). This single flow is coherent, lacks pillowing, and is vesiculated at the top, indicating extrusion in a subaerial environment. The surfaces on which this flow, and the basal flow of the overlying main pile of lavas, were extruded were smooth and firm, indicating the substrates were not soft, unconsolidated, fluid-rich muddy sand. Hematitic alteration of the sediment resulting from baking at the contact was minimal, amounting to a few centimeters or less. Basalt flows above the basal flow in the main pile (which is similar to the isolated flow in the Dox) form steep slopes and ledges rather than sheer cliffs, and they are characterized by a bottle-green color (Walcott, 1894). The bottle-green, slope- and ledge-forming interval, about 100 m thick, consists of multiple flows, most of which are laterally traceable for only short distances. Thin discontinuous beds of sandstone separate a number of the flows. Above the bottle-green interval, the Cardenas consists of a cliff-forming series of flows, at least some of which (chapter 10) were erupted subaerially. Walcott (1894, p. 517) explained the environment of deposition as follows: "The wide distribution of thin layers of sandstone, shale, etc., of uniform thickness over considerable.areas, indicates a relatively smooth sea bed at the time of the spreading of the the first sheet of lava over it; and that the sea bed was shallow is shown by ripple marks and the filling of sun cracks. The occurrence of the latter just beneath the basal flow and the presence of beds of sandstone between the flows prove a gradual subsidence of the sea bed during the deposition of the lava. Portions of some of the lava sheets, especially the basal one, bear evidence of having been broken into fine, rounded fragments, apparently by the hot lava pouring into the sea."

The environment of deposition of the lower, bottle-green interval of the Cardenas has been the subject of controversy. Emplacement in a shallow marine environment has been inferred for the puzzling bottle-green unit because of reported spilitic alteration, secondary chlorite, epidote, talc, and the presence of small concretion-like structures originally thought to be pillow structures (Lucchitta and Hendricks, 1974, 1983). The best of these concretionary features can be observed very locally in the east fork of Basalt Canyon (chapter 1, mile 69.5), and they clearly indicate the presence of water at the time of emplacement. However, clear-cut pillows indicative of submarine emplacement of the flows are not present. Nonetheless, water must have played an important role during eruption, if not emplacement, of the bottle-green, slope-forming unit. The lack of lateral continuity of the individual flows and sandstone interbeds, the irregular but rather pervasive bottle-green alteration, and the lack of clear-cut pillow structures, perhaps can be best explained by appealing to an initial, vigorous series of eruptions from a vent area located in a nearby shallow marine environment (to the west in the area of Hance Rapids?). During this time, sea water presumably was incorporated with the products of eruption, and discontinuous flows and sandstone beds of the bottle-green interval accumulated at and near sea level adjacent to the vent area. With the construction of a volcanic pile and the exclusion of seawater from the vent area, the eruptions became subaerial, allowing the laterally continuous flows and sandstone beds noted by Walcott to accumulate.

Nankoweap Formation

The Nankoweap Formation (Van Gundy, 1951; Maxson, 1967) totals slightly more than 100 m in thickness, and now consists of two informal members that are separated and enclosed by unconformities (Elston and Scott, 1976). Greatly differing paleomagnetic directions (fig. 12.1) indicate that the members accumulated at greatly different times and are separated by a major unconformity. Key relations are diagrammed in Figure 9.2.

Lower (ferruginous) member. The lower member of the Nankoweap Formation is preserved at two places adjacent to the trace of the north-south-trending Butte fault: 1) in the eastern side of the Basalt graben at Tanner Canyon Rapids (mile 68.5, chapter 1) immediately north of the Colorado River, and 2) two kilometers due south and high above the Colorado River, near and at the southern limit of preservation of the Cardenas Basalt and Nankoweap Formation.

The lower member in the Basalt graben at Tanner Canyon Rapids (fig. 1.19) is a highly resistant, hematite-cemented sandstone at the top of the Cardenas Basalt. It has a maximum thickness of about 15 m and wedges out about 200 meters west of the Butte fault beneath the unconformity that underlies the upper member of the Nankoweap Formation. South of the Colorado River, the stratigraphic and structural relations are more complex, and they have been diagrammed by Elston and Scott (1976). The units here include a 10 m-thick ferruginous weathered zone developed across an erosionally truncated section of the upper two-thirds of the Cardenas Basalt, and an adjacent pocket of highly ferruginous sandstone that is unconformable on and that was derived from highly altered detritus of the upper part of the lower, bottle-green unit of the Cardenas. The ferruginous weathered zone, a zone of pervasive hematization, was developed on an east-facing erosional scarp. The small pocket of reworked detritus occupies a step in the erosion surface near the contact between the lower and upper units of the Cardenas

The pocket of sandstone represents the lower discernable limit of the zone of ferruginous alteration. The upper limit of the ferruginous weathered zone is faulted against unaltered basalt of the Cardenas and unaltered strata of the upper member of the Nankoweap Formation by a northwest-trending fault. The unaltered and altered Cardenas, and the Nankoweap, in turn are smoothly truncated by the great unconformity that underlies the Cambrian Tapeats Sandstone. The field relations thus indicate that development of the ferruginous weathered zone, and deposition of the ferruginous sandstone in the erosional pocket near the base of the weathered zone, predated deposition of the upper member of the Nankoweap Formation. Additionally, the relations indicate that the ferruginous alteration was not related to alteration that might have occurred between the end of deposition of the Grand Canyon Supergroup and deposition of Cambrian strata.

The ferruginous weathered zone on the Cardenas, and ferruginous sandstone assigned to the lower member of the Nankoweap Formation north and south of the Colorado River, were sampled for their paleomagnetic directions, which have proven to be similar. The paleomagnetic pole for the lower member of the Nankoweap Formation, shown in Figure 12.1, includes directions from the ferruginous weathered zone as well as the two outcrops of sandstone. The pole lies near the mean pole from flows and sandstone interbeds from the Cardenas Basalt, suggesting that relatively little time passed between the end of eruption of the Cardenas, and the episode of faulting, erosion, and weathering that led to formation of the ferruginous weathered zone and deposition of the lower member of the Nankoweap. The source for the abundant hematite in the lower member of the Nankoweap almost certainly was the weathering of the basaltic rocks of the Cardenas. Lucchitta and others (1983), however, have not accepted the foregoing scenario, and argue that the sandstone beds below the ferruginous weathered zone south of the Colorado River belong to the Cardenas rather than the Nankoweap. Even if this were the case, this would not negate an episode of ferruginous alteration following faulting on the Cardenas, and the accumulation of a ferruginous sandstone above the Cardenas in the Basalt graben.

The field relations described above lead to the following sequence of events. Faulting occurred on and near the Butte fault shortly following eruption of the Cardenas Basalt. South of the Colorado River, the basalt was tilted along the Butte fault system, and a (fault?) escarpment was developed across the upper two-thirds of the 330-m-thick section of flows; across this scarp, a 10-m-thick ferruginous weathered zone was developed as consequence of exposure to the atmosphere. Ferruginous beds, either of the Cardenas or of the lower member of the Nankoweap, are preserved in erosional bench on the Cardenas beneath the scarp. In the area of Tanner Canyon Rapids, near-horizontal iron-rich beds of the lower member of the Nankoweap accumulated unconformably on the Cardenas in the Butte graben adjacent to the Butte fault. South of the Colorado River, prior to regional planation, faulting on a northwest-trending fault subsidiary to the Butte fault served to juxtapose unaltered flows near the top of the Cardenas on the south against the ferruginous weathered zone, following which the uplifted segment of the ferruginous weathered zone was eroded from the upthrown block prior to deposition of the upper member of the Nankoweap Formation.

Although only remnants of the lower member of the Nankoweap remain in the area, a reconstruction (Elston and Scott, 1976) suggests that the lower member once may have been considerably thicker, perhaps of the order of a hundred meters or more. In response to an episode of uplift accompanied by faulting, the lower member and a once more extensive ferruginous weathered zone on the Cardenas were nearly everywhere removed during a long interval of erosion that resulted in planation. The resulting mature surface of low relief is best seen west of the Basalt graben where it underlies the upper member of the Nankoweap and truncates the Cardenas at a low angle. This erosion surface also can be examined in detail (but with great caution) in the Basalt graben above Tanner Canyon Rapids.

Upper member. Red-bed deposits of the upper member of the Nankoweap Formation accumulated to a thickness of more than 100 m following development of a regional surface of low relief marked locally by lag deposits of gravelly chert. Deposition appears to have occurred in sand and mud flats near the sea, with accumulation occurring in shallow water. The upper unit of the upper member of the Nankoweap is a cross-bedded sandstone of probable beach origin, and its upper part is bleached. This unit is sharply and unconformably overlain by coarse dolomite that marks the base of the Tanner Member of the Galeros Formation (Chuar Group). The sandstone and dolomite form a single continuous cliff in the Basalt graben at Tanner Canyon Rapids, and the contact between the Nankoweap and Chuar is in the cliff face. No stratigraphic evidence exists to suggest a large hiatus at this contact, and paleomagnetic directions and poles from the upper member of the Nankoweap and the Galeros Formation (Chuar Group) are closely similar (fig. 12.1), which also suggests no large hiatus at this contact. Temporal extents of the three hiatuses that bound and separate members of the Nankoweap Formation (fig. 9.2) have been estimated on the basis of paleomagnetic correlations.

Chuar Group

Accumulation of the Late Proterozoic Chuar Group marked a major change from the preceding red bed environments of deposition. Strata of the Chuar Group, described by Walcott (1894) and Ford and Breed (1972a, 1972b, and 1973), are generally gray, fine-grained, and laminated to thin-bedded. Characteristics of the strata are consistent with sedimentation in a shallow lacustrine environment (Reynolds and Elston, 1986; chapter 11) inland from a sea coast. This is an alternative to a predominantly marine environment of deposition that has been inferred in the past. Dark shale predominates, recording accumulation under general reducing conditions apparently produced and maintained by an abundant planktonic microbiota. Surprisingly complex organic compounds (Reynolds and others, 1988; chapter 11) could serve as a potential source of Precambrian oil.

Widely spaced beds of dolomite and a few beds marked by distinctive structures and textures are found in the shales. Chuar strata also locally contain stromatolites, which commonly have been thought to represent tidal-flat conditions of deposition. Intervals of red beds reflecting oxidation under shallow water or subaerial conditions also are present in the Chuar; they contain sedimentary structures of subaerial and intermittently flooded shallow-water environments of deposition. Diverse studies are continuing (Vidal, 1986; Horodyski, 1986; Feng and others, 1986; Elston, 1986).

The Chuar is subdivided, in ascending order, into the Galeros and the Kwagunt Formations (fig. 9.2, table 9.1). Because black shales appear much alike across the Chuar Group, subdivision of the Chuar has been based on the presence of distinctive beds of other lithology. The lower unit of the Tanner Member of the Galeros Formation is a coarsely crystalline dolomite. The overlying Jupiter Member is almost entirely dark shale. Above this, the Carbon Canyon Member contains a number of tuffaceous beds (which have yielded no zircons for dating), and in its upper part, a number of red-bed intervals. The Duppa Member at the top of the Galeros Formation is dominantly dark greenish gray shale, with subordinate grayish red shale.

The base of the Kwagunt Formation is marked by cliff-forming red sandstone at the base of the lower Carbon Butte Member. A ledge-forming biohermal (stromatolite) reef marks the base of the overlying Awatubi Member, which otherwise is dominantly black shale. The upper Walcott Member, also dominantly black shale, contains several distinctive beds. A bed of flaky dolomite is found at the base, and about 17 m above this is a thin, distinctive bed of black, pisolitic chert. Vase-shaped microfossils, suggestive of heterotrophic protists (Gonzalo Vidal, 1986, written communication), have been obtained from shale underlying this pisolite bed, and from the pisolite bed itself, from which diverse microfossils also have been reported (Bloeser and others, 1977). The highest marker beds in the Walcott Member are a pair of dolomite beds that are in part disrupted. They lie about 70 m below the top of the highest shale of the member. This pair of dolomite beds thins and pinches out from the axis of the Chuar syncline in Sixtymile Canyon toward the Butte fault to the east, suggesting that minor movement on the fault occurred during deposition of the uppermost strata of the Chuar Group (Elston, 1979).

Paleontology. The relatively rich fossil record preserved in strata of the Chuar Group includes a variety of microfossils and stromatolites. Walcott (1899) was the first to study and report on the Precambrian biota of the Grand Canyon. The fossil record in the Chuar Group can be divided into two categories: 1) the actual organisms preserved as microfossils in cherts and rarely in carbonates; and 2) organosedimentary structures produced by communities of microorganisms (in particular, stromatolites). A brief review of the fossil record, ages, and inferred correlations, has been given by Elston and McKee (1982, p. 693). Bloeser and others (1977, fig. 1) summarized the stratigraphic distribution of microfossils and stromatolites across the Chuar Group. Included among the microfossils are vase-shaped microfossils, acritarchs (among them, *Chuaria circularis, Melanospherillia*), algal filaments, and unicells. The greatest abundance and diversity occurs in the Walcott Member of the Kwagunt Formation, the uppermost unit of the Chuar Group. The identification and distribution of stromatolites and *Chuaria* from the Chuar Group has been discussed by Ford and Breed (1972a, 1972b, and 1973). Stromatolites they identified included *Boxonia, Baicalia, Inzeria,* and *Stratifera*. The first-mentioned form was reported from the Awatubi Member of the Kwagunt Formation, and the last three from the underlying Galeros Formation. *Boxonia*, identified from the biohermal reef at the base of the Awatubi Member, has since been identified as *Baicalia* (Cloud, 1988, fig. 11.5, p. 262). Horodyski and Bloeser (1983) reported possible eukaryotic algal filaments from the Awatubi Member, the middle member of the Kwagunt.

The planktonic biota of the Chuar Group hold promise for the correlation of "late Riphean" and "Vendian" (late Middle and early Late Proterozoic, and Late Proterozoic, respectively) sections of Canada and the United States, Greenland, and Fennoscandia. Acritarch correlations, summarized later, indicate that the Chuar Group should be assigned to the Late Proterozoic.

Sixtymile Formation

The Sixtymile Formation, a unit identified by Walcott (1894), named by Breed and Ford 1973), and further described by Elston (1979), is a 61-m-thick red-bed unit that caps the Grand Canyon Supergroup. Its accumulation records uplift and faulting of the Grand Canyon disturbance that brought deposition of dominantly gray Chuar Group strata to an end. On Nankoweap Butte, one or two very thin beds of tuff (totaling <1 cm in thickness) mark the contact between black shale of the Kwagunt Formation and overlying red beds of the Sixtymile Formation.

The Sixtymile is subdivided into three members of approximately equal thickness, and the members are separated by unconformities. The lower member consists of red sandstone preserved only locally at the very base, overlain by breccia and landslide deposits that accumulated in a depression closely adjacent to the Butte fault. At the type section in Sixtymile Canyon, landslide blocks and detritus derived from the Chuar and Nankoweap were shed from an actively developing scarp into a developing Chuar syncline as a consequence of at least two increments of movement on the Butte fault (Walcott, 1890; Elston, 1979). Much of the structural offset across the Butte fault, which eventually exceeded 3 km, occurred during accumulation of the lower member (stratigraphic offset is diagrammed in Fig. 9.2). At Sixtymile Canyon, the middle member is a quartzite that appears to have accumulated in standing water (a lake?), following which the beds were folded and crenulated toward the axis of the syncline as a consequence of renewed deepening of the structure. Breccia again was shed into the syncline, accumulating above the unconformity that separates the middle and upper members. The upper member, above its basal breccia, is characterized by fluvial sandstone, and only vestiges of this member are preserved along the axial parts of the Chuar syncline on Nankoweap Butte and in Sixtymile Canyon. Sandstone of the upper member may have formed the basal part of a once much thicker section of basin fill that accumulated adjacent to the fault-block "Butte mountain" produced during the Grand Canyon disturbance.

AGE AND CORRELATION

Radiometric Geochronology

Results of Rb-Sr and K-Ar isotopic dating of igneous rocks of the Unkar Group have been reported by Elston and McKee (1982). These include virtually identical 1,070 Ma Rb-Sr whole-rock isochrons from a sill that intrudes strata of the lower Unkar Group and from flows of the Cardenas Basalt. Analytical errors of ±30 and ±70 Ma, respectively, were obtained from the sill and flows. Re-set whole-rock K-Ar dates also have been obtained from flows of the Cardenas Basalt, from several sills, and from the crystalline basement. These record varying degrees of Ar loss with deep burial. Additionally, a seven-step $^{40}Ar/^{39}Ar$ incremental heating diagram for a sill has given ages that are markedly younger than the Rb-Sr isochron ages, also indicating a loss of Ar with burial to depths of >2-~4 km. Ar retention in the flows, intrusions, and basement appears to have begun as a consequence of cooling that attended faulting, uplift, and erosion accompanying the Grand Canyon disturbance. Reset K-Ar ages from five Cardenas flows have provided an apparent age range of 855 to 790 Ma for the time of cooling, following burial by ~2 km of strata; a mean age of about 823 Ma has been calculated from the five determinations. It is possible that the minimum age (790 Ma), which represents the maximum loss of Ar with burial from cryptocrystalline matrix of the basalt, might best represent the time of onset of the cooling accompanying regional uplift, faulting, and erosion related to the Grand Canyon disturbance.

Paleontologic Correlations

The evidence for Proterozoic planktonic life has been reviewed by Vidal and Knoll (983), who argued that Proterozoic plankters, as Phanerozoic microplankton, have environmental and stratigraphic distributions that are both delimitable and useful. Of equal importance, planktonic microfossils in Proterozoic rocks document evolutionary events, the record for which is damped or not observable in the restricted carbonate facies where silicification of stromatolitic microbiotas was most common. Acid-resistant, organic-walled microfossils occur in the Proterozoic siltstones and shales. Most of the fossils obtained from the Proterozoic rocks are morphologically and ecologically comparable to Paleozoic microfossils included in the group Acritarcha (Evitt, 1963), a nomenclaturally informal category established to serve as an "umbrella" for organic-walled microfossils of problematical biological affinities. Gross morphological features, and their problematical nature, allow the spheroidal remains from Proterozoic clastic (and, in some cases, carbonate) facies to be placed among the acritarchs, but Vidal and Knoll (1983) note that no clear phylogenetic lines of connection have been established between Proterozoic and Paleozoic acritarch taxa.

Vidal and Knoll (1983) also have reviewed the nature and distribution of Proterozoic acritarchs, including *Chuaria* and vase-shaped chitinizoanlike microfossils. They noted that distinctive acritarch assemblages are found at several places in the northern hemisphere, and that those assemblages appear similar to the assemblage obtained from parts of the Chuar Group. Microbiota from the Chuar and the Uinta Mountain Groups (Vidal and Ford, 1985; Vidal, 1986) also have been reported to be demonstrably identical to form-taxa previously reported from Late Proterozoic (upper Riphean and lower Vendian) rock sequences in the North Atlantic region (Sweden, Norway, Greenland, Svalbard), and the Russian Platform and southern Urals of the U.S.S.R. The correlations indicate that deposition predated the Varangerian glacial event in the North Atlantic and that glacigenic units in North America thus may be of Vendian age. From paleontologic (and also paleomagnetic - fig. 12.1) correlations in western North America, Chuar strata predate glacigenic deposits of the Windermere Supergroup in Canada, and the Chuar thus appears assignable to the late Riphean and Late Proterozoic.

Boundary between Middle and Late Proterozoic Eras

The boundary between strata of Middle and Late Proterozoic age in the Grand Canyon Supergroup is here drawn at the unconformity that separates the upper member of the Nankoweap Formation from the Chuar Group (fig. 9.2; table 9.1). This assignment serves to separate red-bed strata from strata that record a time of flourishing of plankters and a general, though not complete, suppression of red beds. No structural disturbance is recognized at this horizon. The boundary previously had been drawn to correspond with a structural disturbance (the Grand Canyon-Mackenzie Mountains disturbance) at the top of the Chuar Group, to which a nominal 800 m.y. age had been assigned (Elston, 1979). This structural disturbance preceded onset of glacial conditions recorded in the Late Proterozoic Windermere Supergroup of Canada.

The boundary between the Middle and Late Proterozoic Eras has recently has been assigned a general age of 900 m.y.. (Harrison and Peterman, 1982). It is a geochronometric boundary chosen to escape the effects of orogenic activity. The boundary between the Nankoweap Formation and Chuar Group meets this general requirement, and on the basis of pole postions and correlations inferred therefrom (fig. 12.1), also seems to correspond reasonably well with the nominal 900 m.y. age for the geochronometric boundary.

CHAPTER 10: PETROLOGY AND CHEMISTRY OF IGNEOUS ROCKS OF MIDDLE PROTEROZOIC UNKAR GROUP, GRAND CANYON SUPERGROUP, NORTHERN ARIZONA

John D. Hendricks
U. S. Geological Survey, Flagstaff, Arizona

INTRODUCTION

The Unkar Group of the Grand Canyon Supergroup is comprised of five formations and unnamed diabasic rocks. In ascending order, the named units are the Bass Limestone, Hakatai Shale, Shinumo Quartzite, Dox Sandstone, and Cardenas Basalt. Diabasic intrusive rocks occur as dikes and sills within all units below the Cardenas. Sills greater than 65 feet (20 m) thick intrude the Bass Limestone and Hakatai Shale, whereas dikes intrude all formations of the Unkar above the thick sills.

Sills range in thickness from about 65 feet (20 m) near Hance Rapids in the east (fig. 9.1) to more than 655 feet (200 m) near Dubendorff Rapids to the west (fig. 4.2). With the exception of the sill near Shinumo Creek, all of the sills consist of a main body of medium to coarse-grained olivine-rich diabase having fine-grained, porphyritic chilled margins less than 1 m in thickness. In-place differentiation of the sills is marked by lenses and dikes of syenite and felsite. The sill near Shinumo Creek is unique in that it displays distinct layers of differentiation and segregation products. Chemically the sills are of an alkali-olivine magma. CIPW normative calculations indicate the parent magma was quartz normative.

The flow sequence of the Cardenas Basalt at Basalt Canyon is nearly 985 feet (300 m) thick and can be separated into two distinct topographic units. The lowermost 305 feet (93 m) of this section is a highly altered series of slope-forming basalt flows and thin sandstone interbeds. The basalt of this unit may have undergone spilitic alteration (Lucchitta and Hendricks, 1972) occurring during the course of eruption. Relatively unaltered patches of the basalt, present in the uppermost part of this unit, have a mineralogy very similar to that of the underlying sills. Above the 305 feet (93 m) level the igneous rocks of the Cardenas are cliff-forming basalts and basaltic andesites. Petrographically they are fine-grained with an intergranular to intersertal texture. These rocks change from basaltic andesite to basalt to basaltic andesite upward through the section.

The interpretation favored here is that the lavas of the Cardenas were erupted into shallow, possibly hypersaline sea water, probably in a tidal flat or deltaic environment. Sediments of the Dox Sandstone immediately below the flows were solid but not lithified prior to eruption. The region at the time of eruption was a submerging coastline or possible basin. Subsidence generally kept pace with accumulation of the lavas as witnessed by the sandstone beds within the flow sequence. Features in the upper part of the lava pile, however, suggest a subaerial environment.

The genetic relation, if any, between the sills and dikes, and flows is not straightforward. Chemical variation diagrams indicate a potential common parentage but a direct physical connection cannot be observed in the field. Paleomagnetic observations (chapter 12) and isotopic age determinations (Elston and McKee, 1982; chapter 9) suggest that the sills may be slightly older than the flows (perhaps as much as 40-50 m.y.). If the sills and flows were derived from a common parent, the sills would represent an early phase of igneous activity followed by a length of time during which the magma underwent differentiation at crustal levels before eruption of the lavas. Isotopic (Rb-Sr) determinations yields an age of 1,070±70 Ma for the flows of the Cardenas (McKee and Noble, 1974), whereas one of the sills (Shinumo Creek) has a five point isochron of 1,070±30 Ma (Elston and Mckee, 1982). Although these ages are basically identical, the flows and sills have different $^{87}Sr/^{86}Sr$ ratios (0.70650±0015 and 0.70420±0007 respectively) and apparently distinctly different paleomagnetic pole positions.

INTRUSIVE ROCKS

Petrography

Sills intruding the Bass Limestone and Hakatai Shale crop out in seven general locations in the depths of the Grand Canyon. All of these bodies are composed of olivine diabase and show varying amounts of in-place differentiation. The interior of all of the sills is medium to coarse-grained, containing plagioclase, olivine, clinopyroxene, magnetite-ilmenite, and biotite, with accessory apatite and sphene. The texture is diabasic to subophitic, although a crude alignment of feldspar laths can be seen in many places. Plagioclase laths (An_{45-60}), averaging 1.5 mm in length, are partially to completely altered to sericite and show both normal and reversely zoned crystals within a given rock. The more calcic plagioclase is concentrated toward the center of the sills. Anhedral to subhedral olivine crystals up to 1 mm in diameter are partially altered along borders and fractures to chlorite, talc, magnetite, iddingsite, and serpentine. The fresh interiors of the grains give optic axial angles that indicate compositions of approximately Fa_{20}, and have interference colors that suggest normal zoning.

Plagioclase and olivine grains are enclosed by large, optically continuous, poikilitic pyroxenes, giving the rock its subophitic texture. These pyroxenes have z-C = 45°, $2_{v\partial}$ = 50° and are brownish pink, non-pleochroic, and very fresh. The pyroxene type does not appear to vary in optical properties between the diabasic rocks of the different sills. Large irregular grains of magnetite altered to hematite and biotite, along with primary pleochroic brown biotite, partially altered to chlorite, also occupy interstices between plagioclase and olivine grains.

Toward the center of the sills olivine tends to increase (in percentage) whereas clinopyroxene decreases. In other sills these relationships have been explained by a process of flow differentiation (Simkin, 1967; Drever and Johnston, 1967; Bhattacharji and Smith, 1964; Bhattacharji, 1967), which involves the movement of early formed particles (olivine) away from the margins of a sill or dike during flow of the magma. In this model, crystals of olivine are formed early in the cooling history of the magma (before emplacement of the sill). As the magma moves up and outward, intruding as a sill or dike, hydrodynamic forces concentrate existing crystals toward the center of the moving mass. The principle is known as axial migration, and particles move toward an equilibrium position (i.e., no velocity gradient, radial migration, or rotation) that exists in the center of the moving mass. As the intrusion moves laterally, gravity acts on the crystals to produce a gradational change in olivine content from the lower contact upward while causing an abrupt change in olivine from the upper contact downward. Following emplacement of the sill "crystallization of the surrounding liquid then yields the remaining minerals in relatively constant proportions" (Simkin, 1967).

The chilled margins of the sills in the Unkar are on the order of 1-1.5 feet (30 to 40 cm) in thickness, porphyritic, and show extensive deuteric alteration. These margins consist of phenocrysts of plagioclase and olivine set in a groundmass of plagioclase microlites (averaging An_{50}), opaques, chlorite, epidote, and hematite. Plagioclase phenocrysts (averaging about 30 modal percent) are up to 3 mm in length and have been completely altered to sericite. Olivine phenocrysts (about 5 modal percent) occur as anhedral to subhedral crystals up to 0.5 mm in length and are partially altered to chlorite, talc, and magnetite. The groundmass of the chilled margins show flow textures formed by alignment of microlites

Shinumo Creek Sill

The Shinumo Creek sill was chosen for detailed petrographic and chemical analysis because it represents a classic example of in-place differentiation of a basaltic magma. A section was measured about 0.3 mile (0.5 km) east of Shinumo Creek (chapter 9) where the sill is composed of 280 feet (85 m) of diabasic rock capped by 19 feet (6 m) of granophyre. At this location the sill is well exposed although the lower contact is covered. Textural variations from the normal diabase are of two types: 1) lumps or balls of ophitic intergrowths of augite and plagioclase, and 2) pegmatite veins consisting of plagioclase and augite (Noble, 1914). The mafic minerals of the Shinumo Creek sill vary in modal percentage, whereas plagioclase is uniform at about 60 percent. A transition zone, 2-3 feet (0.6 to 1.0 m) thick, exists between the granophyre and diabase.

The granophyre at the top of the Shinumo Creek sill consists predominantly of potassium feldspar (60%) and quartz (12-20%), with biotite, plagioclase, and magnetite making up the remaining 20-28 percent. The rock is holocrystaline, coarse-grained, and displays a well developed granophyric texture. Quartz, plagioclase, biotite, and magnetite fill interstices between orthoclase grains.

The transition between the granophyre and diabase in the Shinumo Creek sill occurs over a vertical distance of less than 3 feet (1 m) and is a zone rich in biotite and accessory minerals. Apatite makes up as much as 5-10 percent of the rock. Ilmenite and sphene are prominent, and zircon with reaction halos occurs within the biotite grains.

The main body of this sill is mineralogically and texturally similar to the other sills in the Unkar. The entire diabasic portion of the sill, however, contains textural "lumps or balls" mentioned above and described by Noble (1914). Pegmatite veins are confined to the upper part of the diabase. Noble (1914) believed that these features represent a segregation phenomenon. The mineralogy of these "lumps" is similar to that of the "normal" diabase (i.e., olivine and plagioclase with pyroxene filling interstices). Plagioclase laths in the lumps are up to 7.5 mm in length, and olivine occurs as large embayed crystals, with plagioclase filling embayments. The pegmatite has a very similar texture and mineralogy. One sample collected near the base of the Shinumo Creek sill contains nearly 50% olivine, and probably represents a layer in which the heavy olivine settled during cooling.

Contact Metamorphism

Where the sills have intruded the Bass Limestone, metasomatism and recrystallization has produced chrysotile asbestos both above and below the intrusion. The Bass in the Grand Canyon in most places is a siliceous dolomite with common sandstone interbeds. Hot fluids emanating from the intrusions penetrated the Bass along bedding planes and fractures, creating the asbestos. Asbestos, with fibers up to 10 cm in length, commonly occurs within 9 feet (3 m) of both the upper and lower contacts. Although the asbestos is of high quality, mining attempts in the early 1900's failed primarily due to logistical problems involved with removal of the ore.

Where sills intrude the Hakatai, the shale was altered into a knotted hornfels. In a series of samples collected near Shinumo Creek, below the sill, alteration of the Hakatai is clearly evident as much as 16 feet (5 m) below the contact. Within 5 cm of the contact porphyroblasts of biotite reach a maximum size of 0.25 mm. Ten centimeters away from the sill, the rock becomes a knotted hornfels containing porphyroblasts of andalusite and cordierite(?) that have been pseudomorphically replaced by muscovite and green chlorite, respectively.

A high nucleation rate here appears to be reflected in the size (<1 mm) and abundance of porphyroblasts. A sample collected 76 cm below the contact contains well-developed chlorite and muscovite pseudomorphs of cordierite and andalusite porphyroblasts. These porphyroblasts are larger (0.2 to .03 mm) than those nearer the contact and also are not as numerous, apparently reflecting a slower rate of nucleation. Recrystallization in the Hakatai is more noticeable adjacent to fractures along which hot fluids from the intrusion probably moved. This is recorded by a sample collected 13 feet (4 m) stratigraphically below the sill in which rounded porphyroblasts composed of feathery muscovite are concentrated along micro-fractures. These porphyroblasts are as much as 3 mm in diameter and impart a spotted appearance to the rock. The mineralogy suggests that contact metamorphism at the location of the suite of samples is of low to medium grade.

Major Element Chemistry

Major oxide chemistry has been determined for chilled margins of six of the sills in the Unkar and for two dikes that are stratigraphically above the sills. These analyses are presented in Table 10.1. In addition, a suite of eleven samples taken sequentially across the Shinumo sill were analyzed in order study the chemical variation within this differentiated intrusion (table 10.2). The minor chemical differences between the chilled margins of the different sills are believed to result mainly from deuteric alteration, whereas the larger variations between interior samples in the Shinumo Creek sill reflect in-place differentiation. Very low FeO/Fe_2O_3 ratios may represent post-magmatic alteration of iron-bearing minerals. Samples from the interior of the Shinumo Creek sill show higher Al_2O_3, CaO, and Na_2O values than the chilled margins. The higher Al_2O_3 and CaO values may result from concentration of the more calcic plagioclase in the center of the sill. The relatively low Na_2O values of the chilled margins may reflect leaching of sodium by deuteric processes, as shown by the highly altered nature of the plagioclase. Comparison of analyses of the two dike rocks with the other diabasic rocks shows that the dikes are chemically similar to the sills, although the sills tend to have higher magnesia values.

EXTRUSIVE ROCKS

Cardenas Basalt

All Middle Proterozoic igneous rocks of the Cardenas Basalt overlying the Dox Sandstone are extrusive. These volcanic rocks crop out in two locations in the eastern part of the Grand Canyon (Maxson, 1967; Huntoon and others, 1986). The main exposure is along the northeast rim of Unkar Valley from a point near Cape Royal eastward to the main canyon of the Colorado River, then northward to Chuar Lava Hill, a distance of approximately 6.5 miles (11 km). The Basalt Canyon section that was studied for this report is included in this exposure. Outcrops on the opposite (south) side of the Colorado River are stratigraphically equivalent. The other location of extrusive rocks is in Nankoweap Valley, 2.5 miles (4 km) above the mouth of Nankoweap Creek.

	SILLS						DIKES	
	U HN7	L CC13	U BA5	U CR12	L SH17	U TP4	Tid1	Tid 2
SiO_2	44.60	47.00	47.10	43.20	45.50	48.00	46.60	48.10
Al_2O_3	14.84	16.10	16.93	16.49	16.49	15.72	16.70	16.10
Fe_2O_3	2.44	3.31	2.78	2.78	5.48	4.04	13.60	10.90
FeO	8.15	7.46	6.62	5.74	6.03	7.60	0.50	2.88
MgO	6.94	6.12	8.70	8.89	7.33	6.16	2.80	5.10
CaO	13.64	8.78	7.74	4.87	4.87	8.48	7.50	5.90
Na_2O	1.87	2.72	2.89	2.83	3.28	3.85	2.80	2.70
K_2O	1.37	1.10	1.00	3.04	1.91	1.07	2.20	1.90
TiO_2	1.89	2.30	2.23	2.51	2.59	2.64	1.62	1.67
P_2O_5	0.48	0.33	0.35	0.39	0.40	0.41	0.51	0.51
MnO	0.55	0.13	0.13	0.14	0.14	0.21	0.09	0.09
$+H_2O$	2.85	2.42	2.67	4.17	3.71	1.05	1.57	1.99
$-H_2O$	0.38	0.53	0.73	0.69	0.97	0.53	0.47	0.56
CO_2	0.12	2.08	0.51	4.53	1.72	0.12	2.83	1.49
Total	100.12	100.38	100.38	100.27	100.42	99.88	99.80	99.90

U is for upper contact: L is for lower contact
HN7 - Hance Rapids
CC13 - Clear Creek
BA5 - Bright Angel
CR12 - Crystal Creek
SH17 - Shinumo Creek
TP4 - Tapeats Creek

TABLE 10.1. Chemistry of chilled margins of sills and dikes.

	\multicolumn{11}{c}{Distance Below Top of Sill (in meters)}
	0	3.7	5.8	6.1	7.9	13.7	21.3	36.6	51.8	67.1	83.8
SiO_2	66.10	65.70	57.20	48.10	47.50	45.90	47.60	47.20	46.80	47.00	47.70
Al_2O_3	14.90	15.60	15.60	13.70	14.00	18.30	17.20	17.90	19.20	18.90	18.10
Fe_2O_3	3.38	2.94	5.53	7.40	6.30	5.05	4.13	2.25	1.70	2.40	3.73
FeO	0.50	0.60	3.12	7.16	7.25	4.82	6.30	7.07	5.81	6.66	6.55
MgO	3.20	3.08	5.15	7.72	7.00	10.02	7.72	9.09	9.28	7.90	7.76
CaO	0.66	0.52	1.03	2.95	5.20	2.36	5.24	9.00	9.59	9.78	5.75
Na_2O	1.42	1.02	3.11	3.91	4.40	2.55	3.40	2.43	2.43	2.27	2.26
K_2O	7.66	7.89	5.38	0.87	1.30	2.54	2.14	1.20	1.20	1.00	1.81
TiO_2	0.65	0.61	0.98	3.40	3.04	1.88	2.24	1.34	1.03	1.48	1.84
P_2O_5	0.18	0.18	0.50	0.63	0.41	0.28	0.25	0.19	0.17	0.23	0.29
MnO	0.02	0.02	0.05	0.10	0.25	0.11	0.18	0.16	0.12	0.15	0.26
$+H_2O$	1.38	1.52	2.61	3.55	2.37	4.96	3.33	2.20	2.47	2.29	3.13
$-H_2O$	0.32	0.34	0.33	0.77	0.37	1.81	0.81	0.28	0.34	0.29	0.90
CO_2	0.02	0.02	0.09	0.10	0.08	0.06	0.06	0.09	0.02	0.10	0.13

TABLE 10.2. Chemistry of rocks from the Shinumo Creek sill.

Basalt Canyon Section. The lava flows observed at Basalt Canyon comprise one of the thickest and most complete sections of Middle Proterozoic extrusive rocks belonging to the Unkar in the Grand Canyon. At least 14 (and possibly many more) individual flows are present (fig. 10.1). Units in the columnar section do not necessarily represent individual flows but are units defined by distinctive field characteristics such as color, weathering, structure, textures, and mineralogy. Unquestionable flow boundaries are recognized by interlayered sandstone beds and irregular vesicular zones. All flow boundaries are not easily recognized.

Interbedded within the eruptive sequence are at least seven sandstone units that range in thickness from less than 1 foot (30 cm) to as much as 15 feet (4.5 m). Although relatively thin, most of these beds are remarkably continuous in the area studied. Thin sections show that the sandstones are texturally immature, with a poorly sorted coarse fraction of quartz and feldspar set in a matrix of mica and clay minerals. The coarser grains fall within the medium-sand to silt size. Locally derived fragments, up to one centimeter in diameter, of basalt are common in the lower parts of individual sandstone units. Sandstone beds higher in the sequence tend to contain more erosion-derived lava flow fragments. This may be a result of a higher energy environment during deposition, or the surfaces of the flows below the sandstone beds may have contained more loose material than did flows lower in the section. Induration by baking from overlying flows is evident in the sandstones and tends to be more pronounced in the higher beds. Original planar cross-bedding is preserved, and where observed has an amplitude of less than 1.5 feet (0.5 m).

Although propylitization is extensive, the lava beds exposed in Basalt Canyon display a variety of structures. A description of these structures and their relative stratigraphic position above the basal contact follows (fig. 10.1).

Nodular basalt occurs from 0 to 197 feet (60 m), is discontinuous to the 318-foot (97-m) level, and occurs again at the 600-foot (83-m) level. Whereas some of this nodular basalt is clearly a result of spheroidal weathering or exfoliation, there are occurrences where alteration products or exfoliation plates are absent and the individual nodules are in contact with one another. Walcott (1894) believed the nodules represented thermal shattering of the basalt by rapid quenching in water during eruption.

Fan-shaped sheet joints occur from 328-377 feet (100-115 m) above the base of the section. Individual sheets are from two to eight centimeters in thickness, are nearly vertical, and show no preferred strike. The lava in which the sheeting occurs is very dense and free of amygdules.

Ropy lava was observed at the 570-foot (174-m) level. This structure is preserved in a cavity, 5 cm below the base of a sandstone bed. The structure is identical to ropy structures found on the surfaces of present-day pahoehoe lava flows.

Lapillite, consisting of particles from 1 inch (2.5 cm) to more than 1 foot (30 cm) in diameter, set in a matrix of fine-grained ashy material, is preserved near the 755-foot (230-m) level. Volcanic bombs are conspicuous due to differential weathering and the associated bomb sags are still preserved. Vesicles in the scoria are filled with quartz, calcite, chlorite, epidote, and copper minerals.

Petrography. Eight samples were collected from the flow sequence for petrographic analysis. These rocks are petrographically similar to those described by Iddings (1894) from Walcott's Chuar Lava Hill collection. The major difference between the two collections seems to be in the degree of alteration; Walcott's show less alteration. Descriptions of the two collections are similar enough to allow correlation of units of the Basalt Canyon and Chuar Lava Hill sections. The lowermost flow or series of flows (0- 305 ft; 0-93 m) form a green rubbly slope that is quite prominent in the section and is a good marker horizon. Petrographically, this layer is a highly altered basalt. Originally, the rock was hyalocrystalline and contained randomly oriented laths of plagioclase, up

PETROGRAPHIC DESCRIPTION

Sandy siltstone. Laminated to flaggy, moderately to well sorted. Laminations are lenticular. Sorting between layers much poorer than sorting within individual layers. Overall composition: Quartz 75%; feldspar 17%; opaques 5%; muscovite 3%. Adjacent layers vary markedly in composition. Quartz is angular and concentrated in the fine sand and silt. Feldspar, primarily plagioclase, is angular, prismatic, and in the fine sand size. Muscovite is in platy grains as much as 1.15mm in size arranged parallel to lamination. Opaque minerals, primarily magnetite, are widely distributed but especially abundant at the boundaries of laminations. Hematite coats most grains, giving the rock its dark red color. Authigenic quartz fills fractures and large pores. Volcanic rock fragments are concentrated in plagioclase rich layers, as are minor ammounts of heavy minerals.

Basalt, composed of plagioclase, olivine, pyroxene (?) and magnetite, with large patches of altered glsss. Plagioclase (oligoclase) laths are up to 1mm in length. Mafic minerals are completely altered to talc, antigorite (?) and chlorite. Magnetite is present on boundaries of original mineral grains. Patches of devitrified glass are composed of clay, chlorite, epidote, and zeolites (?). Original mineral percentages are: plagioclase 60%; glass 20-25%; mafic minerals 15-20%; minor magnetite.

EXPLANATION

- BASALT OR ANDESITE
- SANDSTONE
- CROSS BEDDED SANDSTONE
- AUTOCLASTIC BRECCIA
- LAPILLITE WITH BOMBS
- VESICLES OR AMYGDULES
- SPHEROIDAL BODIES
- FAN JOINTING

METERS

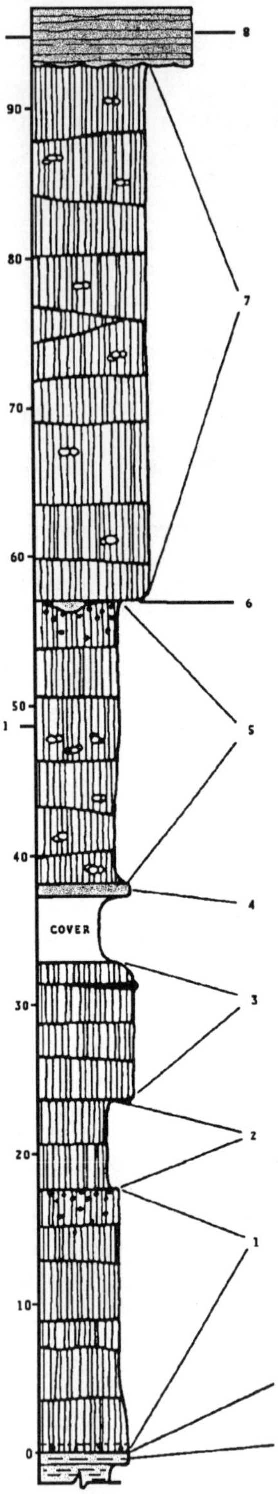

LITHOLOGIC DESCRIPTION

Sandstone, brown to maroon, medium to fine grained, laminated, jointed. Forms prominent sharp cliffs. Upper part moderately baked, hard. Lower contact is planar in gross aspect, but has local irregularities typically 30cm or less in relief, locally as much as 90cm. Some of the irregularities are wave-like in cross section and may represent fluting caused by currents

Basalt, similar to unit 5, but less greenish, more tan and more cliff forming. Common amygdule and fracture fillings of chlorite, epidote, and locally quartz. Abundant spheroidal bodies as much as 60cm in diameter. Most probably are the product of spheroidal weathering, some may represent termal shattering. Several flow units are present, defined by chilled lower contacts as well as vesicles and amygdules near top. Wedging of flows and angular unconformities common within unit. Upper 4.5m of unit is relatively resistant and purplish. Uppermost 7 to 10cm is red and probably represents a weathering zone. Unit forms cliffs with minor slopes

Sandstone, very fine grained, purplish. Discontinuous. Locally fills channels

Basalt, gray to gray-brown, fine-grained. Extensive chlorite-epidote alteration. Lower 45cm has abundant chlorite and epidote filled amygdules. Top of unit is very amygdaloidal and rich in chlorite and epidote, disseminated and in amygdule fillings. Spheroidal bodies, as much as 15cm, typically 8 to 13cm in diameter are common, especially near the base of unit and near the 47m level. Unit is more resistant and less greenish than units below, and forms a steeper slope

Sandstone, maroon to purple, medium to fine grained, platy. Lower contact covered. Contact with overlying basalt has several cm of relief

Basalt, dark greenish gray, fine-grained. Epidote common in groundmass, as blebs, and in shear zones. Thin lenses of fine-grained purplish sandstone near top of unit. More resistant than units below, forms subdued ledge. Weathers into coarser rubble than units below; some fragments are as much as 3cm in size. Composed of at least two flow units

Basalt similar to unit 1, but less resistant and more green. Forms conspicuously greenish slope, typically covered by talus

Basalt, dark greenish gray, locally purplish. Fine grained. Extensive chlorite-epidote alteration. Fractures common, typically filled with chlorite, epidote, and quartz. Quartz amygdules locally common. Basal 23cm is highly amygdaloidal, tan and crumbly-weathering; overlain by 10cm of tan-grey basalt. Vesicle-rich zone at top. Unit is massive, weathers into granular rubble, forms slopes. Probably consists of more than one flow unit

CONTACT. About 60m exposed along strike near line of section. Contact is smooth, with relief of 10cm or less. No weathered zone, sand or pebble lag at top of Dox Formation. Contact appears conformable

baked zone

DOX FORMATION Interbedded maroon, reddish and tan medium to very fine grained sandstone, siltstone, and shale. Baked zone at top

FIGURE 10.1. Measured Section of Cardenas Basalt at Basalt Canyon including petrographic and lithologic descriptions.

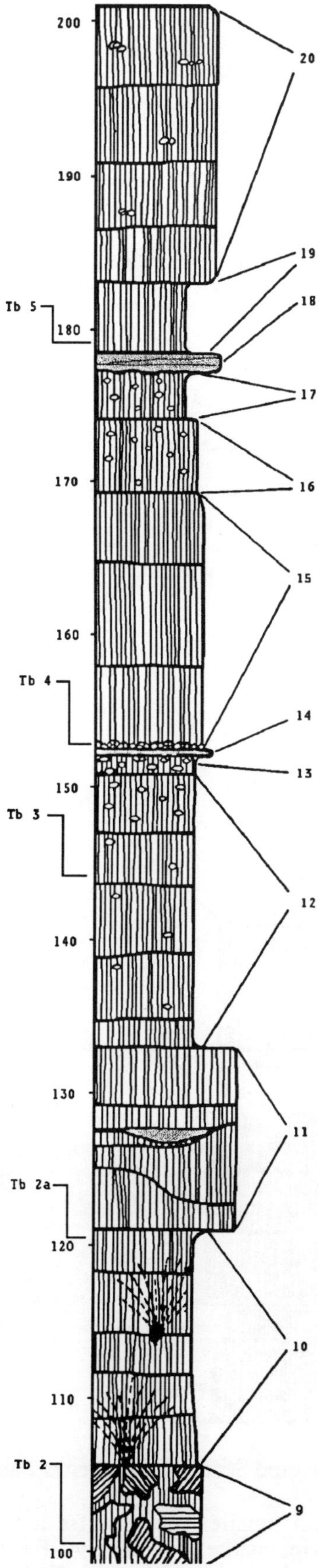

Basalt(?). Similar to Tb 4, but olivine content is less than 5%, chlorite is more abundant as an alteration product of groundmass and large plagioclase phenocrysts are altered to sericite, chlorite and epidote. Sphene(?) is present as an alteration product of augite.

Basalt, amygdaloidal. Intergranular to intersertal texture. Consists of randomly oriented plagioclase laths, altered olivine crystals and aggregates of pyroxene and chlorite-epidote up to 0.3mm in size. Chlorite-epidote may be an alteration product of glass. Plagioclase laths are up to 0.5mm in length, are normally zoned, and have albite twinning typical of oligoclase. Terminations are feathery;, some are slightly bent. Plagioclase is partially altered to sericite. Olivine phenocrysts are anhedral, rounded, and up to 0.15mm in size. Magnetite-hematute fills radial cracks and coats grains. Pyroxene (augite) is present as aggregates of grains; individual grains are up to 0.1mm in size. Magnetite coats mafic minerals and forms large grains associated with these minerals. Chlorite and epidote occur as vesicle fillings and as alteration of groundmass. Chlorite also occurs within altered grains of olivine. Original mineral percentages were: plagioclase 60-65; pyroxene 25; olivine 5; magnetite <5; vesicle filling and groundmass 5-10.

Andesite (?) Randomly oriented laths of plagioclase (An 30-oligioclase, andesine) up to 0.5mm in length and a few completely altered pyroxene crystals, all in a matrix of chlorite and ffine grained magnetite, ilmenite and hematite. Matrix may have been crystalline and/or glassy before alteration.

Andesite, strongly propylitizied. Intergranular to intersertal texture. Randomly oriented laths of plagioclase (AN 30) , phenocrysts of magnetite and hornblend, and aggregates of clinopyroxene (augite) all set in a hyalocrystalline matrix of fine grained plagioclase, orthoclase, quartz, and glass. Plagioclase laths are up to 1mm in length and show normal zoning. Aggregates of augite are composed of 5 or 6 stubby crystals and colored by alteration products (green by chlorite, brown by hematite). Magnetite is associated with concentrations of mafic minerals. Hornblende crystals are euhedral to subhedral, pleochroic green-brown and poikilitic with quartz. Quartz occurs in irregular patches and as intergrowths with orthoclase filling interstices between plagioclase and pyroxene. Pockets of glass as much as 2mm in size are devitrified and now are composed of brown-green pleochroic amphibole (?) and chlorite, some of the original spherolitic structures are still preserved. Original mineral precentages were: plagioclase 55:;, augite 25; primary magnetite 5; quartz and orthoclase 5; hornblende 2. Alteration products of unknown origin and devitrifed glass make up the remainding 8%

Porphyritic andesite. Randomly oriented laths of plagioclase (andesine) up to 0.5mm in length and minor phenocrysts of augite set in a matrix of fine grained pyroxene and alteration products. Orthoclase and quartz occur as large irregular patches up to 0.5mm in size. Ilmenite and/or rutile needles are concentrated in the orthoclase and quartz patches. Chlorite and sericite are abundant. Overall composition: plagioclase 55%, augite 15%, magnetite-ilmenite 10%, quartz 5%, orthoclase <5%, altered groundmass 10-15%.

Basalt, purplish to brownish-gray, pink-gray in outcrop color. Very fine grained, dense, and fractured. Chlorite-epidote alteration. Sphereoids present locally. Unit is massive, cliff-forming, with massive columnar jointing. This unit, together with units 19, 21, 22, probably represents one very thick flow that forms a distinctive outcrop composed of, in assending order, a) a well jointed, rubbly, slope forming, tan-gray zone (unit 19): b) a prominent cliff forming, massive, pink-gray zone with columnar jointing (Unit 20); and c) a well jointed, slope forming, purplish-gray zone (units 21 and 22)

Similar to unit 20, but with closely spaced joints. Rubbly, forms slope

Sandstone, purple-brown, very fine grained, baked. Hard, dense, locally highly fractured and rubbly. Contains fragments of amygdaloidal basalt similar to that of Unit 17. Beds are irregular in thickness; some pinch out. At least two beds are present, locally three. Lower contact is irregular, with relief of as much as 15cm

Amygdaloidal zone in basalt similar to unit 15. Amygdules filled with quartz, chalcedony, epidote, and chlorite. Well developed and preserved ropy structure present near topof unit. Unit weathers into fine rubble (fragments 1.5cm in diam.) and forms slope above cliff held up by units 15 and 16

Vesicular zone in basalt similar to unit 15. Vesicles flattened parallel to bedding, as much as 7.5cm, typically 2-3cm in max. diameter. Quartz common in amygdule fillings, in veinlets, and in geodes. Epidote-chlorite amygdule filling present. Fracturing common.

Basalt, purple to dark brown-gray, very fine grained, dense, fractured. Chlorite - epidote alteration. Shears and fractures mineralized with quartz, chlorite, epidote. Rubbly and amygdaloidal zone 30cm thick at base of unit. Units 15, 16, and 17 probably form one flow

Sandstone, very fine grained purplish, hard, Contains fragments of underlying unit. Overlies with irregular contact the weathered zone developed in unit 13. Appears baked

Basalt, very vesicular, otherwise similar to and continuous with Unit 12. Vesicles elongate in direction of bedding; as much as 5cm, typically 1 - 2.5cm in diameter. Filled with fine ghrained, greenish-black material, probably in part chlorite. Upper 60cm of unit are rich in amygdules filled with calcite and/or zeolites, as well as in bands subparallel to bedding and composed of hard, purplish and greenish felsite (hornfels ?). Uppermost part of unit is probably a weathered zone

Basalt, gray to brown, fine grained dense, hard, fractured. Scattered vesicles and amygdules, become more abundant upward. Amygdule-rich zones 30-60cm thick are common in upper part of unit

Basalt, purple to brown-black. Fine grained dense, hard. Few vesicles or amygdules. Forms cliffs on valley sides . Unit composed of several flows that are very irregular in cross section. Some are lentiform, others fill channels. Unit includes a sandstone bed that is discontinuous and typically fills sharp channels

Basalt, similar to that of unit 8, but without breccia structure. Conspicuous and well developed sheeting, locally fan-like.

Basalt, purplish - gray to brown, dense, fractured. In blocks laced with bowwork of very fine grained, dense basalt of similar appearance: autoclastic breccia?, few to no vesicles and amygdules

FIGURE 10.1 (cont.). Measured Section of Cardenas Basalt at Basalt Canyon .

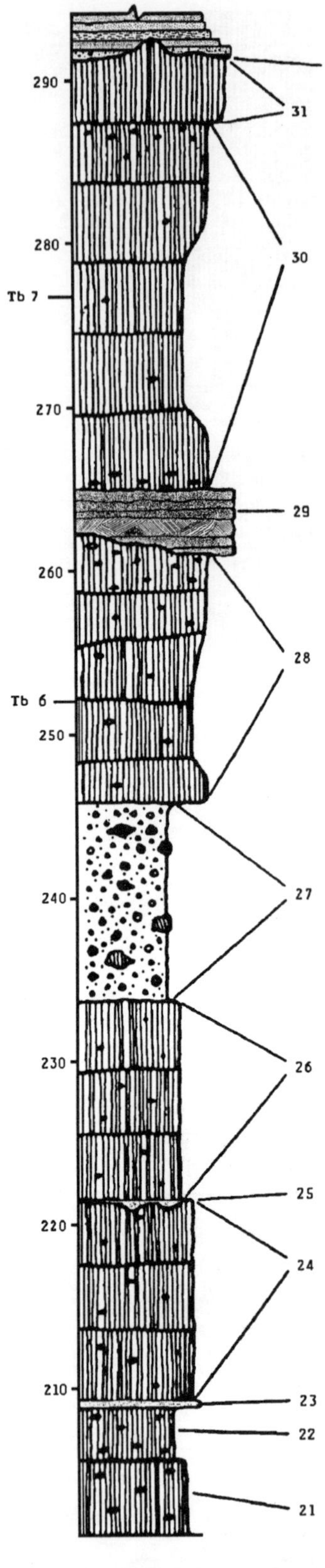

Andesite. Highly altered. Includes phenocrysts of plagioclase, pyroxene, hornblende, minor apatite, and large patches of orthoclase and quartz. Plagioclase laths, up to 1.5mm long, are highly altered to sericite (composition unknown). Clinopyroxene forms small (<0.1mm) grains that are highly altered to talc (?) and magnetite. Euhedral to anhedral pleochroic light green to brown hornblende forms <5% of rock. Concentrations of calcite and talc with rims of magnetite and hematite suggest olivine crystals that have been completely altered

Basalt. Porphyritic, vesicular, amygdaloidal. Randomly oriented plagioclase laths, up to 2mm in length, and minor clinopyroxene (augite) phenocrysts are set in a brown (hematite) groundmass. Opaque material (magnetite-hematite) forms 20% of the rock. Amygdules are filled with quartz and serpentine(?). Alteration of plagioclase to sericite is nearly complete. Pyroxene is fresh. Vesicles originally occupied 10 - 20% of rock

NANKOWEAP FORMATION. Interbedded purplish sandstone, greenish shale and conglomerate beds. Mostly thin-bedded. Unit forms stair-like slope.
CONTACT. Unconformable. Local relief of 1-1.5m.
Basalt, similar to that of unit 30, but with fewer amygdules. Forms pinkish cliff

Basalt, very fine grained dark red to maroon in outcrop color, black to green-black on fresh surfaces. Composed of plagioclase, altered olivine and pyroxene, and magnetite-ilmenite. Chlorite-epidote alteration. Amygdules present near base and top of unit. Those near base are elongate parallel to bedding. Unit is massive, weathers into sharp crags bounded by fractures

Sandstone, fine grained, red to reddish purple on weathered surface, well indurated. Thickness ranges from 3 to 6m. Flaggy bedding. Large-scale cross bedding

Basalt, fine grained, vesicular, amygdaloidal. Vesicles are 1-10mm in size. Some are elongate parallel to bedding. Weathers into steep slope, the lower part of which is littered with nodules. Lowermost part of unit forms a cliff 1-3m high

Lapillite. Scoriaceous lapilli, about 7.5cm in average size, and bombs as much as 1m in size, are set in a matrix of ash. Amygdules are filled with chlorite, epidote, quartz, and calcite. Forms rubbly slope. Bombs weather out in relief

Basalt, similar to that of unit 24, but with greenish outcrop color. Weathers into angular chips and forms a steep slope or subdued cliff. Units 24, 26, and 27 are distinctive from a distance, forming a triplet characterized by the following colors, in assending order: tan, green, pinkish

Sandstone, fine grained, purplish, baked. Thickness is very irregular; forms lenses

Basalt. Very fine grained weathering brown to red-brown. Abundant vesicles and amygdules. Amygdules are filled with quartz, calcite, chlorite, epidote, serpentine, and talc. Chlorite-epidote alteration is common. Upper contact is very irregular, with relief of 3-6m. Unit forms a crumbly cliff

Sandstone, purplish, fine grained, baked

Amygdual zone. Abundant amygdules are filled with quartz, epidote, chlorite, and calcite. Unit weathers into crumbly slope

Vesicular zone. Vesicles flattened parallel to bedding

FIGURE 10.1 (cont.). Measured Section of Cardenas Basalt at Basalt Canyon.

to one millimeter in length, together with smaller phenocrysts of olivine(?), pyroxene(?), and magnetite. Interstices are filled with chlorite, epidote, and opaque material. Large patches of devitrified glass make up the remainder of the rock. Plagioclase is the only primary mineral fresh enough to allow determination of optical properties. Extinction angles of the plagioclase are low (about 0-5°), suggesting oligoclase. Olivine(?) and pyroxene(?) are completely altered to talc, chlorite, calcite, and magnetite, with tentative identification based on crystal shape. Patches of devitrified glass consist of chlorite, epidote, and clay minerals.

Except for variations in the amount of vesicles, all of the flows above the 305-foot (93-m) level show similar textural features. These flows have a smaller grain size than the lower flow unit and an intergranular to intersertal texture. Laths of randomly oriented plagioclase (An_{25-35}), generally less than 0.5 mm in length, make up 55 to 70 percent of the rocks. Interstices between the plagioclase grains contain stubby clinopyroxene crystals, altered glass, orthoclase(?), and minor quartz. Magnetite grains are also associated with the interstitial material. Grain size of the interstitial crystals averages 0.2 mm.

Mineralogically however, the lavas above the 305-foot (93-m) level show distinct variations. Pleochroic green-brown poikilitic hornblende is present in flows Tb2 and Tb7 (fig. 10.1), but is absent in the other units. Altered olivine occurs only in flows Tb4 and Tb5. The plagioclase changes from oligoclase in Tb2 to andesine in Tb4 and Tb5, whereas Tb3 has about An_{30} composition. In flows Tb6 and Tb7, the plagioclase composition cannot be determined because of sericitization. Unaltered plagioclase crystals show slight normal zoning. Vesicles are filled with epidote, quartz, and calcite, and in the most vesicular rock (Tb6), amygdules occupy as much as 20 percent of the rock. Clinopyroxene and hornblende are fresh although olivine phenocrysts are altered to chlorite and magnetite.

Chemistry. Chemical analyses for major oxides of nine samples from the lava series are shown in Table 10.3. The ferric-ferrous ratios have a maximum value of 1.4 and most probably represent the high degree of alteration.

The basal unit, Tb1, differs chemically from the other flows, having the highest silica content, the lowest total iron and potassium, and the highest sodium content of all the lavas. Although the rocks are highly altered, analyses of rocks above the 305-foot (93-m) level are quite similar to one another.

Tb2, Tb3, and Tb7 are classified as basaltic andesites (Coates, 1968) owing to their relatively high silica and low alumina values along with the occurrence of hornblende, orthoclase, and quartz. The rocks in the center of the Basalt Canyon section are siliceous basalts. The trend in the lavas upward through the section therefore, is from intermediate to basic to intermediate.

DISCUSSION

The lowermost sequence of flows in the Cardenas Basalt represents a distinct unit. Differences from overlying lavas are noted on field characteristics, mineralogy, texture, and chemistry. Walcott (1894) suggested that the Unkar lavas were extruded into a shallow sea and states: "Portions of some of the lava sheets, especially the basal one, bear evidence of having been broken into fine, rounded fragments, apparently by the hot lava pouring into the sea". Some of the anomalous chemical values, i.e., high Na_2O and normative albite of this lower unit can be explained by a process of spilitization. The coarser grain size and type of phenocrysts of the lower unit versus finer grain size and interstitial clinopyroxene of the overlying flows suggests a different crystallization history of the magmas that fed the lower (305 ft; 93 m) and upper (305-985 ft; 93-300 m) units. Ropy lava and the deposit of lapillite suggest that the flows higher in the section were extruded in a subaerial environment.

Extrusions of major flow units were separated by sufficient time for the accumulation of the sandstone beds. These sandstones were deposited in quiet water, as shown by the presence of fine silt and clay sized particles. Although varying in thickness due to the irregularity of the upper surface of the underlying flows, the sandstones are apparently continuous over the 7 mi^2 (11 km^2) area of exposure. The thinness of these beds combined with this continuity suggests that deposition was not in localized lakes ponded by the lava flows.

The area of volcanism appears to have been a shallow subsiding sea into which lavas poured occasionally building up deposits above sea level. Flows extruded above sea level were submerged and covered with

	Tb1	Tb2	Tb2a	Tb3	Tb4	Tb5	Tb6	Tb7	Tb7b
SiO_2	56.30	52.30	54.50	52.10	52.10	49.00	53.50	53.20	55.30
Al_2O_3	13.20	14.3	3.10	13.60	12.30	13.90	14.50	13.60	12.10
Fe_2O_3	6.60	7.10	7.40	8.70	6.20	6.50	11.20	10.60	12.20
FeO	3.83	7.29	7.43	4.98	8.51	8.01	3.31	4.50	3.04
MgO	6.90	4.40	3.50	6.30	5.00	6.60	3.30	2.70	2.30
CaO	1.40	2.50	3.90	2.00	4.90	5.90	4.10	4.80	5.20
Na_2O	5.90	4.30	3.60	4.10	4.20	3.30	2.90	3.30	2.90
K_2O	0.20	2.70	3.00	2.20	2.30	1.80	2.30	2.90	2.40
TiO_2	1.10	1.36	1.48	1.55	1.56	1.52	1.52	1.57	1.59
P_2O_5	0.21	0.09	0.25	0.24	0.26	0.29	0.28	0.34	0.36
MnO	0.10	0.12	0.16	0.10	0.17	0.15	0.11	0.15	0.15
$+H_2O$	3.32	2.33	1.62	3.36	1.89	2.07	1.86	1.30	0.99
$-H_2O$	0.90	0.94	0.57	1.29	0.50	0.89	1.16	0.91	1.11
CO_2	0.36	0.24	0.18	0.18	0.18	0.18	0.25	0.31	0.18

TABLE 10.3. Chemical Analyses of flows at Basalt Canyon.
(See figure 10.1 for stratigraphic positions of samples.)

detritus before extrusion of the next sequence of flows. The build-up of the lava pile was at a greater rate than subsidence.

Relations between Intrusive and Extrusive Rocks

A fundamental problem remains concerning the Unkar igneous rocks: do the flows, dikes, and sills represent a single period of volcanism and/or have a common parental magma? No feeders for the flows were found during this study. A dike exposed in Unkar Valley, however, is 200-300 feet (60-90 m) stratigraphically below the base of the flows. This dike was sampled for comparison with the rocks of the flow sequence. Mineralogically, this dike contains plagioclase laths three to four times as long as any from the flows. Phenocrysts of olivine comprise an estimated 10 to 20 percent of the dike rock, whereas the lowermost flow unit is the only extrusive rock containing over 5% olivine. In addition, this lower flow differs chemically from the dike (table 10.3, Tb1 and table 10.1, Tid1). These data suggest that none of the flows were fed by this particular dike. All dikes above the sills that were sampled were found to be mineralogically, texturally, and chemically similar to the sills and different from the flows.

Radiometric age determinations indicate that the intrusive and extrusive rocks were formed at approximately the same time. The occurrence of the flows, sills, and dikes within the same geographical area, together with their similarity in age, suggest a possible genetic relation. If the extrusive and intrusive rock did have a common parent magma, then the textural, mineralogical, and chemical data suggest that these two occurrences of igneous rocks represent two or more episodes of activity. Isotopic ages suggest that the length of time represented between episodes is less than the experimental error of the dating method (~30-70 Ma).

Differentiation of a parental basaltic magma may have produced both the olivine diabase of the sills and dikes, and the more intermediate rocks of the flows. If this is the case, chemical variation diagrams would show trends developed by plotting analyses from all of the rocks studied. Figures 10.2, 10.3, and 10.4 are graphs designed to show a possible chemical relation between the Unkar igneous rocks. Although the analyses show some scatter, a trend is apparent in the diagrams. The samples from the granophyre and granophyre-diabase transition zone in the Shinumo Creek sill are the only anomalous samples.

Kuno (1968) suggests that different chemical trends followed by magmas during differentiation may be a result of different times of crystallization of the iron and titanium oxides. The stage of magnetite crystallization is probably controlled by the oxygen pressure of the magma, which in turn, is mainly a function of the water content of the magma. A higher water content would result in a higher oxygen partial pressure resulting in an earlier crystallization of the magnetite. If the oxygen pressure of the magma emplaced as the Shinumo Creek sill was higher than that of the parent magma, possibly due to assimilation of water during emplacement, and to lower confining pressure, then differentiation within the sill should follow a different chemical trend. The AMF diagram (fig 10.2) shows a trend of the sills and flows in which iron increases as magnesia decreases (i.e., an increase in iron as differentiation progresses). This trend is produced by a constant composition or decreasing PPO_2 (Osborn, 1959). The trend suggested by the plots

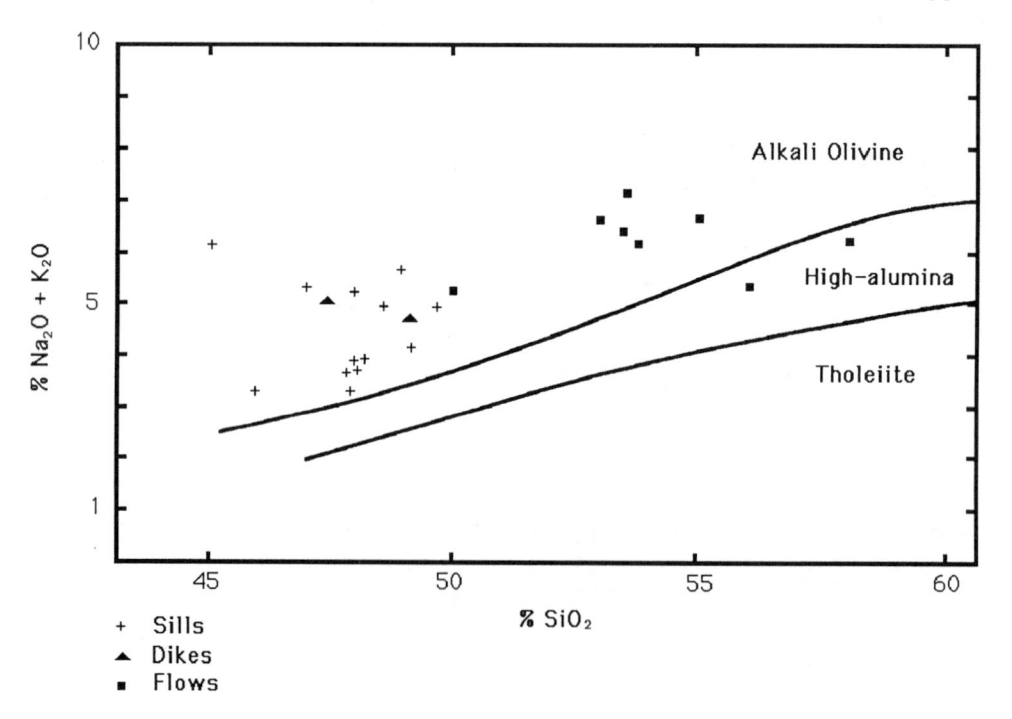

FIGURE 10.2. Silica Versus Alkalai Diagram for Unkar Igneous Rocks. Magma types from Kuno (1968).

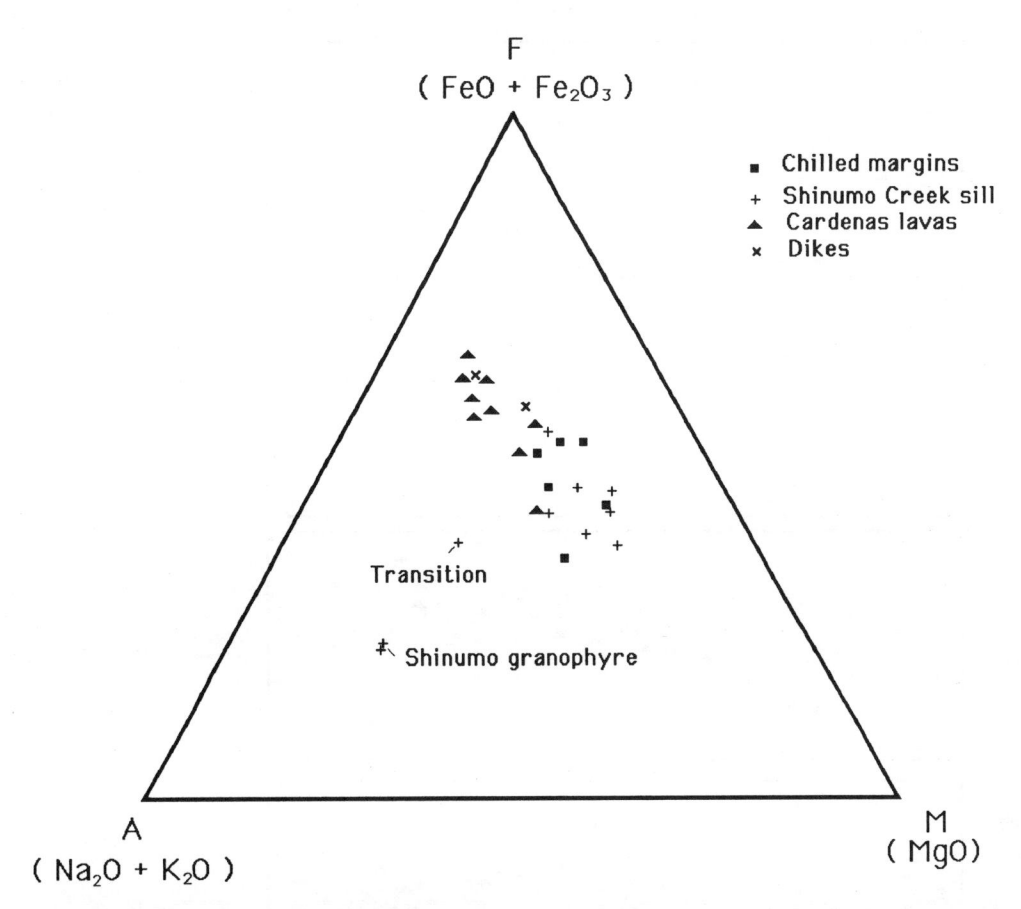

FIGURE 10.3. Alkalai-Magnesium-Iron diagram for Unkar Igneous Rocks.

of the chemical analyses from the Shinumo Creek sill shows that the iron percentage drops as the alkalis increase (i.e., as differentiation progresses). This relationship is expected if the magma emplaced as the Shinumo Creek sill had a higher oxygen partial pressure. The mineralogy of the granophyre and transition zone supports the interpretation of a higher water and other volatile content during crystallization.

Kuno (1968) has suggested that a diagram of solidification index plotted against major oxides is a good indicator of progressive magmatic differentiation. The solidification index would decrease in value with progressive differentiation. Such a diagram (fig. 10.3) also suggests a trend defined by analyses of both intrusive and extrusive rocks.

The relation in time and space combined with the apparent affinity in chemistry favors the suggestion of a common parent magma for the Unkar igneous rocks. If this is the case, the intrusion of the sills and dikes would represent an early phase of volcanism, followed by extrusion of the flows after differentiation of the original melt at depth.

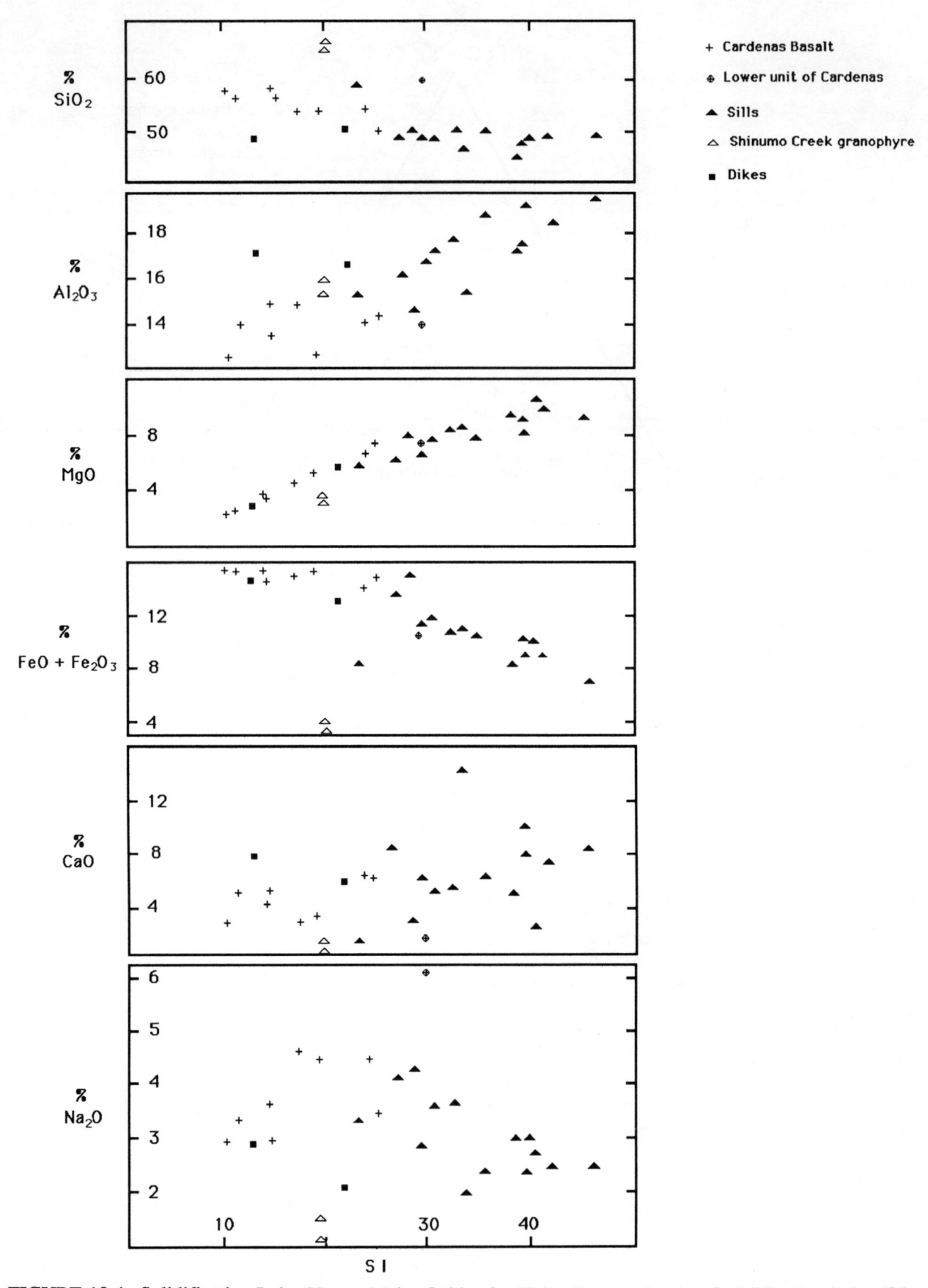

FIGURE 10.4. Solidification Index Versus Major Oxides for Unkar Igneous Rocks. Solidification index (SI) = (MgO × 100) / (MgO + FeO + Fe$_2$O$_3$ + Na$_2$O + K$_2$O).

CHAPTER 11: POTENTIAL PETROLEUM SOURCE ROCKS IN THE LATE PROTEROZOIC CHUAR GROUP, GRAND CANYON, ARIZONA[1]

Mitchell W. Reynolds[2], James G. Palacas[2], and Donald P. Elston[3]

The Chuar Group, upper part of the Grand Canyon Supergroup in northwestern Arizona, is a 1637-meter-thick succession of predominantly very fine grained siliciclastic rocks that contains thin sequences of sandstone and stromatolitic and cryptalgal carbonate rocks (chapter 9). More than half the succession consists of organic-rich gray to black mudstone and siltstone; fossil microorganisms are abundant to common throughout successions of dark mudstone and siltstone.

Strata of the Chuar Group likely accumulated in a succession of environments that included a sediment-starved basin rich in organic material, a coastal or alluvial plain, and mixed coastal or paludal swamp and alluvial plain environments. Stromatolitic, cryptalgal, oolitic, and pisolitic carbonate rocks demonstrate nearshore aqueous environments. Depositional sequences of aqueous and subaerial settings alike seem to be markedly cyclic.

In the past, the site of deposition has been interpreted as a foundering marine embayment on the passive edge of the continent. The character and succession of the strata, in conjunction with new possible correlations with Late Proterozoic rocks of probable marine origin in adjacent parts of the Western United States (chapter 12), suggest the alternate hypothesis that Chuar strata accumulated in a lacustrine setting in a subsiding region within the continent.

Preliminary analyses indicate that dark mudstone from the Chuar Group contains as much as 5 percent total organic carbon. Rock-Eval T_{max} values, generally ranging from 430 to 440 °C, indicate that the rocks, particularly in the upper part of the the succession, are within the principal oil-generating window. Hydrogen index (HI) values of as much as 190 mgHC/gC(org), and genetic potentials ($S_1 + S_2$) of as much as 6kg/ton (6,000 ppm) demonstrate that the rocks still have potential for generating sufficient amounts of gaseous and liquid hydrocarbons for commercial accumulations.

Chuar Group strata, now exposed on the north side of the Colorado River in the eastern part of the Grand Canyon, plunge north beneath the Kaibab Plateau of Arizona and southwestern Utah. In the Grand Canyon the strata are truncated on the east by a normal fault that predates the Cambrian Tapeats Sandstone. We suggest that if the Chuar Group is present more widely beneath Phanerozoic rocks of the Kaibab Plateau or is preserved in pre-Phanerozoic grabens in northern Arizona or southern Utah, the strata may have served as source rocks for potential petroleum accumulations in lower Paleozoic or Proterozoic units yet to be prospected in that region.

---

[1] From U.S. Geological Survey Circular 1025.

[2] U.S. Geological Survey, Federal Center, Denver, Colorado

[3] U.S. Geological Survey, Flagstaff, Arizona

View to north of Basalt Canyon (A) and south end of Chuar suncline (B), and Basalt graben (C), from Tanner Trail. D - Dox Sandstone; E - Cardenas Basalt; F - Nankoweap Formation; G - Galeros Formation; H - Tapeats Sandstone (Cambrian) with Great Unconformity at base. Photograph courtesy of Ivo Lucchitta.

CHAPTER 12: PRELIMINARY POLAR PATH FROM PROTEROZOIC AND PALEOZOIC ROCKS OF THE GRAND CANYON REGION, ARIZONA

Donald P. Elston
U. S. Geological Survey, Flagstaff, Arizona

INTRODUCTION

Approximately two dozen paleomagnetic poles have been obtained from red beds and igneous rocks of the Middle and Late Proterozoic Grand Canyon Supergroup, and red beds of Cambrian, Devonian, Pennsylvanian, and Permian age in the Grand Canyon. An apparent polar wandering (APW) path obtained by connecting the Proterozoic poles in stratigraphic order has allowed provisional correlations to be made on the basis of pole position and polarity with poles reported from Proterozoic strata and intrusive rocks elsewhere in North America. A rather simple composite polar path thus derived appears interpretable in terms of rotations and changes in paleolatitude of the North American plate in turn related to episodes of tectonism.

Paleomagnetic studies of Proterozoic and Paleozoic rocks of the Grand Canyon, begun in 1971, have led to the Arizona poles shown in Figure 12.1. The poles and apparent polar wander path result from detailed stratigraphically controlled sampling across red bed strata and associated mafic extrusive and intrusive igneous rocks. Most samples responded to thermal or alternating field demagnetization treatment, and yielded stable, apparently original magnetizations. Ovals drawn at the 95 percent confidence level are not shown on Figure 12.1, but for the Arizona and Montana poles mostly range from about 1.5° to 7°. Results of studies leading to the Arizona and Montana poles, and studies are described elsewhere, including two major reports that currently are in review (Elston and Scott, 1973; Elston and Grommé, 1974, 1979, 1984, in review; Elston and Bressler, 1977, 1980, 1984, in review; Bressler, 1981; Elston and McKee, 1982; Elston, 1984a, 1986, 1988a, 1988 b).

PROTEROZOIC POLES

The ~4-km-thick Middle and Late Proterozoic Grand Canyon Supergroup (chapter 9) has yielded a series of stratigraphically controlled paleomagnetic field directions and intervals of normal, reverse, and mixed polarity. Oriented samples were collected at ~1/3 m to ~1 m stratigraphic intervals in the Unkar Group, Nankoweap Formation, from several widely spaced red bed intervals in the Chuar Group, and from the Sixtymile Formation. Paleomagnetic poles have been calculated from the paleomagnetic field directions. Poles from reversely polarized rocks have been inverted to plot as if they were of normal polarity. When connected in stratigraphic order, the poles from the Grand Canyon Supergroup form a high-resolution apparent polar wandering path for part of Middle and Late Proterozoic time (fig. 12.1).

Normal polarity paleomagnetic poles from the Grand Canyon Supergroup plot in the area of the present Pacific Ocean. Apparent correlations have emerged where poles and polar paths obtained from successions elsewhere in North America correspond and overlap. The provisional correlations have led to a composite polar path that is temporally controlled by mutually supporting, nonconflicting U-Pb and Rb-Sr isotopic ages from igneous rocks.

The wandering of the pole in an apparent polar wandering path is *apparent* because it is not the Earth's geographic rotation axis (i.e., spin pole) that is believed to have moved. Rather, the spin pole and the magnetic field direction are generally acknowledged to coincide when viewed in the context of geologic time and with the averaging-out of secular variation. Apparent polar motion is thus most commonly interpreted to reflect rotations and changes in paleolatitude of the continent through time with respect to a fixed pole of rotation for the Earth. Thus, in Figure 12.1, where the pole traces from south to north reflecting a clockwise change in declination, the continent can be interpreted to have been rotating in a counter-clockwise sense. Additionally, where the pole moves toward the site direction (in this case, the Grand Canyon) reflecting a steepening of the normal polarity inclination, the continent can be interpreted to have been moving in a northerly direction.

Correlations

Correlations between the Grand Canyon Supergroup and other Proterozoic sequences in North America can be provisionally inferred from correspondences in pole position and the character of polarity zonations (fig. 12.1). On this basis, the Middle Proterozoic Belt Supergroup of western Montana, differing in both pole position and polarity zonation, appears entirely older than all but the lowermost beds of the Grand Canyon Supergroup. Additionally, parts of the Missoula Group of the Belt Supergroup would seem to correlate with the Pioneer Shale of central Arizona and the Sibley Series of western Ontario. Furthermore, lower Keweenawan rocks of the Lake Superior region would seem to correlate with the Shinumo Quartzite; the middle members of the Dox Sandstone would appear to correlate with middle Keweenawan rocks of Lake Superior; and the upper member of the Dox, the Cardenas Basalt, and the lower member of the Nankoweap Formation would appear to correlate with lower to middle upper

Keweenawan rocks. Intrusion of middle Keweenawan mafic sills at ~<1120 Ma in Lake Superior region appears to have been contemporaneous with the intrusion of diabase sills in northern and central Arizona, occurring at the apex of the Unkar loop (fig. 12.1). The upper member of the Nankoweap Formation appears to be somewhat younger than the uppermost upper Keweenawan Chequamegon Sandstone. Lastly, the Chuar and Uinta Mountain Groups appear coeval, and their upper parts appear to correlate with the Little Dal Group of the Mackenzie Mountains Supergroup of northern Canada.

PALEOZOIC POLES

The polar path for the Paleozoic of North America is currently not well defined. Paleozoic red beds in Marble Canyon and the Grand Canyon, and in central as well as northern Arizona, have been sampled with the objective of obtaining new paleomagnetic poles that could aid in defining a reliable polar path for the Paleozoic (Elston and Bressler, 1984; Elston, 1988b). Though not yet complete, the study is sufficiently advanced to allow preliminary poles from Cambrian, Devonian, Pennsylvanian, and Permian strata to be shown on Figure 12.1. The Devonian pole from the Grand Canyon is poorly defined, but a better defined pole has been obtained from the Devonian of western Montana (Elston and Bressler, 1984). Poles from Silurian and lower Carboniferous strata from elsewhere in North America also are shown on Figure 12.1. A Cambrian pole from the Tapeats Sandstone of the Grand Canyon, reported previously by Elston and Bressler (1977), has been refined from additional studies. Data supporting the Paleozoic poles from the Grand Canyon, and central Arizona, shown on Figure 12.1 remain to be reported formally.

SUMMARY

Paleomagnetic poles and the polar path from Middle and Late Proterozoic and Paleozoic rocks of the Grand Canyon of northern Arizona are serving to control much of the apparent polar wandering path for North America. The south-to-north and north-to-south apparent motions represent, respectively, counterclockwise and clockwise rotations of the continent. Paleolatitudes during the interval of time shown on Figure 12.1 were generally within 20° of the equator, with the southernmost positions corresponding with poles at about 800 m.y. (pole for Sixtymile Formation), and the Grand Canyon disturbance that ended deposition of the Grand Canyon Supergroup. The "northern" limit of rotation during the Middle Proterozoic at ~1120 m.y. corresponds with the Keweenawan disturbance of the Lake Superior region, a time of intrusion of mafic sills in the mid-continent region and in Arizona. Places of change in apparent polar motion thus may be interpreted to have arisen from plate interactions that gave rise to several discrete episodes of tectonism..

FIGURE 12.1. Paleomagnetic poles and preliminary apparent polar wandering path for North America. Solid symbols - normal polarity; open symbols - reversed polarity; half solid/half open symbol - mixed polarity. Circles - indicate sites in Proterozoic Grand Canyon Supergroup and overlying Paleozoic strata of northern Arizona.

Proterozoic Rocks: Poles and magnetostratigraphically controlled polar path from Middle and Late Proterozoic successions of western North America (generalized from Elston and Bressler, 1980 and in review; Elston and Grommé, 1974, 1979, 1984, and in review; and Elston, 1984a). B-B', Belt Supergroup, Montana and Idaho. Grand Canyon Supergroup, northern Arizona: U-U', Unkar Group (b - upper member of Bass Limestone; h - Hakatai Shale, lower (hl) and upper (hu) members; s - Shinumo Quartzite; d - Dox Sandstone, lower member (dl), lower middle member (dlm), upper middle member (dum; interval of asymmetric reversals not shown), sills (intruded during deposition of dum), and upper member (du) of Dox Sandstone; c - Cardenas Basalt, flows and sandstone interbeds; N-N', Nankoweap Formation, Nl - lower (ferruginous) member and Nu - upper member; Cg - C'k, Chuar Group, g- Galeros Formation, k - Kwagunt Formation; and S - Sixtymile Formation. Js - Jacobsville Sandstone of Keweenaw Peninsula, northern Michigan (Roy and Robertson, 1978) (overlaps polar path for Nonesuch and Freda Formations); Cs - Chequamegon Sandstone of northern Michigan (McCabe and Van der Voo, 1983). UM - UM', poles 1 - 6, Uinta Mountain Group, Utah and Colorado (Bressler, 1981). LDa and LDb, Little Dal Group, Mackenzie Mountains, northwest Canada (Park, 1981a, 1981); Ts - ~770-Ma sills in Tsezotene Formation beneath Little Dal Group (Park, 1981b; Armstrong and others, 1982; Park and Aiken, 1986). Rz - pole from Z component of magnetization in glacigenic Rapitan Group, Windermere Supergroup, western Canada (Morris, 1977); F - Franklin intrusions (~600-650 Ma), northern Canada (Palmer and Hayatsu, 1975). Ages along apparent polar wandering path are from U-Pb and nonconflicting Rb-Sr isochron ages determined from igneous rocks of Lake Superior, central and northern Arizona, Montana, and northwest Canada.

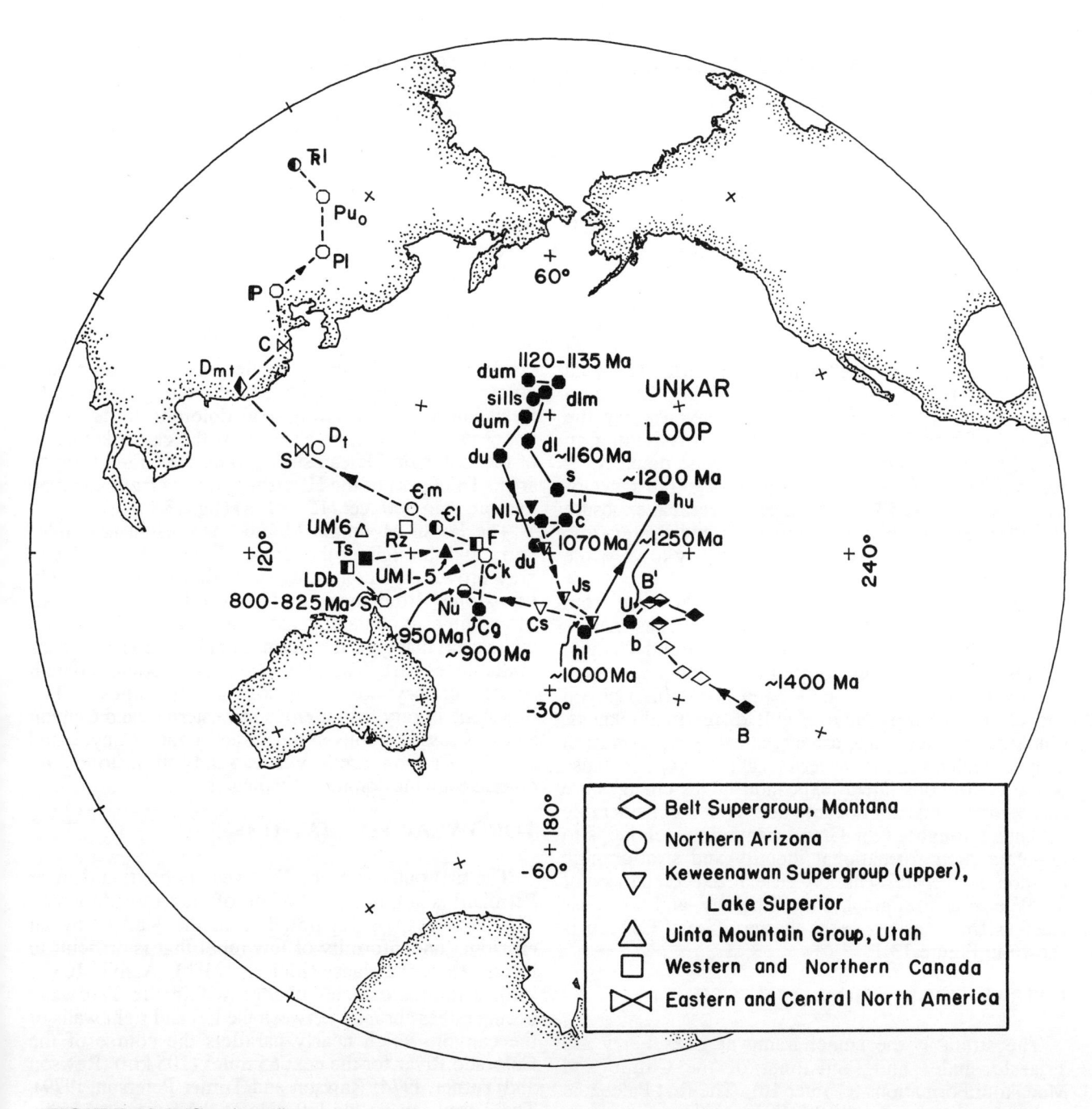

FIGURE 12.1 (Continued)

Paleozoic strata: Poles and provisional polar path for Cambrian to Permian rocks. €l - Lower to Middle Cambrian Tapeats Sandstone, eastern to west-central Grand Canyon; €m - Middle Cambrian Bright Angel Shale, eastern Grand Canyon; S - Silurian Rose Hill, Hersey and Wabash Formations, eastern and central U.S. (French and Van der Voo, 1979; Kent and Opdyke, 1980; McCabe and others, 1985); D_t -poorly defined pole from upper Middle to lower Upper Devonian Temple Butte Limestone, Marble Canyon; D_{mt} - Devonian Maywood Formation, western Montana (Elston and Bressler, 1984); C - Carboniferous strata of New Brunswick (Roy and Robertson, 1968; Roy and Park, 1969, 1974; Roy, 1977); IP -preliminary pole from Pennsylvanian strata of Supai Group, Marble Canyon; Pl - pole from six sites in Lower Permian strata of Supai Group of northern and central Arizona [Chino Point (Steiner, 1988) and Sedona (this report), and Lower Permian Abo Formation of New Mexico (Steiner, 1988)]; Pu_o- Upper(?) Permian overprint pole on Cambrian and Devonian strata, northern and central Arizona; T l - Lower and Middle(?) Triassic Moenkopi Formation, northern Arizona (Purucker and others, 1980).

CHAPTER 13: PALEOZOIC STRATA OF THE GRAND CANYON, ARIZONA

Stanley S. Beus,
Department of Geology, Northern Arizona University, Flagstaff, Arizona

George H. Billingsley
U.S. Geological Survey, Flagstaff, Arizona

From Lees Ferry to the Grand Wash Cliffs, Paleozoic sedimentary rocks (Lower Permian to Middle Cambrian) are exposed in the walls of Grand Canyon along the Colorado River for 277 miles (446 km) providing an effective cross-section over an east-west distance of about 180 miles (290 km) and a north-south distance of about 78 miles (125 km). An average vertical exposure of 4,000 feet (1,220 m) of Paleozoic and Proterozoic rocks and provide an excellent opportunity to view the lateral facies changes.

Most of the formational names were derived from type sections in the eastern Grand Canyon. These names are maintained throughout the entire Grand Canyon because of the lateral continuity of the exposures. However, a few formations viewed in Marble Canyon and eastern Grand Canyon, will differ in thickness, lithologic character, and topographical expression from their equivalents in the western Grand Canyon. Because of the continuous lateral exposure, facies changes are easily identified and documented over a considerable distance throughout the Grand Canyon, which helps to maintain their formational identity and stratigraphic position in a general sense. A generalized comparison of the Paleozoic stratigraphy, the lithology, and thickness changes from eastern to the western Grand Canyon is shown in Figure 13.1.

KAIBAB FORMATION (LIMESTONE)

The strata at the launch ramp at Lees Ferry are Triassic shales and sandstones of the Chinle and Moenkopi Formations (chapter 16). The first Paleozoic rock unit occurs near mile 0.8 (1.3 km) downriver from Lees Ferry. It is the Kaibab Formation (Lower Permian) (Darton, 1910; Noble, 1922; McKee, l938b), a cliff-forming unit separated by an erosional unconformity from the overlying red shale and sandstone of the Moenkopi Formation (Middle? and Lower Triassic) (fig. 1.4).

The Kaibab Formation is subdivided into the upper Harrisburg and lower Fossil Mountain Members (James Sorauf and George Billingsley, work in progress). The Harrisburg Member in the Lees Ferry and Marble Canyon area is a thin-bedded, cherty, ledge and slope-forming, sandy limestone interbedded with a few gray calcareous siltstones. The Harrisburg gradually increases in thickness to the west and southwest of Marble Canyon becoming a ledge and slope-forming sequence of light-red to pale-gray limestone and dolomite, siltstone, sandstone, and white gypsum beds. Resistant sandy limestone and dolomite units in the upper part of the Harrisburg forms the caprock surface of the Colorado Plateau surrounding the Grand Canyon area. Thickness of the Harrisburg ranges from about 50 to more than 300 feet (15 - 91 m) (fig. 13.1).

The Fossil Mountain Member is present near Lees Ferry and throughout the Grand Canyon as a persistent gray-brown cliff consisting of medium-bedded, sandy limestone, limestone, and dolomite. The Fossil Mountain also contains lenses and nodules of gray and white chert that typically forms intraformational breccias in the upper part. The chert lenses and nodules contain fossils of bryozoans, corals, and sponges. The limestone is very fossiliferous in western Grand Canyon and less fossilifereous in the eastern Grand Canyon and Marble Canyon areas where sandy limestone and dolomite are the dominant lithologies.

TOROWEAP FORMATION

The first outcrop of the Toroweap Formation (Lower Permian) is at mile 2.1 (3.4 km) of the Colorado River. The Toroweap is separated from the Kaibab by an erosional unconformity of low relief that is difficult to locate at a casual glance (McKee, l938b). At mile 10 (16 km), a dramatic facies change within the Toroweap occurs rather abruptly between the left and right walls of the canyon which nearly parallels the course of the Colorado River for the next 65 miles (105 km) (Rawson and Turner, 1974; Rawson and Turner-Peterson, 1979). The Toroweap on the left side of the River in Marble Canyon forms a light gray, cliff of thin-bedded, dolomitic sandstone and sandstone. On the right side, it is a light gray and light red, slope and cliff forming sequence of sandstone, limestone, dolomite, siltstone, and gypsum.

Typically, the Toroweap Formation is divisible into three distinct members west of Marble Canyon. They are, from top to bottom, the Woods Ranch, Brady Canyon, and Seligman Members (James Sorauf and George Billingsley, work in progress; fig. 13.1). All three members change to a dolomitic sandstone facies of the Toroweap east of Marble Canyon.

The Woods Ranch Member is typically a red and gray, slope-forming, gypsiferous siltstone and sandstone, interbedded with minor thin-bedded

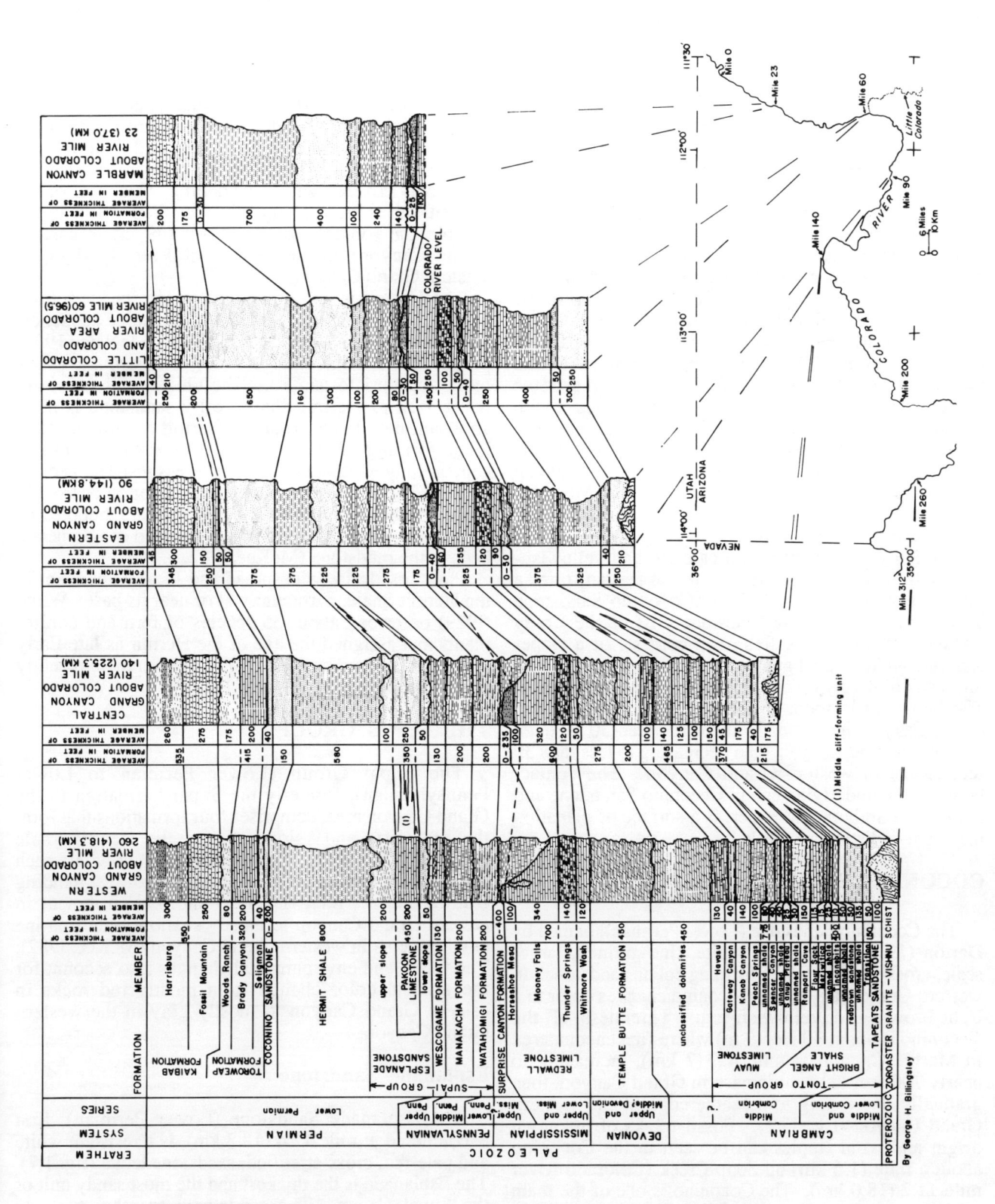

FIGURE 13.1. Paleozoic columnar sections of Grand Canyon, Arizona.

limestone and dolomite in most of Grand Canyon. These lithologies change eastward to a cliff-forming dolomitic sandstone in Marble and eastern Grand Canyon. The gypsiferous siltstone and sandstone of the Woods Ranch are commonly disrupted in thickness, displaying evidence of slumping and solution erosion (Wenrich and others, 1986a). Thickness for the Woods Ranch varies from a few feet (meters) in eastern Grand Canyon to nearly 200 feet (61 m) in central Grand Canyon; decreasing to about 40 feet (12 m) in extreme western Grand Canyon (Wenrich and others, 1986a, 1986b). A gradational contact separates the Woods Ranch and underlying Brady Canyon Members; this contact is arbitrarily placed at the first lithologic change from slope forming siltstone to cliff forming limestone.

The Brady Canyon Member consists of a gray, thick-bedded, fossiliferous, cliff-forming limestone in western Grand Canyon to a light-gray, cliff-forming, dolomitic sandstone and dolomite in extreme eastern Grand Canyon and Marble Canyon. Thickness of the Brady Canyon is about 30 feet (10 m) in eastern Grand Canyon, increasing to nearly 300 feet (91 m) in extreme western Grand Canyon. The Brady Canyon contact with the underlying Seligman Member is gradational and arbitrarily placed below the thick limestone beds of the Brady Canyon Member (fig. 13.1). The Brady Canyon merges with the Woods Ranch and Seligman Members and becomes the sandstone facies of the Toroweap east of Grand and Marble Canyons.

The Seligman Member is a pale-red and yellowish-brown, thin-bedded, calcareous sandstone, that forms a recess separating the underlying Coconino Sandstone cliff from the overlying limestone cliff of the Brady Canyon Member. The Seligman consists of a slope-forming, white to pink gypsum, interbedded with several thin beds of earthy dolomite, limestone, and sandstone. The Seligman is about 35 feet (11 m) thick in eastern Grand Canyon, increasing to about 100 feet (30 m) thick in central Grand Canyon, then decreasing to less then 30 feet (9 m) in western Grand Canyon. The contact between the underlying eolian Coconino Sandstone and the marine Seligman Member, is a surface of planation that forms a distinctive break.

COCONINO SANDSTONE

The Coconino Sandstone (Lower Permian), named by Darton (1910), is a tan to white, fine-grained, large-scale, cross-stratified, cliff-forming eolian sandstone. In western Grand Canyon the Coconino changes color to a light-brown and yellowish-red. Thickness of the Coconino is about 30 feet (9 m) where first encountered in Marble Canyon at mile 4.5 (7 km), increasing to nearly 700 feet (213 m) in eastern Grand Canyon, then gradually decreasing to a featheredge in the western Grand Canyon (fig. 13.1). Fossil tracks of unknown origin and wind ripples can be seen in the Coconino about a mile (1.6 km) up Soap Creek (Colorado River mile 11.2 (18.0 km). The Coconino is one of the main aquifers of the southwestern Colorado Plateau. A few seeps and springs occur at the base of the Coconino cliff throughout most of the Grand Canyon area. The Coconino overlays relatively impermeable red siltstones and shales of the Hermit Shale with a sharp, flat disconformity. Locally, cracks up to 20 feet (6 m) deep and several inches wide in the Hermit Shale are filled with wind blown sand of the Coconino.

HERMIT SHALE

The Hermit Shale (Lower Permian), named by Noble (1922), is encountered at mile 4.9 (8 km) as a slope-forming, red-brown, siltstone and sandstone. Throughout the Grand Canyon the Hermit slopes are commonly covered with talus debris. Thickness of the Hermit varies from about 700 feet (213 m) in upper Marble Canyon to about 160 feet (49 m) in eastern Grand Canyon, and more than 800 feet (244 m) in western Grand Canyon.

The Hermit is predominantly an interbedded, thin-bedded, silty sandstone and sandy mudstone sequence, and has almost no true fissile shale. McNair (1951) proposed the name change to Hermit Formation as a more appropriate name but the unit is still officially recognized as Hermit Shale. Silty sandstone is most abundant near the base of the Hermit as structureless, ledge-forming beds up to 3 feet (1 m) thick locally contain limy nodular concretions. Less resistant beds of silty sandstone may exhibit ripple lamination or trough, cross stratification. Locally, sedimentary pebbles of limestone and sandstone form lense shaped conglomerate beds in the mudstone (McKee, 1982, p. 203).

Plant fossils are widespread and locally abundant in the Hermit Shale, particularly at or near its base. White (1929) described about 35 species of fern and conifer plants and assigned the age of the Hermit as late Early Permian (Leonardian). This age has been more recently confirmed by McKee (1982, p.111).

THE SUPAI GROUP

The Supai Group (Lower Permian to Lower Pennsylvanian), formerly the Supai Formation in the Grand Canyon area, comprises four formations that form the red cliffs and slopes below the Hermit Shale (Mckee, 1975). Several facies changes occur within each formation of the Supai Group from east to west trending from a marine, tidal flat, near shore environment in eastern Grand Canyon to mostly shallow water marine environments in western Grand Canyon (McKee, 1982). The facies and environmental changes also account for the gradual color change from mostly red rocks in eastern Grand Canyon to mostly gray in the western Grand Canyon.

Esplanade Sandstone

The Esplanade Sandstone (Lower Permian), first encountered at mile 11.4 (18.3 km), is a resistant white and pale-red, cross-stratified, sandstone ledge (fig. 1.7). The Esplanade is the thickest and the most sandy unit of the Supai Group. The sandstone, together with its western correlative, the Pakoon Limestone which was placed in the Supai by Billingsley (1978), thickens from about 300 feet (91 m) in eastern Grand Canyon to more than 450 feet (137 m) in western Grand Canyon.

The Pakoon Limestone is a gray, thick-bedded, cliff-forming, fossiliferous marine limestone recognized in the western Grand Canyon. It is not considered to be part of the Supai Group but is equivalent to the lower slope- and lower-middle cliff-forming units of the Esplanade. Between miles 209-260 (km 336-418), the intertonguing Esplanade/Pakoon is an alternating sequence of red sandstone and gray limestone beds that form a series of ledges and small cliffs. The Esplanade and Pakoon have a similar basal conglomerate.

In eastern and central Grand Canyon, the base of the Esplanade is marked by a widely distributed basal conglomerate consisting mainly of limestone and siltstone pebbles and cobbles in a sandy matrix. The conglomerate is rarely more than 3 feet (1 m) thick, but locally, where it fills channels cut into the underlying Wescogame Formation, it may be up to 17 feet (5 m) thick. The lower slope-forming unit of the Esplanade is composed of about equal parts of mudstone and sandstone with some limestone-pebble conglomerate.

The middle cliff-forming unit of the Esplanade, in the eastern and central Grand Canyon, is mainly a fine-grained, well sorted sandstone with beds, 5 to 50 feet (2 to 15 m) thick (McKee, 1982). The dominant and characteristic structure is medium-scale, wedge-planar, to tabular-planar crossbedding. Stratification cosets (as defined by McKee and Weir, 1953, p. 383) are up to 33 feet (10 m) thick. Most of the cross-stratified units are characterized by climbing translatent strata indicative of eolian environments (Ronald C. Blakey, personal communication, 1987).

The uppermost Esplanade thickens westward from a thin bedded sequence of sandstone and siltstone in eastern Grand Canyon to a red-brown, slope-forming unit of sandy siltstone similar to the Hermit Shale, capped by a pale-red to white, fine-grained sandstone. Erosional stream channels, nearly 200 feet (61 m) deep, are cut into this upper slope and ledge unit in the central and western Grand Canyon areas.

The fossil record of the Esplanade Sandstone is limited to western Grand Canyon where its lateral marine equivalent, the Pakoon Limestone, contains abundant corals, fusilinids, and smaller foraminifers together with subordinate brachiopods, gastropods, and echinoderms (McKee, 1982, p. 89).

Wescogome Formation

The Wescogame Formation (Upper Pennsylvanian) is encountered at or near mile 15.0 (24.1 km) in Marble Canyon. It ranges from about 100 feet (30 m) to 225 feet (69 m) thick throughout Grand Canyon. The dominant lithology is sandstone in central Grand canyon, but mudstones increase to the east and limestone to the west (McKee, 1982). The base is marked in some places by a limestone, siltstone, and chert pebble conglomerate, about one meter thick. Locally the basal conglomerate fills channels up to 15 feet (5 m) thick and is absent in about half the localities examined by McKee (1982).

McKee (1982) recognized a lower cliff unit composed of fine-grained, cross-stratified sandstone and an upper slope unit composed of alternating red-brown, siltstone beds and cross-stratified sandstone. Both units are traceable by their topographic expression throughout most of the Grand Canyon. Large scale, tabular planar and wedge planar, cross-strata are common in the lower unit.

Marine fossils are rare but widely distributed in the Wescogame Formation. Several species of mollusks and corals as well as fusilinids, indicative of Virgilian (Late Pennsylvanian) age are reported by McKee (1982, p. 85). Trackways of four-footed vertebrates, probably reptiles, occur on cross-bedding surfaces in some sandstones of the Wescogame at several localities (McKee, 1982, p. 91-93).

Manakacha Formation

The Manakacha Formation is first exposed at river level near mile 20.2 (32.5 km) as a gray, slope-forming sequence of limestone and calcareous mudstone. Unlike most other Paleozoic units, the Manakacha is of relatively uniform thickness of about 200 to 275 feet (61-84 m) throughout Grand Canyon. Like the overlying Wescogame it is predominantly fine-grained, cross-stratified sandstone with lesser amounts of limestone and red mudstone. Crossbedding is medium-scale tabular planar to wedge planar (McKee, 1982). Thin wind-ripple laminations having reverse graded bedding are common in cross-beds of the Manakacha, and are considered diagnostic of eolian deposition (Ronald C. Blakey, personal communication, 1987).

The Manakacha crops out with a distinctive lower cliff unit and an upper slope unit throughout much of Grand Canyon. The sandstone grades westward to predominantly limestone west of Whitmore Wash (mile 188; 302.5 km) (McKee, 1982). A conglomerate (or zone of conglomerate beds), occurs at or near the base of the upper slope unit within the Manakacha. The beds are rarely more than one or two meters thick and contain primarily subangular chert or jasper pebbles with minor rounded limestone clasts. The base of the Manakacha is placed at the base of the lower cliff unit of resistant sandstone beds which overlie slope-forming mudstone and limestone of the Watahomigi Formation. It is not associated with a basal conglomerate as all other formations of the Supai Group.

Marine invertebrate fossils are sparse in the Manakacha but include several brachiopod species and a large bellerophontid gastropod. Fusilinids indicative of Atokan (early Middle Pennslyvanian) age occur as far east as the Marble Canyon area (McKee, 1982, p. 84). The upper part may be Atokan or Desmoinesian in age (McKee, 1975, p. J4).

Watahomigi Formation

The Watahomigi Formation (Lower and Middle Pennsylvanian, or Morrowan and early Atokan; McKee, 1975, p. J4) is the oldest formation of the Supai Group. It is first visible at about mile 21.2 (34 km) where it crops out as a slope-forming sequence of gray limestone and calcareous shale but it is often covered by talus. The formation consists of red and purple mudstone, siltstone, and gray limestone or dolomite with only a minor amount of sandstone. It ranges in thickness from about 80 feet (24 m) in eastern Grand Canyon to as much as 300 feet (91 m) in western Grand Canyon (McKee, 1982).

Three lithologic units, having distinctive topographic expression, are recognized in the Watahomigi throughout the Grand Canyon (McKee, 1982, p. 39). The lower slope unit is predominantly red or purple mudstone and siltstone with some thin limestone beds. In many localities, there is a basal chert pebble conglomerate. The Watahomigi unconformably overlies the Mississippian Redwall Limestone, or locally, the Upper Mississippian-Lower Pennsylvanian(?) Surprise Canyon Formation. The middle cliff unit of the Watahomigi is a limestone tongue that thickens westward and grades from a micritic texture in the east to grainstone and packstone in the west. The upper slope unit consists of alternating beds of limestone and siltstone or mudstone. At the base of the upper unit is a chert pebble conglomerate about 1-10 feet (0.4-3 m) thick that marks the Morrowan (Lower Pennsylvanian)-Atokan (lower Middle Pennsylvanian) boundary within the Watahomigi Formation (McKee, 1982, p. 191-193).

The Watahomigi Formation is the most fossilifereous unit of the Supai Group. Invertebrate marine fossils are abundant in the limestone beds of all three units. Brachiopods typical of Morrowan (Early Pennsylvanian) age occur in both the lower slope and middle cliff units. Fusilinids diagnostic of Atokan (early Middle Pennsylvanian) age occur in the upper slope unit (Mackenzie Gordon in McKee, 1982, p. 122-123.)

SURPRISE CANYON FORMATION

The Surprise Canyon Formation (Upper Mississippian and Lower Pennsylvanian(?)) is mostly covered by talus in Marble Canyon at about mile 23 (37 km). It can be seen on the right bank at mile 23.3 (37.5 km) (fig. 1.9) and again at mile 24.5 (37.5 km). The Surprise Canyon forms isolated, lens-shaped outcrops of dark red-brown strata that fill shallow to deep erosional channels up to 400 feet (122 m) deep cut into the top of the Redwall Limestone (Billingsley and Beus, 1985).

In eastern Grand Canyon and Marble Canyon the outcrops are small and rarely more than 30 to 70 feet (9-21 m) thick. These outcrops typically consist of a basal chert-pebble conglomerate overlain by red-brown siltstone, fine-grained sandstone, and pisolitic ironstone, that grade upward into a pale red-brown sandstone or mudstone with a few conglomerate lenses. The basal contact with the underlying Redwall Limestone is everywhere sharp and well defined. The upper contact with the overlying Watahomigi Formation is, in many places, obscure but is placed where pale red-brown mudstone is overlain by purple-gray mudstone more typical of the Watahomigi Formation.

In central and western Grand Canyon, the Surprise Canyon lenses are thicker and more widespread. They typically consist of a basal slope of chert pebble conglomerate with occasional limestone pebbles, to cobble conglomerate that grades upward into medium-grained sandstone; a middle cliff unit of fossiliferous, skeletal, grainy, yellow-brown, silty limestone; and an upper slope and ledge unit of siltstone and silty, sandy, or algal limestone. Terrestrial plant fossils, including several forms of *Lepidodendron*, occur locally in the lower unit. Abundant marine invertebrate fossils, including diagnostic conodonts and the brachiopod *Rhipiodomella nevadensis*, indicative of late Chesterian (latest Mississippian) age, occur in the middle unit. Ripple marks, ripple laminations and algal fossils are typical in the upper unit.

REDWALL LIMESTONE

The Redwall Limestone (Lower to Upper Mississippian) forms a red-stained, gray limestone cliff about 450 feet (137 m) thick in eastern Grand Canyon, and up to about 700 feet (213 m) or more in the western Grand Canyon. The Redwall is first exposed at about mile 23 (37 km) in Marble Canyon with the overlying Surprise Canyon Formation. The name Redwall Limestone was proposed by Gilbert (1875, p. 177). McKee (1963), subdivided the Redwall into four distinct members. In ascending order, they are the Whitmore Wash, Thunder Springs, Mooney Falls, and Horseshoe Mesa Members. They are described in detail by McKee and Gutschick (1969). All four members are recognized throughout the Grand Canyon area.

Composition of the Whitmore Wash Member is mainly a fine-grained limestone or dolomite. Common textural carbonate types are pelleted, skeletal or oolitic limestones. The Whitmore Wash thickens from about 50 feet (15 m) in eastern Grand Canyon to about 120 feet (36 m) in western Grand Canyon. Brachiopod and foraminiferid fossils indicative of an early Mississippian (late Kinderhookian to middle Osagean) age are found in this member.

The Thunder Springs Member is the most distinctive unit in the Redwall owing to its light and dark banded appearance imparted by alternating gray carbonate and white chert beds. The gray crinoidal limestone beds are typically thin bedded whereas the darker weathering white chert beds and lenses are commonly silicified former bryozoan limestones and mudstones. Thickness of the Thunder Springs increases from about 100 feet (30 m) in eastern Grand Canyon to about 140 feet (43 m) in western Grand Canyon. Invertebrate fossils are abundant in the Thunder Springs and include diagnostic conodont microfossils of Osagean (late Early Mississippian) age (Racey, 1974; Ritter, 1983).

The thickest unit of the Redwall is the Mooney Falls Member, ranging from about 250 feet (76 m) in eastern Grand Canyon to about 340 feet (104 m) in western Grand Canyon. It is nearly a pure, light-gray limestone containing pellets, oolites, or crinoid skeletal grains. It forms a major part of the red-stained sheer wall, to which the name Redwall refers. Invertebrate marine fossils are abundant throughout this member and include conodonts and foraminifera indicative of a late Osagean to Meramecian (early Late Mississippian) age (Skipp,

1969; Ritter, 1983).

The uppermost and thinnest unit of the Redwall is the Horseshoe Mesa Member. It is composed of thin-bedded, light-gray, fine-grained limestone that typically forms a series of weak receding ledges in contrast to the massive cliff of Mooney Falls below. Thickness is variable and ranges from about 0 to 100 feet (0-31 m) throughout the Grand Canyon. It is locally absent near the vicinity of Surprise Canyon deposits. Invertebrate megafossils are present but rare. Foraminifera indicative of Meramecian (early Late Mississippian) age have been identified by Skipp (1979, p. 298).

TEMPLE BUTTE FORMATION (LIMESTONE)

The Temple Butte Formation (Middle? and Upper Devonian), originally designated Temple Butte Limestone by Walcott (1890) and subsequently called the Temple Butte Formation by Beus (1973), is exposed as small, intermittent, lens-shaped channel-fill outcrops cut into the underlying Cambrian unclassified dolomite and Muav Limestone in the Marble Canyon area. The first lense-shaped outcrop is at mile 37.7 (60.7 km) (fig. 1.12). Most of the channels are less then 100 feet (31 m) deep and up to 400 feet (122 m) wide. The purple-gray dolomite and sandstone deposits, that fill the channels, are unconformably overlain by the Redwall Limestone. The Devonian channels and their deposits become more widespread and the channels shallower in the western Grand Canyon. Starting near mile 60 (97 km), the Temple Butte becomes a continuous deposit of dark gray, dolomite ledges overlying local purplish, channel-fill deposits. The unit thickens to more than 450 feet (137 m) farther west.

Lithology of the Temple Butte Formation is predominantly light- to dark-gray, fine- to medium-grained, crystalline dolomite with minor sandy dolomite and quartz sandstone beds. The basal channel-fill strata are commonly sandy to silty reddish-purple dolomite. Bedding is locally irregular or gnarly, in contrast to the thin, at places more uniform bedding of the continuous dolomite outcrops to the west. Walcott (1883, p. 221) reported brachiopods, gastropods, corals and "placoganoid" fish plates from lower Kanab Canyon (mi 143.2; 230.4 km), and Noble (1922), collected fish plates of *Bothriolepis* from Sapphire Canyon (mi 101; 162.5 km). Rare conodont microfossils are reported from the Temple Butte in Matkatamiba Canyon, (mi 148; 238 km), are the most time-diagnostic fossils obtained (Elston and Bressler, 1977, p. 423). Conodont forms identified by Dietmar Schumacher (written communication, 1978) indicate a probable age of late Givetian (latest Middle Devonian) through early Frasnian (earliest Late Devonian).

THE CAMBRIAN ROCKS

The Cambrian consists of three lithologic rock types. From top to bottom, they are the upper unclassified dolomite (upper Cambrian) and Muav Limestone (Middle Cambrian); Bright Angel Shale (Middle Cambrian); and Tapeats Sandstone (Middle and Lower Cambrian). The Muav, Bright Angel, and Tapeats belong to the Tonto Group. The unclassified dolomite (Noble, 1922; McKee and Resser, 1945) is excluded from the Tonto Group and is considered to be equivalent to the Nopah Formation (chapter 15). The first Cambrian rocks exposed by the Colorado River are at river mile 35.1 (56.5 km). Here the Redwall Limestone disconformably overlies the unclassified dolomite (fig. 1.11). The erosional unconformity between the Redwall and the Cambrian rocks is of a much greater magnitude than appears. Rocks of Devonian age probably covered the Cambrian rocks in this area as they do over much of Grand Canyon, but have been mostly eroded from the Marble Canyon area. The disconformity between the Devonian and Cambrian rocks is widespread and strata of Ordovician and Silurian age are not recognized in the Grand Canyon region (fig. 13.1).

The unclassified dolomite is disconformable on the Muav Limestone throughout the Grand Canyon region (fig. 15.3). This unit consists of white to dark gray, thin- to medium-bedded dolomite (dolostone); shale partings between beds are common in the lower part. This dolomite unit forms a series of ledges, cliffs, and slopes.

The Muav Limestone is a thin-bedded, gray, medium- to fine-grained, mottled dolomite; a coarse- to medium-grained, grayish-white, sandy dolomite; and a fine-grained limestone that weathers to a dark gray or rusty-orange color and forms cliffs or small ledges. The Bright Angel Shale is a green and red-brown, micaceous, fissile shale and siltstone that weathers to a slope of the same colors. The Tapeats Sandstone is a conglomeratic, very coarse- to medium-grained, thin-bedded, sandstone; it exhibits small-scale cross-stratification and weathers to a tan or reddish-brown cliff. These three formations, as originally described by Noble (1922) from the Bass Trail in the eastern Grand Canyon (mi 107.7; 173.6 km), are recognized on the basis of their distinct lithologies. McKee and Resser (1945) documented that the three formations intertongue over a considerable lateral distance and represent cyclic transgressions and regressions of the Cambrian sea. Therefore, all the dolomite and limestone lithology in the Cambrian sequence, as defined in this manner, belongs to the Muav Limestone; the shale and siltstones belong to the Bright Angel Shale; and the sandstone and conglomerate belong to the Tapeats Sandstone. Because of the interbedded nature of these mixed lithologies, it is difficult to place mappable stratigraphic boundaries between the three units of the Tonto Group (chapters 14 and 15).

CHAPTER 14: CAMBRIAN STRATIGRAPHIC NOMENCLATURE, GRAND CANYON, ARIZONA - MAPPERS NIGHTMARE

Peter W. Huntoon
Department of Geology and Geophysics,
University of Wyoming, Laramie, Wyoming

McKee and Resser (1945) published a pedagogic classic when they described the Cambrian section in the Grand Canyon. Their work shows cyclic transgressions and regressions of the Cambrian sea in impeccable detail. The work was flawed, however, by their desire to honor Noble's (1922) pre-existing Cambrian formational nomenclature. Noble defined three formations, in ascending order the Tapeats Sandstone, Bright Angel Shale, and Muav Limestone, on the basis of each having a common lithology, -- a perfectly appropriate framework for the superimposed sequence that he described along the Bass Trail in the eastern Grand Canyon. McKee and Resser, however, documented that the three units were intertonguing entities over scores of miles, a fact that is dramatically exposed in the walls of the western Grand Canyon district (figs. 14.1, 1.32).

By adopting Noble's lithology-based nomenclature, McKee and Resser (1945) inadvertently violated the principal rule in defining formations - mappability. As one views the Cambrian section as defined by McKee and Resser (1945) in the western Grand Canyon, all the limestone and dolomite units are members of the Muav Limestone, the shale units are members of the Bright Angel Shale. Consequently one can climb through a series of rocks that alternate back and forth between the Bright Angel Shale and Muav Limestone.

This quirk in the nomenclature led to one costly series of water well prospecting failures on the Hualapai Plateau of western Grand Canyon (fig. 4.1) documented by Huntoon (1977a). The problem developed when Twenter (1962) correctly identified the Rampart Cave Member of the Muav Limestone as the best ground water exploration target under the Hualapai Plateau based on the stratigraphic position of springs discharging from the base of the unit in the Grand Canyon. He recommended that the drilling for ground water supplies penetrate the entire thickness of the Muav Limestone. However he did not make it clear that in order to reach the Rampart

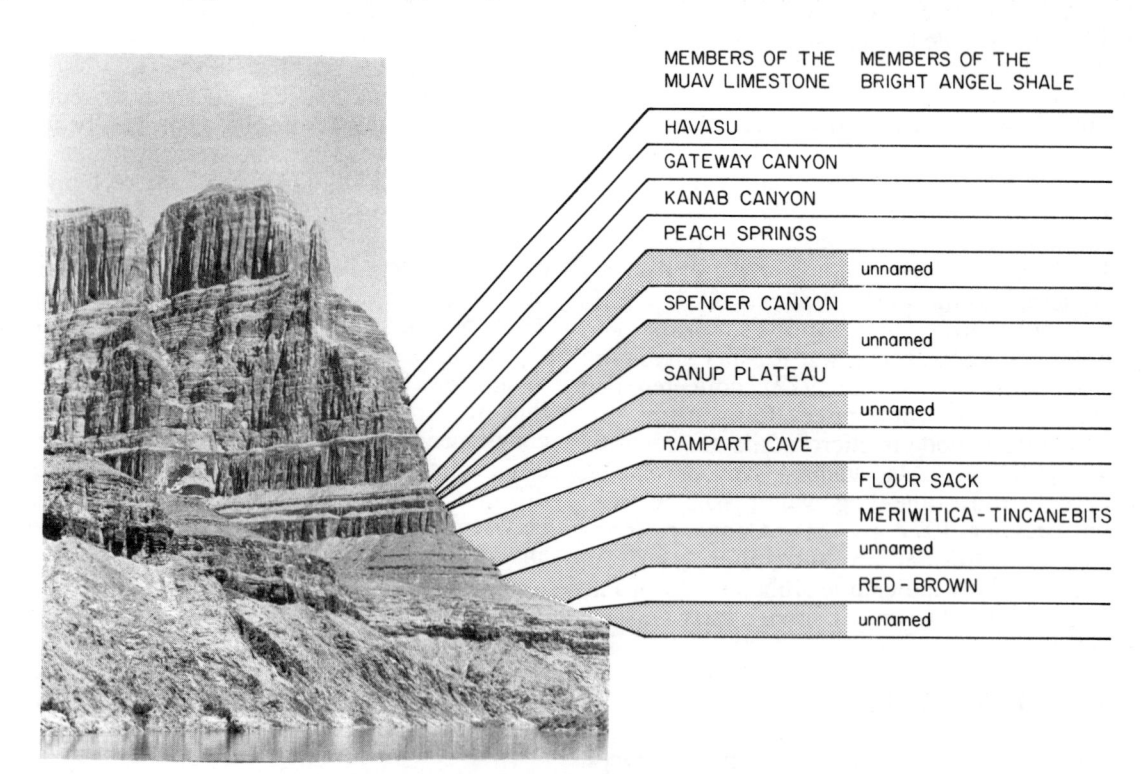

FIGURE 14.1. Intertonguing of the Muav Limestone (unshaded) and Bright Angel Shale (shaded) in the western Grand Canyon, Arizona. Nomenclature and spelling of Meriwitica and Tincanebits from McKee and Resser (1945).

Cave Member, the holes would have to penetrate three thin units of the overlying Bright Angel Shale (fig. 14.1). When the drilling program commenced, the first two holes were terminated at depths of 720 and 800 feet (219-244 m) when classic Bright Angel Shale cuttings were retrieved from the holes. The holes were bottomed in a Bright Angel Shale unit lying between the Peach Springs and Spencer Canyon Members of the Muav Limestone, 200 to 300 feet (61-91 m) short of their targets.

McKee and Resser (1945) were not totally consistent in subdividing and naming the Cambrian section on the basis of lithology. They included in the lower Bright Angel Shale the red-brown sandstone, a unit which is lithologically similar to the Tapeats Sandstone and which was deposited in an environment to that of the Tapeats Sandstone. Huntoon and others (1981, 1982) and Billingsley and Huntoon (1983) arbitrarily used the base of the Rampart Cave Member of the Muav Limestone as their mappable Muav-Bright Angel contact on these three western Grand Canyon geologic maps. Unfortunately the Rampart Cave Member does not extend as a limestone unit into the eastern Grand Canyon but undergoes a facies change to Bright Angel Shale lithology. Thus, the base of the Peach Springs Member of the Muav Limestone (the base of the Muav Limestone originally defined by Noble, 1922) was used for this contact on the eastern Grand Canyon geologic map. This has resulted in a discontinuity between the geologic maps of the western and eastern Grand Canyon. Work is progressing on redefinition of the Cambrian stratigraphic nomenclature, an effort that will honor McKee and Resser's findings that most of the Cambrian section represents a common depositional environment, and that it could be subdivided into formations that more closely honor the criterion of mappability.

View north from south rim to Unkar Creek debris fan and Unkar Rapids (mile 73, km 118; A). Gravels at B and at Hilltop Ruins (C); see chapters 1, 21, and 24 for descriptions and discussions.

CHAPTER 15: CORRELATIONS AND FACIES CHANGES IN LOWER AND MIDDLE CAMBRIAN TONTO GROUP, GRAND CANYON, ARIZONA

Donald P. Elston
U.S. Geological Survey, Flagstaff, Arizona

INTRODUCTION

This report builds on the classic study by McKee and Resser (1945) of the Cambrian Tonto Group of the Grand Canyon (Gilbert, 1874, 1875, p.184; Noble, 1914, p. 61; Noble, 1922). The correlations and interpretations of McKee and Resser (1945) are shown in Figures 15.1 and 15.2. This report presents additional sections of the Tonto Group in the Grand Canyon and Marble Canyon, refines previous correlations, and introduces a revised nomenclature for the group (fig. 15.3).

The revised nomenclature proposed in Figure 15.3 is preliminary and open for interchange and discussion during the International Geological Congress Grand Canyon field trips scheduled for June and July, 1989. For this reason, the nomenclature of McKee and Resser (1945) for the Tonto Group has been retained throughout the other parts of this guidebook. The informal nomenclatural scheme proposed in Figure 15.3 and used in this chapter is introduced as a working solution that has not been formally proposed.

HISTORY OF WORK

Edwin D. McKee (personal communication, about 1974) recounted to me his efforts involved with the completion of his field studies that allowed his classic report on stratigraphy of the Tonto Group to be prepared and published. Field work was begun in the western Grand Canyon in 1936. By 1939 studies had progressed eastward into the area of Noble's (1922) sections at the Bass and Hermit Trails. With the imminence of World War II, a decision was made to complete the study as expeditiously as possible to allow the Carnegie Institution of Washington to publish the results before onset of war. (As it turned out, publication followed the war, in 1945.) Two sections of Bass and Bright Angel Trails, and a section measured by Wheeler and Kerr (1936) along the Tanner Trail, were incorporated in the McKee's study. McKee measured his final section north of Palisades Creek and opposite Lava Canyon (mi 65.5; km 105; identified as "Little Colorado" on McKee's diagrams). As seen in McKee's diagrams (figs. 15.1 and 15.2), the Bright Angel Shale and Muav Limestone of the Tonto Group in the Big Bend area of the eastern Grand Canyon are dominated by sandstone; therefore, some correlations with the type section, a relatively short distance to the west, were uncertain.

During a continuing series of boat trips through Marble and Grand Canyon in 1970's, it was possible to measure additional sections in the Tonto Group north of the easternmost section of McKee, as well as to measure additional sections in the central and western Grand Canyon. Correlation of Cambrian units and the character of facies changes in the Tonto Group has been the subject of scrutiny during more than 25 river trips through the canyon. The results are summarized in Figure 15.3

UNCLASSIFIED DOLOMITE

The type section of the Tonto Group, measured by Noble (1922) along the Bass Trail, consists of the Tapeats Sandstone, Bright Angel Shale, and Muav Limestone. The Muav Limestone of Noble (1922) included a dolostone unit at the top, which McKee and Resser (1945) excluded from the Tonto Group and called "unclassified dolomite." Following the practice of McKee, the unclassified dolostone is excluded from the Tonto Group, and that term is retained here. The unclassified dolostone was recognized by F.G. Poole and D.P. Elston (reconnaissance study, March 1981) as correlative with dolostone underlying the Dunderburg Shale Member of the Nopah Formation (Upper Cambrian) exposed in the Virgin River gorge along Interstate Highway I-15 in the northwestern corner of Arizona. Thus, this dolostone is shown as "Nopah equivalent" in Figure 15.3.

The "unclassified dolomite" is disconformable on the Havasu Member of the Muav Limestone in Marble Canyon (chapter 1; mi 35-38, km 56-61). Not surprisingly, this dolostone unit does not exhibit a southerly facies change observed in underlying strata of the Muav (chapter 1). Recently, this unit has been called the "Grand Wash Dolomite" in an unpublished MS thesis from a study carried out in the western Grand Canyon (Brathovde, 1986), but the the name "Grand Wash" cannot be used because it is preempted for a different stratigraphic unit in the same area (Grand Wash Basalt). Additional work is needed before a decision is made on whether the name Nopah is to be extended into the Grand Canyon region or whether an entirely new name is to be established.

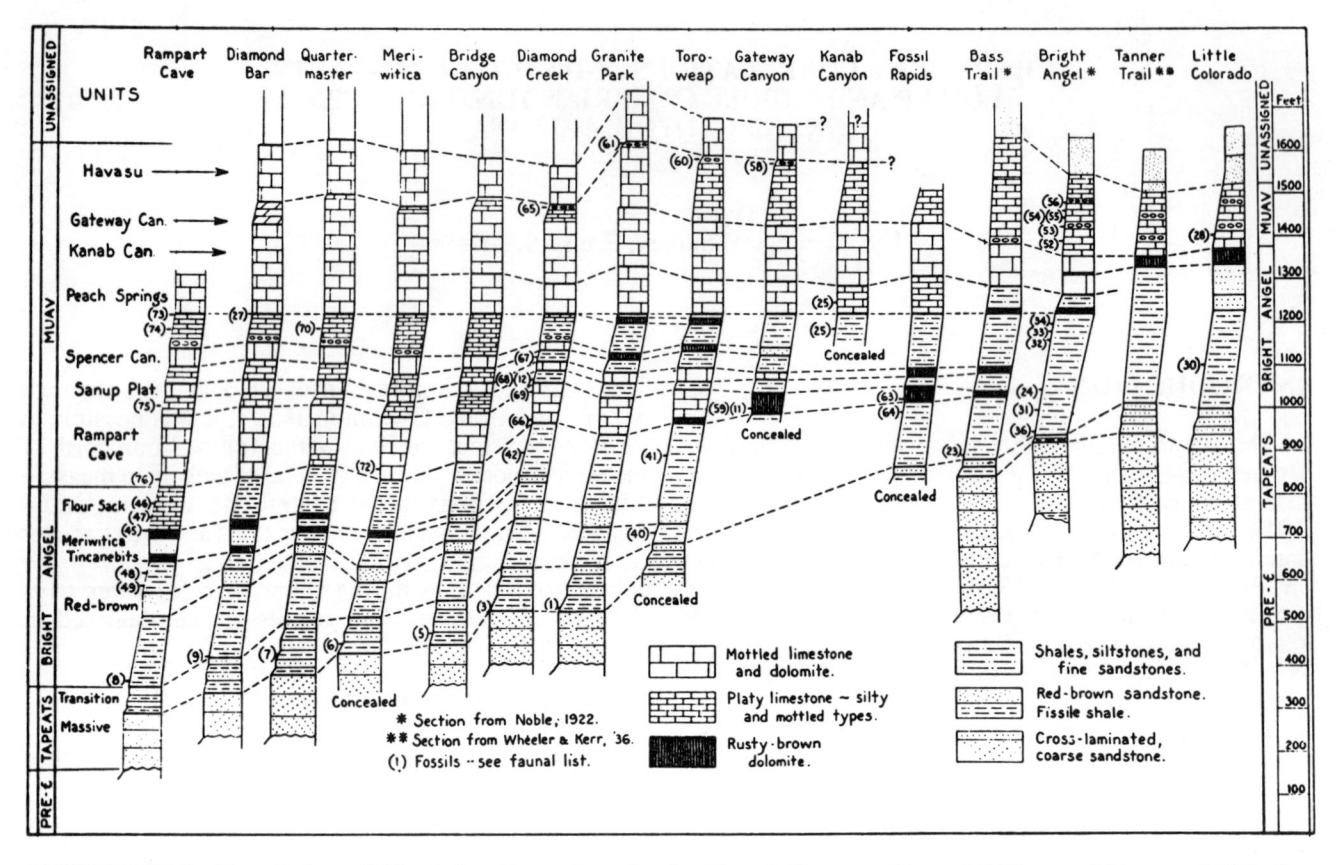

FIGURE 15.1 Correlation of Cambrian formations in the Grand Canyon. (from McKee and Resser, 1945, fig. 2B).

TONTO GROUP

Tapeats Sandstone

Central and Eastern Grand Canyon. At the type section of the Tonto Group along the Bass Trail (Noble, 1922), and throughout much of the central and eastern Grand Canyon, the Tapeats Sandstone consists principally of a single depositional unit. However, the lower part at places (for example, the Deer Creek section) is dominantly parallel bedded and cross laminated, in contrast to the more highly cross-bedded overlying beds. Additionally, a cross-bedded unit near the top of the formation at Deer Creek is underlain by a shaly and sandy unit that is recognized eastward as far as the Bass Trail. This cross-bedded unit is the easternmost extension of the red-brown sandstone of the Bright Angel Shale of McKee and Resser (1945) in the western Grand Canyon. The parallel-bedded sandstone at the base of the Tapeats in the central and, locally, in the eastern Grand Canyon probably correlates with part of the unnamed shaly and sandy unit that lies between the massive sandstone member and the red-brown sandstone in the western Grand Canyon.

Western Grand Canyon. Three units of the Tapeats Sandstone are recognized in the western Grand Canyon: 1) a lower massive sandstone; 2) a middle unnamed unit consisting of shale and interbedded sandstone of Tapeats lithology; and 3) an upper red-brown sandstone, which also has Tapeats lithology. A lithologic transition zone from sandstone to shale overlies the red-brown sandstone in the western to eastern Grand Canyon. The loss of citrine quartz and the appearance of abundant glauconitic sand mark the top of the transition zone. Zircons from an ash-fall tuff near the base of the red-brown sandstone at mile 205.7 (km 329) was dated at 563 ± 49 Ma from fission track analysis (table 17.3). This tuff appears to approximately correlate with the *Zacanthoides* cf *walpai* horizon of the Toroweap section, considered by McKee and Resser (1945) to be of early Middle Cambrian age.

The massive sandstone member is a grayish-red to purple, cross-bedded, quartzitic sandstone that has a laterally persistent bleached zone at the top. The *Olenellus* horizon is in shale just above the top of the member. The easternmost exposure of the massive sandstone member along the Colorado River is at mile 205.7 (km 329), and it persists to the Grand Wash Cliffs to the west. This member also is present on Whitney Ridge, Nevada, west of the Grand Wash Cliffs, where it is called the Prospect Mountain Quartzite (of Proterozoic and Early Cambrian age) and closely underlies the *Olenellus* horizon in the lower part of the Pioche Shale (F.G. Poole and D. P. Elston, reconnaissance, March 1981). Because the red-brown sandstone also is present within the Pioche Shale on Whitney Ridge, the lower part of the Pioche Shale correlates with the unnamed shaly middle unit of the Tapeats Sandstone in the western Grand Canyon. This unnamed interval appears

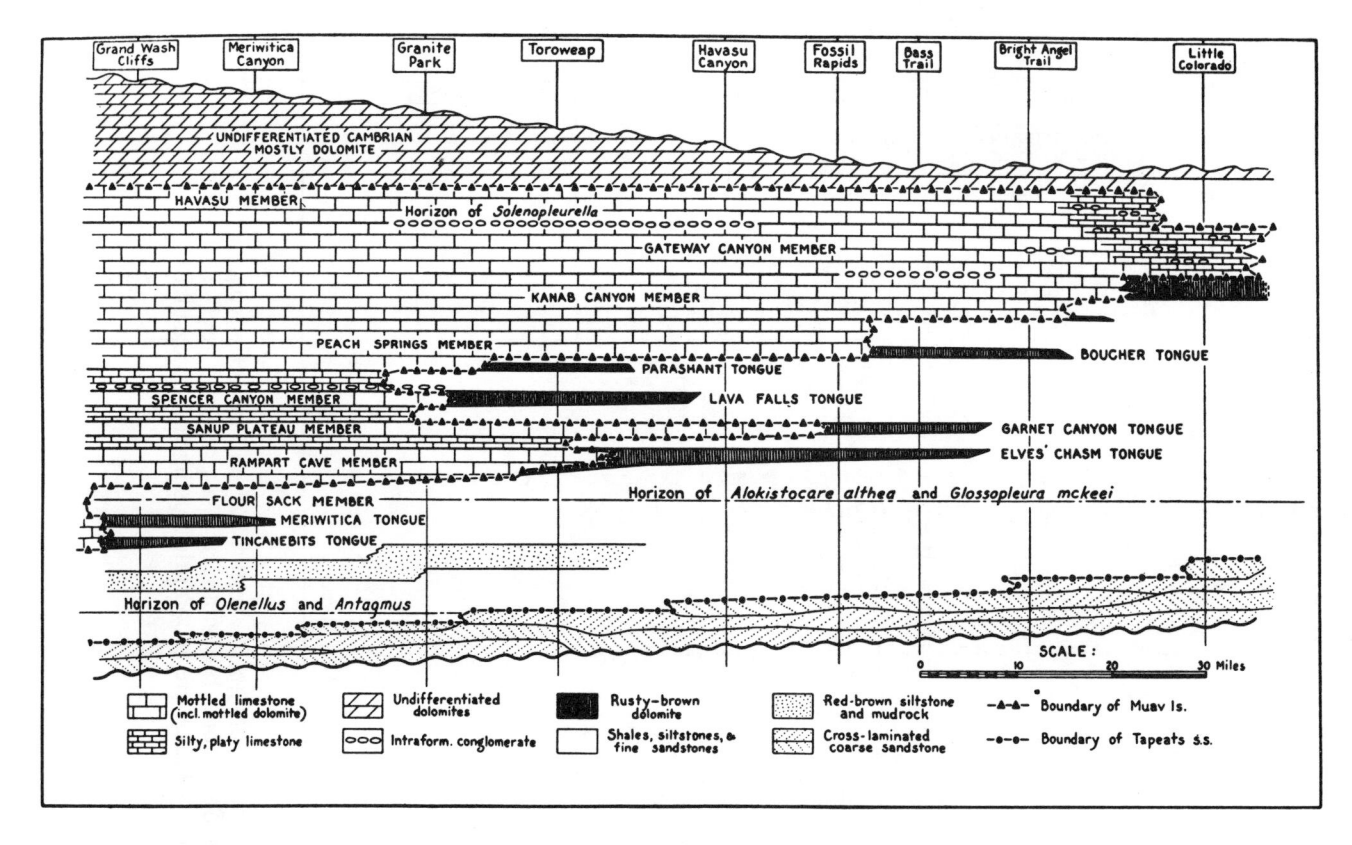

FIGURE 15.2 Diagrammatic section of Cambrian deposits in Grand Canyon, showing stages in transgression and regression and distribution of facies from east to west. Time planes are horizontal; actual thickness ranges from 1,500 feet (457 m) in west to 800 feet (244 m) in east. (from Mckee and Resser, 1945, fig. 1).

to become more sandy in the vicinity of mile 190 (km 304), where only the middle and upper members are recognized in the Tapeats. The massive sandstone member is inferred to wedge out by onlap onto the crystalline basement east of mile 205.7 (km 329). A major episode of eastward marine transgression thus appears to have occurred at about the level of the *Olenellus* horizon. The somewhat reduced thickness of the Tapeats east of mile 205.7 supports this supposition. However, west of Granite Park (mi 208; km 335), the massive sandstone member locally wedges out against a basement "high" in the vicinity of mile 211 (km 339).

Bright Angel Shale

A number of distinctive members occur in an interval correlative with the Bright Angel Shale of the Bass Trail section. Near the Grand Wash Cliffs on the west, the section as shown by McKee and Resser (figs. 15.1 and 15.2) is mostly limestone, and the proportion of shale increases in an easterly direction. As seen in Figures 15.1 and 15.2, limestone beds on the west pass laterally into dolostone beds in the area of the central Grand Canyon, and these in turn become glauconitic sandstone beds in the eastern Grand Canyon (fig. 15.3). Exposed at river level from mile 50 to 60 (km 80-97), the Bright Angel consists dominantly of silty sandstone. From miles 50 to 60 (km 80-97), intervals of slightly more resistant, magenta-colored, glauconitic sandstone are seen to be distributed across a sandstone section of the Bright Angel similar to the distribution of dolostone and sandy dolostone members in the Bright Angel Shale to the west. A provisional correlation of these beds in the Bright Angel section at mile 60 (km 97) with members of the Bright Angel Shale to the west is proposed in Figure 15.3.

The Bright Angel Shale of the west-central and western Grand Canyon perhaps should be called the Bright Angel Formation, rather than Bright Angel Shale, because of its diverse lithology and an abundance of carbonate rock; the designation Bright Angel Sandstone might also be applied in the easternmost Grand Canyon where the Bright Angel is a sandstone. The change of the rank term on the west should correspond to the place where the Rampart Cave Member of the Muav Limestone of McKee and Resser (1945) grades from a dolostone to a limestone just east of Kanab Canyon (mi 140; km 225).

As documented by McKee and Resser (1945; figs. 15.1 and 15.2), facies changes in the Bright Angel are seen in an east-to-west direction (fig. 15.3). The purplish or "magenta-colored" sandstone beds of the Bright Angel appear to correlate with beds that become the yellowish-brown to reddish brown, "snuff-colored" beds of dolostone in the central Grand Canyon, and these dolostones grade laterally into limestone members of the Bright Angel to the west. A westward increase in thickness accompanies the facies changes in the Bright Angel, and sandstone and mudstone-siltstone decrease as limestone increases.

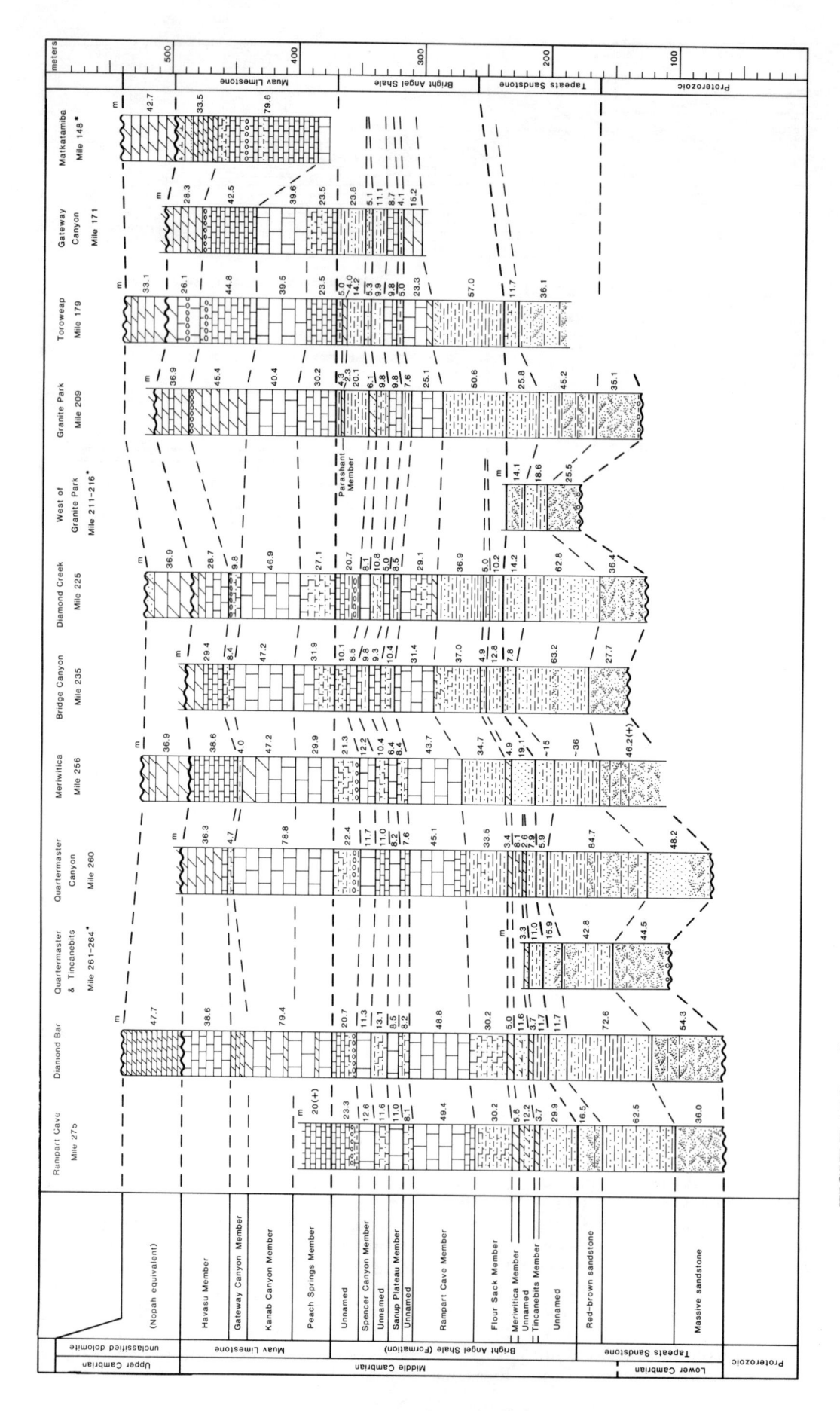

FIGURE 15.3a Correlation of Cambrian Tonto Group showing facies changes in western half of Grand Canyon, wedge out of lower massive sandstone member of Tapeats Sandstone by onlap, and a proposed revised nomenclature. Lithology of the sections of McKee and Resser (1945) were recompiled using their measured sections.

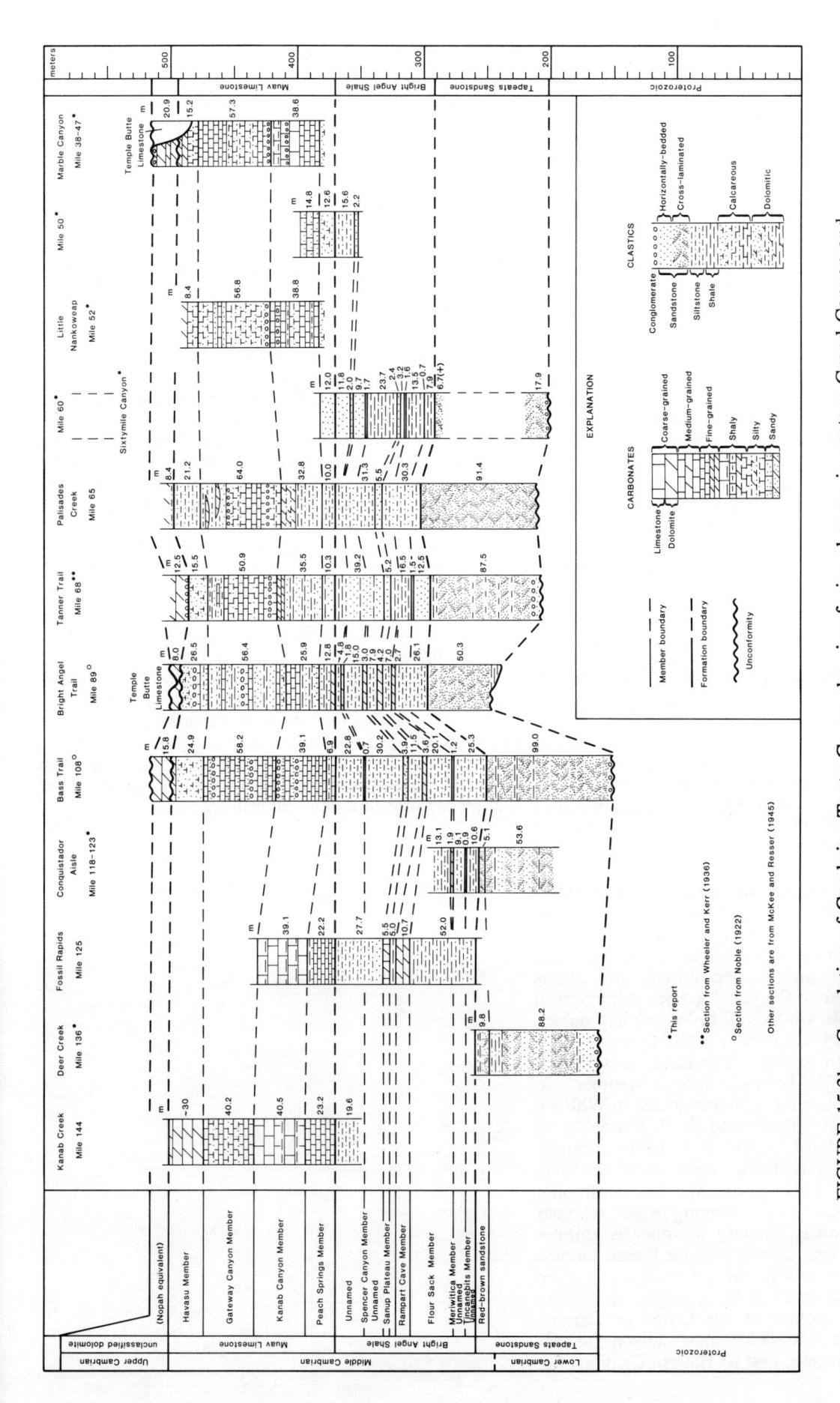

FIGURE 15.3b Correlation of Cambrian Tonto Group showing facies changes in eastern Grand Canyon and Marble Canyon and a proposed revised nomenclature. Measured sections of Noble (1922), Wheeler and Kerr (1936), and McKee and Resser (1945), supplemented by sections measured by D. P. Elston and a number of associates, 1974-1978, including E.M. Shoemaker, S.L. Gillett, and G. Mercado. M. L. Dennis is responsible for the careful recompilation of the sections and the drafting of Figure 15.3.

Underlying the Peach Springs Member of the Muav Limestone and overlying the Flour Sack Member of the Bright Angel on the west are four members or tongues named by McKee and Resser (1945) that here are assigned to the Bright Angel rather than the Muav Limestone. This assignment and the lateral continuity of the members has removed the need for an upward shingling base of the Muav Limestone, recognized by McKee and Resser (1945) in their lithology-based nomenclature. The Parashant Tongue at the top of the Bright Angel Shale has limited distribution, but appears to be present locally in the area of Conquistador Aisle (not shown on section) as well as in the Granite Park-Toroweap area.

Muav Limestone

The four members of the Muav Limestone as defined by McKee and Resser (1945) -- Peach Springs, Kanab Canyon, Gateway Canyon, and Havasu Members -- (figs. 15.1 and 15.2) are retained in this report (fig. 15.3). Following the precedent of McKee and Resser (1945), the base of the Peach Springs Member of the Muav has been used as a datum in construction of Figure 15.3. A north-to-south facies change from carbonate to sandstone is observed in the Muav Limestone in Marble Canyon and the easternmost Grand Canyon (fig. 15.3). Marker carbonate beds persist into the dominantly sandstone sections of the Big Bend area measured by McKee ("Little Colorado River" or Palisades Creek section) and by Wheeler and Kerr (1936; Tanner Trail section). The dominantly sandy nature of the Bright Angel and Muav in this area indicates that the shore line was near the Big Bend area in the Grand Canyon. Away from this area, the laterally traceable members of the Bright Angel and Muav appear to be contemporaneous units that reflect slightly changing environments of marine deposition rather than major transgressions and regressions of the epeiric sea.

Members of the Muav Limestone remain fairly constant in lithologic character following the transition from sandstone and sandy carbonate to carbonate. The Peach Springs Member most commonly is represented by a single relatively uniform depositional unit, and its contact with the Kanab Canyon Member is marked in many places by a thin ash-fall tuff bed which has yielded a zircon fission-track age of 535 ± 48 Ma (table 17.3) from a site a short distance below Lava Falls Rapids (chapter 1; mi 179.8). The Peach Springs Member, and the tuff bed at the top, have been recognized in Whitney Ridge, Nevada (F. G. Poole and D. P. Elston, field reconnaissance, March 1981) and in Frenchman Mountain, Nevada (F.G. Poole, field reconnaissance, July 1981). The Kanab Canyon Member commonly crops out as two resistant, cliff-forming ledges, whereas the overlying Gateway Canyon Member is thinner bedded, tends to be less resistant than the Kanab Canyon Member, and commonly is characterized by a steep slope/cliff/slope rather than by a single cliff. An abnormally thick section of the Gateway Canyon Member was measured at Matkatamiba Canyon (mi 148; km 238), a short distance east of Havasu Canyon (fig. 15.3). The reason for its greater thickness is unknown. The Havasu Member at the top of the Muav is cliff-forming, and characteristically consists of a lower unit of limestone overlain by a unit of dolostone.

CONCLUSIONS

The correlations proposed in Figure 15.3 indicate that following deposition of the massive sandstone member (= Prospect Mountain Quartzite of Nevada) of the Tapeats Sandstone in the western Grand Canyon, an eastward transgression of the epicontinental sea across the central and eastern Grand Canyon area occurred at or near the *Olenellus* horizon, which lies a few feet above the top of the massive sandstone member. The red brown sandstone member in the west traces into the upper part of the Tapeats Sandstone in the central and eastern Grand Canyon, and the underlying shaly interval in the west passes into parallel-bedded, cross-laminated sandstone east of Granite Park. Correlations above the Tapeats Sandstone indicate that a series of marker beds, identified as members of the Bright Angel Shale, record facies changes reflecting slight shifts in environments of marine deposition rather than major transgressions and regressions of the epeiric sea as concluded by McKee and Resser (1945). The latter conclusion arose from the use of a lithologic-based nomenclature for the the Tonto Group in the western Grand Canyon, and the upward stair-stepping of the base of the "Muav" eastward, across stratigraphic horizons where limestone units pass laterally into dolostone in an easterly direction (figs. 15.1 and 15.2). The stair-stepping correlation gave rise to the mapping problem recounted in Chapter 14 (an upward shingling of the base of the Muav Limestone from west to east), and to problems involving the identification of target horizons in water-well drill holes.

CHAPTER 16: MESOZOIC STRATA AT LEES FERRY, ARIZONA

George H. Billingsley
U.S. Geological Survey, Flagstaff, Arizona

Most of the Colorado River trips through Grand Canyon begin at Lees Ferry, Arizona, which is nearly surrounded by beautiful red rock strata of Mesozoic age: the Vermilion Cliffs to the west and Echo cliffs cliffs to the north and east (fig. 4.1). This is the first and last place that rocks of this age are viewed while on a trip through Grand Canyon. The Colorado River flows southward from Lees Ferry in Lower Triassic strata for about one mile (1.5 km) before encountering Paleozoic strata (fig.1.4).

Mesozoic rocks once covered most, if not all, of the Grand Canyon area to an unknown thickness, perhaps as much as 2,000 feet (610 m), but perhaps much less as the margin of the Colorado Plateau is approached. Erosional processes began removing the Mesozoic from the Grand Canyon area about 70 million years ago (chapter 18). Only a few isolated remnants of Mesozoic strata are found at various places surrounding the Grand Canyon.

The lowest and oldest Mesozoic unit is the Moenkopi Formation of Middle? and Early Triassic age (figs. 1.1, 1.3). The Moenkopi consists mostly of red siltstone and sandstone, about 400 feet (122 m) thick, that forms a slope above a light-gray cliff of Permian limestone (Kaibab Formation). The Moenkopi is seen on the left bank of the river just downstream from the boat landing (fig. 1.3). The Moenkopi and overlying strata dip upstream towards the northeast reflecting the Echo Cliffs monocline which trends to the northwest through the boat landing area (fig. 4.1).

The Upper Triassic Chinle Formation overlies the Moenkopi Formation. At Lees Ferry, the Chinle is divided into three members; the basal Shinarump, the middle Petrified Forest, and the upper Owl Rock Members (figs. 1.2, 1.3). The Shinarump Member is a light-brown, coarse-grained sandstone and conglomerate that forms a cliff. The Shinarump contains abundant petrified logs and wood fragments. It fills erosional channels cut into the Moenkopi Formation, and thins and locally pinches out away from the stream channels. The Shinarump averages about 45 feet (14 m) in thickness,and has a maximum thickness of about 150 feet (46 m) (Phoenix, 1963). The cable for the water gaging station at Lees Ferry is anchored to the Shinarump on the left bank of the Colorado River near the boat landing (fig. 1.3).

The Shinarump grades upward into multicolored, soft shale, mudstone, and sandstone of the Petrified Forest Member (fig. 1.3). These sediments were deposited by meandering rivers in a floodplain environment and contain many petrified logs, plant fossils, and early dinosaur remains. The Petrified Forest is about 800 feet (244 m) thick in the Lees Ferry area.

The Petrified Forest Member grades upward into green and red-brown, mudstone, sandstone, silty limestone, and conglomerate of the Owl Rock Member. The Owl Rock forms a series of slopes and ledges and averages about 200 feet (61 m) in thickness. The three members of the Chinle Formation total about 1,050 feet (320 m) thick at Lees Ferry (Phoenix, 1963).

An erosional unconformity separates the Chinle Formation from the overlying Moenave Formation. The Moenave (Upper Triassic) is divided into the basal Dinosaur Canyon Sandstone Member and the upper Springdale Sandstone Member (fig. 1.2). The Dinosaur Canyon Sandstone Member consists of reddish-orange and dark-red, fine-grained, friable sandstone, siltstone, and mudstone that form a series of slopes and ledges. Thickness of the Dinosaur Canyon ranges from about 90 to 220 feet (27-67 m); it grades upward into the Springdale Sandstone Member. The Springdale consists of dark to pale reddish-brown, fine-grained sandstone and minor siltstone that form a series of ledges and small cliffs. Average thickness of the Springdale is about 220 feet (67 m).

The Springdale Sandstone Member of the Moenave is difficult to distinguish from the overlying Kayenta Formation (Upper Triassic?) because of similar color and lithology. However, the Kayenta generally is lighter colored, less jointed, and is thicker- and parallel-bedded rather than the more lenticular character of the Springdale. The Kayenta consists mainly of pale-red, cross-bedded sandstone layers separated by thin beds of red-brown siltstone which together form a series of ledges and slopes similar to those of the Moenave. The upper part of the Kayenta and the lower part of the Navajo Sandstone are similar in color (red-brown), and cross-bedded sandstone beds of the Kayenta intertongue with those of the lower Navajo. The top of the Kayenta Formation is arbitrarily placed at the top of the uppermost, horizontally bedded sandstone (Phoenix, 1963). The Kayenta can be distinguished from the Navajo by a topographic break, where the ledge- and slope-forming Kayenta grades into the cliff-forming Navajo (fig.1.2). Because of the arbitrary upper boundary of the Kayenta, the Kayenta has a variable thickness ranging from 120 to 180 feet (37-55 m). Because of the arbitrary boundary between the Kayenta and Navajo, and the lack of fossil evidence in the Lees Ferry area, the boundary between the formation is considered to be Late Triassic(?) and Jurassic age. Most of the Navajo Sandstone is Jurassic age.

The Navajo Sandstone forms the upper sheer cliff of the Vermilion and Echo Cliffs that nearly surrounds Lees

Ferry and is the highest Mesozoic rock unit exposed in the area. The Navajo consists of pale-red, medium- to fine-grained, well-rounded, eolian quartz sandstone. Several large-scale, cross stratified sets of sandstone beds form an impressive sheer cliff nearly 1,600 feet (427 m) thick; the upper 400 feet (122 m) is eroded from the Echo Cliffs behind (north of) Lees Ferry (Phoenix, 1963).

View to southwest of north end of Marble Canyon. A - Lees Ferry; B - Marble Plateau; C - Shinarump Member of Chinle Formation (Upper Triassic) capping bench; D - Vermillion Ciffs. The Upper Triassic strata of the Vermillion Cliffs once covered much of the Grand Canyon area. The Colorado River flows southward (arrow), incised in Paleozoic strata of the Marble Plateau. Photograph courtesy of Richard Hereford.

CHAPTER 17: FISSION-TRACK DATING: AGES FOR CAMBRIAN STRATA AND LARAMIDE AND POST-MIDDLE EOCENE COOLING EVENTS FROM THE GRAND CANYON, ARIZONA

C. W. Naeser[1], I. R. Duddy[2], D. P. Elston[3], T. A Dumitru[4] and P. F. Green[4]

INTRODUCTION

This progress report on fission-track ages from rocks collected in the Grand Canyon, Arizona, combines two separate studies. One study (CWN and DPE) involves the dating of rocks collected at river level along the Colorado River from Lees Ferry to Diamond Creek. The location for these samples can be found on Figure 4.2. The second study (IRD, TAD, and PFG) involved rocks sampled along a traverse from the rim of the canyon at Grand Canyon Village to the bottom of the canyon at Phantom Ranch. The purpose of these studies is to determine the thermal history of the Grand Canyon through fission-track analysis. If the thermal history can be constrained, it may be possible to better understand the timing of events that led to uplift of the Colorado Plateau and by inference the excavation of the Grand Canyon.

Compressional events of Late Cretaceous and early Tertiary (Laramide) age, originating from the west, led to formation of the major monoclines and plateau upwarps of the Grand Canyon region (Hunt, 1956, 1969, 1974; Lucchitta, 1979; Young, 1979, 1987), but the central Colorado Plateau had yet to be subjected to strong regional uplift. Although the southwest margin of the Colorado Plateau was uplifted late in Cretaceous time, the central Plateau remained low compared to the country to the southwest at least until Eocene (Young, 1987). The central Colorado Plateau then was eventually uplifted and became an area of positive relief relative to the country to the south (chapters 18, 19, and 20). Two periods of Cenozoic uplift are now recognized in the geologic record in the Grand Canyon region.

Hunt (1974) reviewed two major theories concerning the development of the Colorado River and the Grand Canyon. One theory proposed that the Colorado River and the Grand Canyon are as old as middle Tertiary (early Miocene; Hunt, 1956, fig. 59), whereas a Pliocene age for the time of canyon cutting is proposed in the second theory (see discussion, chapter 18). These models can be examined in light of a tectonic/thermal history of the Colorado Plateau deduced from cooling records obtained from fission tracks.

---

[1]U. S. Geological Survey, Denver, Colorado.

[2]Geotrack International Pty. Ltd, Melbourne University, Parkville, Victoria, Australia.

[3]U. S. Geological Survey, Flagstaff, Arizona.

[4]Department of Geology, University of Melbourne, Parkville, Victoria, Australia.

FISSION-TRACK DATING

A fission track is a cylindrical zone of intense damage formed when a fission fragment passes through a solid. In natural minerals and glasses, fission tracks are produced by the spontaneous fission of ^{238}U. When an atom of ^{238}U fissions, the nucleus splits into two lighter nuclei, one averaging about 90 atomic mass units (amu) and the other about 136 amu, with the liberation of about 170 MeV of energy. The two highly charged nuclei recoil in opposite directions, and as they travel through the host material they strip electrons from atoms along their paths. This produces positively charged ions that are mutually repulsed, forming linear damage zones or latent tracks (Fleischer and others, 1975). A latent track is tens of angstroms in diameter and about 10-20 μm in length.

A latent track can only be observed with a transmission electron microscope; however, a suitable chemical etchant can enlarge the damage zone so that it can be observed with an optical microscope at intermediate magnifications (x200-500). Common etchants used to develop tracks include nitric acid (apatite), hydrofluoric acid (mica and glass), concentrated basic solutions (sphene), and alkali fluxes (zircon) (Fleischer and others, 1975; Gleadow and others, 1976).

Because ^{238}U fissions spontaneously at a constant rate (in a statistical sense), the accumulation of fission tracks in a mineral or glass is an indication of its age. A fission-track age of a mineral or glass can be calculated from the surface density of spontaneous fission tracks and the concentration of ^{238}U that produced those tracks. Because the relative abundance of ^{238}U and ^{235}U is constant in nature, the easiest and most accurate way to determine the amount of ^{238}U present in the sample is to create a new set of fission tracks by irradiating the sample with a known dose of thermal neutrons in a nuclear reactor, which induces fission of ^{235}U. The reader is referred to Naeser and Naeser (1984) for a more detailed discussion of the theory and procedures of the method.

Apatite fission-track ages can be used to constrain times and rates of uplift (for example, Naeser and others, 1983; Dokka and others, 1986). Fission-tracks disappear at elevated temperatures through a process known as *track annealing*. Heating a mineral containing fission tracks allows ions displaced along the damage zone to move back into normal crystallographic positions in the mineral. This repair of the damage zone leads to shortening and ultimately to the total disappearance of the fission track. The annealing of fission-tracks is a time-temperature function. Short heating times at high temperature can have the same effect on fission tracks as long heating times at lower temperatures. Over the periods of time that are usually required for geological processes (>10^5 years), total annealing will take place at temperatures between 100 °C and 150 °C (Naeser, 1979a, 1981). The longer the heating time, the lower the temperature required for total annealing.

The apatite fission-track age is generally regarded as a cooling age that records the time since the grain last cooled through its "closure/blocking" temperature. Although the apatite closure temperature commonly is given as 100 °C (Naeser, 1979b), the closure temperature in reality can range from 100 to 150 °C depending upon the rate of cooling (Naeser, 1981). Slow uplift and cooling results in closure temperatures close to 100 °C, whereas the closure temperature for rapid uplift and cooling could be closer to 150 °C. An apatite at the surface will yield a cooling age when it has been uplifted from depths where the combination of time and temperature did not allow tracks to be preserved.

Unfortunately, fission-track ages cannot always be interpreted simply as cooling ages because annealing is not a simple process where the tracks are either totally retained or totally lost. There is a zone of partial annealing (Naeser, 1979a), which spans a temperature interval of about 30 to 40 °C (Naeser, 1981). Apatite that has resided in the zone of partial annealing will yield a mixed age that is less than the primary age but greater than the cooling age. Cooling ages can be distinguished from mixed ages by means track length and the frequency distribution of track lengths (Gleadow and others, 1983, 1986a, 1986b). A prolonged residence in the zone of partial annealing shortens fission tracks and gives a skewed track-length distribution, with some tracks <12 µm in length. Such short lengths are not seen in apatites with a rapid cooling histories.

Measured apatite fission-track ages will always be either equal to (for samples that cooled quickly through their closure temperature) or greater than (for unannealed or mixed-age samples) the age of the last uplift or cooling event. Thus, the youngest apatite age obtained from of a series of samples from the same structural block is a maximum age for the event that brought those samples to the surface.

The fission-track annealing properties of zircon are less well known. As established through numerous studies (for example, Harrison and others, 1979; Zeitler, 1985), fission tracks in zircon are more resistant to annealing than those in apatite. Several estimates have been made for the closure temperature for zircon; Harrison and others (1979) estimated ≈175 °C and Hurford (1986) estimated 240 ± 50 °C. Hurford's estimate is based on data generated in the Lepontine Alps, Switzerland, in an area of relatively rapid cooling. Because of the rapid cooling in these rocks Hurford's value should be considered as a maximum temperature. Thus, in areas of less rapid uplift, zircon ages can probably be considered to indicate the cooling of the rock below ≈200 °C.

In addition to providing information on the thermal history of a region, fission tracks can also be used to date volcanic events. For example, zircons separated out of a bentonite can be used to date that horizon and provide a numerical age assignment for sedimentary strata that enclose the bentonite bed (Ross and others, 1982). During several trips down the Colorado River, three bentonites in Cambrian strata were found and collected. Zircons were separated and dated from two of these bentonites (one from the top of the Peach Springs Member of the Muav Limestone and the other from near the base of the red-brown sandstone of the Bright Angel Shale of western Grand Canyon (chapter 15).

RESULTS AND DISCUSSION

Apatite Data.

Twenty-four apatite concentrates were dated by the fission-track method. Figure 17.1 summarizes the fission-track ages of apatite separated from rocks collected at or near the river from Lees Ferry to Diamond Creek, a river distance of 226 miles (364 km). Table 17.1 lists the miles down river for the apatite samples and either the formation name or rock type collected. Figure 17.2 shows the apatite fission-track ages determined on samples collected along a vertical traverse along the Bright Angel Trail extending from Grand Canyon Village to just beyond Phantom Ranch (near the junction of Phantom Creek and Bright Angel Creek). The sample closest to the river from this second data set is also shown on Figure 17.1. The rocks sampled for each data set include Phanerozoic and Middle Proterozoic sedimentary rocks, and igneous and metamorphic rocks of late Early Proterozoic age.

The data collected at river level (fig. 17.1) show a general trend of increasing age with distance down river. The youngest ages are found between Lees Ferry and the junction of the Colorado and Little Colorado Rivers. The ages along this reach of the river average 35 Ma. The oldest ages are found in the lower part of the canyon beyond mile 219 (km 352). In this part of the canyon, between the Hurricane fault and Diamond Creek, the apatite ages average 82 Ma. Within this overall trend of increasing age with distance down river, deviations occur when some of the major structures are crossed. A significant age discontinuity occurs at the Palisades fault, a bounding fault of the East Kaibab monocline. The apatite age from strata a few meters east (upstream) of the fault is 38.9 ± 17 Ma (± 2σ), and all ages upstream of the fault are <45 Ma. In contrast, five samples immediately down river from the fault have apatite ages that average 63.4 ± 11 Ma (± 2σ). A similar shift in ages appears to occur when the Bright Angel and Hurricane faults are crossed, though the data are sparse. The apatite ages are consistently younger on the upstream side (downthrown blocks at the Palisades and Bright Angel faults; upthrown block at the Hurricane fault), and older on the downstream side. The apatite age at Phantom Ranch (GC-8) is significantly older than the ages determined on apatite from outcrops several miles upstream and downstream from this locality. The dated

TABLE 17.1. Approximate distance down river from Lees Ferry with formation name or rock type collected for apatite and zircon dated in the river-level sampling part of this study.

Sample number	Miles down river	Formation/rock type
78N3	0.0	Chinle Formation
78N4	12.5	Esplanade Sandstone
78N5	17.0	Supai Group.
78N6	58.8	Tapeats Sandstone.
80N1	63.0	Dox Sandstone.
80N3	64.5	Dox Sandstone.
78N8	65.4	Dox Sandstone.
78N9	65.7	Dox Sandstone.
82N1	66.0	Dox Sandstone.
82N4	72.0	Dox Sandstone.
78N12	74.0	Dox Sandstone.
78N13	83.8	Vishnu Schist
78N15	105.0	Ruby Creek pluton
80N6	113.3	Elves Chasm Gneiss
78N19	189.6	Granitic rock
80N10	219.2	Granitic rock
78N20	225.7	Granitic rock

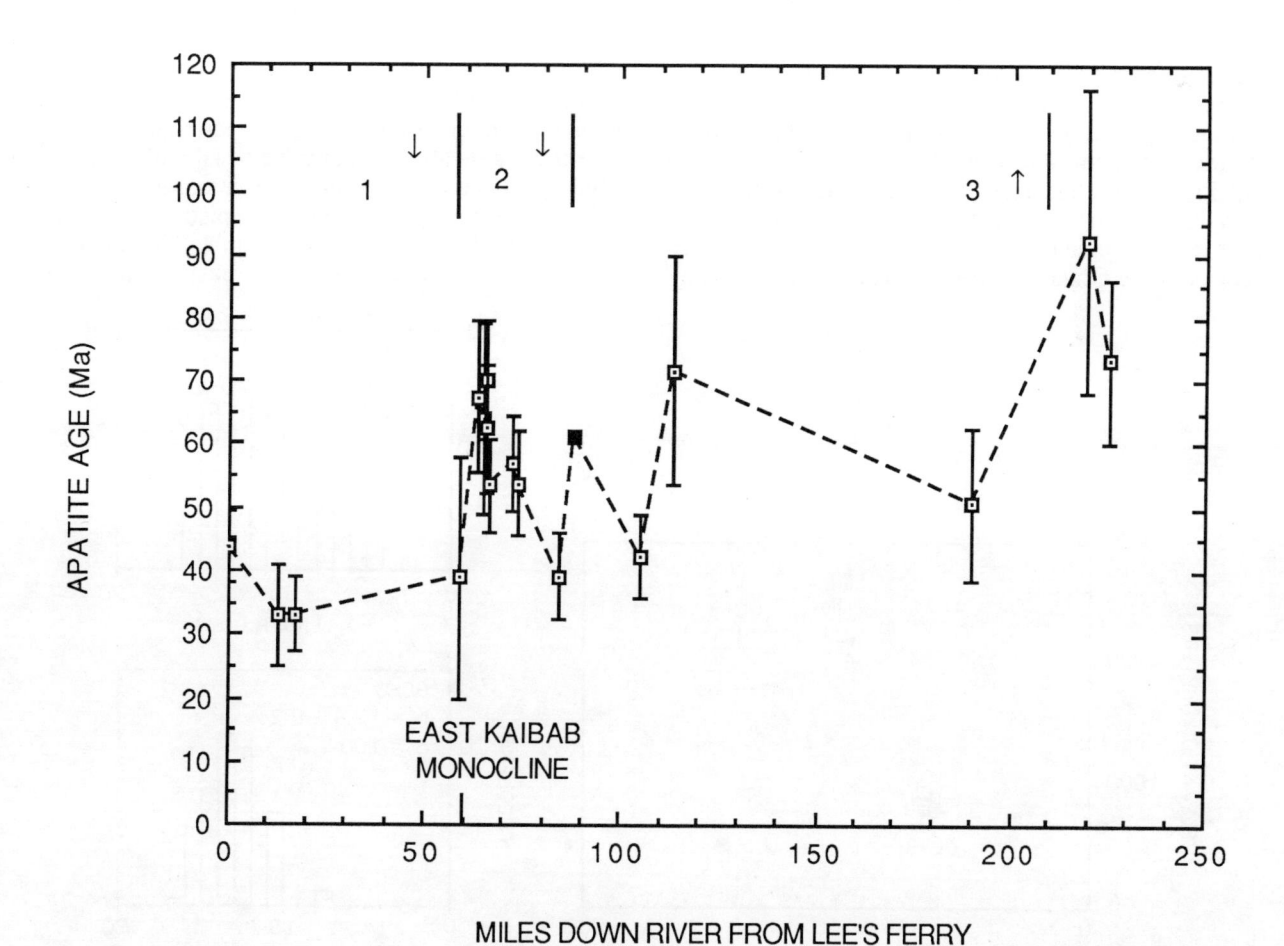

FIGURE 17.1. Fission-track ages of apatite collected in the Grand Canyon at river level from Lees Ferry to Diamond Creek. The error bars are ± 2σ. Solid square indicates the sample closest to river level from the vertical traverse (fig. 15.2). The following faults are identified by number: Palisade fault (strand of the Butte fault) (1), Bright Angel fault (2), and Hurricane fault (3). Arrows indicate sense of last movement on faults.

samples from the Phantom Ranch area (fig. 17.2) are on the up-thrown (downstream) side of the Bright Angel fault. The old, ≈60 Ma age from upthrown Proterozoic metamorphic rocks near the river with respect to the ~40 Ma age from downthrown Vishnu Schist approximately 4 miles (6.4 km) upstream, parallels age relations seen across the Palisades fault at the East Kaibab moncline. The discordant ages (fig. 17.1) across the Palisades fault, the Late Cretaceous-early Tertiary ages obtained from the eastern side of the East Kaibab moncline, and a series of progressively older ages obtained downstream and west of the moncline, indicate that the East Kaibab moncline existed as a topographic high at the time of Laramide folding and faulting. The basement and overlying strata on the west sides of the Bright Angel and Hurricane faults also are inferred to have been topographically elevated at this time relative to rocks on the east sides of these faults (Young and Huntoon, 1987). These data indicate differential uplift (tilting) within each block.

A definite change in age with elevation is shown in the traverse that extends from the rim to the river (fig. 17.2). The apatite ages from the upper Paleozoic strata average 124 ± 34 Ma ($\pm 2\sigma$). The average age for the four apatite concentrates separated from the Proterozoic basement at Phantom Ranch is 62.4 ± 5.2 Ma ($\pm 2\sigma$). A classical interpretation of this limited data set would be that detrital apatites dated from the upper Paleozoic strata had undergone partial track annealing and that apatites from the basement had been totally annealed prior to Laramide upwarping. Then, with moderate uplift of the Colorado Plateau and development of the monoclines during the Laramide (latest Cretaceous-early Paleocene), the rocks of the Grand Canyon region cooled, "freezing-in" the age of the Laramide uplift. The younger ages (<50 Ma) then indicated a second period of uplift and cooling in the Eocene.

The fission-track length data from the vertical traverse (figs. 17.2, 17.3; table 17.2), however, reveal a different story. Two of the six fission-track length distributions that were measured are shown on Figure 17.2. One is from the Proterozoic basement (GC-8) and the other is from a sample of Kaibab Limestone (GC-9). The distributions are similar to the patterns determined for apatites from the other basement and sedimentary rocks. The fission-track length data are summarized on Figure 17.3 and Table 17.2. The short mean track lengths in all of the samples (≈12 μm) and the presence of short tracks (<10 μm) indicate a complex thermal history for these rocks. There is little doubt that the apatites from the Paleozoic section underwent partial annealing prior to Laramide uplift and that the apatites from Phantom Ranch were totally annealed at the same time, because the ages of individual grains from the Paleozoic samples show excessive scatter (all failed the Chi-Square test), indicating the possibility of a non-uniform age population. If these apatites had been totally annealed, the ages from the individual grains would represent a single uniform population, not one with excessive scatter.

The presence of tracks in apatite from the Proterozoic basement with lengths between 4 and 10 μm indicates that the basement remained at a temperature that allowed some track annealing to continue to take place after the initial uplift and cooling during the Laramide. The final cooling occurred during a second episode of uplift and erosion in middle or late Tertiary time. The increase in the mean track length with decreasing elevation is

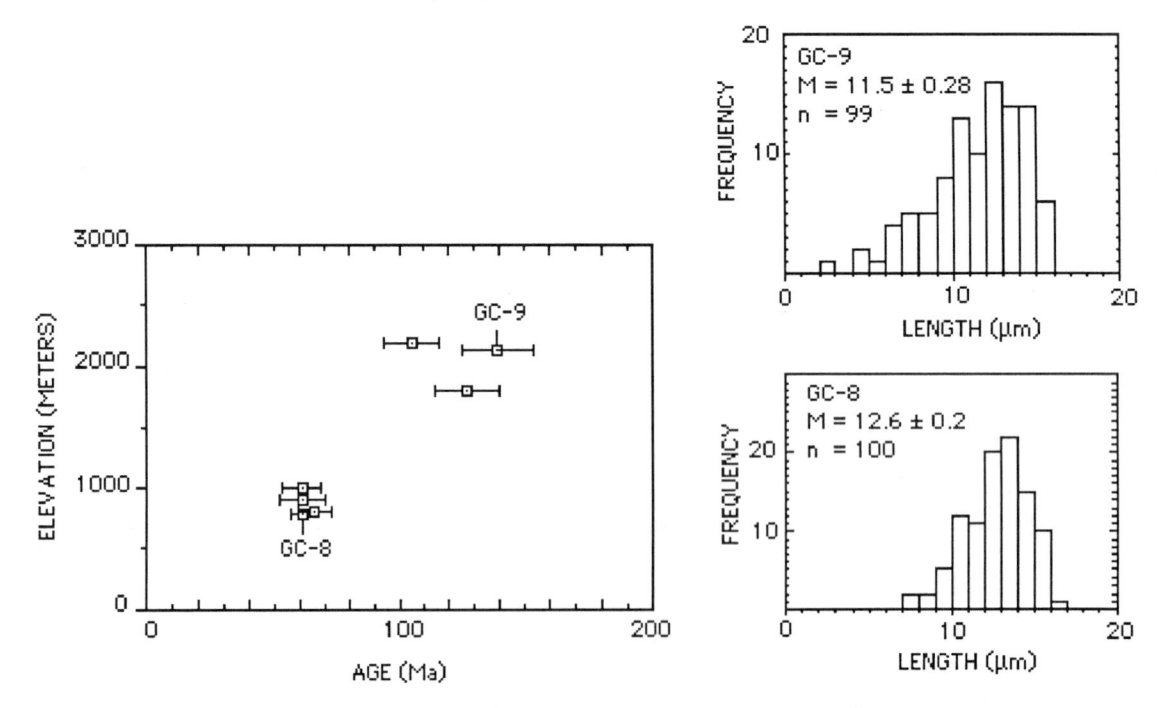

FIGURE 17.2. Fission-track ages and length data for apatite collected along a vertical traverse along the Bright Angel Trail from Grand Canyon Village to the junction of Bright Angel and Phantom Creeks, plotted as a function of elevation. The error bars are $\pm 2\sigma$.

TABLE 17.2. Confined fission-track lengths in apatite, Bright Angel Trail, Grand Canyon, Arizona.

Sample number[1]	Elevation meters	Mean length μm	Standard error of mean	Standard deviation	Range μm
GC-10	2192	11.66	0.33	3.27	2-18
GC-9	2134	11.47	0.28	2.82	2-16
GC-1	1814	11.86	0.24	2.59	2-16
GC-5	1006	12.40	0.29	1.77	7-16
GC-6	896	11.98	0.22	2.21	4-16
GC-8	792	12.64	0.20	1.95	7-17

[1] GC-10 and -9 from Permian Kaibab Limestone; GC-1 from Permian and Pennsylvanian Supai Group; GC-5,-6, and -8 from Proterozoic basement.

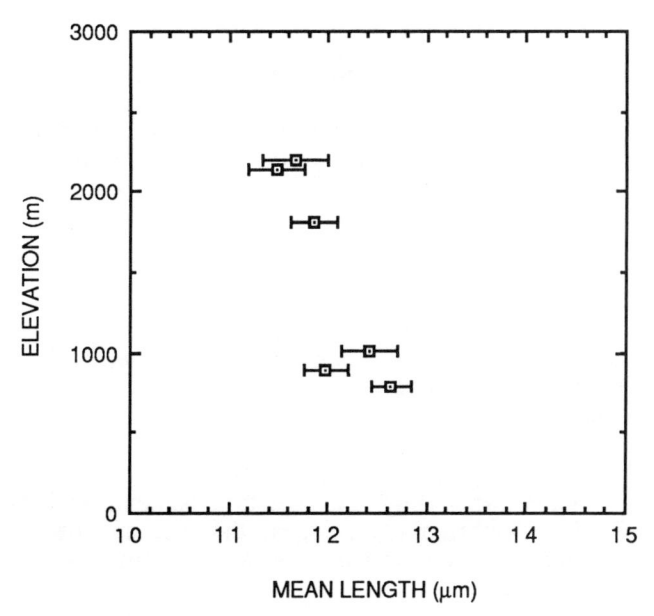

FIGURE 17.3. The mean lengths of confined fission tracks in apatite collected traverse down the Bright Angel Trail from Grand Canyon Village to the junction of Bright Angel and Phantom Creeks (fig. 17.2). The error bars are ± one standard error of the mean.

These ages indicate that the maximum temperature reached by rocks presently exposed at river level in the Grand Canyon has not exceeded ≈200 °C in the last 1,000 m.y.

The remaining two zircon suites (table 17.3) were separated from Middle Cambrian bentonite horizons in the Bright Angel Shale of the western Grand Canyon (correlative with the Tapeats Sandstone of the central Grand Canyon; chapter 15), and with the Muav Limestone. Zircon 80N8 (563 ± 49 Ma [± 2σ]) is from a bentonite bed underlying the red brown sandstone at mile 205.7 (km 331), above Granite Park. The horizon is at the top of an interval of shale and subordinate sandstone of Tapeats lithology that overlies the Early Cambrian *Olenellus* horizon; this faunal horizon, in turn, closely overlies the lower cliff-forming massive sandstone unit of the Tapeats, which can be traced to the Grand Wash Cliffs to the west. The bentonite horizon beneath the red-brown sandstone appears to correlate with the *Zacanthoides cf. walpai* horizon of the Toroweap section to the east (chapter 15). McKee and Resser (1945, fig. 2B, p.19) considered this horizon to be of early Middle Cambrian age.

The second zircon concentrate, 80N7 (535 ± 48 Ma [± 2σ]), comes from a bentonite layer, 2 - 3 cm thick, in a thin, thin-bedded interval of limestone that separates the Peach Springs Member from the overlying Kanab Canyon Member of the Middle Cambrian Muav Limestone. The sample locality is a ledge at river level on the left (south) bank of the river at mile 179.85 (km 289), immediately below the small rapids that follow Lava Falls Rapids.

CONCLUSIONS

Fission-track age and track-length data provide information on the thermal history of the Grand Canyon. Fission-track ages of ≈1,000 Ma obtained from zircons from Proterozoic rocks now exposed at river level indicate that these rocks have been at temperatures of <≈200 °C for the last 1,000 m.y. Fission-track data on detrital apatite crystals separated from Pennsylvanian and Permian strata near Grand Canyon Village on the south rim of the Grand Canyon indicate that these rocks have never been buried to sufficiently high temperatures for

consistent with this interpretation of the data. If apatites from near the river at Phantom Ranch had been only partially annealed during or prior to Laramide uplift, the mean fission-track lengths would decrease rather than increase with depth. Apatite fission-track ages of ≈40 Ma from other parts of the Grand Canyon indicate that this second event is late Eocene or younger in age.

Zircon Data

Six zircon separates from the Grand Canyon were dated by the fission-track method (fig. 17.4). Three separates were from Middle Proterozoic sedimentary rocks collected between miles 65 and 70 (km 105 - 113) and one was from a sample of a Proterozoic granitic rock collected near mile 220 (km 354). The fission-track ages of the zircons from the four Proterozoic rocks are concordant, and they average 1,038 ± 80 Ma (± 2σ).

TABLE 17.3. Fission track ages of zircons from Middle Cambrian bentonites, western Grand Canyon, Arizona.

Sample no.	Mineral	Number grains	Fossil density X 10^6 t/cm^2	Fossil tracks counted	Induced density X 10^6 t/cm^2	Induced tracks counted	Neutron Dose X 10^{15} n/cm^2	tracks counted	Ma ±2σ
80N7	ZIRCON[1]	15	24.38	3251	12.14	809	4.64	2268	535 48
80N8	ZIRCON[2]	12	39.98	3392	19.50	827	4.80	2500	563 49

[1] Peach Springs Member, Muav Limestone
[2] Base of red-brown sandstone, Bright Angel Shale

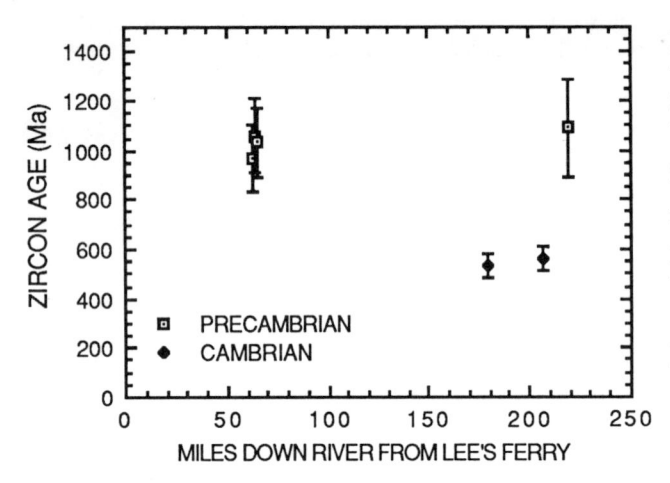

$\lambda_f = 7.03 \times 10^{-17}$ yr^{-1}

FIGURE 17.4. Fission-track ages of zircon collected at river level in the Grand Canyon between Lees Ferry and Diamond Creek.

the fission tracks to have been totally annealed, suggesting maximum temperatures of <100 °C. Rocks presently exposed at river level between Lees Ferry and the Hurricane fault were all in the zone of total track annealing for apatite prior to the onset of Laramide tectonism at the end of the Cretaceous Period. This suggests temperatures >100 and <200 °C for these rocks prior to the initiation of Laramide uplift, following which they underwent an initial cooling. Because apatite ages from near Phantom Ranch in the Grand Canyon are mixed ages (average 62.4 ± 5.2 Ma), initiation of the Laramide uplift and folding occurred shortly before 62.4 Ma. Moderate Laramide uplift only served to cool these rocks to a temperature where some annealing still was taking place. In contrast, abruptly different cooling ages across the East Kaibab monocline and the Bright Angel and Hurricane faults suggest the monclines and structural blocks on the west sides of the faults were elevated during Laramide compression, allowing these rocks to cool. The present opposite sense of displacement on the Hurricane fault suggests that the observed reversal of this Laramide offset occurred following a second episode of uplift.

A second episode of uplift occurring after middle Eocene time brought the rocks up to a temperature zone where no further significant annealing took place. The minimum cooling ages obtained in Marble Canyon and the eastern Grand Canyon, in the range of 35-40 m.y., generally correspond with a late Eocene-early Oligocene time of regional uplift and tilting of the plateau inferred for the time of drainage reversal and integration of an ancestral Colorado River on the Paleocene and Eocene Rim gravels (chapter 19), followed by removal of Rim gravels and the inferred initiation of cutting of the Grand Canyon (chapter 18). Rocks at river level west of the Hurricane fault were not totally annealed prior to the episodes of Laramide uplift and cooling, suggesting an attenuated section of Paleozoic strata near the present western margin of the Colorado Plateau.

Fission-track dating of zircons from bentonite beds of Middle Cambrian age suggest that the age of the boundary between the Early and Middle Cambrian lies near 565 Ma.

CHAPTER 18: DEVELOPMENT OF CENOZOIC LANDSCAPE OF CENTRAL AND NORTHERN ARIZONA: CUTTING OF GRAND CANYON

Donald P. Elston
U.S. Geological Survey, Flagstaff, Arizona

Richard A. Young
Department of Geological Sciences, SUNY College at Geneseo, Geneseo, N.Y.

INTRODUCTION

The most controversial aspects of the geology of the Grand Canyon continue to be concerned with the history of development of the landscape, particularly as it pertains to the time of cutting of the Grand Canyon. This is not surprising because only a fragmentary depositional record, poorly dated or of unknown age, is preserved on and near the Colorado Plateau and within the Grand Canyon. In this chapter, the character and distribution of deposits, structures, and morphologic features related to evolution of the landscape are described, and they are placed within a temporal and climatic framework that serves to constrain aspects of the Cenozoic history of the region. For the locations of place names, features, and deposits mentioned in this article, the reader is referred to Figures 4.1 and 4.2, and 19.1, and to various figures in Chapters 20, 21, 23, and 24.

More than thirty years ago, C. B. Hunt (1956, fig. 59) postulated from regional relations that the Little Colorado River and lower Colorado River of the Grand Canyon, and the Mogollon Rim, were in existence during early Miocene time. Ten years later, an assumption that led to a different conclusion was made concerning the age of the informally designated "Rim gravels" (chapter 19), which underlie Miocene basaltic flows on the high plateaus. McKee and others (1967, p. 33) assumed that the gravels, which contain no local basaltic detritus, were not much older than the overlying basalt flows; subsequently, ~7-14 Ma K-Ar dates were obtained from the flows (McKee and McKee, 1972). This assumption of near-equivalency in age thus served to place the ancestral Colorado River, and associated tributary drainages, near the level of the (presently high) plateau surface during late Miocene time. It fixed the time of onset of entrenchment of the Colorado River system to form the Grand Canyon accompanying an inferred uplift of the Colorado Plateau, and resulted in a Grand Canyon cut during Pliocene time (McKee and McKee, 1972). (This inference also implied a Pliocene cutting of the Mogollon Rim). In the 1960's, the Pliocene Epoch was considered to include the interval 12 Ma to 2.5 Ma (Arizona Highway Geologic Map, 1967, Cooley, M.E., compiler), so adequate time was believed available for cutting of the Canyon.

The same assumption, calling for a coeval age of gravel on the Shivwits Plateau and overlying basalt, is employed in a current model that proposes a Pliocene cutting of the western Grand Canyon. Gravels underlying ~7 Ma basalt on the Shivwits Plateau of the western Canyon are considered by Lucchitta (1984, 1987, 1988) not to be much older than the overlying basalt. This has led Lucchitta to propose that much of the western Grand Canyon was cut during the Pliocene (from ~5-3.8 m.y.) from the level of the Shivwits Plateau (equivalent to the high plateau surface in the area of the central and eastern Grand Canyon).

The specific interval of time involved (~5-3.8 m.y.) is related to the age and incision of the Muddy Creek Formation in the Grand Wash Trough (Lucchitta, 1987, 1988). This proposed episode of erosion is postulated to have resulted in capture (by headward erosion) of an ancestral Colorado River that was flowing northwesterly between the Kaibab and Shivwits Plateaus. The inferred older Miocene stream course left no trace of a specific channel and no unique deposits on the Shivwits Plateau. Moreover, the place of capture is not specifically recognized, and the older Miocene river is inferred to have flowed toward an unspecified destination in the Basin and Range Province near the Arizona-Utah border.

Models that propose a Pliocene cutting of the Grand Canyon have been concerned with solving the "Muddy Creek problem" (Longwell, 1936). Deposits of the Muddy Creek Formation, which may range in age from late middle Miocene to early Pliocene, accumulated across the course of the Colorado River against and west of the Grand Wash Cliffs, where the Colorado River enters the Basin and Range Province. As shown in the diagrams of Lucchitta (1984), none of the solutions proposed during the past 20 years for the time of cutting of the Grand Canyon have considered evidence bearing on the time of regional uplift of the Colorado Plateau that may be gained from the area of the Mogollon Rim, and the adjacent Verde Valley. Rather, the problem of the time of presumed regional uplift required to initiate cutting of the Grand Canyon in the plateau north of the Mogollon Rim has been satisfied by inferring that the Colorado River flowed on the currently high, smooth plateau surface during late Miocene time. Although a reasonable case has been made for this model, frustrating uncertainties remain concerning the specific sequence of events leading to the timing and initiation of Grand Canyon erosion. This is especially true in view of newly developed, more refined evidence for multistage Laramide uplift and depositional events (chapter 17), and the evidence for the early Tertiary definition of a Mogollon Rim having several thousand feet of relief (i.e., elevation). Given these basic uncertainties and the recently shortened time scale for post-Muddy Creek incision (Damon and others, 1978; Shafiqullah and others, 1980; Metcalf, 1982), we question whether all viable scenarios for canyon development have been exhausted.

LARAMIDE EVENTS AND FEATURES

During Late Cretaceous and Paleocene time, a broad cuestaform landscape formed across central and northern Arizona as a result of Laramide-initiated uplift (~80-64 m.y. ago; episode 1, table 18.1). Accompanying and subsequent to this widespread episode of erosion, dominated by more or less uniform cliff recession of the northwest-striking Paleozoic and Mesozoic rocks, arkosic sediments containing crystalline clasts filled canyons on the margin of the uplift and then were spread across the erosion surface as "Rim gravels" (episode 2, table 18.1). These deposits are presently found at elevations higher than 6,400 feet (1,950 m) on the Coconino Plateau, south of the Grand Canyon (Young, 1987; Dickinson and others, 1988). This erosional episode cut canyons in excess of 4,000 feet (1,200 m) deep along the uplifted southwestern Plateau margin, and subsequent deposition buried parts of the present Coconino and Hualapai Plateaus in the western Grand Canyon region under several hundred feet of early Tertiary sediments.

The ages of the Rim gravels are discussed at somewhat greater length in Chapters 19 and 20. The Rim gravels have now been generally constrained as ranging from late Paleocene to late Eocene (~64-37 m.y.) in age across Arizona on the basis of K-Ar ages of basal volcanic clasts, freshwater gastropods, magnetic polarity studies, scanty pollen data (presence of grasses), and dated volcanic ash beds. Additionally, similar sediments recently have been recognized north of the Grand Canyon on the west side of the Kaibab Uplift where they occur at elevations near 5,700 feet (1,735 m). In part, these more northerly deposits contain imbrication indicative of reworking off the west side of the Kaibab uplift, presumably as a result of a late pulse of Laramide deformation (chapter 17). The original gravels deposited on the Kaibab uplift may have been similar to the Rim gravels south of the Colorado River, but they appear to have been derived from a more westerly source consistent with lithologic differences (Richard A. Young, work in progress).

The antiquity of the southerly-derived Rim gravel deposits is further indicated by their intimate association with hanging channel segments bordering monoclinal structures on the Hualapai Plateau, especially along the Hurricane fault. At an earlier time, this late Miocene(?)-Pliocene fault was the site of an eastward-verging Laramide monocline, which deflected major trunk drainages converging along its eastern flank (Young and Huntoon, 1987). Early Tertiary episodic movement on this structure (chapter 17) produced a series of disjointed, abandoned meandering channel segments now preserved on the upthrown (east) side of the younger Hurricane fault near Peach Springs Canyon. These disconnected, hanging channel segments are contemporaneous with more extensive Laramide channels on the Hualapai Plateau west of the Hurricane fault, previously described by Young (1985, 1987).

The positions and geomorphology of these channel systems demonstrate that a complex, incised drainage system was cut at least 4,000, feet (1,200 m) into the Colorado Plateau margin by late Paleocene time (~<64 m.y.). This required an equivalent amount of uplift of the southwestern plateau margin and a corresponding minimal gradient sufficient to transport 18-inch (45 cm) boulders northward as far as the area of the present Colorado River. If one assumes (from the hydraulic literature) a reasonable gradient of .01 for a slope needed to transport such clasts, it can be argued that the central Colorado Plateau could have remained near sea level while the southern margin of the Plateau was being strongly uplifted by several thousand feet (at least 1 km). The sediments and associated drainages of northern Arizona created during this depositional episode must be temporally correlative with the thick fluvial and lacustrine formations of south-central Utah (e.g., North Horn, Flagstaff, Claron, Cedar Breaks, Wasatch, and Caanan Peak Formations).

The most important conclusion to be drawn from the early Tertiary erosional and depositional episodes is that the basic cuestaform (stair-step) paleogeomorphology of the modern plateau surface in northern Arizona was inherited from a major early Laramide orogenic event of Late Cretaceous and early Paleocene age. Significantly, this early episode of erosion, and the subsequent deposition of the Paleogene Rim gravels, are in no way related to the younger episode of Grand Canyon erosion. Furthermore, development of the Grand Canyon had a relatively minor effect on the broad, regional elements of the landscape inherited from the early episode of uplift, erosion, and deposition.

Following the Paleocene-Eocene episode of deposition, the Paleogene deposits were deeply weathered. This is evident in the best preserved Rim gravel sections (episode 3, table 18.1). The deep weathering would appear to correspond in time with formation of the Eocene (post-Laramide) surface of Epis and Chapin (1975) and Gresens (1981). However, little evidence is recognized that documents late Eocene to late Oligocene (37-27 m.y.) history across central and northern Arizona, nor for that matter has such a record been recognized across much of the remainder of the Colorado Plateau. This general lack of a stratigraphic record for much of the Plateau is interpreted as indicative of regional erosion following an episode of Laramide uplift, followed by the establishment of an integrated drainage system across the Plateau and removal of the products of erosion from the region (e.g., Hunt, 1956). Alternatively, the period might have been characterized by general tectonic quiescence and by an absence of any well-integrated regional drainage system. However, it can be argued that deposits of even such a "quiet" tectonic interval should be recognizable at many places across the Colorado Plateau.

TABLE 18.1. Erosional and depositional episodes, and inferred climate, proposed during development of landscape of northern and central Arizona.

AGE	EPISODE — EVENTS AND EVIDENCE
Late Cretaceous-Paleocene ~80 to 64 m.y. (early Laramide tectonism)	1. Laramide uplift, erosion, and development of Mogollon Rim; 4,000 ft. (1,200 m) of relief on Hualapai Plateau; canyons filled with Paleocene and Eocene deposits. Climate wet.
Paleocene-Eocene ~64 to 37 m.y. (later Laramide tectonism)	2. Deposition of Rim gravels north and south of Mogollon Rim, with one or more increments of later Laramide regional uplift and intermittent tectonism (fission track data; gastropods at Long Point; 37 Ma ash date in Apache Reservation; clast ages; paleomagnetics). Climate wet.
Oligocene ~37 to 27 m.y.	3. Evolution poorly constrained (few rocks or ages). Erosion dominates? Increasing(?) aridity with time. North of Mogollon Rim, plateau drainage possibly began to remove Rim gravels and to flow on Kaibab surface south of Kaibab Uplift, exiting to west or northwest. South of Mogollon Rim, inferred drainage reversal with erosion of Rim gravels and final stages in development of topography on which lower Miocene deposits accumulated. Ancestral Verde Valley with east and south-flowing drainage developed beneath lower Miocene Hickey sediments that underlie basalt of Hickey Formation. Correlative(?) surface with west and northwest drainage developed in Little Colorado Valley beneath late Miocene-Pliocene Bidahochi Formation. On south edge of Transition Zone, upper Oligocene-lower Miocene beds are unconformable on Rim gravels. Inferred time of erosion and beginning of drainage integration(?) on Colorado Plateau as well as in Transition Zone.
Late Oligocene - late middle Miocene ~27 to 12 m.y.	4. Increased aridity; deposition dominates? Beginning of Basin and Range extension and volcanism. Major Basin and Range tectonism, 20 to 12 m.y., disrupts regional drainage. In west, regional northeast dip reduced (18 Ma Peach Springs Tuff datum). Early plateau drainage continues(?) to develop, flowing west or northwest(?) (pre-Bidahochi Formation valley). South of Mogollon Rim, aggradation of lower Miocene deposits in ancestral Verde Valley and in central Arizona to south (Hickey sediments and equivalents) suggest at least local interruption of south-flowing drainage system. Hickey basalts (14.5-11.25 Ma) formed a continuous north-to-south ramp across Transition Zone. Deposition of thick sedimentary section and volcanics on Hualapai Plateau make incision of western Grand Canyon unlikely (Separation Canyon; chapter 20). Major pre-Muddy Creek erosional relief along Grand Wash Cliffs on west, including area of present course of Colorado River, indicates deep dissection of high-standing terrain prior to deposition of the Muddy Creek Formation.
Late middle Miocene - Pliocene ~12 - 4 or 2.5 m.y.	5. Time of maximum(?) aridity; region-wide aggradation and (Basin and Range?) interruption of drainage systems, followed by re-establishment of drainages and important drainage changes. Accumulation of upper middle and late Miocene-Pliocene Muddy Creek, Verde, and Bidahochi Formations between ~12 and ~3 m.y. Complex, sedimentary record in eastern Grand Canyon records a pre-Pleistocene(?) episode of aggradation (Nankoweap Canyon and basin). On Hualapai Plateau, change from deposition to erosion. Begin erosion of Muddy Creek Formation near Grand Wash Cliffs ~5 m.y. ago. Open (re-open?) western Grand Canyon drainage by integration and capture(?) by 3.8 m.y. ago (basalt of Grand Wash). Basin and Range events wane, followed by extensional faulting of western plateau, and volcanism (Shivwits-Hualapai Plateaus).
Pliocene-Holocene ~4 or 2.5 m.y. to present	6. Increased precipitation and re-establishment(?) of perennial streams, including Verde Valley and Little Colorado River drainages. Multiple damming events by lavas in Grand Canyon, <1.2 or <0.7(?) m.y. ago.

From this early Tertiary perspective, the central questions concerning events controlling the origin of the modern Colorado River (and Grand Canyon) would appear to include: 1) What are the amounts and timing of the late Laramide uplift of central and northern Arizona and the central Colorado Plateau leading to post-Laramide drainage reorganization? 2) What was the timing of post-Laramide drainage reorganization leading to renewed erosion across central Arizona, and presumably formation of the Grand Canyon in northern Arizona? 3) Is it reasonable to expect a very rapid cutting of the western Grand Canyon in early Pliocene time as proposed in one hypothesis? and 4) Could there be some validity to an alternate hypothesis that suggests a somewhat older, pre-Basin and Range (Oligocene-early Miocene) drainage system may have developed as a precursor to the modern canyon, remaining relatively dormant during at least part of Miocene and Pliocene time due either to climatic conditions or to inadequately documented sedimentation episodes

EARLY POST-LARAMIDE SETTING

The first well-dated regional geologic event following widespread Laramide erosion and deposition of Rim gravels (episodes 1 and 2, table 18.1), and weathering and erosion(?) (episode 3, table 18.1) was the onset of late Oligocene volcanism around the Plateau margin about 27 to 24 m.y. ago. In several areas in Arizona, volcanism continued until about 11-12 m.y. ago accompanying Basin and Range extension to the south and west of the Grand Canyon. Development and maintenance of a well-integrated regional drainage system that flowed south across the Transition Zone of central Arizona and southwesterly across the Basin and Range Province during this time of significant structural disruption and volcanism, would appear to be difficult, if not impossible.

EVOLUTION OF LANDSCAPE OF HUALAPAI PLATEAU AND WESTERN GRAND CANYON: CONSTRAINTS

In middle Tertiary time (at least part of episode 4, table 18.1), locally derived gravels filled the remaining Laramide canyon relief in the western Grand Canyon and spread over the intervening divides. These sediments, interbedded with lower to middle(?) Miocene volcanic rocks (~19-16 m.y. old), document regional fluvial aggradation on the plateau extending up to the modern canyon rim in the western Grand Canyon region (Buck and Doe Conglomerate; chapter 20).

On the Hualapai Plateau, formed by Laramide scarp retreat of the Pennsylvanian-Permian strata, the geomorphology and Tertiary stratigraphy make it difficult to allow for a significant episode of canyon cutting prior to the widespread volcanic episode about 19 to 16 million years ago (chapter 20). Gravel aggradation continued locally in the vicinity of the western Grand Canyon burying these early to middle Miocene volcanic rocks beneath 300 feet (100 m) of post-volcanic gravels (Willow Springs Formation of Young, 1966). These relations argue against any emerging ancestral Colorado River incision on the Hualapai Plateau of the western Grand Canyon at that time. The cessation of the depositional episode represented by these post-volcanic gravels in the western Grand Canyon is poorly constrained in time. Deposition of related sediments in the nearby Truxton Basin (fig. 4.1) continued into Pleistocene time (Twenter, 1962), but it appears that Grand Canyon erosion on the Hualapai plateau immediately adjacent to this closed basin had begun somewhat earlier. An unresolved question is whether the Grand Canyon erosion episode was controlled mainly by tectonic events (epeirogenic uplift), climatic change (increased runoff), slow drainage integration (capture), or by a complex sequence of episodic entrenchment, aggradation, and drainage re-exhumation. The most realistic solution (possibly unknowable) is that some combination of these events occurred consistent with the regional post-Laramide tectonics, the timing of which is poorly constrained on the Plateau.

Essential arguments in support of the different scenarios can be summarized as follows:

1) The simplest tectonic model assumes that canyon cutting reflects contemporaneous uplift. Although rapid downcutting accompanying uplift has been documented in many areas of the world, the story of the Colorado Plateau uplift has been poorly documented and misinterpreted. Naeser and others (chapter 17) report evidence of two distinct episodes of Laramide uplift from fission track data in the Grand Canyon, evidence supported by the emerging early Tertiary sedimentation history. Because the Hualapai Plateau has canyons with 4,000 to 5,000 feet (1,200-1,500 m) of Laramide relief, and because its present surface lies near an elevation of 5,000 feet (1,500 m), little <u>significant</u> late Tertiary uplift appears necessary in the <u>western</u> Grand Canyon to explain the youngest erosion. However, it should be noted that north-trending structures in the region, which display older down-on-the-east monoclinal displacements of Laramide age, have had such displacements wholly mitigated by down-on-the-west normal fault displacements of late Tertiary age. Thus, the present elevation of the western Grand Canyon area may be considerably lower now than during Paleogene time. The magnitude of regional warping (uplift) is difficult to assess.

2) A climatic model could invoke sporadic or cyclic Oligocene-Miocene erosion followed by undocumented aggradation and re-exhumation, thereby lengthening the canyon-forming process. The Hualapai Plateau Tertiary stratigraphy presents a strong argument against this sequence of events for late Oligocene through Miocene time (episodes 4 and 5, table 18.1). Deposits spanning this interval (chapter 20) record no evidence of significant downcutting (unconformities) in close proximity to the Grand Canyon, and datable relics of such deposits have yet to be found within the Canyon. The concept of a depositional sequence accumulating rather continuously through Miocene time appears to adequately fit the thick Tertiary sequences exposed in Milkweed and associated canyons, above and below the Miocene volcanic rocks exposed there. Lastly, the Muddy Creek sediments of the Grand Wash Cliffs and Lake Mead area to the west (5-12 Ma, Metcalf, 1982) correspond temporally with some of the deposits on the the Hualapai Plateau, implying deposition throughout the area of the western Grand Canyon from early Miocene time to about 5 million years ago (ages of regional volcanics capping basin fills).

3) Slow drainage integration by capture presumably

would involve north and northeasterly headward erosion all along the physiographic margin of the Colorado Plateau beginning at some unspecified time. This would lead to capture and draining of local basins on the Plateau following the gradual disintegration of the Laramide paleodrainage. It seems likely that such a slow, gradual process must be one factor in any model of drainage reversal, whether or not it represents the major mechanism of change. Headward erosion from the west is currently excavating basins such as Truxton Valley, a small basin isolated from the Grand Canyon drainage (fig. 4.1). Drainage from the western edge of the plateau has headwardly eroded in the vicinity of Route 66 near Valentine, Arizona, and has begun re-excavating this broad Laramide valley in Pleistocene time by breaching an interior basin.

4) The old Laramide canyons on the Hualapai Plateau that converge on Peach Springs Canyon are themselves examples of drainage burial and re-excavation on a fairly large scale. Hells Canyon (eastern tributary to Peach Springs Canyon; fig. 20.1) is an excellent example of a sinuous Laramide channel whose outline has just begun to be destroyed by incision below the former canyon floor profile cut into the Muav Limestone. The partial burial of any hypothetical ancestral Grand Canyon drainage during some part of Miocene time might be plausible if such an ancestral drainage had become established earlier, but conclusive proof of such an event is lacking in the western canyon.

Pre-Pleistocene(?) gravels and related deposits (chapter 21) that aggraded along the course of the Colorado River from Marble Canyon to the west-central Grand Canyon, and their possible equivalents in the western Canyon, are of uncertain age. They may be no older than late Miocene in age.

An ancestral Grand Canyon of pre-late Miocene (pre-11 m.y.) age would not necessarily have to conform exactly to the modern course, but might be expected to include some of the basic elements seen in the modern drainage pattern, i.e., a similar course, in places at perhaps higher topographic levels. The course postulated by Lucchitta (1984) around the Kaibab Uplift and into a northwest outlet from the plateau would offer such a solution.

It is extremely difficult to proceed from a well-established Laramide and early Paleogene channel system to a Pliocene(?) Grand Canyon without considering a reasonable mechanism for the gradual evolution of the new drainage, which flows in an opposite direction. Any reasonable explanation must deal with the drainage system(s) that must have evolved between late Eocene time and the modern Grand Canyon episode. It is unlikely that the landscape ceased to evolve for this length of time, a minimum of 30 million years.

EVOLUTION OF LANDSCAPE OF CENTRAL ARIZONA AND CENTRAL AND EASTERN GRAND CANYON: CONSTRAINTS

Erosion of the Transition Zone of central Arizona and the Mogollon Rim apparently occurred in consequence of regional uplift of the Colorado Plateau during early Laramide time (episode 1, table 18.1), suggested by evidence in the Hualapai Plateau. Early formation of the topographic elements of the region was followed by deposition of Paleocene-Eocene Rim gravels on an irregular terrain across central and northern Arizona (episode 2, table 18.1). A pulse of uplift of late Laramide age (post-middle Eocene age; chapter 17) then may have initiated a reversal of drainage south of the Mogollon Rim, leading to removal of most of the Paleocene-Eocene Rim gravels and other products of erosion from the area of central Arizona by a southward flowing ancestral Verde River drainage system prior to the accumulation of Miocene deposits in irregular basins (episode 3, table 18.1). North of the Mogollon Rim on the Colorado Plateau, a contemporaneous drainage reversal presumably led to establishment of a westerly-flowing, ancestral Colorado River on the Rim gravels, contrasting with north-flowing drainages that existed during accumulation of the Rim gravels. Following removal of much of the Rim gravels, the inferred ancestral Colorado River may have become incised in early Mesozoic and Paleozoic strata (Cooley and Akers, 1961) to form at least the initial stages of the Grand Canyon, and topographic relief comparable to that developed in central Arizona. In central Arizona, evidence for the second, later episode of erosion exists at the unconformable contact between early Miocene deposits and an underlying erosional landscape that locally contains Paleogene Rim gravels, seen at places across central Arizona from the Verde Valley on the north to Cave Creek on the south (fig. 19.1). However, direct evidence for this episode of erosion is lacking north of the Mogollon Rim.

The Mogollon Rim in the area of the Verde Valley marks an approximate erosional boundary of the Colorado Plateau, one that long has been recognized on purely physiographic grounds. The Rim here lies directly south of a deeply entrenched Colorado River (fig. 4.1; fig. 19.1). The Mogollon Rim and south rim of the Grand Canyon are at similar elevations and are connected by a virtually unfaulted high plateau, episode 1 surface (table 18.1) that locally is overlain by remnants of Rim gravel. This ancient surface can not be related to any episode of Miocene erosion by the Colorado River or the Verde River, and, creation of the scarp of the Mogollon Rim clearly pre-dates Miocene deposition in the Verde Valley.

Cretaceous-Paleogene erosion to produce the mountainous terrane of the Transition Zone of central Arizona and to form the high-standing Mogollon Rim that bounds the mountain region on the north apparently resulted from two general episodes of regional uplift (chapter 17). Abundant rainfall and vigorous perennial stream flow were required, first to move clastic materials northward across the region to deposit the Paleocene-Eocene Rim gravels, and then, following drainage re-organization and reversal, to erode the Rim gravels and move products of erosion southward out of the Transition Zone of central Arizona. Because similar climatic conditions should have prevailed north of the Mogollon Rim, a contemporaneous drainage reorganization, and entrenchment, also should have occurred on the Colorado Plateau. The Transition Zone (or mountain region) of central Arizona is a physical and structural extension of the Colorado Plateau, --an area in which Paleozoic strata have been largely removed and basement rocks exposed by erosion. No evidence exists that allows the Transition Zone to be divorced from the early and later episodes of uplift and erosion on the Colorado Plateau. A major, early Laramide episode of uplift, erosion, and deposition, followed by a younger, less severe episode of Laramide uplift, tilting, and drainage derangement (reversal), seems to best explain the development of the landscape and the distribution of Cenozoic deposits in central Arizona. A similar scenario may also explain the far less well documented relations in the Grand Canyon.

Models that call for a very young (Pliocene) Grand Canyon have not addressed problems posed either by requiring a very late Cenozoic uplift of the region (consequences of which are not observed in depositional sequences on and off the Colorado Plateau), or by delaying the main canyon-cutting episode and calling for the cutting to have taken place in slightly more than a million years (from ~5-4 m.y.). This rapid cutting is postulated to have occurred at a time when aridity dominated, and when late Miocene-Pliocene deposits were accumulating regionally in basins both on and off the Colorado Plateau.

Studies in the Verde Valley have indicated that the Mogollon Rim was eroded to its present height and near its present position by early Miocene (~24-17 m.y.) time, and that the last significant erosion could have been no younger than Oligocene (37-24 m.y.) (Elston, 1978, Gomez and Elston, 1978; Peirce and others, 1979; Elston 1984b). Evidence from the area of the Hualapai Plateau to the west, already summarized, indicates that a major part of the erosion of the Mogollon Rim occurred during the early Laramide.

Lower Miocene deposits lie on Paleozoic rocks at the base of the Mogollon Rim. They are at the level of Oak Creek, south of Sedona, which also is the level of the ancestral Verde Valley. Erosion that developed the Mogollon Rim thus had taken place prior to accumulation of the early Miocene deposits (Hickey sediments underlying 14.5-11 m.y.-old Hickey basalts), which were derived from the Black Hills to the west (Anderson and Creasey, 1958; Levings, 1980; Elston, 1984b). Fluvial, lacustrine, evaporite, and interbedded volcanic deposits of early to late Miocene and Pliocene age accumulated in the Verde Valley and in irregular basins across the Transition Zone. Deposits of aggradation also accumulated in the Little Colorado River basin on the Colorado Plateau (Miocene-Pliocene Bidahochi Formation). The deposits of aggradation were largely the products of flash flooding in the absence of perennial streams capable of removing the products of erosion from the region.

The climate was drier during Miocene and Pliocene time than earlier in Cenozoic time and at present (e.g., Smiley, 1984). Intermittent lakes were shallow, and most water presumably resided in the subsurface, moving down the regional gradient in the subsurface across the Transition Zone to lowland areas of the desert region of Arizona to the south. North-south diagrammatic sections in Elston (1984b) show the distribution of early to late Miocene and Pliocene deposits across the Transition Zone from the Mogollon Rim of the Verde Valley to Cave Creek, Arizona.

LANDSCAPE DEVELOPMENT: EPISODES, DEPOSITS, AND CLIMATE

Stages in development of the present landscape are assignable to six general episodes. These episodes, deposits related to them, and inferred climates, are summarized in Table 18.1. In this synthesis, the history of the southern and southwestern Colorado Plateau of northern Arizona, and the adjacent Transition Zone of central Arizona, are treated as a single entity, interrelated as to timing of uplift, amounts of erosion, environments of accumulation of sediments, and similarity of climate. The model is presented with the object of stimulating review and developing an improved perspective on problems bearing on the time of cutting of the Grand Canyon, and the roles that climate and tectonic history may have played in development of the Cenozoic landscape.

Episode 1 (~80-64 m.y.)

Episode 1 was marked by development of a regional surface of erosion, presumably beginning about 80 m.y. ago with the onset of Laramide deformation bordering the area of the Colorado Plateau. The erosion surface extended at least from the southern margin of the Transition Zone of central Arizona into southern Utah (chapter 19). At the southern margin and in the south-central parts of the Transition Zone, the surface of erosion was on crystalline rocks of Early Proterozoic age. A wedge of Paleozoic strata in the northern half of the Transition Zone is truncated by this regional unconformity, and the wedge thickens to the north. North of the Mogollon Rim, the erosion surface lies close to and generally corresponds with the boundary between the Permian Kaibab and Triassic Moenkopi Formations, forming an extensive planar surface of erosion that extended from the Mogollon Rim on the south to the vicinity of the Utah State line on the north. Farther north, this erosion surface appears to truncate progressively younger Mesozoic strata preserved in the high plateaus of Utah.

The end of the episode of planation is estimated from the youngest age (~64 Ma) determined on Late Cretaceous and early Paleocene volcanic clasts found in gravel at Long Point and in Peach Springs Canyon (chapters 19 and 20). A relatively wet environment is inferred to have accompanied the regional planation and transport of products of erosion into Late Cretaceous-early Paleocene basins of deposition in Utah.

Folding and faulting associated with northeast-directed Laramide compression served to moderately, or perhaps even substantially, elevate the terrane that was to become the Colorado Plateau in the area of the Grand Canyon, setting the stage for the accumulation of deposits of Episode 2. Substantial uplift (4,000 ft/1,200 m) apparently occurred in the area of the Hualapai Plateau of the western Grand Canyon, adjacent to higher-standing terrane to the west that was to become the Basin and Range Province (chapter 20). Strong uplift also may have occurred along the southern boundary of the Transition Zone. Definition of a step-bench topography by erosion and development of the Mogollon Rim was initiated during Episode 1 by streams flowing northeastward on the regional dipslope.

Episode 2 (~64-37 m.y.)

Clastic deposits (Rim gravels) hundreds of feet thick and locally containing fresh water limestone accumulated as a near-continuous apron across the area of central and northern Arizona (Young and Hartman, 1984; chapters 19 and 20). In the Lees Ferry area of northern Arizona, quartzitic gravels on the surface of the Paria Plateau (fig. 19.1) rest on the Jurassic Carmel Formation (Phoenix, 1963). The Rim gravels of the Coconino Plateau and the gravels near Lees Ferry would appear to correlate with Paleocene and Eocene deposits of the Canaan Peak, Pine Hollow, and Wasatch Formations of the Bryce Canyon, Utah, area.

In the Long Point area of the Coconino Plateau (fig. 19.1), deposition appears to have begun during late Paleocene time. Deposits preserved here may represent only part of an originally much thicker section of gravel and limestone. From paleontologic and paleomagnetic considerations, the Rim gravel preserved on the Coconino Plateau may be no younger than middle Eocene in age (chapter 19). Regionally, however, the episode of gravel and limestone deposition appears to have extended to the end of Eocene and possibly into early Oligocene time. A moderately wet environment, a continuation of a Late Cretaceous-early Paleocene wet climate that gave rise to Episode 1 planation, is suggested by the following: 1) transport of coarse- to fine-grained detritus to form fanglomeratic and conglomeratic deposits; 2) common to locally abundant beds of carbonate at a number of places; and 3) extensive chemical alteration marked by a pervasive precipitation of hematite pigment.

Episode 3 (~37-27 m.y.)

Deposits recording episode 3 are scarce and mostly poorly dated in central and northern Arizona, and southern Utah. This episode is marked chiefly by an unconformity that appears to include much of Oligocene time. During this interval, a drainage reversal (to the south) appears required to remove most of the Rim gravel deposits from the Transition Zone (mountain region) of central Arizona prior to the accumulation of extensive Miocene and Pliocene sedimentary deposits and volcanic rocks. Erosion, and extensional tectonism perhaps beginning late in Oligocene time, led to development of an irregular topography in central Arizona, which included an ancestral Verde Valley bounded by a Mogollon rim on the north having its present relief. Although much of the relief on the Mogollon Rim may have predated the Oligocene episode of erosion, some additional regional uplift and a continuing generally wet climate appears required to remove the Laramide Rim gravels from the region and to develop a topography that would receive early Miocene deposits of aggradation. Erosion of the older Rim gravels from plateaus north of the Mogollon Rim presumably also occurred contemporaneously. This interval of time seems to be critical for the establishment of an ancestral, south-flowing Verde Valley drainage system in central Arizona. It also is a prime candidate for the time of drainage integration on the Rim gravels of the Colorado Plateau north of the Mogollon Rim, and the establishment of a west-flowing ancestral Colorado River.

Integration of regional drainage systems in central and northern Arizona perhaps began during late Eocene or early Oligocene time, ending the episode of interior drainage that saw Paleocene and Eocene deposits accumulate in several large basins across the present area of the Colorado Plateau. These basins on the Plateau may have drained to the sea to the northwest (Dickinson and others, 1988). The inherited Laramide drainage, joined by an evolving ancestral upper Colorado River deflected(?) by the Laramide Kaibab uplift, found its way around the south end of the structure. In this area, a new river system presumably became established on the Paleocene-Eocene Rim gravels. The younger of two cooling events revealed from fission track dating on apatite is of post-middle Eocene age (chapter 17), and may record the episode of uplift that led to the postulated post-Laramide drainage adjustment. This episode of uplift, accompanied by unspecified regional tilting, perhaps was responsible for the differing drainage reversals in central and northern Arizona, as well as for re-organization of the drainage system and entrenchment of the new stream courses. If stream courses on the Colorado Plateau and in the Transition Zone were downcut contemporaneously to broadly similar base levels, the Plateau drainage ("ancestral Colorado River") in the area of the central and eastern Grand Canyon would have had perhaps 2,000+ feet (600+ m) of relief by the end of Episode 3, similar to relief on Mogollon Rim in Verde Valley (Elston, 1984b) and to pre-Bidahochi relief inferred in the Little Colorado River drainage (Cooley and Akers, 1961). In this scenario, drainage reversal and integration, accompanied by erosion and removal of most of the Rim gravels on the plateau and partial entrenchment of the ancestral Colorado River, occurred during an interval of about 10-13 million years. This is an interval substantially greater than has been appealed to for a Pliocene cutting of the western Grand Canyon.

Episodes 4 and 5 (~27-12 and ~12-4 or 2.5 m.y.)

A particularly controversial part of the Cenozoic history concerns the general environment of accumulation of Miocene and Pliocene deposits in the region, and the accumulation of potentially correlative but undated (pre-Pleistocene?) deposits in the Grand

Canyon and its tributaries (chapter 21). The Miocene-Pliocene depositional history of the Transition Zone of central Arizona has been pieced together from scattered dated deposits. Miocene-Pliocene deposits in central Arizona appear to record a general episode of accumulation of deposits of aggradation. Re-establishment of a through-going drainage system in the Verde Valley, and presumably also elsewhere in the Transition Zone, apparently occurred no earlier than late Pliocene time (~<2.5 m.y. ago).

If the general environment of deposition and the depositional history inferred for Miocene and Pliocene deposits in the Transition Zone of central Arizona is extrapolated to the area of the Grand Canyon, perennial stream flow can be inferred to have become reduced to the point where aggradation in the main and tributary stream courses occurred and gravel increasingly choked the channelways. At some point, surface flow no longer transported detritus downstream, except during times of flash flooding, and underflow presumably prevailed. With time, deposition of detritus derived from tributary stream courses came to dominate over detritus derived from upstream (trunk) sources (a relationship seen in the gravel deposit at the mouth of Nankoweap Canyon; chapter 21). Such mixtures of local and exotic materials appear to have once choked the course of Marble Canyon and the eastern Grand Canyon with perhaps 400 feet to locally as much as 800 feet (120-240 m) of gravel. Such deposits, which would have required appreciable time to accumulate, appear to have accumulated along much (if not all) of the length of the Grand Canyon after the canyon had been cut to very near its present depth. A gravel-choked channelway would have allowed reduced runoff and groundwater from the headwater areas of the Colorado River to move in the subsurface through the Grand Canyon and into Miocene-Pliocene deposits in the Basin and Range Province. Except for sporadic flash floods, the surface of deposits of aggradation along the river courses would have remained mostly dry during a period dominated by sediment accumulation.

The climate in central and northern Arizona during Miocene and Pliocene time appears to have been drier than it was during Oligocene and Pleistocene time. Ephemeral streams with reduced carrying capacity (except during times of flash flooding) apparently prevailed, leading to the accumulation of a variety of deposits, including poorly sorted clastic deposits, and locally, lacustrine deposits and evaporites indicative of aridity. As deduced from upper Miocene-Pliocene strata in the Verde Valley, this situation persisted until near the end of Pliocene time. Similar environments of deposition appear to have marked accumulation of the upper Miocene-Pliocene Muddy Creek Formation, deposited west of and high against the Grand Wash Cliffs of the western Grand Canyon, and the upper Miocene-Pliocene Bidahochi Formation in the Little Colorado River basin as well.

The Bidahochi Formation in the Little Colorado River drainage system fills an older, northwest-sloping valley defined by bedrock contours as low as 5,800 feet (1,770 m) and having more than 1,000 feet (300 m) of relief (Cooley and Akers, 1961, fig. 237.3). Sutton (1974) reported that the middle volcanic member of the Bidahochi contains trachybasalt dated at 6.7 Ma. These relations strongly suggest pre-Bidahochi (pre-late Miocene or older) establishment of northwest-flowing drainage into an ancestral(?) Colorado River, a conclusion reached by Cooley and Akers (1961) and Sutton (1974).

Present structural, topographic, and age constraints between the ancestral Little Colorado River valley and the south rim of the Grand Canyon in the vicinity of the Laramide Kaibab uplift imply a minimum canyon depth of 1,200 feet (365 m), assuming no westward gradient, to allow any drainage to cross the southern end of the uplift in pre-Bidahochi time. Conceivably, this would appear to require a significant canyon-cutting episode beginning in middle Miocene time, or possibly earlier (uncertainty of early Bidahochi age). Coincidently, the elevation of the floor of such an hypothetical canyon crossing the southern Kaibab uplift would correspond closely with the elevation of the top of the Redwall Limestone, a logical position for canyon incision to be impeded. However, this amount of incision would have to be a minimum value for the main stem of the master stream, and would only indicate an intermediate stage of development, not the onset of regional downcutting, nor the controlling, external base level.

A major episode of reduced stream flow during episodes 4 and 5 (table 18.1), accompanied by a choking of existing stream courses with thick fills of gravel, would serve to mitigate the "Muddy Creek problem" of the western Grand Canyon and the Basin and Range Province beyond. No longer a problem, accumulation of the Muddy Creek deposits would become part of the solution. Accumulation of the Muddy Creek Formation across the Grand Wash Cliffs would not preclude the existence of either an ancestral Grand Canyon upstream or an ephemeral Colorado River whose flow had disappeared into the subsurface through deposits of late middle Miocene to Pliocene age. Accumulation of the Muddy Creek Formation athwart the Colorado River only precludes the existence of a Colorado River actively carrying significant volumes of sediments into this late middle Miocene-Pliocene basin, and a river carrying detritus through to a lower base (sea?) level.

Episode 6 (~4 or 2.5 m.y. - present)

The sixth and final episode is inferred to have begun during Pliocene time, apparently at about 2.5 m.y. ago in the Verde Valley and about 4 m.y. ago in the area of the western Grand Canyon. The greater age for the end of deposition and onset of erosion derives from Muddy Creek basalt K-Ar dates near Lake Mead (Metcalf, 1982) and an ~3.8 m.y.-date for the basalt of Grand Wash (mi 285, km 458, chapter 1), which is unconformable on an eroded section of the Muddy Creek Formation. The younger age is from Verde Formation magneto-stratigraphy (Bressler and Butler, 1978). Accumulation of a thick section of white limestone in the uppermost Verde Formation at the north end of the Verde Valley suggests an increase in rainfall and the end of ephemeral lacustrine conditions. This was followed by re-establishment of the Verde River as a perennial stream, and fluvial entrenchment of the Verde Formation on the order of a few hundred to several hundred feet. A perennial Colorado River presumably also became re-established at this time, accompanied by incomplete removal of products of aggradation from the trunk and tributary stream courses. Lastly, multiple lava dams appear to have been rapidly emplaced (and rapidly destroyed) in the western Grand Canyon during the past 1.2 m.y. (or 0.7 m.y.) (chapter 23). Temporary lakes behind the lava dams, some of which are inferred to have reached Lees Ferry and beyond, perhaps were responsible for the accumulation of terrace deposits of unconsolidated alluvium on relict deposits of older gravel.

SUMMARY

An entirely satisfactory, comprehensive explanation for the Tertiary evolution of the Grand Canyon has yet to be devised. Although some interesting models have been presented and some important constraints have been documented in a number of areas (table 18.1), several problems are unresolved. These problems center around the following issues.
1) What specific drainage adjustments accompanied the change from the well-established northeastward (Laramide) paleoslopes to the southwestward Colorado River physiography?
2) How can remnants of Miocene or older buried channels (Little Colorado River Valley) be reconciled with the apparent age constraints posed by the geologic relations preserved in the Lake Mead area prior to, during, and after Muddy Creek (late middle Miocene to Pliocene) time?
3) If the plateau margin had thousands of feet (4,000 ft/1,200 m) of relief during Laramide time (indicating substantial early uplift), why is it not likely that a significant amount of fluvial incision occurred during Oligocene and part of Miocene time, as can be inferred for central Arizona?
4) What was the timing and amount of late-Laramide uplift that led to inferred regional drainage reversal and incision (as distinct from later Basin and Range deformation)?
5) How was the erosional history of the Grand canyon related to, or affected by, apparent climatically controlled cycles of aggradation and erosion in Miocene basins in regions on and bordering the Colorado Plateau.
6) What effect did Basin and Range structural adjustments have on the regional drainage system that must have existed at that time; did they merely serve to enhance a climatically controlled interruption of drainage systems in the region?
7) Why do some parts of the Colorado River system appear to be older than others?

Until some of these unanswered questions, and contradictory lines of evidence, are better resolved, explanation of the Grand Canyon's complex origin will remain incomplete.

View west from the top of the Redwall Limestone (Mississippian) towards Royal Arch Creek (Elves Chasm; A), which cuts massive travertine terrace. The travertine cliff is about 400 feet (122 m) thick near the Colorado River from mile 115.2-116.7 (km 185.4-187.8). See chapter 21 for discussion.

CHAPTER 19: PALEONTOLOGY, CLAST AGES, AND PALEOMAGNETISM OF UPPER PALEOCENE AND EOCENE GRAVEL AND LIMESTONE DEPOSITS, COLORADO PLATEAU AND TRANSITION ZONE, NORTHERN AND CENTRAL ARIZONA

Donald P. Elston[1], Richard A. Young[2], Edwin H. McKee[3]
and Michael L. Dennis[1]

STATEMENT OF PROBLEM

The age of gravel and interbedded limestone deposits preserved discontinuously on a high erosion surface developed on the lower part of the Moenkopi (Triassic) and upper part of the Kaibab (Permian) Formations in northern Arizona was unknown until a few years ago. These deposits, long recognized and commonly referred to informally as "Rim gravels" for the Mogollon Rim region (Wilson, 1962; Moore, 1968; McKay, 1972; Peirce and others, 1979), are widely distributed across the southern margin of the Colorado Plateau and Transition Zone of central Arizona (fig. 19.1). The true antiquity of the deposits was first revealed from the identification of *viviparid* gastropods of early Eocene or greater age from limestone interbedded in the gravel in the area of Long Point on the Coconino Plateau (Young and Hartman, 1984). In addition, the limestones contain pollen from grasses (which did not evolve until Paleocene time) (Leslie Sirkin, written communication, 1975), and *charophytes*, presently known from rocks ranging from early Eocene to middle Oligocene in age (Richard Forester, written communication, 1984). This combination of forms is consistent with a Paleocene to late Eocene age for these sedimentary deposits. Young (1987) noted that the upper Paleocene to lower Eocene Flagstaff Formation of the high plateaus of southern Utah may be most nearly equivalent in age to the paleontologically dated Rim gravels south of the Grand Canyon. Similar fluvial deposits found in the Fort Apache Indian Reservation of eastern Arizona contains an airfall tuff near the top of a thick Rim gravel section from which biotite dated by K-Ar analysis yielded as age of about 37 Ma (Potochnik, 1987).

Some 20 years ago, the Rim gravels of the Coconino Plateau at Long Point had been assumed to be not much older than the age of overlying ~7-14 Ma basalt flows (McKee and McKee, 1972). Lucchitta (1984, p. 278, p. 294-295; 1987; 1988) noted the existence of similar gravels on the Shivwits Plateau north of the Colorado River. Because the gravel on the Shivwits Plateau was inferred to be not much older than than an overlying ~7 Ma flow, Lucchitta concluded that the regional drainage

---

[1]U.S. Geological Survey, Flagstaff, Arizona
[2]Department of Geological Sciences, State University College of Arts and Science, Geneseo, New York
[3]U. S. Geological Survey, 345 Middlefield Road, Menlo Park, California

(base level) was near the level of the Shivwits Plateau during late Miocene time. (See chapter 18 for a review of models for the time of cutting of the Grand Canyon.)

The Eocene and possibly Paleocene age for Rim gravels underlying Miocene basalt on the Coconino Plateau, as determined from paleontology, indicates that the subjacent erosion surface was of Paleocene or even Late Cretaceous age, and thus a product of the Laramide orogeny. The gravel deposits and underlying surface of erosion must have been contemporaneous with the thick lacustrine and fluvial sequences of early Tertiary age in southern and central Utah, which formed when that region may have been nearer sea level.

To obtain additional information bearing on the age of the Rim gravels, radiometric age studies were initiated. Eight volcanic clasts from near the base of several gravel sections were dated by whole rock K-Ar analysis to obtain an internally derived estimate of a minimum age for the time of accumulation of the deposits. To complement this, a paleomagnetic study of sandstone, siltstone, and limestone in the gravels was carried out at seven sections at five localities to obtain paleomagnetic directions and polarity information that might serve to correlate the deposits, and to perhaps place them in the polarity time scale. These studies have provided information that generally accords and supports the paleontologically determined estimate of age and age range. The paleontologic and supporting paleomagnetic data suggest a general late Paleocene to early Eocene age for the Rim gravels on the high plateaus south of the Grand Canyon. The surface on which the gravels rest thus would appear to have been cut during Late Cretaceous to early Paleocene time, as previously surmised by Young (1985, 1987), and the gravels would be of the order of 45 m.y. older than had been previously assumed by some workers. The early Tertiary events represented by the erosion surface and Rim gravels consequently appear to be much older than and clearly distinct from events related to formation of the Grand Canyon.

EXTENT AND DISTRIBUTION OF GRAVELS

The Rim gravels have been long known and studied (e.g., Robinson, 1913; Koons, 1948, 1955; Price, 1950; Young, 1966, 1979; McKee and McKee, 1972; Peirce and others, 1979). The most detailed studies have been undertaken on the Coconino Plateau (fig. 19.1), and exposures in this area continue to provide useful information. The gravels recently have been remapped

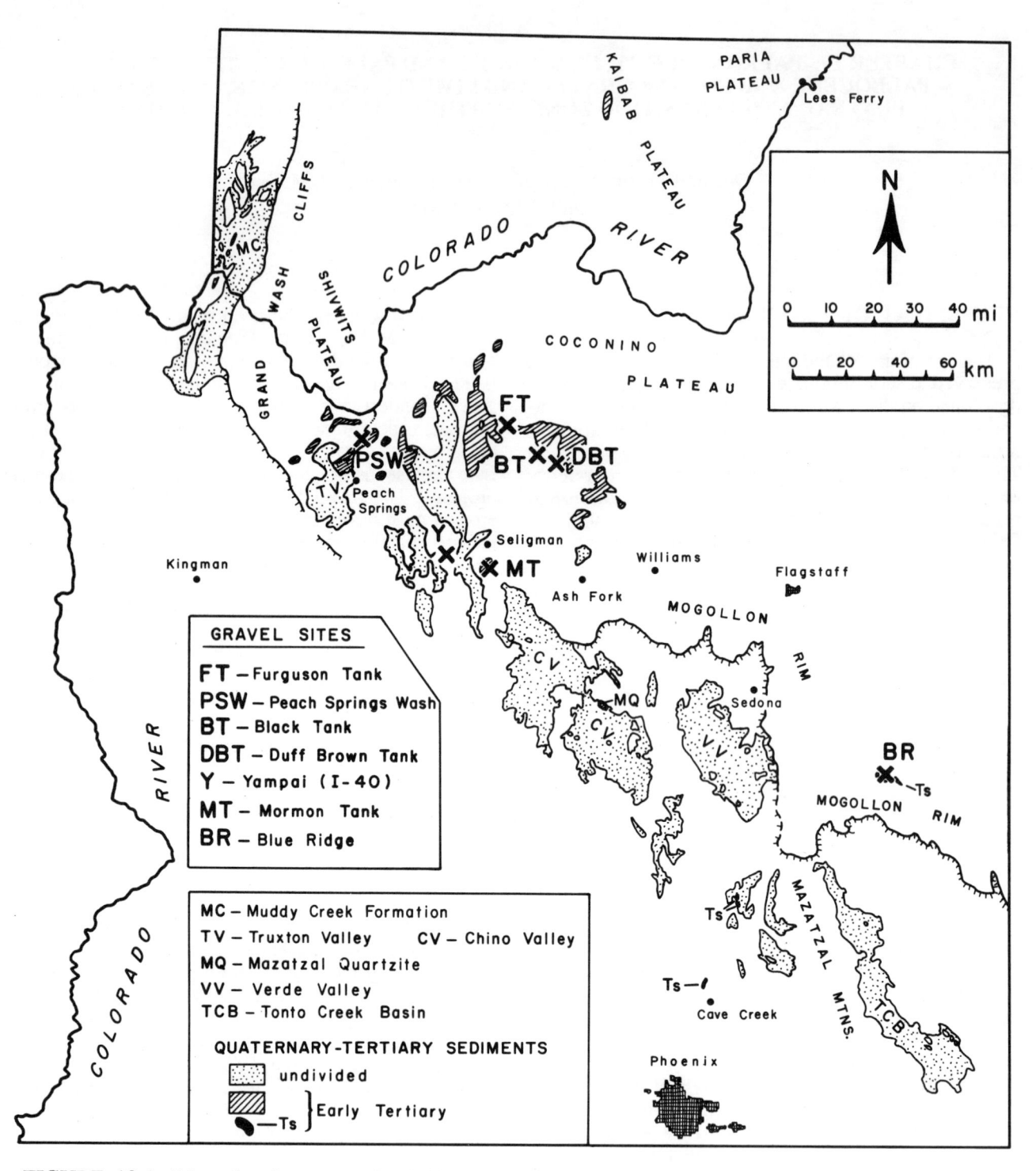

FIGURE 19.1 Map showing generalized distribution of Paleocene-Eocene Rim gravels and undivided Quaternary-Tertiary deposits in northern and central Arizona. Sampling site for paleomagnetism and stratigraphic study indicated by **X**. Mormon Tank site not sampled for paleomagnetism.

and subdivided geologically from the west side of Cataract Creek to the area of Peach Springs Canyon and Peach Springs Wash (Billingsley and others, 1986a, 1986b; Wenrich and others, 1986a, 1986b), where correlative deposits had been the subject of earlier studies (Young, 1966, 1979; Young and Brennan, 1974).

In the Peach Springs area, the Rim gravels occupy multiple channel segments cut into the plateau surface at several discrete elevations, and they also occupy an erosionally embayed margin of the Colorado Plateau at Truxton Valley (chapter 20). As concluded by Young (1987), the Rim gravels in the vicinity of Peach Springs and Truxton Valley had their sources to the south and west, and transport directions were to the north and northeast.

The ancient, high-level Rim gravels consist

predominantly of white or pale reddish to locally dark reddish brown arkosic sandstone, and they commonly are moderately well cemented. At some localities the basal beds contain many more clasts of Paleozoic rocks than overlying beds. Thus the gravels record the gradual unroofing of the Precambrian basement during Laramide uplift. Recognized at widely separated localities on the Colorado Plateau of northern Arizona, they also are present discontinuously across the Transition Zone of central Arizona. Several representative localities are shown on Figure 19.1. The deposits are well exposed not only in the Long Point and Peach Springs Wash areas, but also in exposures along a 5 mile (8 km) stretch of Interstate Highway I-40, about 10 to 15 miles (16-24 km) west-southwest of Seligman, and near Mormon Tank, about 6 miles (9.7 km) south of Seligman. Additionally, similar deposits mapped as Ts (Tertiary sediments) on the 1967 Geologic Highway map of Arizona (Cooley, 1967) are found 1) in the Blue Ridge area, 48 miles (77 km) southeast of Flagstaff, 2) in two exposures in Bloody Basin in the Transition Zone of central Arizona south of the Verde Valley, and 3) in several exposures of apparently correlative deposits north of Cave Creek at the southern margin of the Transition Zone (fig. 19.1). Presumably equivalent deposits are found near and south of the Mogollon Rim of the Fort Apache Indian Reservation of east central Arizona, east of the area of Figure 19.1. Lastly, during preparation of this guidebook, a new gravel locality of probable Tertiary age has been verified north of the Colorado River, on the west side of the Kaibab Uplift (fig. 19.1; Peter W. Huntoon and Richard A. Young, oral communication). However, these gravels contain abundant black chert, which has been assumed to indicate a more westerly provenance (Cooley, 1960).

Exposures of well-cemented red conglomerate (or fanglomerate) found at and near the southern margin of the Transition Zone are preserved near the foot of the New River Mountains near Cave Creek, north of Phoenix (Gomez, 1979). The conglomerate here is unconformably overlain by a lithic tuff containing an Oligocene oreodont near its base (Lindsay and Lundin, 1972; Lander, 1977). The lithic tuff is overlain by a basalt flow dated at 22.4±2.6 Ma by K-Ar whole rock analysis (Lindsay and Lundin, 1972). A thick section of interstratified lower Miocene fluvial, shallow water lacustrine and volcanic strata, and middle Miocene basaltic flows, unconformably overlie the well-cemented red conglomerate to form the New River Mountains (Gomez, 1979). The lithologic character, unconformable relations, and age constraints in the Cave Creek area thus indicate that the well cemented conglomerate at the base of the section correlates with Rim gravel deposits elsewhere in the Transition Zone and on the Colorado Plateau.

Gravels on the Paria Plateau, west of the Colorado River near Lees Ferry (fig. 19.1), may be correlative with the lower Tertiary gravels of the Coconino and Kaibab Plateaus. Phoenix (1963, p. 38) encountered high-level, quartzite-bearing gravels resting on Jurassic strata more than 3,500 feet (1,070 m) above the Colorado River during his mapping of the Lees Ferry Quadrangle (Phoenix, 1963, p. 38, "Qg" deposits, Plate 1). These gravels are at an elevation above 6,500 feet (1,980 m), similar in elevation to Rim gravels of the Coconino Plateau and those west of the Kaibab uplift. The regional distribution of the Rim gravels (fig. 19.1) suggests that deposits as old as Paleogene may have once blanketed much of central and northern Arizona.

STRATIGRAPHY AND AGE

The generalized stratigraphy of Rim gravels examined in the vicinity of Long Point, Peach Spring Wash, Highway I-40, Mormon Tank, and Blue Ridge is depicted in Figure 19.2. Source areas for the gravels near Seligman and at Blue Ridge appear to have been to the south. Clasts apparently derived from the Proterozoic Mazatzal Quartzite found at Mormon Tank and Blue Ridge presumably came from exposures of Mazatzal Quartzite preserved in Chino Valley and in the Mazatzal Mountains, respectively. The Rim gravels contain no volcanic detritus characteristic of Miocene basaltic flows that were extensively erupted across the region. The gravels therefore appear predate this episode of regional volcanism. Upper Oligocene volcanic units, present in the Chino Valley-Juniper Mountains region south of Seligman, overlie the gravels; clasts from them have not been observed in localities such as the gravel of the Mormon Tank site.

Vivaparid gastropods indicative of an early Eocene or greater age as well as *Physa* and *Lioplacodes* (Young and Hartman, 1984) have been collected from limestone at the Black Tank locality of the Long Point area (figs. 19.1 and 19.2). Additionally, volcanic pebbles from three sites in the Long Point area and from three sites in Peach Springs Wash now provide K-Ar dates indicative of a Late Cretaceous to early Paleocene age for a time of volcanism in the source area for the gravels (fig. 19.2). The youngest of the dates (63.7 Ma) also establishes a maximum (early Paleocene) age limit for the time of deposition of the Rim gravels in the Long Point and Peach Springs Wash areas.

PALEOMAGNETIC STUDIES

Sampling for paleomagnetic analysis has been carried out at six of the seven localities shown on Figure 19.1, and across the stratigraphic intervals diagrammed in Figure 19.2. These studies were undertaken with the object of providing information that might allow the deposits to be correlated with one another and perhaps with the polarity time scale. Fine- to medium-grained sandstone, silty sandstone, and limestone at five of six sections responded to thermal cleaning. Coarse-grained sand at the sixth section (Peach Springs Wash) provided only scattered paleomagnetic directions. Polarity obtained following thermal cleaning is dominantly reverse, and most paleomagnetic directions exhibited anomalously low inclinations. The distribution of polarity in the sections studied is diagrammed in Figure 19.2.

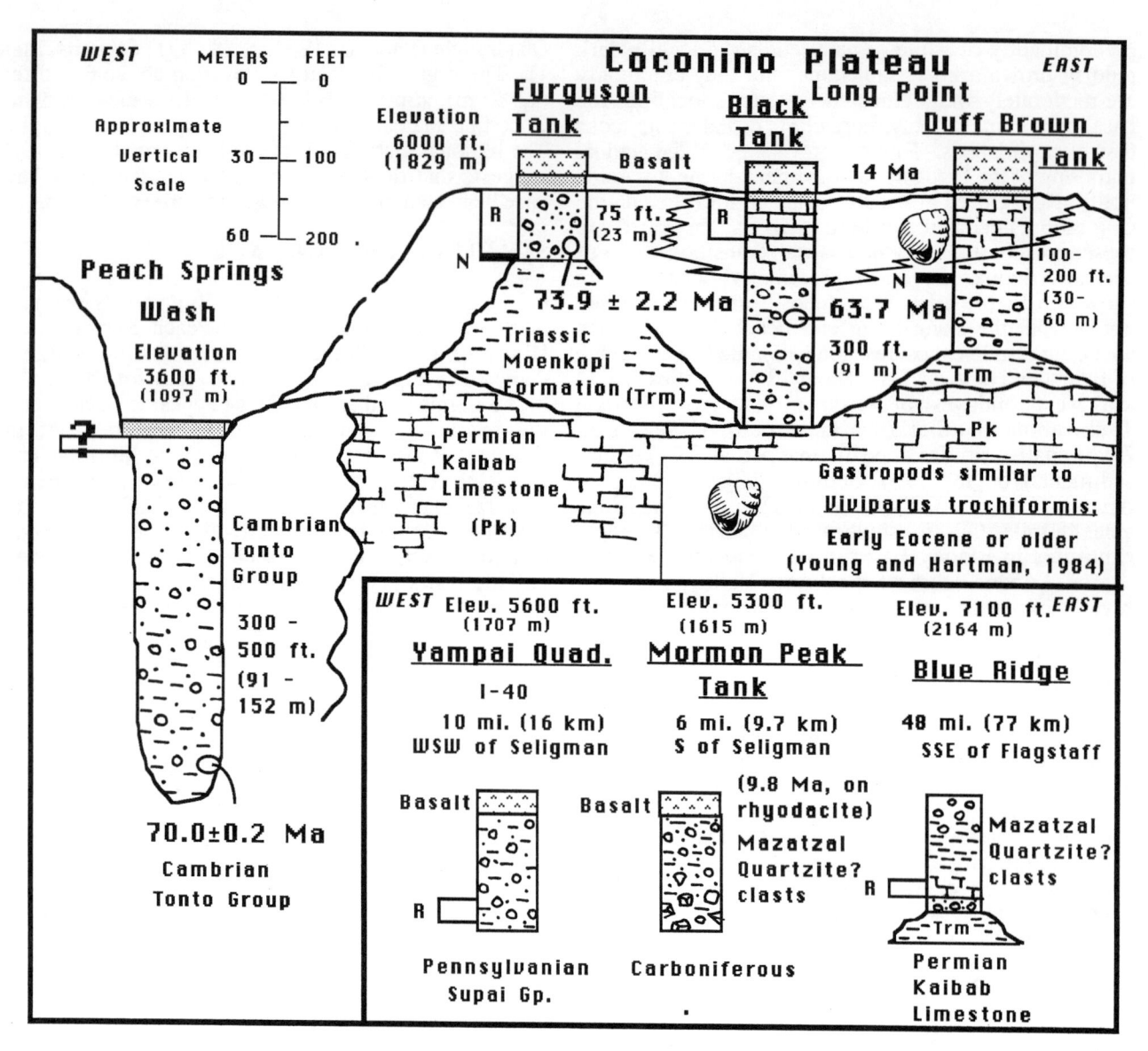

FIGURE 19.2. Diagrammatic sections showing approximate intervals in Rim gravel sections sampled for paleomagnetic analysis; white - reverse polarity, R; black - normal polarity, N. Also shown are approximate positions of volcanic clasts that yielded whole-rock K-Ar ages, and location of paleontologically dated gastropods. 1) Rim gravels; 2) interbedded limestone.

Stability of Magnetization

Paleomagnetic directions obtained from individual samples from the six sections, measured prior to and following progressive, stepwise thermal cleaning, are shown in Figure 19.3. Directions measured prior to cleaning are labelled as NRM directions (natural remanent magnetization). Characteristics observed during stepwise progressive demagnetization treatment for representative samples from the individual sites are shown in a series of equal area plots of directions and on orthogonal demagnetization diagrams (fig. 19.4).

Mean Directions

Furguson Tank. The most stable magnetization obtained from any of the sites came from two thin (~15-20 cm-thick), dark red silty sandstone beds found within a half meter of the base of the Furguson Tank section, a few kilometers west of Long Point. These beds, derived from red beds of the closely subjacent Triassic Moenkopi Formation, provided tightly grouped, normal and reversed polarity directions. The directions remained virtually unchanged from the NRM step to the 670 °C cleaning step. The declinations are not exactly anti-parallel, and the inclinations (2.5° and 5.9°; fig. 19.3) are much shallower than expected for deposits of Paleocene or Eocene age. An inclination of 51° has been reported from middle Paleocene strata of the Nacimiento Formation of the San Juan Basin, New Mexico, a direction employed for the calculation of a Paleocene pole for North America (inclination = 51.3°, declination = 343.9°; Butler and Taylor, 1978). The shallow inclinations thus appear to be anomalous. Nonetheless, such inclinations can have value for purposes of correlation if they can be demonstrated 1) to be stable, characteristic, and early ("primary"), and 2) to be restricted to a discrete part of the time scale.

Near-antiparallel reversals were obtained from two, thin (~1 ft; 25 cm), closely adjacent beds at the Furguson Tank section (FT 1 and FT 2; fig. 19.3), and these beds closely overlie the Moenkopi Formation. The beds are darker red than overlying strata of the Rim gravel section, and appear to have been derived directly from the subjacent Moenkopi. The directions are very stable to high demagnetization temperatures (670 °C), exhibit univectorial decay and display no apparent secondary components of magnetization (figs. 19.3 and 19.4), indicating that the magnetization is characteristic and early, and thus apparently acquired during the course of accumulation and lithification. The very high stability also indicates that the magnetization resides in detrital specularite grains (black hematite) derived from the Moenkopi Formation, as well as in (less stable) red hematite pigment. Additionally, the anomalously shallow inclinations and non-antiparallel declinations suggest that depositional bedding errors are present. The magnitudes of potential inclination and declination errors have been documented from the Moenkopi Formation elsewhere in the region and from experimental studies employing naturally disaggregated detritus of the Moenkopi Formation. About 12-20° of inclination error and 10° of declination error is expected (Elston and Purucker, 1979; Bressler and Elston, 1980). This can only partly account for the very shallow inclinations and for the skewed declinations obtained from the two near-basal beds in the Furguson Tank section. Even with adjustments that might arise from bedding errors, the directions would far from correspond with ambient field directions anticipated for the Paleocene and Eocene. No ready explanation, including remagnetization, exists for these anomalous directions, which from stratigraphic considerations appear to represent only a geologically brief interval of time.

Strata of the Furguson Tank section, above the basal 1-2 m, are light brick-red to pink in color. Twenty samples were collected from five horizons spaced across an approximately 20 m-thick interval of strata. The magnetization of the pale red beds was found to be considerably less stable than that measured from the basal beds at Furguson Tank. Nonetheless, although NRM directions were rather scattered and tended to plot in northerly directions, directions for all samples moved into southerly positions with cleaning, but failed to cluster tightly. The poorly defined direction obtained at the 590°C cleaning step (implying the remanence resides principally in hematite pigment) is close to mean reverse polarity directions obtained from other reversed polarity sites in the Rim gravels. The stability of magnetization degraded abruptly above 590°C, accompanied by a large increase in the dispersion of directions.

Black Tank. A 30 foot-thick (9 m) section of very light gray fresh water limestone at the top of the Rim gravel section was sampled at Black Tank, a few kilometers west of Long Point on the Coconino Plateau. Magnetization was very weak (~1×10^{-7} gauss); only highly scattered directions, slightly skewed toward a southerly reversed polarity direction, were observed prior to cleaning. The limestone did not respond to alternating field demagnetization treatment, but did respond to thermal treatment to reveal a shallow reversed polarity direction (fig. 19.3). The best grouping occurred at the 400 °C cleaning step, and a more southerly, slightly less well clustered direction was obtained at 450 °C. Above this temperature, intensities of magnetization decreased below the measurement capability of the magnetometer and a coherent grouping of directions was lost. The directions arrived at from the groupings are shallow and may indicate contamination with a secondary normal polarity component of magnetization.

Duff Brown Tank. Six samples were collected east of Long Point from two thin (<1 m-thick) red bed intervals, approximately 1 m and 5 m below limestone that contains the early Eocene (or older) *vivaparid* gastropods. The NRM directions were steeper than the present field direction, suggesting that a reversed polarity direction would be revealed with cleaning. However, the directions moved in an opposite sense, toward a normal polarity direction in the range 620-640 °C. The mean inclination (~35°), although fairly steep, still is 15° shallower than expected. The declination is similar to the declination of the normal polarity bed at Furguson Tank, and is antiparallel to the reverse polarity declination obtained from the 450 °C cleaning step from Black Tank. However, as for the Black Tank samples, an unremoved secondary component of magnetization might account for the shallow mean inclination.

Yampai Quadrangle (I-40). Samples from this site came from discontinuous sandstone interbeds within a few meters of the base of the Rim gravel section exposed in road cuts on the north side of Interstate Highway I-40, about 10 miles (16 km) west-southwest of Seligman. The gravel at this locality contains a high proportion of clasts derived from red bed strata of the Supai Group (Pennsylvanian and Permian). It was hoped that detrital specularite derived from these rocks would carry a depositional remanent magnetization (such as apparently seen in the basal beds at Furguson Tank) to supplement magnetization residing in red pigment. The NRM directions were highly scattered. However, 16 of 22 samples responded to thermal treatment in the range 400-590 °C to provide a shallow reverse polarity direction. Of the 16 samples, twelve also provided an anomalous, north-seeking normal polarity direction at the 620 °C cleaning step. Coherence of direction was lost at higher temperatures. We report a provisional reverse polarity direction from strata at this site, considering it to reside in an early post-depositional pigment, and suggest that a normal polarity field may have been in existence at the time of accumulation of these beds. Neither the normal polarity or reverse polarity direction can be considered to be free of unremoved secondary components of magnetization.

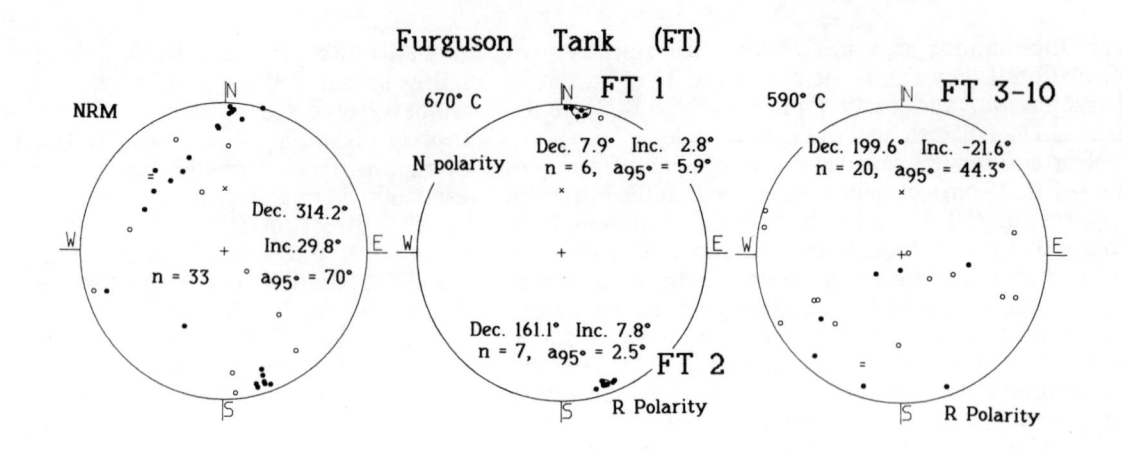

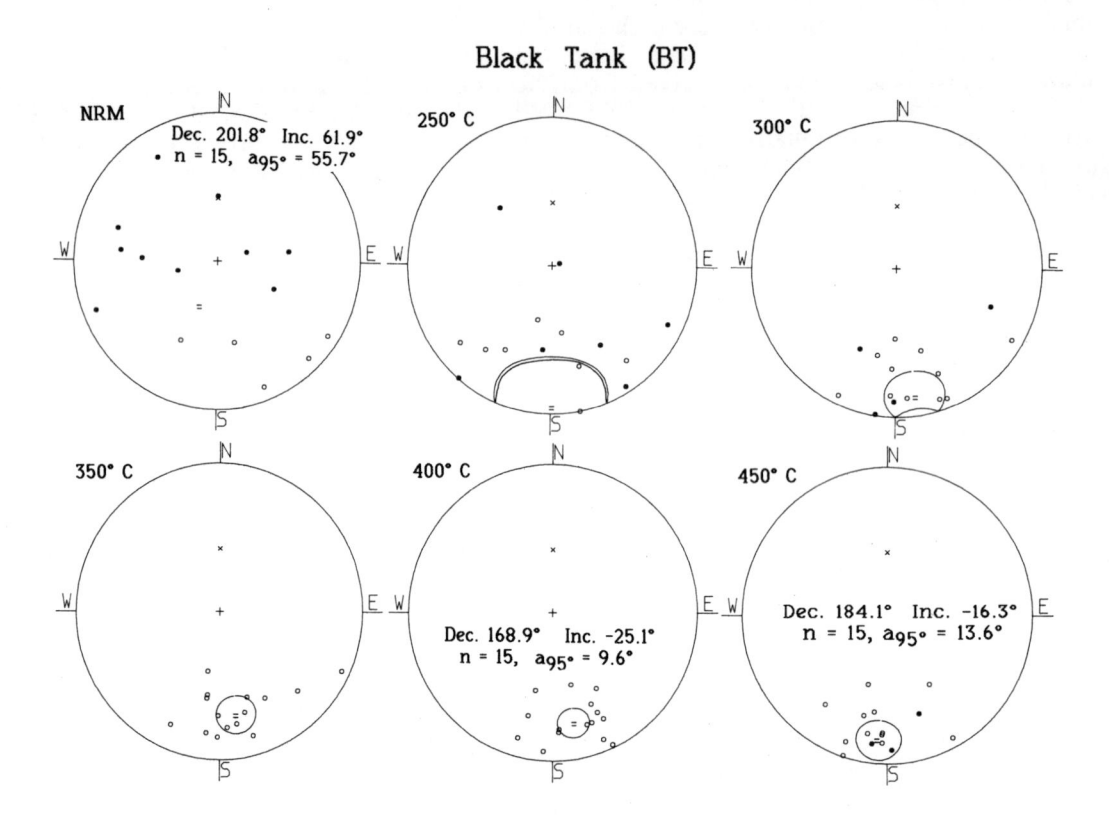

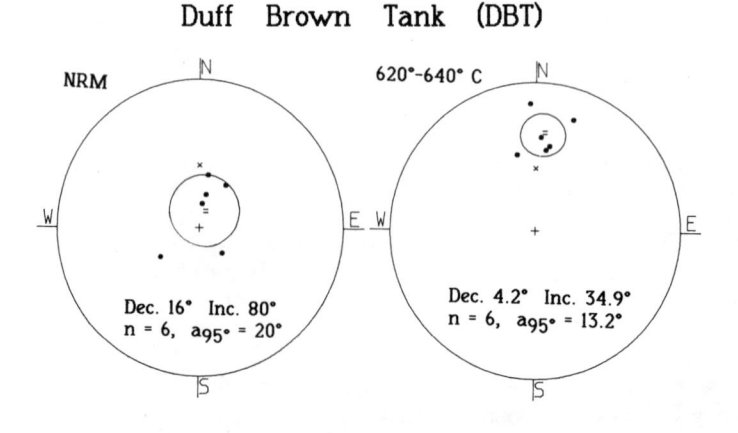

FIGURE 19.3.

Yampai (I-40) (Y)

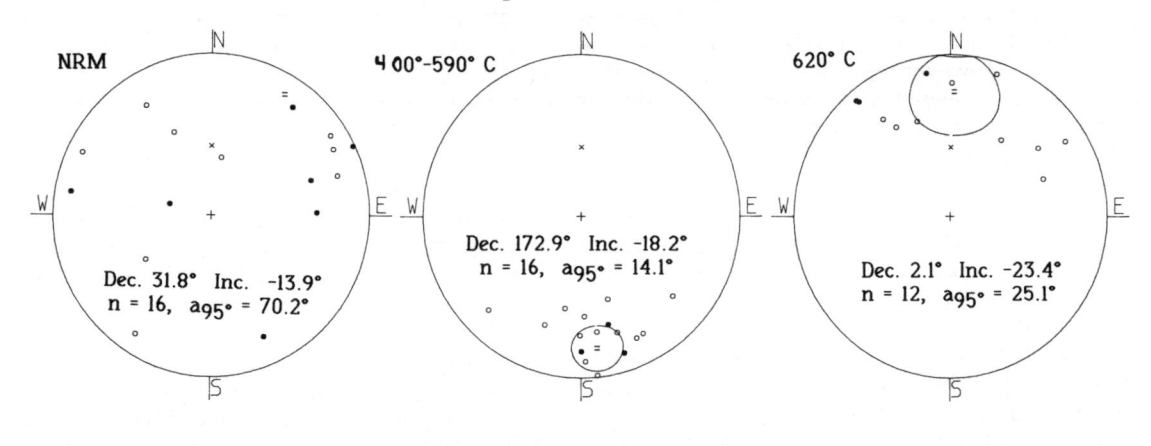

NRM
Dec. 31.8° Inc. -13.9°
n = 16, a₉₅° = 70.2°

400°-590° C
Dec. 172.9° Inc. -18.2°
n = 16, a₉₅° = 14.1°

620° C
Dec. 2.1° Inc. -23.4°
n = 12, a₉₅° = 25.1°

Blue Ridge (BR)

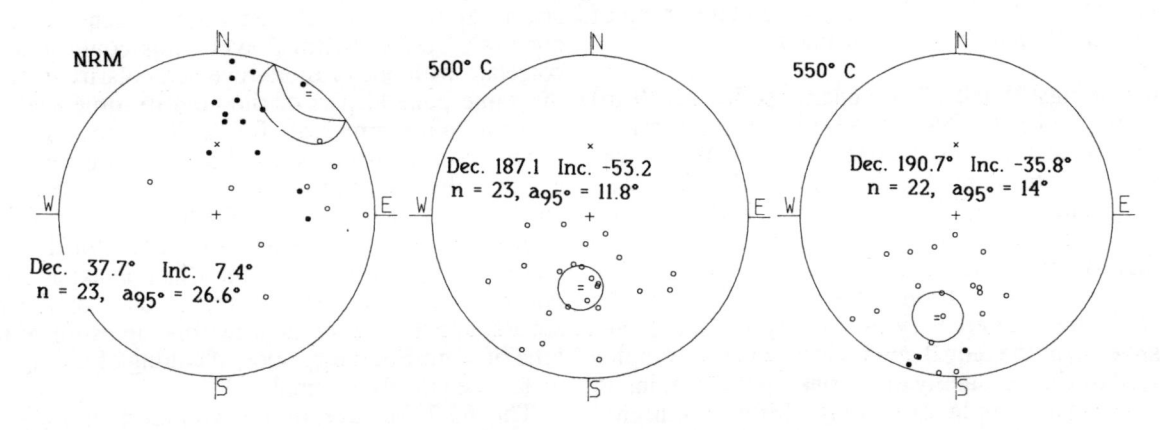

NRM
Dec. 37.7° Inc. 7.4°
n = 23, a₉₅° = 26.6°

500° C
Dec. 187.1 Inc. -53.2
n = 23, a₉₅° = 11.8°

550° C
Dec. 190.7° Inc. -35.8°
n = 22, a₉₅° = 14°

Peach Springs Wash (PSW)

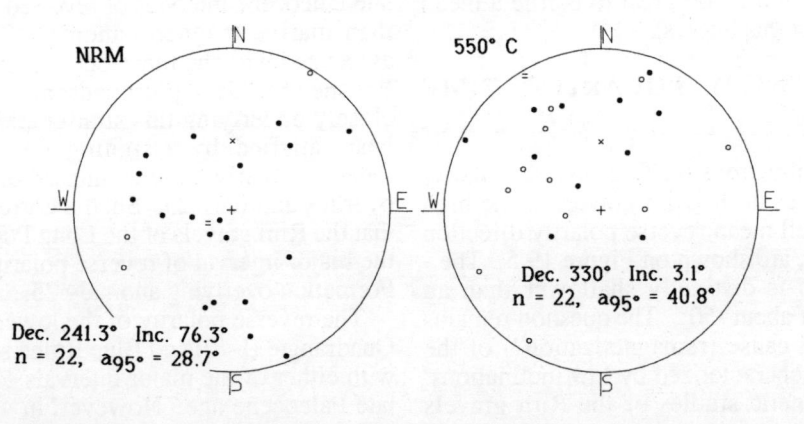

NRM
Dec. 241.3° Inc. 76.3°
n = 22, a₉₅° = 28.7°

550° C
Dec. 330° Inc. 3.1°
n = 22, a₉₅° = 40.8°

FIGURE 19.3. Equal area plots of paleomagnetic directions from six localities showing distributions of sample directions, and mean directions (= sign), before cleaning (NRM), and at thermal cleaning steps that yielded clustered directions. Solid dots - lower hemisphere; open circles - upper hemisphere; Dec. - declination; Inc. - inclination; n - number of samples; a₉₅° - circle drawn at the 95 percent confidence level (Fisher, 1953); X indicates present axial dipole field direction.

Blue Ridge. Strata sampled from this site are exposed along State Highway 87, about 5 miles (8 km) north of Clints Well and the junction with the Forest Service Road that leads to Happy Jack and Flagstaff. The samples come from a 12 foot-thick (3.6 m) interval of red, fine grained carbonate-cemented silty sandstone and silty limestone that overlies a basal conglomerate, 9.8 feet (3 m) thick. The conglomerate, in turn, rests on red beds of the Moenkopi Formation (Triassic). A population of shallow northerly directions and streaked easterly upper hemisphere directions suggested the existence of two polarities at the NRM step. However, all samples moved into a reverse polarity upper hemisphere direction at 500 °C. The mean inclination at this step is close to that expected for rocks of Paleocene and Eocene age. However, cleaning at 550 °C resulted in a shallower mean, reverse polarity inclination, similar in magnitude to the normal polarity inclination at Duff Brown Tank. Dispersion increased markedly at higher temperatures. The question remains as to whether an "end point" mean direction at 550 °C or the mean direction at 500 °C better represents the mean direction at the time of deposition and lithification of beds of this thoroughly oxidized deposit. The great circle "streak" in the demagnetization data clearly shows a strong normal polarity secondary overprint, and this streak passes through the area of the expected reverse polarity direction before arriving at a lower inclination direction at the highest stable temperature. Even if lines and planes analyses indicated a grouping at the steeper inclinations, the cause of the shallower, higher temperature group of directions would remain to be explained.

Peach Springs Wash. The uppermost 30 feet (9 m) of an estimated 300-500-foot (91-152 m) thick gravel section exposed in lower Peach Springs Wash was sampled to determine if pale red, poorly indurated sandstone, unconformably overlain by a well-cemented gravel containing basaltic detritus (and therefore probably of Miocene age) retained an early stable direction of magnetization. NRM directions of the samples were scattered, lying mainly on the lower hemisphere of the equal area plot; several samples displayed distinct southerly directions. With cleaning to 550 °C, most sample directions shifted to highly scattered, northerly positions, about equally distributed between the lower and upper hemispheres of the equal area plot. The dispersion is too great to define a mean direction or polarity for this interval.

CORRELATION WITH POLARITY TIME SCALE

Provisional mean directions for five reverse polarity intervals from four sites in the Rim gravels of the high plateaus, and an overall mean reverse polarity direction for the five directions, are shown on Figure 19.5. The -20° mean inclination is distinctly shallower than an expected inclination of about -50°. The question remains as to the validity and cause (remagnetization?) of the anomalous directions characterized by low inclinations. Additional paleomagnetic studies of the Rim gravels clearly are needed.

Paleomagnetic results from the Rim gravels are compared with results reported from a Paleocene-Eocene sedimentary section that has been studied for its polarity zonation (Black Peaks Formation of the Big Bend area of west Texas; Rapp and others, 1983). The Black Peaks Formation, of late Paleocene and early Eocene age as determined from its vertebrate fauna, was found to be dominantly of reversed polarity. Curiously, many of the reverse polarity directions across the middle part of the section exhibited anomalously low paleomagnetic inclinations. The mean direction from 22 reverse polarity sites provided a mean inclination of about -26° (fig. 19.5), a direction that is statistically identical with the mean direction from the Rim gravels of Arizona. However, as for the Rim gravels of Arizona, the validity and cause of the low inclination directions for the Black Peaks Formation remain to be determined.

The temporal interval embraced by the interval of low inclination reversed polarity directions in the Black Peaks Formation is diagrammed in Figure 19.6. The stratigraphically much less comprehensive polarity zonation from the Rim gravels thus could potentially correlate with one or more reversed polarity intervals in the same general part of the polarity time scale. The position in the time scale for the Rim gravels shown in Figure 19.6, constrained by the middle Eocene or older age for the *viviparid* gastropods, would seem to correspond best with the late Paleocene - early Eocene interval of reverse polarity (and anomalously shallow inclinations) reported from the Black Peaks Formation. Ages of volcanic clasts from the base of the gravel section appear to provide a reasonable estimate for the time of a pre-Eocene episode of cutting of the surface on which the gravels accumulated.

The 63.7 Ma age of the youngest of three dated volcanic clasts indicates that the Rim gravels of the Coconino Plateau may be no older than the oldest of the late Paleocene intervals of reversed polarity, and younger than marine magnetic anomaly 27. Reverse polarity associated with the limestone facies that contain the early Eocene (or older) gastropods, and normal polarity strata closely underlying this stratigraphic level, appear to be best satisfied by assigning this interval to the late Paleocene-early Eocene interval of reverse polarity that overlies anomaly 25. Such a correlation would suggest that the Rim gravels of the Long Point area correlate with the major interval of reverse polarity in the Black Peaks Formation overlying anomaly 25.

The reverse polarity of the lower parts of the Yampai Quadrangle (I-40) and Blue Ridge sections may correlate with either of the major intervals of reversed polarity of late Paleocene age. However, in view of a lack of data supporting an age greater than that inferred for the deposits of Long Point, the age of the basal parts of the Rim gravels at the Yampai Quadrangle and Blue Ridge

FIGURE 19.4. Demagnetization behavior of representative samples from five localities that yielded well grouped directions. The most stable magnetizations came from strata near the base of the Furguson Tank section (sites FT 1 and 2), which are normally and reversely polarized. Less stable magnetizations are found in overlying pale red strata of the Furguson Tank section (sites FT 3-10). Very weakly magnetized limestone from Black Tank (BT) yielded fairly uniform univectorial decay toward the origin of the orthogonal demagnetization diagram. Progressive demagnetization analysis revealed multicomponent magnetizations in samples from the Duff Brown Tank (DB), Yampai Quadrangle (Y), and Blue Ridge (BR) sections, but apparently original polarities were resolved in these sample groups (fig 19.3).

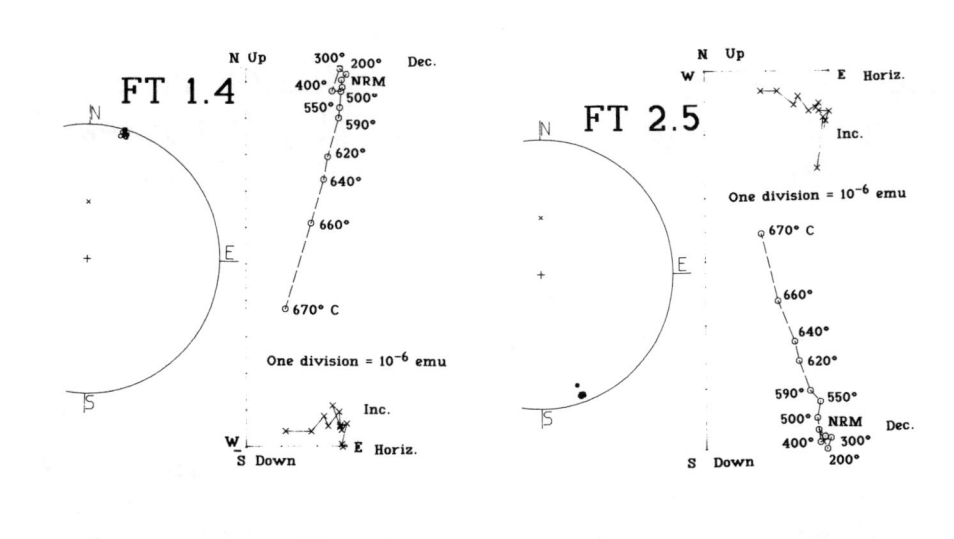

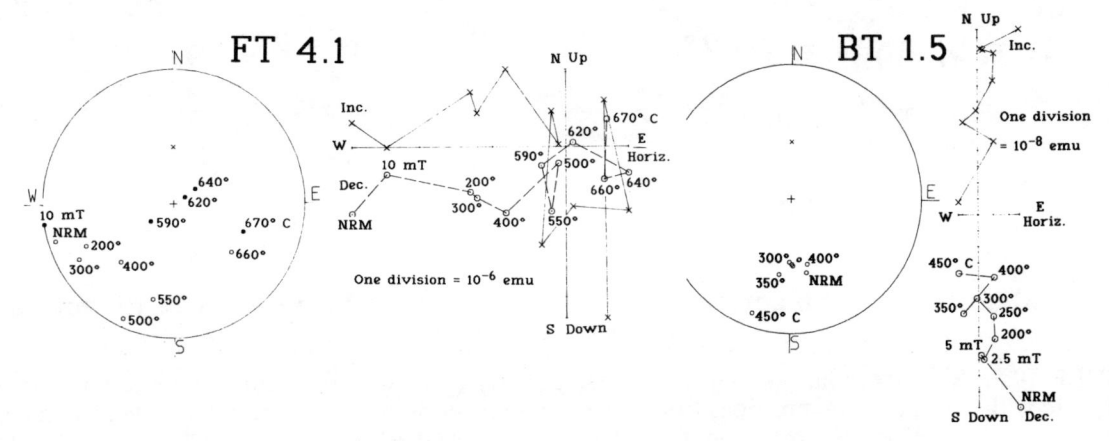

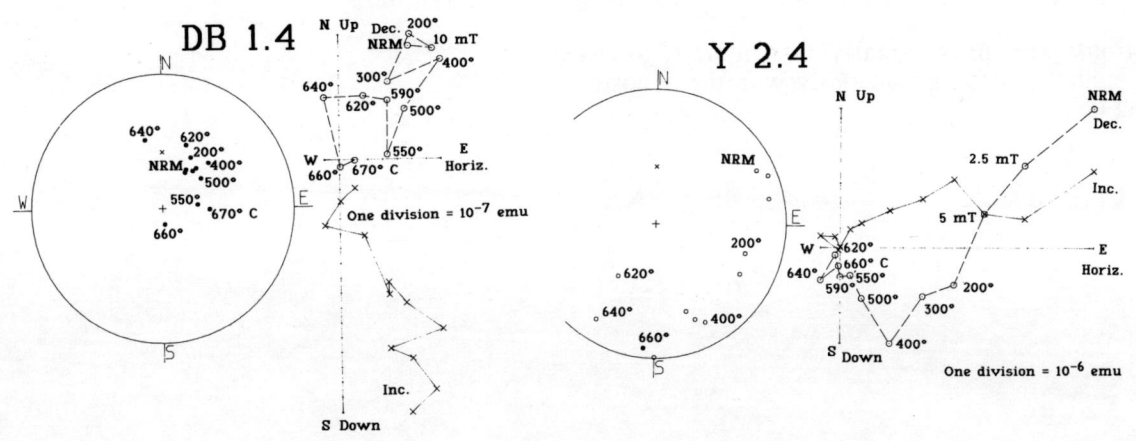

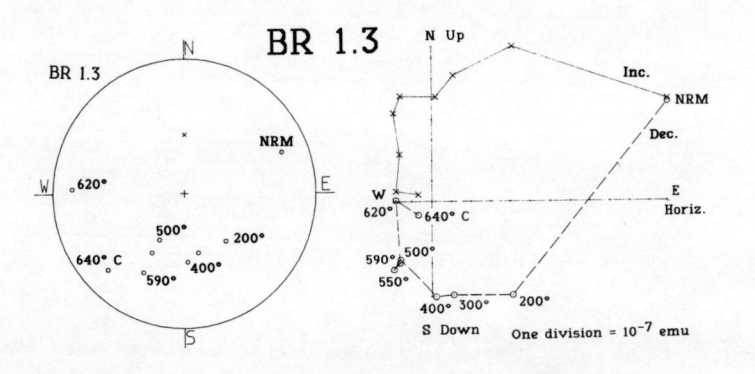

T115/315: 163

Rim Gravels, Northern Arizona

Dec. 181.2° Inc. −17.4°
n = 5; $a_{95°}$ = 20.7°

Mean of site directions

Black Peaks Formation, Big Bend area, Texas

Dec. 358.6° Inc. 44.6°
n = 3; $a_{95°}$ = 16.3°

Dec. 177.6° Inc. −25.8
n = 22; $a_{95°}$ = 6°

Mean of site directions

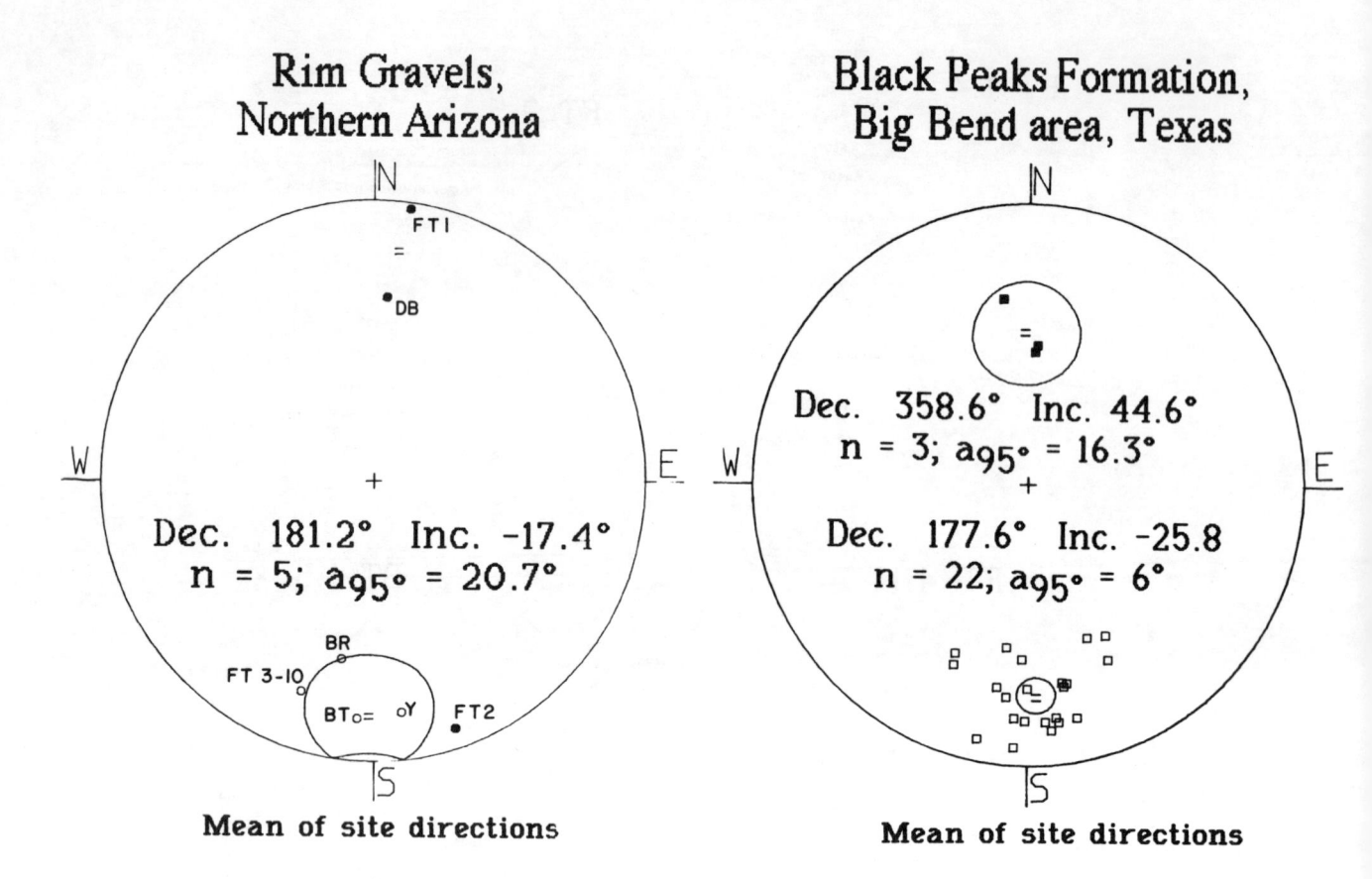

FIGURE 19.5. Site directions and a mean reverse polarity direction from Rim gravels of northern Arizona compared with site and mean directions from Paleocene-Eocene strata of the Black Peaks Formation of Texas. Data from the Black Peaks Formation are from Rapp and others (1982).

sections are provisionally considered to have accumulated contemporaneously with the deposits of Long Point.

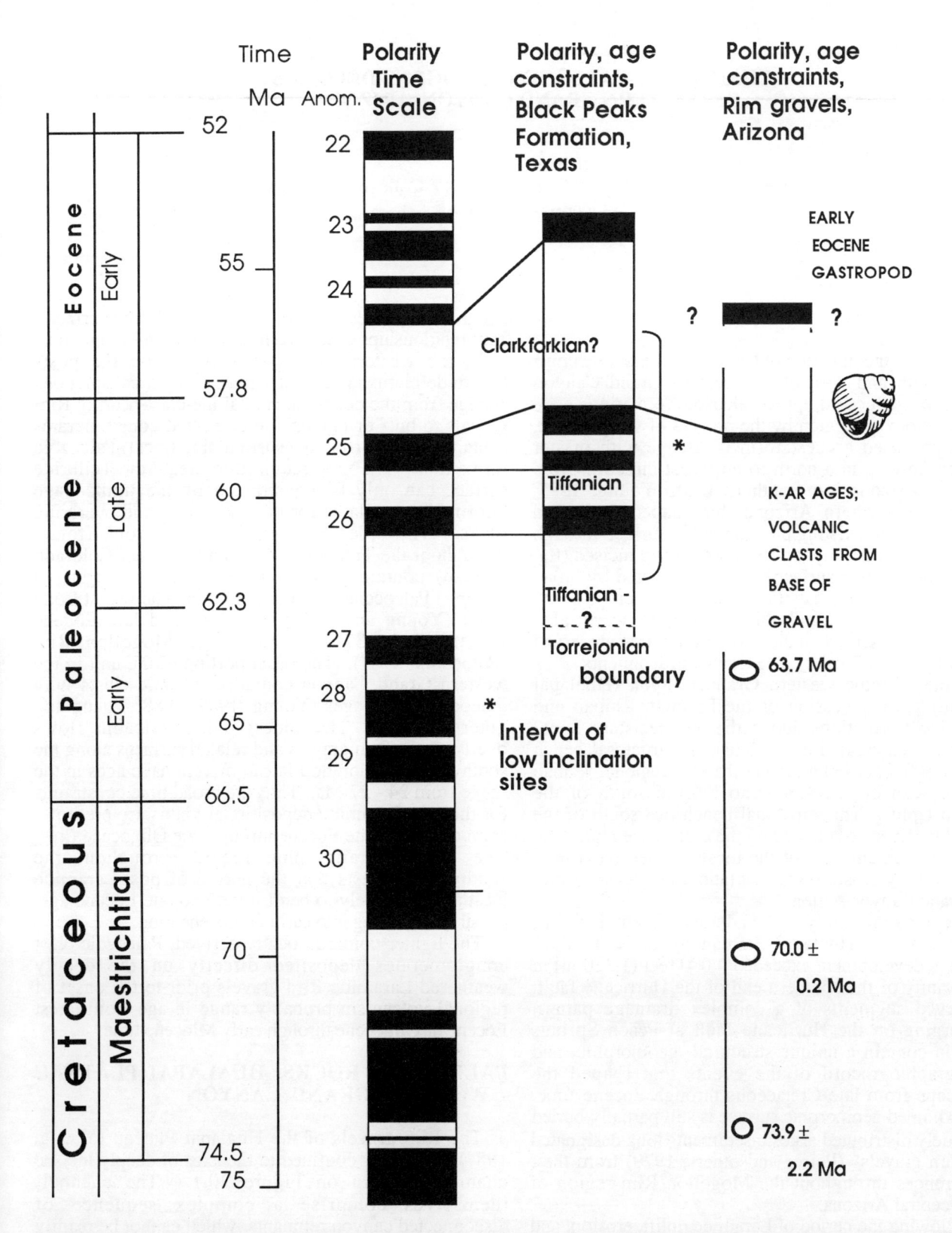

FIGURE 19.6. Provisional correlation of polarity zonation and age of Rim gravels of northern Arizona with polarity zonation and age range of Black Peaks Formation. Polarity time scale from Berggren and others (1985).

CHAPTER 20: PALEOGENE-NEOGENE DEPOSITS OF WESTERN GRAND CANYON, ARIZONA

Richard A. Young
Department of Geological Sciences, SUNY College at Geneseo,
Geneseo, New York

INTRODUCTION

Following the initiation of Cretaceous Laramide uplift to the south and west of the western Grand Canyon region, Mesozoic and upper Paleozoic formations were stripped from the region by the process of scarp retreat, which produced a cuestaform landscape with master streams flowing in a north to northeast direction from northern Arizona into southern Utah (Young, 1985, 1987). In northern Arizona this shaped the broad character of the existing physiographic surface through which the Grand Canyon was subsequently incised (fig. 4.1). Much of this surface is now dominated by broad exposures of the Kaibab limestone, but outcrops of the resistant lower Moenkopi sandstones were also commonly present, as indicated by remnants preserved beneath younger Tertiary lava flows and sediments.

In the extreme western Grand Canyon (Hualapai Plateau) scarp recession of the Permian Kaibab and Coconino Formations above the less resistant Supai Group shales created a subplateau or structural bench about 1,500 feet (460 m) lower than the regional Kaibab surface seen on the Coconino Plateau south of the Kaibab Uplift. This erosional bench lies south of the Colorado River, is bounded on the east by the Hurricane fault, and contains one of the most detailed records of early Tertiary erosion, sedimentation, and tectonism in the Grand Canyon region.

Due to its proximity to the adjacent Laramide uplift, incision of the Hualapai Plateau by early Tertiary drainage development exceeded 4,000 feet (1,220 m) in the vicinity of the southern end of the Hurricane fault. Preserved segments of a complex drainage pattern converging on the Hurricane fault at Peach Springs Canyon contain a unique structural, geomorphic, and stratigraphic record of the events that shaped the landscape from late Cretaceous through Eocene time. The exhumed geomorphic surface is still partially buried by widely distributed arkosic sediments long designated as "Rim gravels" (Peirce and others, 1979) from their occurrences throughout the Mogollon Rim region of north-central Arizona.

Following the period of Laramide uplift, erosion, and deposition of Rim gravel around the southwestern margin of the Colorado Plateau there appears to have been a period of limited tectonism and extensive weathering of these sediments under relatively humid climatic conditions. The resulting regional unconformity is marked by a depositional hiatus or weathered interval throughout much of the western U.S. and Mexico designated by Gresens (1981) as the "Telluride surface" from relationships described in the San Juan Mountains.

In the western Grand Canyon region the post-Laramide hiatus is generally marked by a conspicuous change from the deep red, crystalline-clast-bearing Rim gravels to buff or tan, calcite-cemented conglomerates containing clasts from the surrounding (local) Paleozoic formations. The time required to create the Telluride surface can only be constrained by the limited age information available for the underlying and overlying rocks. The older Rim gravels of Arizona (found throughout the Transition Zone and adjacent Colorado Plateau) contain gastropods of lower(?) Eocene to upper(?) Paleocene age south of Grand Canyon at Long Point (Young and Hartman, 1984) and late Eocene volcanic ash (37 Ma) along the Mogollon Rim (Potochnik, 1987). The basal portion of the unit in the western Grand Canyon contains volcanic clasts with Paleocene K/Ar ages (Young, 1987, 1988; Young and others, 1987). The oldest dated volcanic flows overlying the Rim gravels and related surfaces along the southwestern Colorado Plateau margin have ages in the range from 24 - 27 Ma. Thus the broad time constraints for the post-Laramide depositional hiatus are presently from middle or late Eocene through late Oligocene time. The tectono-stratigraphic record throughout the southwest suggests that the period of post-Laramide weathering is likely to be of middle to late Eocene age, possibly extending into early Oligocene time.

The lighter-colored, locally derived, Paleozoic clast conglomerates deposited directly on the deeply weathered Laramide Rim gravels prior to the onset of regional volcanism probably range in age from latest Eocene or Oligocene through early Miocene time.

PALEOGENE ROCKS, HUALAPAI PLATEAU - WESTERN GRAND CANYON

The Rim gravels of the Hualapai Plateau (Young, 1987) are largely confined to the axes of deeply incised channels, shown on Figure 20.1. The channels themselves comprise a complex sequence of disconnected canyon remnants, which cannot be readily reconstructed into a single, integrated drainage pattern. The deepest channel segments have more than 4,000 feet (1,220 m) of relief preserved from the exhumed margins to the buried channel axes. This paleorelief implies a minimum elevation (uplift) of the region of 4,000+ feet (1,220+ m) above sea level during the early period of erosion that preceded deposition of the Paleocene-Eocene sediments.

The distribution of clast types in the channel gravels indicates both a temporal and areal diversity of rocks in the source terranes to the south and west. The gravel stratigraphy records the gradual stripping of Precambrian quartzites from the source terranes, followed by an increase in the percentage of igneous and metamorphic rock types. A small but persistent component (1 to 5 %) of Laramide volcanic rock clasts is present in most of the Rim gravel exposures on the Hualapai and Coconino Plateaus. The volcanic clast component in the gravels increases locally to greater than 50 percent of all clasts near the tops of some sections east of the Hurricane fault. These areal variations in Rim gravel compositions reflect both the variability in individual paleovalleys draining different source terranes and the gradual unroofing of older Precambrian basement rocks, as well as the selective preservation of incomplete gravel sections produced by the complex Tertiary history.

Eight K-Ar ages obtained on clasts collected from basal or lowermost gravel exposures range from 63 to 80 Ma, with only one clast giving an age as old as 120 Ma (Young and others, 1987; Edwin H. McKee, written communication). Gravel age estimates based on included volcanic ashes (Potochnik, 1987) and on fossil gastropods, pollen, and charophytes found in the Rim gravels from the Coconino Plateau eastward to the Mogollon Rim suggest the deposition of the gravels continued through late Eocene time. The gravel clast variations reflect the late Laramide erosional unroofing of the older source terranes, followed by the increasing influence of randomly distributed, late Cretaceous-Paleocene volcanic centers adjacent to some portions of the Plateau.

The Rim gravels also show much more evidence of structural deformation than the younger (Neogene) sediments and gravels overlying them, as indicated by many small faults that terminate upward at the disconformity separating the Laramide and younger sediments. There is indirect evidence (in the form of small folds and thrusts) in some Rim gravel exposures near monoclines that the final pulse(s) of Laramide compressional folding is recorded in the Rim gravel sequences (Young, 1979).

Paleochannel trends, elevations, and imbrication data preserved in the channel sediments as depicted on Figure 20.1 suggest that a series of channels were progressively incised and then abandoned along the trend of the Hurricane fault, accompanied by significant structural deformation. This could be most readily explained if the early Hurricane "fault" structure was deformed as an east-verging Laramide fold or monocline (Young, 1979; Naeser and others, chapter 17) whose present sense of offset is opposite to its early Tertiary displacement, as documented for other similar Grand Canyon structures (chapters 5, 6, 7). Continued (intermittent?) Laramide deformation along the Hurricane structure could have progressively disrupted the earliest channel segments along the fold and created the apparent hanging valley relationships (Young and Huntoon, 1987). Modern rivers elsewhere in the world that approach the synclinal axes of exposed monoclines from the downdip direction exhibit similar sinuous patterns, reflecting structural influence over the drainage pattern (e.g., Cincinnati Arch region of Kentucky).

This inferred connection between the early (Laramide) drainages and intermittent structural deformation is further supported by the occurrence of lacustrine facies (Young, 1979) near the tops of Rim gravel sections in several paleocanyons (temporary structural blockage of drainage?), as well as by the lateral interfingering of the arkosic (Rim gravel) sediments with coarse, boulder fanglomerates derived from Paleozoic rocks in close proximity to major structures (Young, 1979). These fanglomerates (fig. 20.2) are presumed to reflect the influence of renewed structural relief associated with episodic movement along structures (Young, 1966).

NEOGENE ROCKS, HUALAPAI PLATEAU - WESTERN GRAND CANYON

Locally derived fluvial conglomerates containing clasts derived from Cambrian through Mississippian rocks lie disconformably above the deeply weathered, early Tertiary Rim gravels on much of the Hualapai

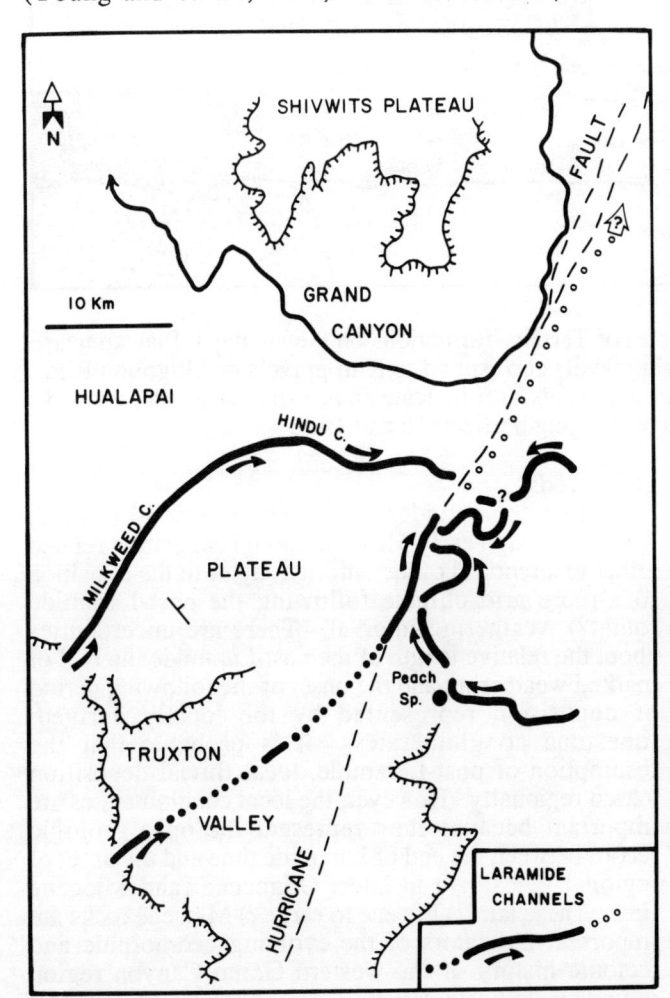

Figure 20.1. Laramide paleochannels converging on the Hurricane fault zone at Peach Springs Canyon. Arrows indicate former streamflow directions demonstrated by gravel imbrication.

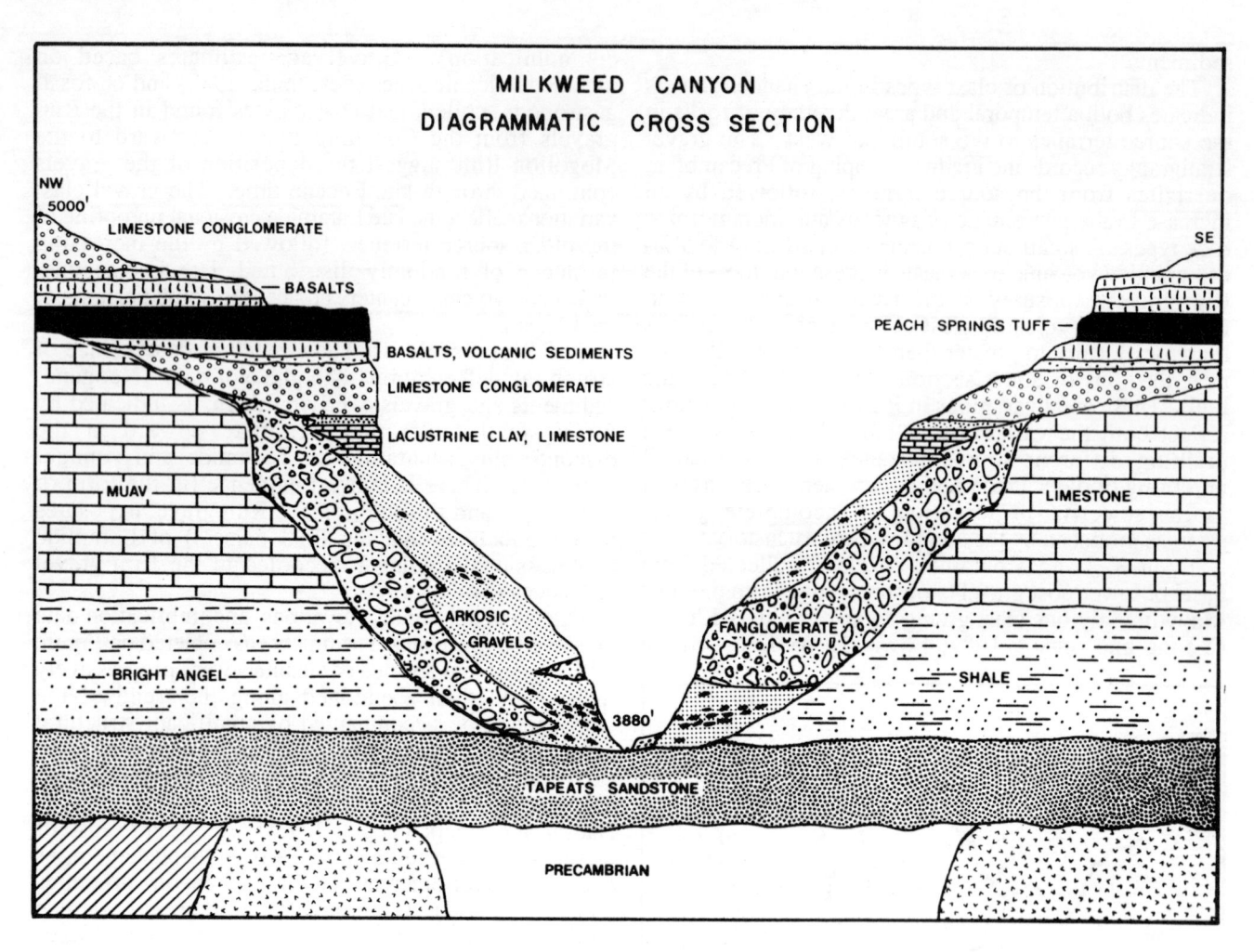

Figure 20.2. Composite, diagrammatic section (not to scale) of Tertiary formations on the Hualapai Plateau near Milkweed Canyon (modified after Young, 1966). Arkosic gravels correspond to Rim gravels of Mogollon Rim area. Numerous analyses of mineral separates from the Peach Springs Tuff indicate an age slightly greater than 18 Ma. Other volcanic rocks in the region have whole-rock ages ranging between 16 and 19 Ma.

Plateau, and they predate the local (early to middle Miocene) volcanic flows in this part of the western Grand Canyon (Fig. 20.2). These buff-colored conglomerates contrast markedly with the older reddish sediments and are up to 250 feet (75 m) thick along the axes of the old Laramide canyons. In some sections along the margins of the canyons the buff-colored conglomerates appear to have been deposited conformably (gradationally) with the red-orange coarse fanglomerates which interfinger laterally with the arkosic Rim gravels.

The upper stratigraphic limit of these locally derived, fluvial conglomerates, occurring between the Laramide sediments and the Miocene volcanic rocks, is often defined by an abrupt transition to dominantly volcaniclastic sediments, which reflect the onset of middle Tertiary volcanism in the adjacent Basin and Range province. In some sections the onset of volcanism is less sharply defined and appears to have been more broadly contemporaneous with the deposition of a significant thickness of the uppermost conglomerate beds, as indicated by the more gradual appearance of increasing numbers of isolated volcanic clasts in the gravel beds.

The geologic evidence suggests a period of late(?) Eocene through early Miocene fluvial deposition related either to a renewal of tectonic activity or to the transition to a more arid climate following the post-Laramide (quiet?) weathering interval. There are uncertainties about the relative length of the post-Laramide interval of marked weathering and the onset of the following period of deposition represented by the locally derived, limestone conglomerates. It is probable that the resumption of post-Laramide, local fluvial deposition varied regionally. However, the local conglomerates are important because they represent the only geologic record between the end of Laramide time and the onset of regional volcanism in latest Oligocene and Miocene time. These late(?) Eocene to early(?) Miocene rocks are important indicators of the evolving geomorphic and tectonic history of the western Grand Canyon region during this critical interval.

GEOMORPHIC SIGNIFICANCE OF THE EOCENE(?) - EARLY MIOCENE(?) ROCKS

The locally derived conglomerates resting on top of the Laramide Rim gravels on the Hualapai Plateau were named the Buck and Doe Conglomerate (fig. 20.2) by Gray (1964), a name which was extended by the mapping of Young (1966), and is included in the U.S. Geological Survey lexicon of geologic names (Keroher, 1970). This conglomerate buried much of the relief remaining on the older Laramide channels which converged on the Hurricane fault zone. Subsequently the Buck and Doe gravels spread across the surface of the Hualapai Plateau, and erosional remnants of this conglomerate define a broad depositional surface at elevations between 4,700 and 4,900 feet (1,430-1,490 m) south of the Grand Canyon from the Hurricane fault to the western margin of the Hualapai Plateau (figs. 20.3, 20.4).

On the rim of south Separation Canyon, a short

The stratigraphic and geomorphic position of this Buck and Doe Conglomerate outcrop at Separation Canyon implies a late Eocene-early Miocene surface of deposition across the Hualapai Plateau up to the very edge of the modern Grand Canyon during this period of time. This could only have occurred prior to the dramatic incision of the modern Colorado River drainage, as demonstrated by the extreme relief on all sides of this tiny Tertiary outlier (fig. 20.4). The most abundant clasts in the Separation Canyon exposure reflect the local Paleozoic rocks cropping out south of the Colorado River, and clast imbrication indicates a northeasterly direction of transport. However, a short distance to the southeast (near Bridge Canyon) stratigraphically older, fanglomeratic gravels occur at a slightly lower elevation and contain a high percentage of clasts derived from the Coconino Sandstone. The nearest Coconino Sandstone outcrops are now to the north on the Shivwits Plateau, 1,500 feet (460 m) higher in elevation and north of the Colorado River in this

Figure 20.3. Nineteen million year old basalt (B) capping Buck and Doe Conglomerate remnant (A) on southeastern edge of south Separation Canyon, tributary to western Grand Canyon (lower left). Conglomerate and Sanup Plateau (middle distance) are equivalent surfaces, 1,100 feet (335 m) below Shivwits Plateau (skyline) to the north. See areal location on Figure 20.4.

Colorado River tributary, a remnant of the Buck and Doe Conglomerate capped by basalt was mapped (Young, 1966) on the very edge of the modern Grand Canyon (fig. 20.3). The capping basalts have recently been shown to have a K/Ar age of 19.0±0.4 Ma (George Billingsley and Paul E. Damon, personal communication). The basalts are thus spatially and temporally related to the other volcanic flows on the Hualapai Plateau with ages close to the Peach Springs Tuff (18.5± Ma), as originally presumed by Young (Young, 1966; Young and Brennan, 1974).

vicinity. (The recession of the Shivwits Plateau scarp was an earlier Laramide event preceding the development of Laramide channels on the Hualapai Plateau).

The gravel distribution, stratigraphy, and clast differences imply that the Hualapai Plateau continued to be a surface of deposition, and not erosion, in late Eocene-early Miocene time. Following the cessation of Laramide channel development it appears that younger drainages spread local gravels northeast across the Hualapai Plateau toward older, south-sloping alluvial fans, which headed along the southern edge of the

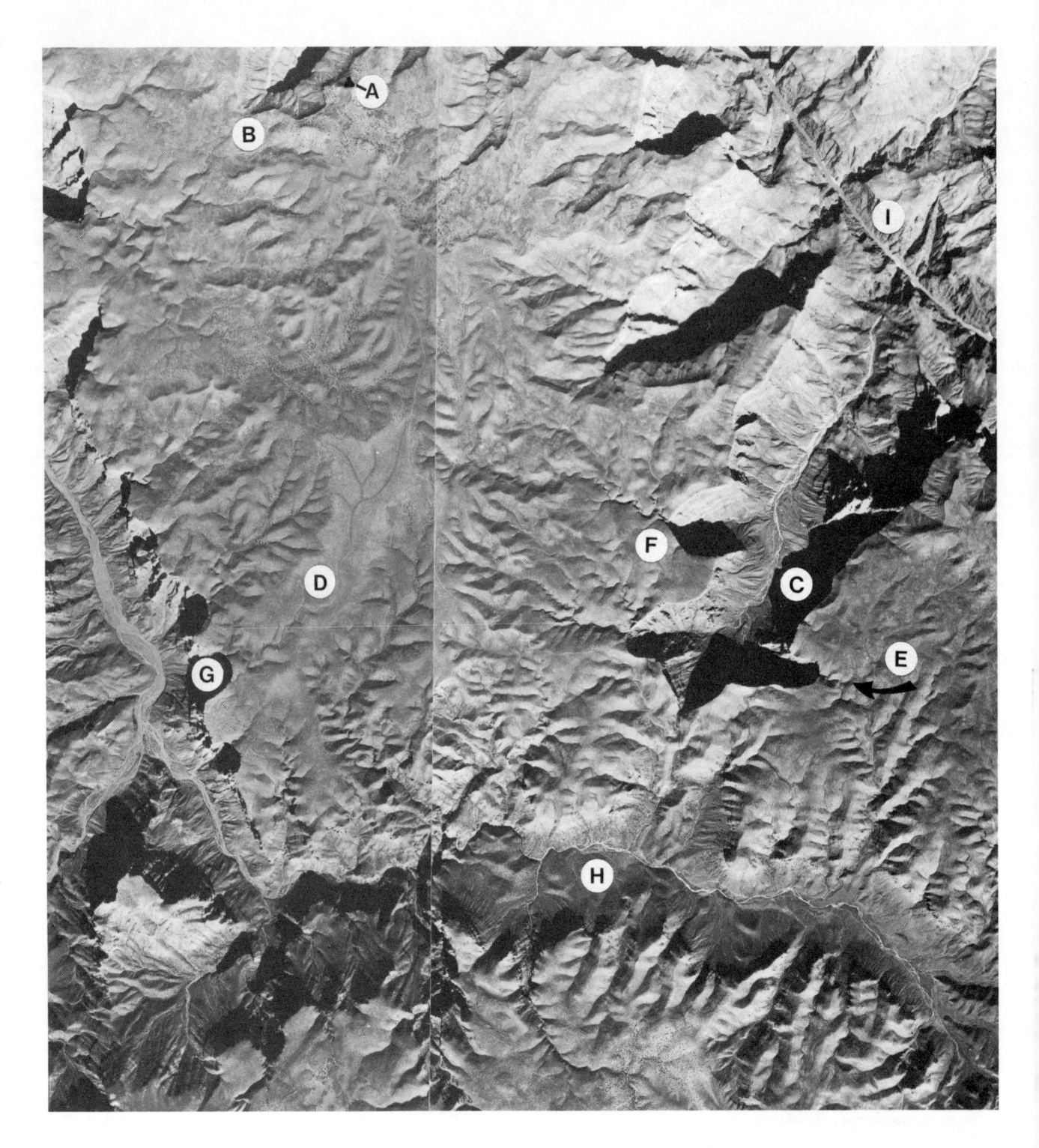

Figure 20.4. U.S. Geological Survey aerial photograph of divide area between Bridge, Separation, Spencer, and Hindu Canyons; view shows Buck and Doe Conglomerate and older surfaces being exhumed and older drainages being captured by headward erosion and development of short, younger Colorado River tributaries. North at top; Bridge Canyon is 3 miles (4.8 km) long. A - Basalt-capped conglomerate of Figure 20.1 (Source of flow is west of Spencer Canyon.); B - Separation Canyon; C - Bridge Canyon; D - Buck and Doe Conglomerate on Supai group near 4,800 feet elevation (1,465 m); E - Stream capture of Hindu Canyon tributary by Bridge Canyon tributary; F - Area of additional stream captures; G - Spencer Canyon (also headwardly eroding into Hindu channel); H - Hindu Canyon (partially re-excavated Laramide channel connecting Milkweed and Peach Springs Canyons as depicted on fig. 20.1); I - Grand Canyon.

Shivwits Plateau (local north rim of the modern Grand Canyon). These geomorphic indicators demonstrate fluvial aggradation up to (across) the very edge of the modern Grand Canyon from both northerly and southerly directions through early Miocene time, with no evidence of any significant erosion episode on the northern Hualapai Plateau between the post-Laramide depositional hiatus and the onset of regional (late Oligocene-middle Miocene) volcanism.

The partially exhumed Laramide canyons containing thick Tertiary sedimentary fills on the western Plateau (Milkweed, Hindu, Lost Man, and Peach Springs Canyons) all contain this obvious record of relatively continuous, post-Laramide deposition above and below Miocene volcanic rocks which erupted from 16 to 19 million years ago. Such a plateau surface immediately adjacent to an evolving canyon would have reflected any episode of significant fluvial incision if the regional erosion relating to the modern Grand Canyon on the Hualapai Plateau had begun during that time. In other words, if early Grand Canyon erosion is proposed to have begun by Oligocene or early Miocene time, the local geomorphic relations would require evidence of drainage incision throughout the Hualapai Plateau during this interval. In contrast, the limited stratigraphic evidence implies continuous deposition to within a few feet of the south rim of the modern Grand Canyon into middle, and possibly late, Miocene time. Likewise, thin volcanic flows, such as those capping the gravels near Separation Canyon (19 Ma; figs. 20.3, 20.4) would not have reached their present location if any significant relief had existed at that time along the courses of Spencer or Separation Canyons, which lie between the lavas and their sources.

MIOCENE VOLCANISM AND RELATED EVENTS: HUALAPAI PLATEAU

The interval of early to middle Tertiary deposition continued following the episode of Miocene volcanism across the Hualapai Plateau. Eventually, when the Grand Canyon was incised across the Hualapai Plateau, it captured those local tributary drainages which had, until then, drained toward the nearly buried, Laramide channel (Hindu-Lost Man Canyons) three miles south of the modern canyon, a drainage reversal process which continues to the present (Young, 1970; 1982, fig. 3).

The uppermost beds of the Buck and Doe Conglomerate generally show a gradual increase in the number of volcanic clasts and cinders, especially along the axes of the major Laramide channels, until some of the strata finally become totally volcaniclastic. This change reflects the onset of volcanism to the west and southwest of the Plateau beginning regionally about 27 million years ago. On the Hualapai Plateau the oldest dated flows are between 16 and 19 million years old, including the Peach Springs Tuff. There do not appear to be any significant erosional hiatuses during this relatively short volcanic episode. However, immediately east of the Hurricane fault the volcanic activity was younger, spanning the interval from 1 to 14 million years ago (Reynolds and others, 1986).

The outpouring of volcanic rocks on the Hualapai Plateau came from some local vents as well as from the region to the west of the plateau. Flows filled in much of the relief that remained following Buck and Doe Conglomerate time. The thicker flow sequences followed the axes of the nearly buried Laramide canyons. This period of volcanism was presumably associated with the onset of the Basin and Range episode of volcanism and tectonism to the west.

POST-VOLCANIC SEDIMENTATION HISTORY

Following the relatively short period of volcanic activity, fluvial sediments continued to accumulate across the Hualapai Plateau surface (fig. 20.5), burying both the Buck and Doe Conglomerate and the Miocene volcanic rocks. These gravels, the Willow Springs Formation of Young (1966), accumulated to a maximum thickness of approximately 330 feet (100 m) on the Hualapai Plateau, depending on local relief (fig. 20.5). They are distinguished from the similar Buck and Doe Conglomerate either by their stratigraphic position above the volcanic rocks or by the inclusion of numerous local volcanic clasts. In addition, this younger gravel unit tends to be less well indurated than the Buck and Doe Conglomerate. The source of the gravels can be traced to the local scarps and divides remaining after the cessation of volcanic activity. Eruption of the Peach Springs Tuff at about 18+ Ma had helped to eliminate the relief that remained in the Laramide paleovalleys.

Figure 20.5. Middle Miocene to Pliocene(?) Willow Springs Formation of Young (1966) resting on basalts on north side of Milkweed Canyon; closest outcrop of similar gravel to Grand Canyon is 6 miles (10 km) south of Separation Canyon (fig. 20.4). Top of unit ranges locally from 4,800 to 5,200 feet elevation (1,460 to 1,585 m) representing top of Tertiary section filling Laramide channel in Milkweed Canyon (Young and Brennan, 1974, fig. 5; fig. 20.2). A - Upper conglomeratic member 170 feet (52 m) thick; B - Middle silt/clay/sand member 50 feet (15 m) thick; C - Lower conglomeratic member 80 feet (24 m) thick.

The geomorphic implications of these late(?) Miocene (or younger) Willow Springs gravels are similar to those recorded by the older Buck and Doe Conglomerate. The Hualapai Plateau surface close to the modern Grand Canyon records a potentially lengthy episode of regional fluvial deposition, with no evidence of significant drainage incision. Unfortunately, outcrops of the Willow Springs Formation (Young, 1966) do not occur as close to the Grand Canyon as the Buck and Doe Conglomerate remnant near Separation Canyon. Thus the geomorphic evidence provided by the post-volcanic sediments is less direct than that seen at the 19 Ma Separation Canyon outcrop. This apparently reflects the fortuitous location of the volcanic rocks that capped and preserved the older south Separation Canyon exposure, as well as the greater erosion of the stratigraphically younger and less-well-indurated deposits as the modern canyon developed. Thus, the outcrop pattern of gravel of the Willow Springs Formation by itself does not absolutely demonstrate the absence of any middle to late(?) Miocene Grand Canyon erosion. However, the relatively continuous deposition of gravels over a wide portion of the Hualapai Plateau in the same relative locations above the Miocene volcanic rocks, combined with the distribution of the Buck and Doe Conglomerate below, suggest that the regional geomorphic setting continued to be one of overall, long-term deposition (not erosion) throughout the Hualapai Plateau (western Grand Canyon) until late(?) Miocene time, perhaps longer.

REGIONAL STRUCTURAL PERSPECTIVE

The Tertiary deposits in the western Grand Canyon must have a direct relation to the tectonic and erosional history of the region. It is clear that the the older Laramide channels developed during an episode of marginal plateau uplift or regional tilting in response to Cretaceous - Paleocene events. The minimum amount of uplift along the southwestern margin of the plateau at that time must have been in excess of 4,000 feet (1,200 m) in the immediate vicinity of the western Grand Canyon. This amount of uplift is documented by the relief preserved within the Hualapai Plateau paleovalleys converging on the southern Hurricane fault and cut into Cambrian through Devonian strata (Truxton Valley). In addition, if one restores the upper Paleozoic rocks that must have also been stripped away during contemporaneous erosion of the Laramide surface, the minimum relief up to the Kaibab Limestone surface would be at least 5,500 feet (1,675 m) This marginal Laramide uplift and regional tilting must have decreased northward so that the central Colorado Plateau was nearer sea level in early Tertiary time. The steepness of the regional paleoslope into Utah (Green River-Uinta Basin area) is unresolved, but an increase in the regional structural dip of 1° would more than suffice to permit the Laramide channels to maintain gradients northward across the younger Grand Canyon.

The newest data on timing of uplift (chapter 17) from fission track studies may resolve some of these questions. Those fission track results for the vertical Grand Canyon traverse could be interpreted to mean that between 124± and 63± m. y. ago, the Laramide uplift near Grand Canyon gradually first brought the rocks up through their critical annealing temperatures (including exposure by erosion). This long period of time may have included a series of unresolved, discrete structural events accompanied by the progressive uncovering of deeper, older rocks through the normal processes of erosion. This uplift interval would have been broadly coincident with the incision of the Laramide channels around the Plateau margin. It seems reasonable to suppose that the uplift is generally older than the erosional stripping, which gradually exposes the rocks to cooling. Thus the fission track ages should be minimum ages for the tectonic events.

A group of younger apatite fission track ages, which cluster near 35 to 40 Ma (chapter 17), may reflect the youngest Laramide tectonic pulse. It correlates well with the plate tectonics-sea level chronology of Haq and others (1987). World-wide sea level changes indicate periods of increased oceanic ridge growth (ridge inflation linked to marine transgressions) with peaks near 141 Ma, 129 Ma, 117 Ma, 90-95 Ma, 79 Ma, 52 Ma, and 35 Ma. Presumably these periods of change in the rates of plate motions are closely related to global tectonic events. Within these periods of changing sea levels, the classical Laramide Orogeny seems to be characterized by somewhat more significant sea level "events" at 90-100 m.y., 48 - 54 m.y., and 32 - 37 m.y.

Young (1979, 1985, 1987) postulated that the early(?) Eocene limestones in Rim gravel sections on the Hualapai and Coconino Plateaus were indicative of a late Laramide disruption of north-northeasterly Laramide drainage, resulting in ponding adjacent to compressional monoclines and upwarps. However, the fission track data (chapter 17) suggest a slightly younger age for the latest Laramide event than the paleontological evidence (Young and Hartman, 1984). The work of Hartman (1984) indicates that the *viviparid* gastropods in the Coconino Plateau Rim gravels are not younger than middle Eocene in age (upper limit near 40 m.y.). Thus the two sets of data are in reasonably close agreement.

SUMMARY AND CONCLUSIONS

It does not appear reasonable to invoke a major "Grand Canyon" incision episode in the western Grand Canyon region (Hualapai Plateau) beginning in middle Tertiary time while about 650 to 1,000 feet (200 to 300 m) of fluvial conglomerates and volcanic rocks were accumulating immediately adjacent to the modern canyon on a surface only now being headwardly eroded by relatively short, "immature" Grand Canyon tributaries. On the westernmost Colorado Plateau, these constraints suggest there was no significant Grand Canyon erosion until middle(?) Miocene time at the earliest, following the well-documented Miocene volcanism and Willow Springs fluvial deposition. This scenario is generally compatible with the prevailing interpretation (Lucchitta, 1988) of the geology in the Grand Wash Trough near Lake Mead where Muddy Creek sediments were also accumulating across the modern course of the Colorado River in late Miocene time.

The erosional and stratigraphic evidence for a multistage Laramide history gradually accumulated by Young (1966, 1979, 1982, 1985, 1987) and by other workers (Young and others, 1987; Young and Huntoon, 1987) is supported by the studies included in this volume (chapters 17, 18, and 19), which indicate an early Laramide interval of significant regional erosion followed by sedimentation and accompanied by one or more secondary pulses of late Laramide tectonism continuing into middle or late Eocene time. Given these more refined time constraints and the resulting improved age estimates for the sequence of Hualapai Plateau gravels and conglomerates, it is difficult to allow for any significant interval of Grand Canyon erosion in parts of the western Grand Canyon prior to middle and late Miocene (15 to 6 m.y.) or Pliocene time. However, the geologic evidence on the Hualapai Plateau is incomplete and subject to alternative interpretations. The geomorpholgy and chronology in other regions is also contradictory and open to alternative interpretations. There is evidence which can be interpreted to suggest significant Miocene erosion and the presence of a related ancestral(?) Colorado River system (chapter 18) in northeastern Arizona.

View upstream from mile 137 (km 220), looking toward "overhang campsite", showing displaced dolomite unit of Bright Angel Shale (base of landslide) on south side of river overlying pre-landslide Colorado River gravel and travertine above the camp. See chapter 21 for discussion.

CHAPTER 21: PRE-PLEISTOCENE(?) DEPOSITS OF AGGRADATION, LEES FERRY TO WESTERN GRAND CANYON, ARIZONA

Donald P. Elston
U.S. Geological Survey, Flagstaff, Arizona

STATEMENT OF PROBLEM

Unconsolidated alluvium, and well-cemented gravel and intercalated talus and landslide that locally underlie the alluvium, are considered in Chapter 24 to comprise an interrelated series of Pleistocene deposits that accumulated during a single progressive entrenchment of the Colorado River. This model is commonly employed during assessing of he age and history of accumulation of surficial deposits that overlie the pre-Mesozoic bedrock within the Grand Canyon.

However, geologic relations in Marble Canyon and the Grand Canyon indicate an alternative incision and accumulation history could be seriously considered. The relations indicate that the well-cemented gravel and associated talus and landslide deposits may be substantially older than suprajacent deposits of unconsolidated alluvium. Distribution of the coarse, crudely stratified deposits suggest they accumulated during a single, general episode of aggradation after the canyon and its tributaries had been cut to very near their present depths. Following aggradation, the master and tributary streams re-excavated the stream courses. A model for the pre-Pleistocene(?) accumulation of the deposits is given in the following section. The distribution and characteristics of the deposits are summarized in sections that follow.

Model for Pre-Pleistocene(?) Accumulation

Field geologic relations lead to the following speculative model. Accumulation of the pre-alluvium deposits in the Grand Canyon and its tributaries record a discrete episode of accumulation of gravel, talus, travertine, and locally landslide. Aggradation appears to have occurred over an appreciable interval of geological time, one presumably encompassing several millions of years. This inference derives from an apparently slow accumulation of crudely stratified talus derived from the canyon walls, and from a presumed equally slow, intermittent accumulation of thick sections of gravel of flash flood (storm) origin that apparently choked the master and tributary stream courses and accumulated as continuous fills in tributary basins. Additionally, blocks associated with apparently ancient debris flows exhibit extensive dissolution and cavernous weathering arising from *in situ* weathering. The accumulation of talus and gravel was accompanied by the precipitation of travertine along the canyon walls, and travertine accumulated even across extensively eroded benches, as ground waters were drained from various horizons in the Paleozoic section in the canyon. None of the foregoing features would seem to be compatible with episodes that might be related to Pleistocene events correlated with glacial and interglacial episodes recognized elsewhere in the western U.S.

The lower parts of the talus and travertine deposits bordering the stream courses contain stream-transported detritus. In the course of the Colorado River, the gravel commonly contains detritus of both up-stream and local origin. However, accumulation and choking of the stream courses by gravel and talus appears to have marked a waning of perennial stream flow and a diminished capacity of the Colorado River to transport detritus both into and out of the region. At one locality (Nankoweap Creek), detritus derived from the headwater areas of the Colorado River is present in the lower part of the gravel section, but only detritus of local origin is observed above this. The gravel deposit at Nankoweap Creek was once part of a more extensive deposit that connected with gravel fill preserved in Nankoweap basin west of the Colorado River. Unconsolidated alluvium accumulated on the gravel deposits before Nankoweap Creek and the Colorado River were re-established and re-entrenched themselves to their former levels.. A few hundred to locally perhaps as much as several hundred feet of gravel accumulated along the course of the Colorado River and its tributaries prior to re-establishment and re-entrenchment of the Colorado River and its tributary streams.

Appeal to a completely dry Colorado River during the time of gravel accumulation is not necessary. With the onset of accumulation of gravel in the master and tributary stream courses, the diminished flow of the Colorado River in its tributaries came to increasingly move in the subsurface through the gravel. A stage soon was reached when surface flow occurred only at times of flash flooding.

Catastrophic landsliding appears to have occurred in the area of Deer Creek Falls (Surprise Valley landslide) in the central Grand Canyon before the end of the episode of accumulation of gravel. If not related to lubrication arising from a wet climate, landsliding alternatively could have been triggered by one or more increments of movement on a near-by fault (Sinyala fault). Two apparently ancient, nested debris flows in Comanche Creek in the Big Bend area of the eastern Grand Canyon, which have their source in a large talus cone that overlies the Palisades fault beneath the Palisades of the Desert, appear to record two abruptly emplaced debris flows. These flows, which merge with gravel along the Colorado River, may correlate with the episode (or episodes?) of landsliding in the Deer Creek area. Similar debris flows are present in a few other drainages in the eastern Grand Canyon, including nearby Tanner Wash.

The gravel and interrelated talus and landslide deposits are provisionally correlated with upper Miocene and Pliocene deposits of the Bidahochi, Verde, and Muddy Creek Formations, which appear to record a prolonged (~>10 m.y.) episode of aggradation elsewhere on and around the Colorado Plateau (chapter 18). This regional episode of aggradation presumably was brought on and maintained by increased aridity that resulted in disappearance of perennial streams and a loss of capability for transport of detritus into and out the region.

Re-establishment of perennial streams and a throughgoing drainage system, presumably in late Pliocene(?) and Pleistocene time (chapter 18), allowed detritus once again to be removed from the region. Following entrenchment of the streams to their former levels in the early and middle Pleistocene, flooding events in late Pleistocene time related to temporary lava dams in the Lava Falls area may have been responsible for the accumulation of minor deposits of unconsolidated alluvium on terraces developed across the gravel deposits.

CHARACTER AND DISTRIBUTION OF DEPOSITS

Undated Neogene deposits of aggradation consist of interbedded and intertonguing gravel, talus, travertine, and landslide deposits. Their characteristics and geologic relations are summarized in the order of their encounter in the canyon to allow the traveler to anticipate the deposits and field relations that may be observed during the course of a trip through the canyon. The generalized distribution of the deposits of aggradation, from Lees Ferry to a short distance above Lava Falls Rapids, is shown on Figure 21.1 The distribution has been derived from geologic maps, modified and supplemented from field observations. A summary of the deposits, placed within the framework of episodes of Table 18.1, is given in Table 21.1.

Gravel of Lees Ferry Area

Two markedly different, high- and low-level gravel deposits that consist of well-rounded, water-worn pebbles are shown on the geologic map of the Lees Ferry area (Phoenix, 1963). Both deposits were designated as Qg (Quaternary gravel) by Phoenix although the high-level gravel obviously could not be related to Pleistocene or Holocene erosion, transport, and deposition related to processes operating at and near the level of the Colorado River. The high-level gravels are on the Paria Plateau, some 3,500 feet (1,067 m) above the present level of the Colorado River. The high elevation and isolated locations of the deposits suggest they may correlate with Paleocene-Eocene Rim gravels of the Coconino Plateau, south of the Grand Canyon (chapter 19). The source area (or areas) for these high-level gravels remains to be elucidated, but if they are correlative with the Rim gravels of the Coconino Plateau they should contain quartzite and other rocks characteristic of the Proterozoic of central Arizona.

The lower-level gravels mapped near Lees Ferry are found at two levels. A main, lower-level deposit appears to have accumulated at least as an unbroken apron extending from an elevation of about 3,400 feet (1,035 m) to near river level (elevation ~3,180 ft; ~970 m). The gravel contains abundant clasts derived from more headward areas of the Colorado River. If the gravel accumulated as a fill across the course of the Colorado River, as can be inferred from relations downstream within the canyon, a thickness of about 200 feet (60 m) or more is suggested. It is a thickness broadly comparable to the thickness of potentially correlative gravel deposits in the Grand Canyon and its tributaries.

Remnants of the same(?) lower-level gravel deposit also are found at higher elevations. One remnant is about 400 feet (122 m) higher (elevation of ~3,800 feet; 1,160 m), preserved on a pinnacle of the Shinarump Member of the Chinle Formation (Upper Triassic), just west of Lees Ferry. Other isolated deposits of gravel at similar and somewhat lower elevations also are found south of Lees Ferry on the stripped surface of the Marble Plateau, adjacent to the rim of Marble Canyon and on both sides of the river. These gravel deposits are found as far south as mile 7.5 (km 12), above Badger Canyon, where they lie as much as 750 feet (230 m) above river level. If the continuous and discontinuous lower-level gravels of the Lees Ferry area aggraded during a single depositional episode (paralleling an inferred single episode of aggradation in the canyon), the Colorado River drainage in the Lees Ferry area would seem to once have been choked with gravel.

Talus Deposits of Upper Marble Canyon

Well-cemented talus occurs along both walls of Marble Canyon downstream from Badger Canyon (mi 7.5; km 12). Cemented cones of colluvium and fan alluvium on the Hermit Shale appear at approximately mile 8.5 (km 13.7), about 200 feet (60 m) above the river on the right. The apexes of a series of cones abut zones of carbonate seepage at the base of the overlying Coconino Sandstone. The cemented talus is crudely stratified, and locally is overlain by unstratified talus that accumulated prior to dissection of the cones and colluvial fans by the current episode of erosion. These deposits extend south to mile 28 (km 45), and beyond, rising ever higher above river level as they follow the contact between the Hermit Shale and overlying Coconino Sandstone.

Gravel, Talus, and Travertine Deposits: Buck Farm Canyon to Little Colorado River

Well-cemented talus developed on the Gateway Canyon Member of the Muav Limestone (Middle Cambrian) makes its appearance at Buck Farm Canyon (mi 41; km 66). From here to the Little Colorado River (mi 61; km 98), cemented talus forms small to extensive

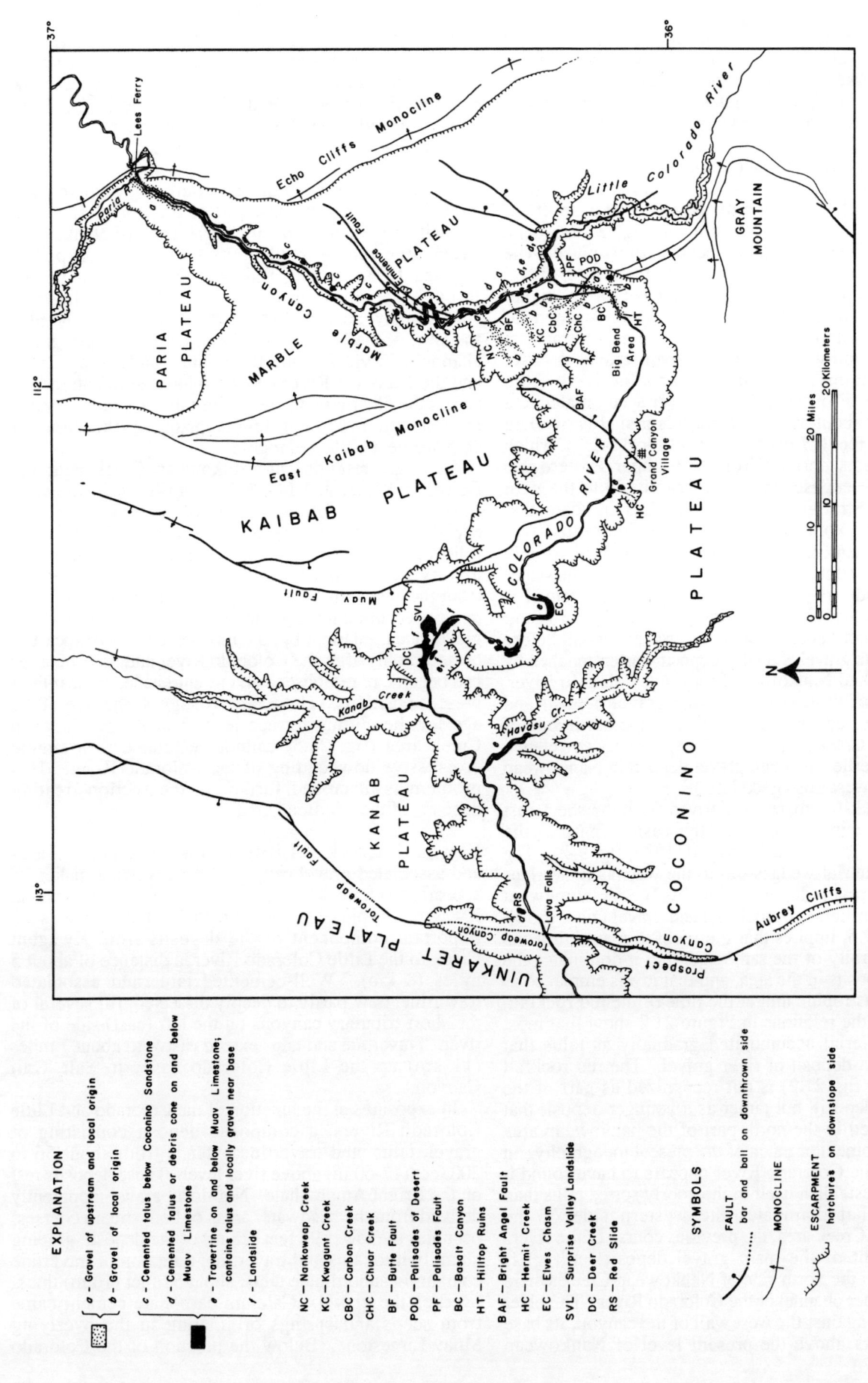

FIGURE 21.1. Map showing distribution of pre-Pleistocene(?) deposits of aggradation, Lees Ferry to west-central Grand Canyon.

deposits at low to moderate elevations on the canyon walls. The abundance and preservation of talus increases markedly with the appearance of travertine on the left side of the river, which cements talus and gravel at and below Kwagunt Rapids (mi 56; km 90).

Nankoweap Creek. A composite gravel and talus deposit, capped by alluvium, is preserved on the west (right) side of the Colorado River at and south of Nankoweap Creek (mi 52; km 84; fig. 21.2). The designation of Grand Canyon begins at this place. The western part of the deposit consists mainly of gravel (unit g), the top of which is about 160 feet (50 m) higher in elevation than river level at the lower end of Nankoweap Rapids. As seen along and near a trail that climbs the southwest side of the main gravel deposit, gravel containing abundant clasts derived from headwater areas of the Colorado River accumulated to a thickness of about 30 feet (9 m); this gravel is overlain by about 55 feet (17 m) of gravel of local origin, which includes clasts derived from the Upper Proterozoic Chuar Group exposed in Nankoweap basin to the west (fig. 9.1). Stratigraphic relations thus indicate that transport of exotic clasts down the Colorado River waned and then ceased, following which clasts of only local origin accumulated during the medial and late stages of the episode of gravel accumulation. The gravel is capped by 8 feet (2.4 m) of fine grained alluvium (unit oa), which accumulated across the gravel deposit before the composite gravel and talus deposit was entrenched by a re-established Nankoweap Creek and Colorado River. A remnant of unit oa overlies the gravel deposit and underlies the red rock fall (unit r) on the north side of Nankoweap Creek (fig. 21.2). An apparently correlative alluvium overlies the main gravel deposit in Nankoweap basin to the west (discussed later).

Talus (rock-fall material) derived from the sheer east wall of the canyon occupies the eastern part of the Nankoweap Creek area (Hereford, 1984, fig. 9-2). The rock-fall material wedges-out to the west as it overlaps the main gravel deposit (fig. 21.2). A somewhat different distribution of rock fall and gravel units, shown in Figure 24.5, implies that the rockfall material of the area is generally of the same age, that it dominates the surficial deposits in the area, and that it was emplaced as a single catastrophic unit at the time of the red rockfall. In contrast, the relations in Figure 21.2 show that most rockfall material accumulated gradually as talus that overlapped a deposit of river gravel. The red rockfall unit (unit r, fig. 21.2) is not recognized as part of the main talus deposit, but rather as a younger deposit that was emplaced in the north part of the nankoweap area after development of much of the present topography. In any event, the Colorado River appears to have found it easier to re-establish itself in the poorly sorted talus than in gravel that dominates the western side of the Nankoweap Creek area, the previous course of the river.

A remnant of the main gravel deposit (unit g) is preserved on the north side of Nankoweap Creek where it fills a former channel of the Colorado River. The older channel lies against the west wall of the canyon, its base a few meters above the present level of Nankoweap Creek and the Colorado River. The gravel deposit north of Nankoweap Creek is overlapped by red rockfall material that has been dated at approximately 210,000 years by the uranium-trend method (table 24.1). From its age, the red rockfall is placed between terrace levels 5 and 7 of the Big Bend area of the eastern Grand Canyon (figs. 24.2 and 24.3; table 24.1). Red rockfall (unit r, fig. 21.2) derived from the east wall of the canyon was deposited against the east end of the gravel and talus deposit on the north side and at the level of Nankoweap Creek; the red rockfall at its easternmost exposure appears to have been fluidized from having passed through an intervening Colorado River to the east. Additionally, an outlier of the red rockfall is found near the level of the Nankoweap Creek west of the pre-gravel channel. In view of the foregoing, Nankoweap Creek and the Colorado River appear to have been entrenched across the older composite gravel-talus-alluvium deposit to near their present levels prior to the time of emplacement of the red rockfall.

Geologic relations at Nankoweap Creek near the Colorado River, and in Nankoweap basin to the west (discussed in a following section), seem to be best explained as arising from an episode of aggradation that followed cutting of the Grand Canyon and its tributaries to very near their present depths. Rock falls derived from the east wall of the Grand Canyon can be inferred to have accumulated sporadically during the episode of gravel accumulation, but the most recent major rock fall occurred only after the Colorado River and its tributaries had become re-established and re-entrenched to near their present positions. The relations suggest that the 40 ka age for the cutting terrace level 3 at the Nankoweap Creek area (fig. 24.5) cannot be related to a simple progressive downcutting of the Colorado River. This problem is discussed further in the section treating deposits of the Big Bend area.

Kwagunt Creek - Little Colorado River. Talus and associated gravel cemented by travertine makes its appearance on the left (east) side of the Colorado River at Kwagunt Rapids (mi 56; km 90). Travertine is an important component of the deposits from Kwagunt Creek to the Little Colorado River, a distance of about 5 miles (8 km). Well-cemented talus and associated travertine, now partly to deeply dissected, fill several of the short tributary canyons on the left (east) side of the river. Travertine and talus extend eastward about 7 miles (11 km) up the Little Colorado River to Salt Trail Canyon.

In exposures at the junction of the Colorado and Little Colorado Rivers, a composite deposit consisting of gravel, talus, and travertine extends from about 56 to 200 feet (17-60 m) above river level. These deposits rest on the Bright Angle Shale. Non-local gravel (apparently derived from the headwater areas of both stream courses) occupies the lower 92 feet (28 m) of the deposit, grading laterally into talus and travertine; a section of travertine containing subordinate talus, about 52 feet (16 m) thick, overlies the gravel. Calcium-carbonate cement came from seeps and springs originating in the overlying Muav Limestone. Below the junction of the Colorado

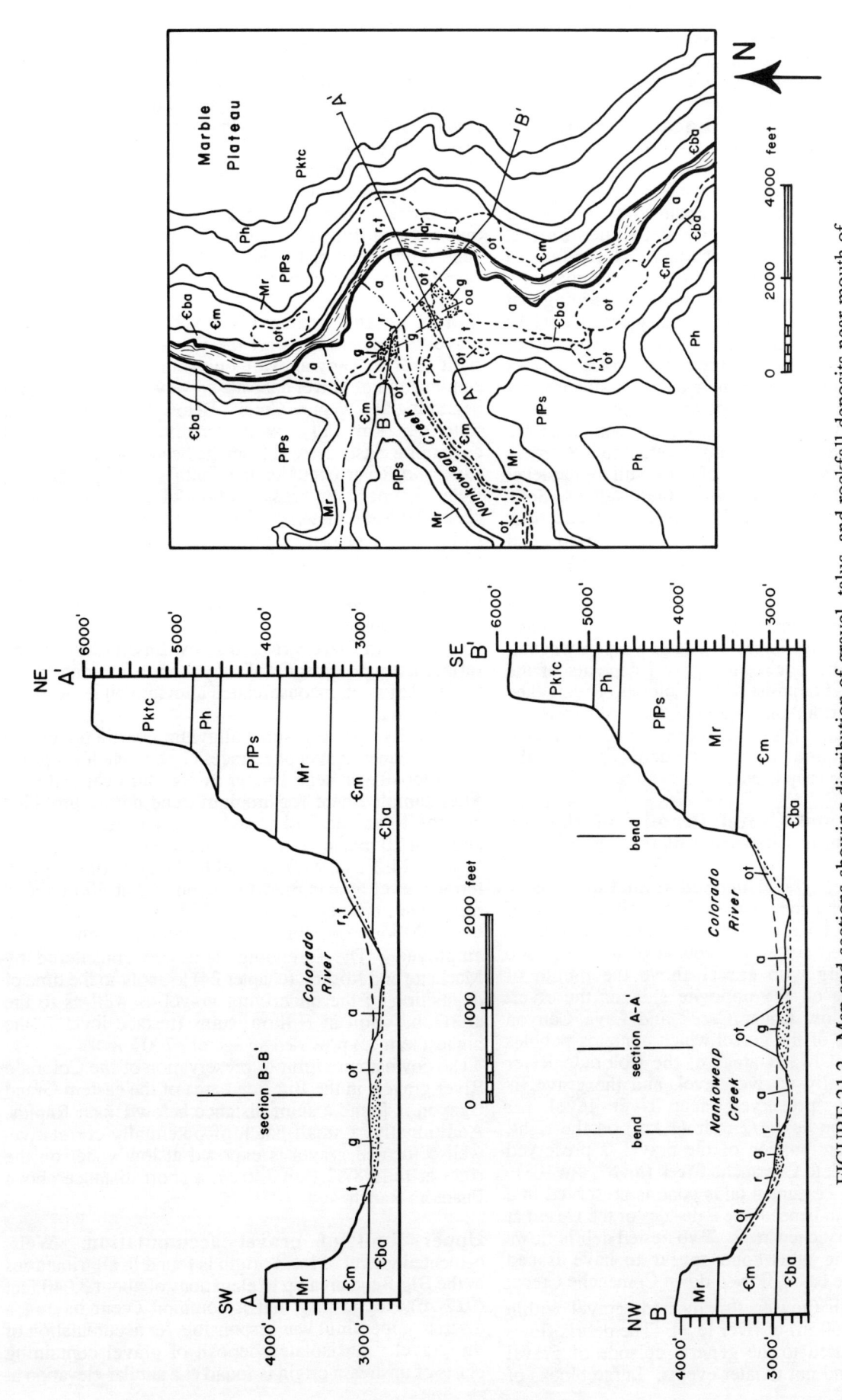

FIGURE 21.2. Map and sections showing distribution of gravel, talus, and rockfall deposits near mouth of Nankoweap Canyon, upper Grand Canyon. Deposits of aggradation: g - river gravel containing clasts of upstream origin in lower part but only clasts of local origin in upper part; oa - older fine grained alluvium, capping unit g; ot - older talus, in part well cemented; r - red rockfall; t - talus, undivided; a - younger fine grained alluvium. Bedrock: Cba - Cambrian Bright Angel Shale; Cm - Muav Limestone; Mr - Mississippian Redwall Limestone; PIPs - Pennsylvanian and Permian Supai Group; Ph - Permian Hermit Shale; Pktc - Permian Kaibab, Toroweap, and Coconino Formations, undivided. Adapted from Maxson, 1967; sections drawn at 2X map scale.

and the Little Colorado Rivers, seeps occur in the lower part of the Tapeats Sandstone, a stratigraphically lower unit, but these manifest themselves as salt seeps.

Gravel Deposits of Local Origin in Nankoweap, Kwagunt, and Chuar Valleys

Gravel deposits of local origin, a few meters to a few tens of meters in thickness, accumulated in the isolated basins of Nankoweap, Kwagunt, and Chuar Creeks, west of and between miles 52 and 65.5 (84-105 km) on the Colorado River The gravels record an apparently prolonged episode of aggradation by flash flooding that followed entrenchment of these tributary stream courses to essentially their present levels. The upper surfaces of the deposits at places display a series of terrace surfaces that step down toward the main tributary drainages, but these stepped surfaces do not appear correspond to different underlying depositional units, and thus are developed across what are interpreted to be single aggradational units. Remnants of a few still higher level gravels also are preserved locally in these valleys. Soils developed on the lower- and higher-level gravel deposits in Nankoweap basin were sampled for uranium-trend dating, but without conclusive results (chapter 24), suggesting the soils may be older than the limits of the dating method (~700 ka).

The main (lower-level) gravel deposits also once occupied the stream courses of Nankoweap, Kwagunt, and Chuar Creeks, connecting gravel deposits of the basins with gravel deposits of the Colorado River. The erosion, aggradation, and re-entrenchment histories of the three basins appear similar to one another, and seem to parallel the three-step history deduced from the deposits at Nankoweap Creek.

Debris Flows and Gravel Deposits of the Big Bend Area, Chuar Creek to Unkar Rapids

Well-cemented gravel derived from Chuar basin accumulated along the course of and at the mouth of Chuar Creek (mi 65.5; km 105). The gravel appears to have accumulated across the course of the Colorado River, connecting with gravel above the mouth of Palisades Creek on the opposite side of the river. Immediately below Chuar Creek and Lava Canyon Rapids, the base of the gravel, which contains pebbles derived from headwater areas of the Colorado River system, is virtually at river level, and the gravel is discontinuously preserved near river level for approximately the next 1/2 mile (1 km) on the right; below this, a continuation of the gravel is preserved along the left bank to Comanche Creek (mi 67; km 108).

A large, well-cemented talus cone is preserved in a steep-walled basin beneath the Palisades of the Desert at the head of Comanche Creek. Two nested debris flows derived from the talus cone appear to have issued abruptly from the basin, flowed down Comanche Creek, and merged with (now well-cemented) gravel within about 200 feet (60 m) of river level. The debris flows thus appear related to the general episode of gravel accumulation, and not to later events. Large blocks of sandstone and limestone derived from Permian strata of the Palisades of the Desert are preserved on the margins of the debris flows. The blocks display cavernous weathering and the effects of extreme dissolution produced *in situ* by rainfall, indicating exposure over a geologically appreciable interval of time. Similar debris flows are present in Tanner Wash (mi 68.5; km 110). Triggering mechanism for the debris flows could have been climatic, but earthshocks from the Palisades fault (extension of Butte fault), which passes beneath the talus cone at the head of Comanche Creek, also could have been responsible for triggering the flows.

Gravel at and near river level. Well-cemented gravel is plastered against the Cardenas Basalt on the right (north) canyon wall opposite Tanner Wash, a continuation of gravel deposits preserved near river level beginning at Kwagunt Creek. Below Tanner Rapids, at mile 69 (km 111), well-cemented fluvial gravel containing clasts derived from the headwater areas of the Colorado River occupies the south side of the present river channel. A cross section of a former fluvial channel system that aggraded at the level of the present Colorado River can be observed by looking upstream after passing the deposit. The relations argue for an episode of accumulation not related to the current Colorado River. Extrapolating from deposits at Nankoweap Creek, gravel preserved along the course of the river from Chuar Creek to below Basalt Canyon may reflect a single episode of aggradation pre-dating the accumulation of unconsolidated alluvium on terraces cut on the gravel.

Deposits of fine-grained alluvium overlie the coarse well-cemented gravel on terraces close to the level of the Colorado River near Tanner Wash and near mile 69. Sites sampled here for uranium-trend dating provided ages of 40 ka and 75 ka, respectively (terrace levels 3 and 4, at 20 m and 36 m above river level, respectively; figures 24.2 and 24.3 and table 24.1). Alluvium on terrace level 5 near Basalt Canyon, about 50 m above river level, provided an age of 150 ka, also well within the ~700 ka upper limit of the uranium-trend method employed. The foregoing dates are considered by Machette and Rosholt (chapter 24) to apply to the time of deposition of the underlying gravel as well as to the alluvium. Soil at Hilltop Ruins (terrace level 7, the highest terrace) provided an age of ≥700 ka.

The downstream limit of preservation of the Colorado River gravels in the Big Bend area of the eastern Grand Canyon is found a short distance below Unkar Rapids. Additionally, a small patch of potentially correlative, well-cemented gravel is exposed at low water on the right at mile 87.2 (km 140.3), a short distance above Phantom Ranch.

Upper limit of gravel accumulation. Well-cemented gravel of local origin is found in all tributaries in the Big Bend area up to elevations of about 3,040 feet (927 m), suggesting that a common event having a discrete upper limit was responsible for accumulation of the gravel. An isolated deposit of gravel containing clasts of upstream origin is found at a similar elevation at

Hilltop Ruins, above Unkar Rapids, away from the canyon wall and on the south side of the Colorado River. The upper limit of the gravel at Hilltop Ruins lies about 425 feet (130 m) above river level. The gravel here also contains clasts derived from the flaky dolomite unit of the Kwagunt Formation, upper part of Chuar Group. The flaky dolomite unit is exposed only in Nankoweap and Kwagunt basins on the north side of the Colorado River. Two scenarios are possible to account for the presence of the clasts in the upper part of the gravel: 1) the class were deposited when the Colorado River was at this high level during a single, progressive entrenchment of the canyon; or 2) the clasts were deposited after a gravel fill some 400 feet (120+ m) thick had accumulated in the Big Bend area. Below the level of Hilltop Ruins, similar gravel is found in a series of small benches preserved along the course of the Colorado River from Basalt Canyon to below Unkar Rapids. These are considered by Machette and Rosholt (chapter 24) to be remnants of gravel left as the Colorado River progressively cut its channel through the bedrock, as their dates would seem to indicate. However, this scenario does not to accord with geologic relations preserved upstream, at Nankoweap Creek and in Nankoweap, Kwagunt, and Chuar basins, which indicate a distinct episode of aggradation followed cutting of the canyon and tributary basins to very near their present depths.

The time of cutting of the terraces here and at Nankoweap, and the time of deposition of alluvium on the terraces indicated from the uranium-trend dates, might be the time that the river re-established itself through a gravel fill. However, at least one other possibility exists. Dated alluvium related to terraces 3-5 in the Big Bend and Nankoweap Creek areas may reflect episodes of flooding associated with temporary lakes backed-up by late Pleistocene lava dams in the Lava Falls area (chapter 23). The eruptions occurred mainly, if not entirely, within the past 720 ka. If this is the case, the Colorado River already had been entrenched to its present level prior to the eruptions, as seen from geologic relations in the the Lava Falls area. This correlation implies that gravels associated with the discontinuous terraces in the Big Bend area are relict, that excavation of the canyon to its present depth had occurred prior to the accumulation of alluvium on the terraces, and that the uranium-trend ages are dating younger events.

Travertine of Hermit Rapids Area

Discontinuous deposits of travertine are observed on the left wall and rim of the inner gorge, a short distance below (west of) Hermit Rapids (mi 95; km 153). Part of a once-continuous deposit, the travertine is within about 100 feet (30 m) of river level, where its lower part contains water-worn pebbles and cobbles, and some lenses of sand. Lying on the crystalline basement within the inner gorge, and on the Tapeats Sandstone and Bright Angel Shale on and above the rim of the inner gorge, a series of isolated deposits of travertine appear to have once been connected with a deposit preserved against the Muav Limestone, about 1-1.5 miles (1.5-2 km) southeast of the Tapeats rim and east of Hermit Creek. The travertine must have once occupied and crossed the headwater area of Hermit Creek Canyon. The head of the deposit is at an elevation of about 4,000 feet (1,220 m), about 1,650 feet (500 m) above river level. The extent and dissected nature of the deposit would seem to be compatible with a pre-Pleistocene age.

Travertine, Talus and Gravel of Elves Chasm-Conquistador Aisle Area

A continuous deposit of travertine, containing talus and locally gravel in its lower parts, once occupied the left (south) wall of the canyon extending from about river-mile 114.4 to 117.5 (km 184-189). Its least-dissected part is centered along the short east-west stretch of river that contains Elves Chasm, which incises the deposit. Dissected travertine deposits also are found farther along the left (west) wall of the canyon along Stephen Aisle. Springs that were the source for the travertine are in the upper and middle parts of the Muav Limestone, more than 1,200 feet (365 m) above river level; the lower margins of the deposits at places lie near river level.

Deposits of cemented talus and gravel are found locally near and close to river level along east-west-trending Conquistador Aisle, beginning at mile 120.5 (km 194). The talus and gravel are more abundant along the bend and north-trending stretch of the river that follows Conquistador Aisle, ending at mile 127.5 (km 205). The gravel and cemented talus are interbedded, and similar to deposits of lower Marble Canyon.

Travertine, Gravel, and Landslide of Surprise Valley Area

Key geologic relations bearing on the history of accumulation of travertine, gravel, and landslide in the Deer Creek area can be observed along the east-west stretch of the Colorado River that lies above and traverses the area of the Surprise Valley landslide. They are summarized in schematic cross sections shown in Figure 21.3. On viewing the landslide from up-river on the east, from about mile 135 (km 217), an abandoned channel of the Colorado River occupied by travertine-cemented gravel, overlain(?) by travertine, is observed on the right (north) bank a few tens of meters above river level (figs. 1.31 and 26.5; fig. 21.3, section at mile 135.2). Landslide occupies the remainder of the channel above the travertine, and tilted, jumbled landslide derived from the Cambrian Tonto Group, Devonian Temple Butte Limestone, and Mississippian Redwall Limestone mantles the bedrock above and away from the channel. Travertine is present on the left (south) side of the river, and farther down river (mile 136.0; km 219), ledge-forming dolomite of the Bright Angel Shale (correlative with the Rampart Cave Member of the Muav Limestone a few miles downstream; see chapter 15) lies at the base of the landslide. The displaced dolomite unit and overlying landslide are seen to abruptly climb section, crossing undisturbed beds of the Bright Angel Shale. The up-

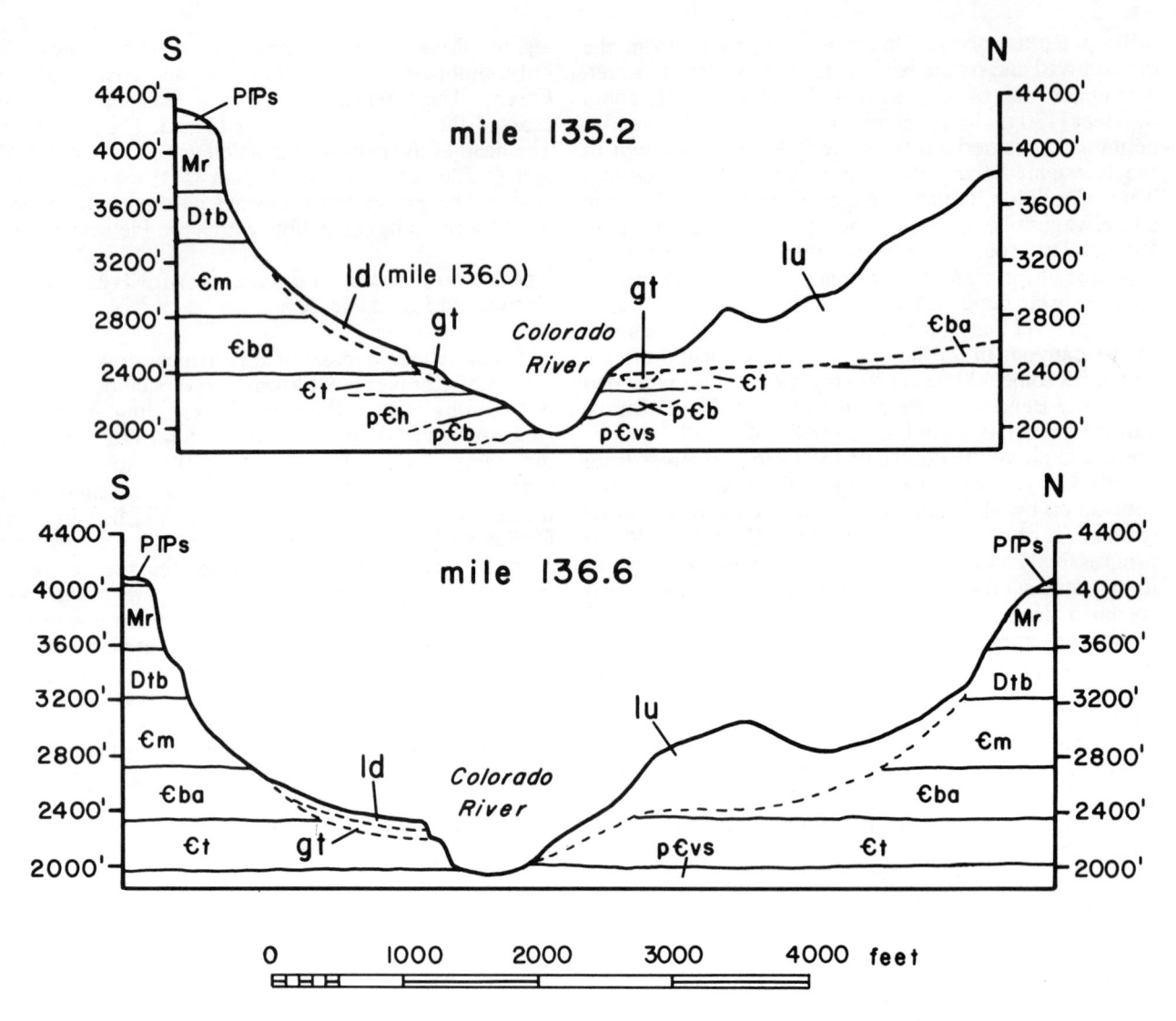

FIGURE 21.3. Generalized sections showing distribution of landslide and travertine and gravel in the Surprise Canyon area, central Grand Canyon. Surficial deposits: ld - landslide with dolomite unit of Bright Angel Shale at base; lu - landslide, undivided; gt - travertine-cemented gravel and travertine. Bedrock: Precambrian, pCvs - Vishnu Schist; pCb - Bass Limestone; pCh - Hakatai Shale; Paleozoic, Ct - Cambrian Tapeats Sandstone; Cba - Cambrian Bright Angel Shale; Cm - Cambrian Muav Limestone; Dtb - Devonian Temple Butte Limestone Mr - Mississippian Redwall Limestone; PIPs - Pennsylvanian and Permian Supai Group. Adapted from Huntoon and others, 1976 (note: edition of 1986 does not show some of the landslide and talus at mile 135.2).

slope end of the dolomite-bearing landslide, which locally overlaps the Muav Limestone, lies high above undisturbed, correlative dolomite in the Bright Angel Shale in the south wall of the canyon. The stratigraphic and structural relations, and the coherence of the dolomite unit at the base of the landslide, indicate that the leading edge of the landslide was emplaced catastrophically as a single, broadly coherent body derived from the north side of the canyon.

The abandoned Colorado River stream course at the upper (east) end of Granite Narrows parallels the present river course on the north for 0.8 mile (1.3 km). The abandoned channel appears to cross the present stream course just upstream from Deer Creek Falls (mi 135.8; km 218.5), where the channel deviates to the left (south) and is partly buried by the Surprise Valley landslide. On the north side of the river, a landslide-covered channel of an original Deer Creek Canyon joins the present river course at mile 136.2 (km 219.2). A short distance below this, at mile 136.6 (km 219.8), one can obtain an excellent view upstream of landslide relations on the south side of the river (diagrammed in fig. 21.3, mile 136.6). Landslide containing the dolomite unit of the Bright Angel Shale at the base, about 450 feet (140 m) above river level, transects undisturbed Colorado River gravel that in turn overlies travertine.

If the time of landsliding is inferred to have occurred during the Pleistocene, the cause might be attributed to lubrication of strata during a wet episode in the Pleistocene. Alternatively, if the landsliding is pre-Pleistocene in age and occurred during a time characterized by general aridity, cause for the landsliding might be attributed more plausibly to earthshock accompanying one or more increments of movement on the near-by Sinyala fault.

Huntoon (1975) considered the entire mass in and above the channel on the right (north) side of the river to be landslide. However, apparently undisturbed travertine-cemented gravel appears to underlie the landslide within the abandoned channel on the north side of the river, and undisturbed travertine-cemented gravel underlies the landslide on the south side of the river, a short distance downstream from Deer Creek (fig. 21.3). The relations thus seem to indicate that gravel had accumulated along the course of the river prior to the episode of catastrophic landsliding, and that it was a time that the Colorado River was not actively removing detritus from the area. The pre-landslide episode of aggradation thus appears to parallel the episode of aggradation seen in the eastern Grand Canyon, and landsliding can be inferred to have occurred during the episode of aggradation. Because of inferred diminished stream flow and subsurface movement of the Colorado River through the gravel, no immediate entrenchment by the Colorado River needs to be assumed following the episode of landsliding.

The Surprise Valley landslide and related deposits of gravel and cemented talus continue downstream to below mile 141 (km 227). Above mile 137.5 (km 221), high on the right (north) wall, a large apron of well-cemented, crudely stratified red talus is observed above the landslide. It bears a strong resemblance to well-cemented, crudely stratified deposits of the talus cone at the head of Comanche Creek (mi 67; km 108), and to deposits at "Red Slide" (mi 175; km 282) a short distance above Lava Falls Rapids.

Deposits Below Surprise Valley

Havasu Creek. Extensive travertine deposits and multiple low- to high-level travertine dams, and waterfalls, are found along the course of Havasu Creek (mi 179; km 288). Alluvium behind the travertine dams is inferred in Chapter 23 to represent Pleistocene silt that accumulated when lava dams blocked the course of the river in the area of Lava Falls Rapids, and the terraces are inferred to represent a series of lake levels formed by a succession of younger Pleistocene lava dams. Alternatively, red alluvium in the valley behind the highest dam appears to differ in no appreciable way from undated (but potentially older) alluvium found in drainages elsewhere in the general area (Gary R. Scott, personal communication, 1988). The accumulation of travertine deposits of Havasu Canyon parallels the accumulation of travertine deposits of the Elves Chasm area in stratigraphic setting and southerly direction of source. However, because of the length of the Havasu drainage system, travertine has accumulated in step-like fashion over the course of several miles of outcrop of source beds, which are well away from the Colorado River. Because travertine continues to be precipitated in Havasu Creek to this day, it is commonly assumed that travertine deposits elsewhere in the region must be very young.

"Red Slide". "Red Slide" (mi 175; km 282), is a poorly cemented, crudely stratified deposit of talus and landslide debris that has gravel at its base a few tens of meters above river level. The talus and gravel horizon can be traced discontinuously downstream toward Lava Falls Rapids, where it is overlain by Pleistocene basalt and associated gravel of the Toroweap area (chapter 23). "Red Slide" is the westernmost of the essentially undisturbed, pre-Pleistocene(?) deposits of aggradation of the central and eastern Grand Canyon.

Deposits Below Lava Falls

Gravel, travertine, and landslide deposits highly similar to deposits found above Lava Falls Rapids also are present in the western Grand Canyon. (The division line between the central and western Grand Canyon nominally is drawn at Lava Falls Rapids). However, these deposits of aggradation not only have been complicated by overlapping Pleistocene lava flows and cascades, and their associated gravels, but also by faulting and complex landsliding that has occurred along the Hurricane and related faults. Late Paleogene(?) and Neogene movements on the Hurricane fault system, and faults that lie to the west, have served to reverse and mitigate at least some of the earlier down-to-the-east Laramide displacements. The younger down-to-the-west displacements presumably occurred as the western margin of the plateau became stepped down toward the Basin and Range Province, probably beginning during Miocene extension. A number of major landslides in the western Grand Canyon, considered to be relatively old on the basis of their degraded appearances, have been mapped as Quaternary-Tertiary in age, and large travertine deposits that block the course of tributary drainages have been mapped as Quaternary in age (Weinrich and others, 1986; Billingsley and others, 1986). Landsliding in the western Grand Canyon conceivably was related to earthshocks accompanying increments of movement on the faults. The general distribution of the landslide and travertine deposits near the river is indicated in the geologic river log (chapter 1).

SUMMARY

The distribution of undated but interrelated gravel, talus, travertine, and landslide deposits in the master and tributary drainages of the Colorado River, extending from upper Marble Canyon to the west-central Grand Canyon, and very likely to the western margin of the Grand Canyon, suggests that a relatively simple, three-step history can account for their accumulation and subsequent partial dissection. The inferred history has been summarized already in the section entitled "Model for pre-Pleistocene(?) accumulation." In this model, the deposits are provisionally correlated with upper Miocene and Pliocene deposits of the Bidahochi, Verde, and Muddy Creek Formation, which record an episode of aggradation in near-by areas on and adjacent to the Colorado Plateau (Table 18.1).

Farther afield, correlative deposits are identified as "T5" units on the 1:1,000,000 scale geologic map of

Utah (youngest of five Tertiary subdivisions; Hintze, 1975), and they include a number of T5 deposits in southwestern Utah that are correlated with the Muddy Creek Formation; to the north, correlative deposits include the Salt Lake Formation, adjacent to the Wasatch Front. T5 deposits also are shown distributed across the high plateaus and basins of the Colorado Plateau. These deposits in Utah, apparently correlative late Miocene and Pliocene deposits in northern and central Arizona, and analogous deposits that might be enumerated from across a much larger part of the western U.S., would seem to record a widespread episode of aggradation that followed definition of fundamental topographic elements by erosion.

The sediments of the "T5" episode appear to have accumulated at a time of general aridity, when the products of mechanical disintegration were transported relatively short distances by flash flooding but when stream flow was insufficient to remove the products of erosion from the region. It was a time when stream flow presumably also would have been insufficient to cut the western Grand Canyon in the interval ~5-3.8 m.y., as is proposed in a model for the time of cutting of the western Grand Canyon (chapter 18, table 18.1). If the pre-Pleistocene(?) deposits of aggradation in the eastern and central Grand Canyon also accumulated in the western Grand Canyon (as appears likely), and if these deposits are truly Miocene and Pliocene rather than Pleistocene in age, the entire Grand Canyon had been cut to essentially its present depth prior to late Miocene time.

The time of cutting of terraces in the eastern Grand Canyon during the past few hundred thousand years would appear to be more easily and simply related to temporary lava damming events in the Lava Falls area than to stages of an initial, progressive downcutting of the Colorado River. The Colorado River was at its current level prior to eruption of the lava flows that apparently dammed the river during late Pleistocene time. The foregoing speculative conclusions remain to be verified.

TABLE 21.1. Summary of pre-Pleistocene(?) and Pleistocene deposits of Grand Canyon, referred to Episodes 5 and 6 of Table 18.1. Deposits are listed in order of their encounter along the Colorado River.

Episode 4. Erosion of Marble Canyon and the Grand Canyon and its tributaries to near their present depths by middle(?) Miocene time.

Episode 5. Accumulation of pre-Pleistocene(?) deposits in Grand Canyon and its tributaries.

Gravel of Lees Ferry area
   Gravel of upstream and local origin, 220 to 750(?) feet thick, choked(?) course of Colorado River at Lees Ferry and occupied(?) upper Marble Canyon to below mile 7.5. Climate dry; aggradation by flash flooding; attenuated Colorado River flowed in subsurface.

Cemented talus below Coconino Sandstone, upper Marble Canyon
   Well-cemented, now dissected, talus cones, begin about mile 8.5 near river level; they rise in elevation to south following increase in elevation of Coconino Sandstone; cementation resulted from seeps issuing from base of Coconino Sandstone.

Gravel, talus, travertine -- Buck Farm Canyon to Little Colorado River
   Well-cemented, dissected talus is developed on the Muav Limestone and Bright Angel Shale from Buck Farm Canyon (mile 41) to the Little Colorado River (mile 61). Gravel is associated with the lower parts of the talus; travertine makes its appearance at Kwagunt Rapids (mile 56) and is locally abundant to the area of the Little Colorado River. Cementation of talus and deposition of travertine related to seeps in the Muav Limestone issuing along both sides of the river.

Gravel and talus at Nankoweap Creek (mile 52), and gravel of local origin in Nankoweap, Kwagunt, and Chuar Valleys.
   Gravel accumulated to thicknesses of the order of 100 feet or more along the Colorado River and its tributaries, and in Nankoweap, Kwagunt, and Chuar basins. With accumulation, stream flow moved predominantly in the subsurface; accumulation of clasts from headwater areas of Colorado River waned; middle and upper part of Colorado River gravel contains only clasts of local origin. Talus was shed sporadically from sheer east wall of canyon at Nankoweap Creek, overlapping the gravel. Alluvium accumulated on the gravel at Nankoweap Creek and in Nankoweap basin prior to re-establishment of perennial streamflow and re-entrenchment of stream courses to former stream level. Red rockfall from east wall, dated at 210,000 years, was deposited against dissected gravel deposit on north side of Nankoweap Creek and in Nankoweap Creek.

Gravel and debris flows of Big Bend area

Gravel may have accumulated to thickness of as much as 400 feet in western end of Big Bend area; two debris flows derived from talus cone at head of Comanche Creek merge with gravel above Colorado River; blocks along margin of flows exhibit deep dissolution from *in situ* weathering. Late Pleistocene alluvium on gravel terraces, having ages of a tens of thousands to a few hundred thousand years, may have accumulated at times of flooding related to temporary late Pleistocene lava dams at Lava Falls.

Travertine of Hermit Rapids area

Formerly extensive, now deeply dissected travertine deposit whose source was the Muav Limestone above the head of Hermit Canyon, south of river.

Travertine of Elves Chasm-Conquistador Aisle areas

High wall of travertine whose source was the Muav Limestone, south of the river, and deposits of travertine and gravel below Elves Chasm.

Travertine, gravel, and landslide of Surprise Valley area

Inter-relations indicate that landsliding occurred catastrophically following the accumulation of travertine and gravel along a pre-existing, higher level course of the river. Landsliding may have been triggered by an increment of movement on the Sinyala fault.

Deposits below Surprise Valley

Travertine of Havasu Canyon is a horizontally stretched-out version of travertine of the Elves Chasm area. "Red slide" is a talus and landslide deposit similar to talus deposits at Comanche Creek and Surprise Valley landslide.

Deposits below Lava Falls

Landslide and talus deposits of the western Grand Canyon are similar to those of the central and eastern Grand Canyon; they have been complicated by Pleistocene lava flows and associated gravel deposits, and by pronounced landsliding presumably triggered by movements on the Hurricane and other faults.

Episode 6. Late Pliocene(?) and Pleistocene re-establishment of perennial stream flow and re-entrenchment of streams to former levels.

Colorado River stream course eroded to present level by about 1 m.y. ago, or less, in Lava Falls area. Eruptions leading to formation of a series of temporary lakes behind lava dams may have occurred mainly since 720,000 years ago.

Alluvium of Nankoweap Creek and Big Bend areas

Late Pleistocene alluvium on gravel terraces, having ages of a tens of thousands to a few hundred thousand years from uranium-trend dating, may record times of flooding related to temporary late Pleistocene lava dams at Lava Falls.

CHAPTER 22: PETROLOGY AND GEOCHEMISTRY OF LATE CENOZOIC BASALT FLOWS, WESTERN GRAND CANYON, ARIZONA

J. Godfrey Fitton
Grant Institute of Geology, University of Edinburgh
West Mains Road, Edinburgh EH9 3JW, U.K.

INTRODUCTION

The basaltic lava flows around Vulcan's Throne (Colorado River mile 179 (km 288) represent some of the most recent volcanic activity in northwestern Arizona. Northwestern Arizona contains three separate volcanic fields, from west to east, the Grand Wash (mile 285; km 458), Shivwits Plateau, and Uinkaret Plateau fields. The latest volcanic activity was concentrated on the Uinkaret Plateau (fig. 4.1) and it was from this field that lavas cascaded into the Grand Canyon to form the lava dams.

Volcanism in the western Grand Canyon area was accompanied by faulting associated with extension and uplift of the Colorado Plateau. This faulting resulted in prominent fault scarps of the Grand Wash and Hurricane Cliffs. The oldest flow yet dated (PED37-66, Table 22.1) is about 7 million years old (Best and others, 1980) but very few radiometric ages are available. A relative time scale has, however, been established by Hamblin (1970) on geomorphological criteria. This, coupled with the few available K-Ar ages, suggest a general eastward migration of volcanism at a rate of about 1 cm/yr (Best and Brimhall, 1974).

The purpose of this contribution is to review the current state of knowledge of the petrology and geochemistry of the western Grand Canyon lavas. Detailed descriptions of the petrography and mineralogy have been given by Best and Brimhall (1970, 1974).

COMPOSITION OF THE BASALTIC FLOWS

The flows are entirely mafic; olivine basalt being the most common rock type. Flows more evolved than hawaiite have not been reported. The most basic flows have abundant small olivine phenocrysts which are joined, in more evolved compositions, first by augite and later by plagioclase. Plagioclase is the dominant phenocryst phase in the hawaiites.

Sample descriptions and localities from the western Grand Canyon area are given in Table 22.1. Chemical analyses of representative flows are given in Table 22.2. The analyses are arranged by volcanic field (Grand Wash, Shivwits, and Uinkaret), and within each field in approximate order of decreasing age based on Hamblin's (1970) geomorphological stages. For comparison, an average of 97 analyses of Basin and Range basalts is also given. All the data were obtained in Edinburgh by X-ray fluorescence techniques and are, therefore, directly comparable.

The data in Table 22.2 indicate that the basalt flows range from transitional to strongly alkaline in character, and that the more undersaturated flows are confined to the Uinkaret Plateau. Best and Brimhall (1974) have shown that the older basalts have compositions close to the critical plane of silica-undersaturation, whereas the younger basalts tend to have more diverse compositions dominated by strongly undersaturated types.

TABLE 22.1. Sample descriptions and localities.

GW 17	Olivine basalt, S of Mud Mountain		36°37.8'N	113°47.7'W
GW10	Olivine basalt, E side of Virgin Mountains		36°46.9'N	113°47.7'W
PED37-66	Plag-ol-augite-phyric basalt. Mt. Dellenbaugh		36°06.6'N	113°32.5'W
PED39-66	Olivine basalt, NW of Poverty Mountain		36°25.8'N	113°33.5'W
AZ99	Olivine basalt, S side of Seegmuller Mountain		36°47.3'N	113°33.0'W
SS11	Olivine basalt, 13 km N of Mt. Dellenbaugh		36°13.2'N	113°30.8'W
AZ98	Plag-ol-augite-phyric basalt. 4 km S of Diamond Butte		36°32.3'N	113°22.8'W
AZ96	Olivine basalt, S side of Mt. Trumbull		36°23.8'N	113°08.8'W
U73	Basanite. Uinkaret Plateau		36°41.1'N	113°08.4'W
AZ92	Olivine basalt. Small lava field S of Antelope Valley		36°33.9'N	112°54.9'W
AZ93	Plag(-ol-augite)-phyric hawaiite, N of "The Hat"		36°30.0'N	113°00.8'W
AZ97	Basanite. Recent flow 4 km SW of Mt. Trumbull		36°22.6'N	113°09.8'W
AZ95	Basanite containing abundant peridotite xenoliths Recent flow by track to W of Vulcan's Throne		36°12.7'N	113°05.2'W
AZ95	Alkali olivine basalt. Vulcan's Throne		36°12.8'N	113°04.5'W

Plag = plagioclase; ol = olivine.

TABLE 22.2. Chemical analyses of late Cenozoic basaltic lavas from the western Grand Canyon area, Arizona. Major elements and normative values in weight percent. Trace elements in parts per million.

	GW17	GW10	PED37-66	PED39-66	AZ99	SS11	AZ98	AZ96	U73	AZ92	AZ93	AZ97	AZ94	AZ95	Basin & Range Mean (1)
Area (2)	GW	GW	S	S	S	S	S	S	U	U	U	U	U	U	
Stage (3)	I	III	I	I	I	I	III	I	III	III	III	IV	IV	IV	
Major elements															
SiO_2	49.64	48.55	48.93	48.53	49.32	48.38	50.40	50.31	44.77	47.44	50.70	45.84	43.41	45.14	46.63
Al_2O_3	14.99	15.76	18.01	15.81	15.75	15.93	17.85	15.47	13.56	13.78	18.69	15.29	13.86	13.67	15.68
Fe_2O_3(4)	11.92	11.26	10.86	11.18	11.41	12.16	10.73	11.34	13.29	11.36	9.81	13.60	13.19	11.00	11.67
MgO	7.43	7.27	5.09	7.47	7.36	7.39	4.38	8.04	10.01	12.07	4.03	7.78	10.31	12.28	8.02
CaO	7.25	9.52	10.14	10.16	9.18	10.01	8.99	9.29	8.96	9.71	8.50	8.09	9.98	10.59	9.06
Na_2O	4.21	3.18	3.45	3.24	3.36	3.16	4.06	3.17	3.59	2.96	3.70	4.35	3.99	3.00	3.62
K_2O	1.77	0.94	1.15	1.11	1.17	0.91	1.37	0.93	1.98	0.81	1.75	2.01	1.58	1.16	1.66
TiO_2	2.06	1.66	1.49	1.53	1.63	1.64	1.66	1.46	2.69	1.36	1.56	2.65	2.65	1.75	2.24
MnO	0.16	0.17	0.18	0.16	0.17	0.18	0.16	0.17	0.17	0.17	0.16	0.16	0.19	0.17	0.18
P_2O_5	0.54	0.29	0.40	0.38	0.35	0.34	0.52	0.30	0.71	0.48	0.53	0.56	0.78	0.65	0.55
Total	99.96	98.60	99.70	99.58	99.71	100.10	100.12	100.48	99.72	100.15	99.45	100.33	99.94	99.40	99.31
CIPW Norms (5)															
or	10.55	5.71	6.88	6.65	7.01	5.41	8.14	5.52	11.83	4.83	10.46	11.97	9.44	6.99	9.97
ab	31.52	27.52	28.40	25.70	28.80	26.97	34.00	26.96	12.34	21.95	31.77	16.62	7.13	12.07	20.26
an	16.93	26.58	30.62	25.67	24.72	26.82	26.63	25.34	15.26	22.08	29.61	16.35	15.38	20.69	21.99
ne	2.43	-	0.61	1.12	-	0.03	0.33	-	9.98	1.78	-	11.10	14.65	7.41	5.89
di	12.90	16.17	14.52	18.64	15.51	17.20	12.25	15.43	20.30	18.74	7.96	16.63	23.85	22.72	16.26
hy	-	3.87	-	-	2.15	-	-	9.82	-	-	6.15	-	-	-	-
ol	17.24	13.17	12.23	15.35	14.76	16.37	11.33	10.40	19.83	23.81	7.13	17.27	19.04	22.23	16.83
mt	3.21	3.07	2.93	3.02	3.08	3.27	2.88	3.03	3.59	3.05	2.65	3.65	3.55	2.97	3.16
il	3.95	3.22	2.87	2.94	3.13	3.15	3.18	2.78	5.18	2.60	3.01	5.06	5.10	3.37	4.32
ap	1.28	0.70	0.95	0.91	0.84	0.80	1.25	0.72	1.70	1.15	1.27	1.34	1.86	1.56	1.32
Trace elements															
Ni	162	93	22	95	120	105	21	115	214	319	18	131	192	276	138
Cr	182	276	27	238	350	286	15	408	286	619	21	135	287	580	208
V	165	217	277	226	211	226	206	217	225	203	184	213	238	234	201
Sc	14	26	31	30	28	23	21	31	15	25	26	14	20	26	21
Cu	51	63	52	60	60	60	37	57	56	66	36	49	55	58	49
Zn	114	90	84	80	85	91	90	87	117	86	85	128	103	78	81
Sr	664	390	502	439	472	402	628	370	779	685	601	663	917	834	666
Rb	18	6	16	19	13	12	15	13	20	9	32	18	16	10	28
Zr	243	142	160	156	155	148	180	138	270	140	199	227	255	191	230
Nb	39	17	27	28	24	26	36	19	58	33	34	42	64	46	52
Ba	478	305	677	514	452	422	747	456	524	1096	773	362	1053	1344	457
La	31	18	31	25	18	18	38	15	40	47	40	31	57	59	35
Ce	60	36	50	47	42	54	74	29	79	90	74	67	121	126	73
Nd	26	18	25	20	19	22	33	16	42	37	33	27	48	48	32
Y	20	22	25	22	23	24	24	24	23	21	27	19	25	22	28
La/Nb	0.80	1.07	1.16	0.90	0.77	0.69	1.06	0.82	0.70	1.43	1.18	0.74	0.88	1.28	0.67
Ba/Nb	12.38	18.35	25.26	18.43	18.92	16.18	20.86	24.41	9.10	33.30	22.99	8.56	16.38	29.27	8.80
Rb/Sr	0.027	0.015	0.032	0.043	0.028	0.030	0.024	0.035	0.026	0.013	0.053	0.027	0.017	0.012	0.042

(1) Average of 97 analyses of basaltic lavas, mostly from Potrillo (NM), Geronimo (AZ), Lunar Craters (NV) and the Mojave Desert (CA) (from Fitton and others 1988).
(2) GW, Grand Wash. S, Shivwits Plateau. U, Uinkaret Plateau.
(3) Relative ages of flows according to geomorphological stages of Hamblin (1970). Approximate absolute ages (Best & Brimhall, 1974); Stages I and II, 4.5 to 7 Ma. Stages II and III, 1 to 3.5 Ma. Stage IV < 1 Ma. K-Ar ages have been published by Best and others (1980) for PED37-66 (7.06 Ma) and PED39-66 (4.75 Ma).
(4) Total Fe expressed as Fe_2O_3.
(5) Norms calculated using $Fe_2O_3/(Fe_2O_3 + FeO) = 0.20$.

NATURE OF THE MANTLE SOURCE

The basalt flows of the Grand Canyon area form part of an arcuate province of dominantly basaltic, late Cenozoic volcanic fields along the western and southern margins of the Colorado Plateau. Other volcanic fields include St. George Basin in Utah, the San Francisco and Springerville fields in Arizona, and the Zuni-Bandera field in New Mexico. These fields occur on or near the physiographic boundary between the Colorado Plateau and Basin and Range.

If the mantle beneath the Colorado Plateau differs from that beneath the tectonically active Basin and Range, chemical differences between Basin and Range basalts and those erupted on the edge of the Colorado Plateau in the Transition Zone should be seen. Fitton and others (1988) have compared suites of basalts erupted in the two tectonic regimes and have demonstrated significant chemical and isotopic differences between them. Basin and Range basalts are, in general, chemically and isotopically similar to ocean island basalt and therefore appear to have a source within the asthenosphere. Interaction with lithospheric mantle or crust has been insignificant. In contrast, Transition Zone magmas, including those from the Grand Canyon area, have relatively high concentrations of barium, low concentrations of niobium and rubidium, radiogenic strontium, and non-radiogenic neodymium when compared with Basin and Range basalts.

This can be illustrated with respect to the Grand Canyon basalts by inspection of the data in Table 22.2. All the Grand Canyon basalts have higher La/Nb than the Basin and Range mean, and nearly all have higher Ba/Nb. This relative enrichment in Ba is not, however, reflected in the abundance of rubidium. With the exception of the hawaiite sample AZ93, which has probably lost plagioclase by fractional crystallization, with consequent increase in Rb/Sr, all the Grand Canyon samples have lower Rb/Sr than the Basin and Range mean. This is surprising in view of the differences in $^{87}Sr/^{86}Sr$ between the Transition Zone and Basin and Range basalts. Eleven $^{87}Sr/^{86}Sr$ determinations on basalt samples from the Grand Canyon area (Leeman, 1974) have a mean value of 0.7038, significantly higher than the mean value (0.7030) for Basin and Range basalts (Ormerod and others, 1988).

That these differences cannot be due to crustal contamination can be easily demonstrated. Assimilation of crustal rocks would increase the silica content of a basaltic magma and result in a correlation between degree of silica saturation and parameters such as Ba/Nb. Inspection of the data in Table 22.2 shows that this is not the case. AZ94, for example, is strongly nepheline-normative and yet has high Ba/Nb. Furthermore, this sample was collected from a flow containing abundant peridotite xenoliths and therefore could not have been stored for long in a magma reservoir at crustal levels.

It is concluded that the geochemical characteristics of the Grand Canyon basalts were inherited from their mantle source. This source must have been depleted in Rb and Nb (both highly incompatible elements) and yet enriched in Ba and ^{87}Sr. Ba is also highly incompatible in mantle phases and therefore this enrichment must have been superimposed on a previously depleted mantle. Because the asthenosphere is well stirred by convection, the only available long-term repository for enriched and depleted mantle domains is the lithospheric mantle. Therefore, Transition Zone basaltic magmas, including those in the Grand Canyon area, have their source within the lithospheric mantle or have interacted with it on the way to the surface.

The nature of the mantle enrichment process is still far from clear. It has introduced Ba, ^{87}Sr, and light rare earth elements (as shown by low $^{143}Nd/^{144}Nd$ in Transition Zone magmas; Fitton and others, 1988), but not Rb and Nb. Pelagic sediment, returned to the mantle during subduction, is a possible agent for this enrichment. Some of this sediment may be introduced into the lithospheric mantle and remain there until remobilized during a later magmatic event. If this is the case, the enrichment may have been related to Mesozoic subduction beneath the western U.S. but need not have been. It may have been introduced during much more ancient subduction episodes, perhaps even those which led to formation of the local continental crust.

Why is lithospheric mantle involved in the genesis of Transition Zone magmas but not Basin and Range magmas? The answer probably lies in the different styles of tectonism shown by the two magmatic provinces. During extension the lithosphere is stretched and thinned, and adiabatic upwelling of the mantle will lead to melting in the thermal boundary layer between the rigid lithosphere and the convecting asthenosphere (McKenzie and Bickle, 1988). As these melts rise through the attenuated lithospheric mantle they will heat it and perhaps lower its viscosity enough to enable it to mix with the convecting mantle below, especially if the convecting mantle has a higher than normal temperature, as in a plume. Melting of this mobilized lithospheric mantle, or interaction between it and asthenosphere-derived melts, will produce magmas with geochemical and isotopic signatures inherited from the lithospheric mantle. In this way lithospheric mantle may be involved in the generation of magmas in the early stages of extension.

As extension proceeds, the lithospheric mantle may eventually be eroded away and replaced by asthenosphere as postulated by Best and Brimhall (1974). This appears to have happened beneath the Rio Grande rift in New Mexico and parts of the southern Basin and Range (Sinno and others, 1986; Perry and others, 1988). Because volcanism and extensional faulting have been migrating into the Colorado Plateau over the last few million years (Best and Brimhall, 1974), it follows that the Transition Zone (including the Grand Canyon area) is in the early stages of extension whereas the central parts of the Basin and Range are at an advanced stage. Hence, the Basin and Range magmas are generated from the asthenosphere alone whereas the Transition Zone magmas have interacted with lithospheric mantle.

SUMMARY AND CONCLUSIONS

1. The late Cenozoic volcanic fields of the Grand Canyon area are composed entirely of mafic lavas ranging from transitional alkali basalt and hawaiite to thoroughly undersaturated basanite.

2. Volcanism was accompanied by faulting associated with extension and uplift of the Colorado Plateau. Both faulting and volcanism migrated eastwards into the Colorado Plateau at a rate of about 1 cm/yr. The oldest basaltic lava flows (7 Ma) occur around the Shivwits Plateau and Grand Wash areas and the youngest around Vulcan's Throne.

3. There was a general trend towards undersaturation with time. The most strongly alkaline basalts, which are confined to the Uinkaret Plateau, often contain peridotite xenoliths as found at Vulcan's Throne.

4. The Grand Canyon basalts, like those erupted by other parts of the Transition Zone, are chemically and isotopically distinct from contemporaneous basalts of the Basin and Range Province. Transition Zone basalts generally have higher La/Nb, Ba/Nb, and $^{87}Sr/^{86}Sr$, but lower Rb/Sr and $^{143}Nd/^{144}Nd$ than do Basin and Range basalts. The Basin and Range basalts resemble ocean island basalt and probably had a source within the asthenosphere whereas the Transition Zone magmas either originated within the lithospheric mantle or interacted with it.

5. The different mantle sources tapped by Basin and Range and Transition Zone magmas reflect the different styles of tectonism experienced by the two provinces. Extension is at an advanced stage in the Basin and Range with consequent erosion and removal of lithospheric mantle, whereas it is still at an early stage in the Grand Canyon area and other parts of the Transition Zone.

ACKNOWLEDGEMENTS

Most of the rock samples (AZ numbers) were collected in the course of field work funded by a research grant from NATO (grant no. 0198/85). Additional samples were supplied by Myron Best and Paul Damon. The author is indebted to Bill Leeman and Myron Best for their help and advice with field work in the western U.S. and to Dodie James for her help with the chemical analyses.

CHAPTER 23: PLEISTOCENE VOLCANIC ROCKS OF THE WESTERN GRAND CANYON, ARIZONA

W. Kenneth Hamblin
Department of Geology, Brigham Young University, Provo, Utah

INTRODUCTION

The late Cenozoic basalt flows and cinder cones of the western Grand Canyon constitute one of the most spectacular displays of volcanism in North America (fig. 23.1). Numerous flows were extruded near the southern margin of the Uinkaret volcanic field, cascading over the outer rim of the Grand Canyon in Toroweap Valley and Whitmore Wash. Others were extruded within the canyon on the Esplanade platform, some of which formed spectacular "frozen" lava falls as they plunged over the rim of the inner gorge into the Colorado River 3,000 feet (900+ m) below. In addition, several volcanic cones are perched precariously on the very rim of the canyon and remnants of others cling to the steep walls of the inner gorge. Dikes, sills, and volcanic necks are also exposed in the canyon and represent part of the complex sequence of late Cenozoic volcanism.

From viewpoints near the Toroweap Campground it may appear at first glance that the lava cascades into the Grand Canyon represent a single period of extrusion in which two great lava dams were built across the Colorado River; one at the mouth of Toroweap Valley below Vulcan's Throne, and the other 3 miles (5 km) downstream. However, the volcanic features within the canyon are much more complex than one might first expect. More than 150 lava flows poured into the canyon during the last ~1.2 million years leaving an incredible record of volcanic extrusions that formed a sequence of great barriers or lava dams across the Colorado River. These dams formed temporary lakes upstream, some as far as the head of Lake Powell. As the lake behind each barrier overflowed, a new gorge was eroded through the dam leaving only small remnants of lava clinging to the walls of the Canyon. Later eruptions formed new dams that were largely destroyed by overflow of the Colorado River leaving only remnants of the dams stacked side by side against the canyon wall. The relative ages of the major dams within the Grand Canyon therefore are clearly expressed by juxtaposition. Remnants of older flows lie adjacent to the bedrock of the canyon walls with succeedingly younger flows stacked in sequence against them.

The results of our work (Hamblin, in press) indicate that at least 12 major lava dams were formed in the Grand Canyon during the last ~1.2 million years. Some were more than 2,000 feet (600+ m) high; one was more than 84 miles (135 km) long. The remnants of the lava dams and the sediment deposited in the lakes behind the dams provide a fascinating record of this unusual and most recent series of events in the history of the Grand Canyon. Table 23.1 summarizes the characteristics of the dams and the lakes that formed behind them. The following section describes the geologic relationships of the most important remnants observable from the river.

DESCRIPTION OF LOCAL AREAS

Toroweap Valley

Toroweap Valley, on the north side of the Colorado River, is near the center of volcanic extrusions in the western Grand Canyon and has been involved, either directly or indirectly, with most volcanic events in the area. It is, also, the most accessible area to view the volcanic features from the ground because it is possible to drive down Toroweap Valley and out onto the Esplanade bench to the rim of the inner gorge.

A cross-section across the mouth of Toroweap Valley is shown in Figure 23.2a. Four major sequences of basalt are found in this area: 1) the Toroweap sequence, which formed a large dam in the Grand Canyon, remnants of which adhere to the walls of the inner gorge and extend far downstream beyond the mouth of Toroweap Valley; 2) a sequence of thin basalt flows that fill Toroweap Valley to the level of the Esplanade; 3) younger intracanyon flows; and 4) lava cascades. Isolated high level remnants of uncertain age also are found at some localities.

1. The Toroweap Valley Sequence. The older flows beneath the Toroweap cascades consist of a thick sequence of nearly horizontal flows that adhere to the canyon walls. Although the Toroweap Valley sequence is located at the mouth of Toroweap Valley, it is clear that the sequence of flows is not confined to the ancestral valley because large remnants are found along the walls of the inner gorge for a considerable distance downstream beyond the limits of the buried Toroweap Valley.

The oldest flows in the Toroweap Valley sequence consist of three basaltic lava units, designated by McKee and Schenk (1942, p. 253) as A, B, and C flows of their "Lower Canyon lavas". Two additional units, recognized in our studies as belonging to this same sequence but superposed on the C flow (Hamblin, in press), are designated as units E and F. The "D" flow of McKee and Schenk is much younger and rests unconformably upon, or is juxtaposed against, the Toroweap Valley sequence. River gravels and channels cut in the lower Toroweap Valley sequence indicate that each flow was partly eroded during the construction of

FIGURE 23.1 Sketch showing the volcanic features in Toroweap Valley. View looking northeast. Toroweap Valley is filled to the level of the Esplanade platform with thin-bedded basalt and intraflow sediments. Vulcan's Throne is perched above the valley fill on the rim of the inner gorge. Remnants of the Toroweap dam can be seen adhering to the canyon wall 1,400 feet (427 m) above the river.

Table 23.1 Summary of age, characteristics, and distribution of Pleistocene basalt flows of Toroweap area and below, and major lava dams, western Grand Canyon, Arizona.

Dam	Elevation	Height Above River	Radiometric Date	Type of Dam	Minimum Length of Dam	Volume of Lava (Km3)	Lake Length Miles	Water Fill Time	Sediment Fill Time
Prospect	4000	2330	---	2 major flows 700-800 ft. thick	?	4.0	324	23 years	3018 years
Ponderosa	2800	1130	---	1 major flow 1000 feet thick	12 miles	2.5	126	1.5 years	163 years
Toroweap	3093	1413	1.2 Ma	5 major flows 100-200 ft. thick	8-10 miles	3.7	177	2.62 years	345 years
Esplanade	2600	960	---	6-8 flows 100-200 ft. thick	8 miles	1.8	109	287 days	92 years
Buried Canyon	2480	850	0.89 Ma	8 major flows 100-200 ft. thick	?	1.7	108	231 days	87 years
Whitmore	2500	900	0.99 Ma	More than 40 thin flows 10-20 ft. thick	18	3.0	108	240 days	88 years
"D" Flows	2295	635	0.57 Ma	More than 40 thin flows 10-20 feet thick	?	1.1	77	87 days	31 years
Lava Falls	2260	600	---	Single flow	22	1.2	77	86 days	30 years
Black Ledge	2033	373	0.55 Ma	Single flow	86 miles +	2.1	53	17 days	7 years
Gray Ledge	1813	203	---	Single flow	13	0.3	37	2 days	10.3 mos.
Layered Dbs.	1938	298	0.64 Ma	Up to 20 thin flows	14 miles	0.3	42	8 days	3 years
Massive Dbs.	1826	226	0.14 Ma	Single flow	10	0.2	40	5 days	1.4 years

the Toroweap dam.

2. Lavas Filling Toroweap Valley. The sequence of lavas that fill Toroweap Valley, exposed in the cliffs below Vulcan's Throne, consist of numerous thin basalt flows, each approximately 20 feet (6 m) thick (fig. 23.2b). Although the entire sequence is not exposed because of a partial cover of younger cascades and talus, each flow originally moved down Toroweap Valley and into the inner gorge of the Grand Canyon. Erosion, however, has removed all evidence of any cascades, if they did exist, so that only the eroded margin of basalt flows filling the ancient valley are exposed. It is thus uncertain if this sequence of thin flows formed a lava dam to the level of the Esplanade. If they did, it was one of the highest in the canyon, more than 2,500 feet (760 m) above the river.

3. Younger Lava Dams. Four remnants of younger lava dams, juxtaposed against the Toroweap Valley sequence, provide a record of events that occurred after the Colorado River excavated a new gorge through the Toroweap Valley flows. Several can be correlated with flow remnants farther downstream. The two largest flows are located at the west end of the Toroweap Valley flow exposure just below Lava Falls Rapids.

4. Cascades at Vulcan's Throne. One of the most impressive sights at the mouth of Toroweap Valley is the "frozen lava falls" that spill over the rim of the Esplanade just west of Vulcan's Throne and cascade into the inner gorge (fig. 23.1). It might appear at first that the cascades originated from the vent at Vulcan's Throne, but geologic mapping indicates that the flows forming the cascades are much younger. They were extruded from vents high on the southern tip of the Uinkaret Plateau. They then flowed to the outer rim of the canyon, cascaded into Toroweap Valley, and then spread out over the Esplanade; following this, they spilled over the rim of the inner gorge, just west of Vulcan's Throne.

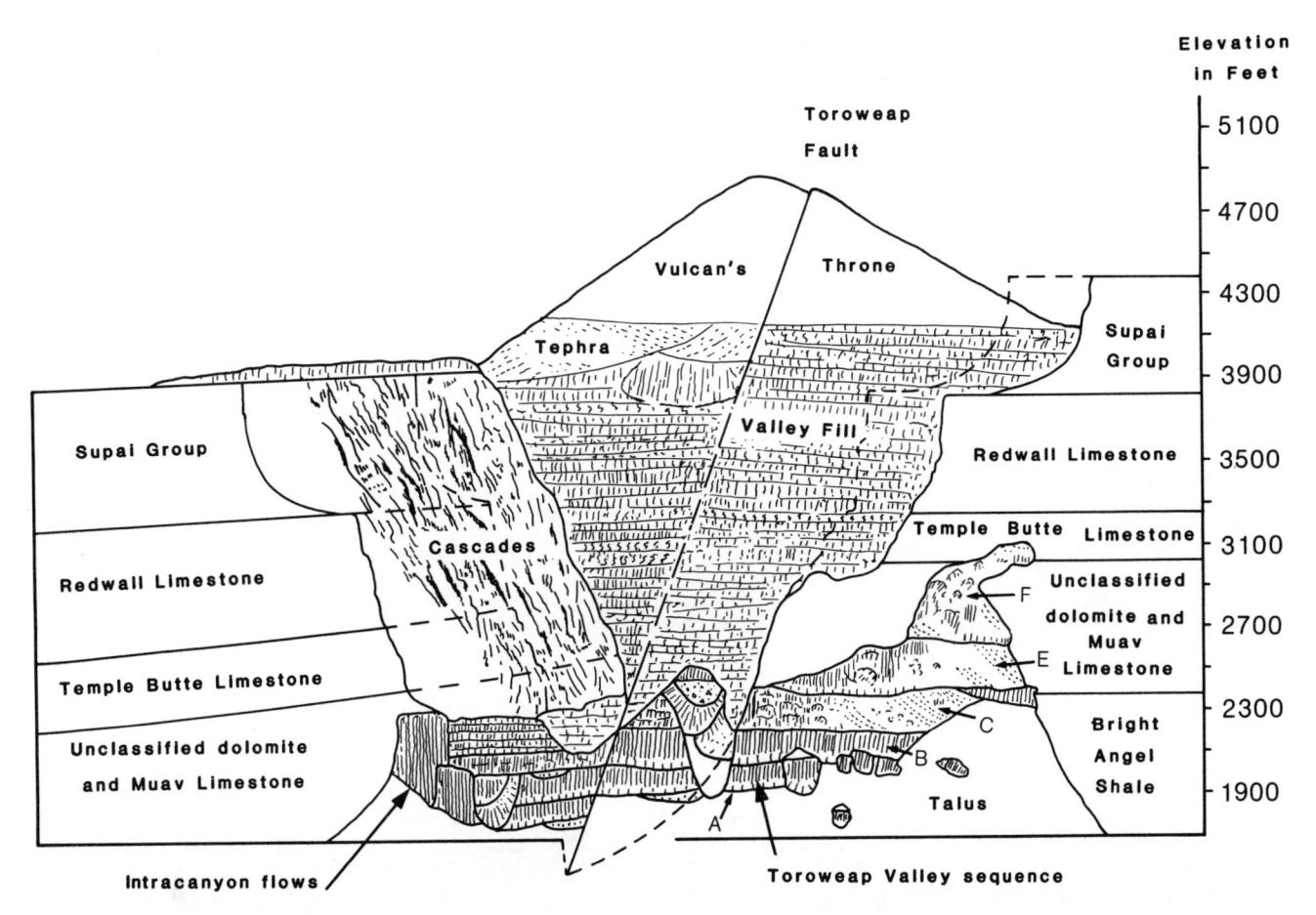

FIGURE 23.2a Cross-section across Toroweap Valley where it enters the Colorado River at mile 170 (273 km). The profile of the valley below Vulcan's Throne before it was filled with lava is shown by a dashed line. The face of the canyon wall is shown in solid lines. Note that most of Toroweap Valley is filled with thin flows which have been displaced by recurrent movement along the Toroweap fault. Remnants of the Toroweap dam (flows A-F) are juxtaposed against the lower units that fill the Toroweap Valley and against the Paleozoic strata east and west of the mouth of the ancient tributary canyon. The cascades on the western side of Vulcan's Throne mask part of the western side of the ancestral Toroweap Canyon. Vulcan's Throne rests upon an older cone west of the fault and the uppermost flows that fill Toroweap Valley east of the fault. Recurrent movement has displaced Vulcan's Throne approximately 15 feet (5 m)

Isolated High Level Remnants. Several small isolated remnants of basalt of uncertain relative age are found high on the canyon walls on both the north and south sides, about 1,400 to 1,600 feet (425 to 490 m) above present river level. The highest remnant, on the south wall at mile 177 (285 km), is protected in a recess of the canyon wall at the base of an amphitheater in the Redwall Limestone. Several other high level remnants are located on the south wall downstream from the Toroweap fault. Each high level remnant is composed of several thin flows separated by tephra.

Prospect Canyon

Prospect Canyon, on the south side of the Colorado River, in many ways is similar to Toroweap Valley. Both are fault-controlled canyons along the Toroweap fault and both have been filled with lava to the level of the Esplanade. In terms of structural control and pre-volcanic geomorphology, Prospect Canyon is nearly the mirror image of Toroweap Valley. The volcanic histories of the two tributary canyons, however, are completely different. Toroweap Valley has been filled with very young flows that have not been subjected to significant erosion. Only a small incipient valley has developed on the east margin of Vulcan's Throne, and there is little or no erosion along the zone of weakness produced by recurrent movement along the Toroweap fault. In contrast, Prospect Canyon, has been filled with lava and is partly exhumed (fig. 23.3a). Headward erosion has cut a new deep canyon more than a mile (1.6 km) long in the basalt flows, which once filled the inner gorge of Prospect Canyon. All of the original cascades in Prospect Canyon have been removed and only remnants of the flows and cinder cones that once filled the canyon remain. With much of the volcanic material removed from Prospect Canyon, the most significant volcanic features are dikes and volcanic necks associated with the older volcanic eruptions and remnants of cinder cones that once filled the ancient canyon.

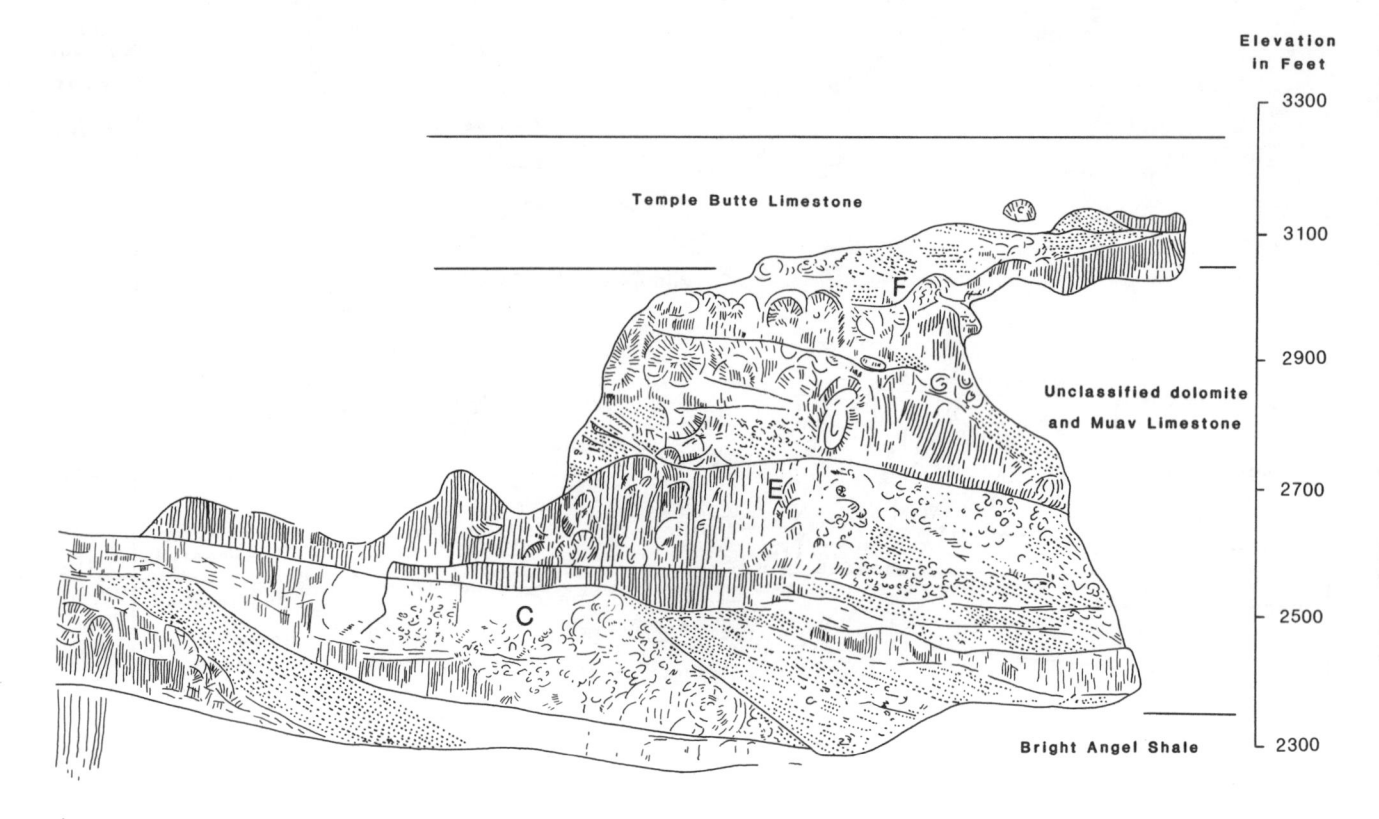

FIGURE 23.2b Sketch showing details of the internal structure of the Toroweap dam exposed in remnants high on the north wall of the inner gorge at mile 178.4 (287 km). The individual flows are 150-200 feet (45-60 m) thick and are characterized by classical colonnade and entablature jointing. Eastward the flows develop elliptical structure which in turn grade into pillow basalts and stratified tephra inclined upstream. These structures suggest that this area was near the head of the lava dam.

As seen in Figure 23.3b, Prospect Canyon is a rugged V-shaped gorge nearly 2,500 feet (760 m) deep. The present canyon has been cut by headward erosion along the Toroweap fault, exposing remnants of a variety of volcanic features on the canyon walls. A large alcove east of Prospect Canyon contains a thick sequence of basalt exposed in a vertical wall 2,000 feet (600+ m) high. The oldest basalts here are capped by a thick horizontal flow (Upper Prospect flow). Remnants of two younger units, the Ponderosa and the Toroweap sequence, are juxtaposed against the face of the Massive Prospect unit.

Upper Prospect Flow. The Upper Prospect flow forms a horizontal cap rock for all the volcanic features that filled ancient Prospect Canyon (fig. 23.3a,b). It appears to be a single flow, approximately 400 feet (120 m) thick, characterized by large thick columnar jointing; however, it locally may include several smaller flow units. The Upper Prospect flow is not confined to the deep, inner gorge of the old Prospect Canyon, but spreads out across the wide upper part of the Grand Canyon and over parts of the adjacent Esplanade platform. The Upper Prospect flow is offset approximately 160 feet (50 m) by recurrent movement along the Toroweap fault.

Dikes and Volcanic Necks

The lava cascades between Toroweap Valley and Whitmore Wash (downstream) are so striking when seen from the air and from the vicinity of Vulcan's Throne that there is a tendency to conclude that they were the source of all intracanyon flows. There are, however, numerous dikes, volcanic necks, and cones within the inner gorge, which indicate that much of the lava that formed the dams was extruded within the canyon itself. One of the largest and certainly most spectacular remnants of a volcanic vent within the canyon is a volcanic neck commonly referred to as Vulcan's Forge, located in the center of the Colorado River at mile 177.8 (286 km). In the canyon walls near Vulcan's Forge, several small dikes intrude the older intracanyon flows and remnants of laminated ash. The dikes are nearly vertical, strike in a general north-south direction, and range from 2 to 5 feet (1 to 2 m) in width. The most prominent intrusion is a large dike located high on the south rim of the canyon near the mouth of Prospect Valley. The dike is 30 to 40 feet (9 to 12 m) thick and projects as a high wall above the surrounding country rock.

The igneous intrusions within the canyon are of unquestionably late Cenozoic (apparently Pleistocene) age and are significant in that they must have been formed during a time that the canyon was filled with one

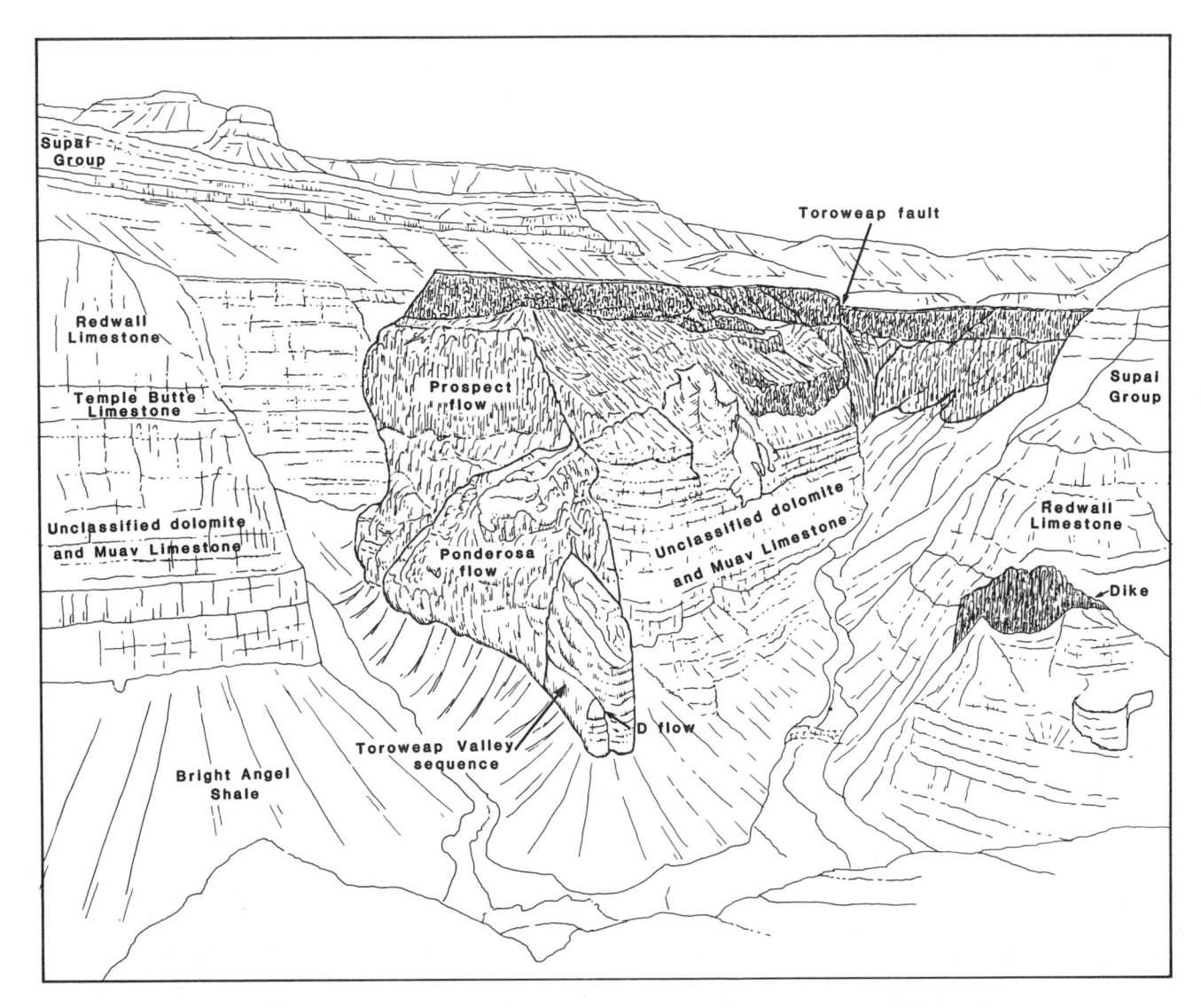

FIGURE 23.3a Prospect Canyon as seen from the west side of Vulcan's Throne. Over a mile (1.6 km) of headward erosion has occurred in Prospect Canyon along the Toroweap fault since the canyon was filled with lava. The thick Prospect flow forms a sheer vertical cliff almost 2,000 feet (610 m) high near the left margin of the sketch. The Upper Prospect flow consists of a horizontal unit nearly 400 feet (122 m) thick which once spread across the top of Prospect Canyon and across part of the Esplanade platform. It formed a cap rock covering the older flows and remnants of cinder cones that filled the ancient canyon. Note the offset of the Upper Prospect flow near the apex of the canyon produced by recurrent movement along the Toroweap fault. Recent movement has also displaced the alluvium covering the Upper Prospect flow.

of the various lava dams. The lava dams had to fill the canyon in order to provide a country rock into which the intrusions could be injected. The thin dikes exposed on the canyon walls in the vicinity of Vulcan's Forge must indicate the minimum elevation of the lava dam into which they were intruded.

The Esplanade Platform

One of the largest exposures of late Cenozoic volcanic material in the Grand Canyon is located on the north wall of the inner gorge between mile 180.6 and 182.9 (290.6 to 294.3 km). In this area, remnants of at least 8 lava dams are preserved in addition to cinder cones, dikes, and stratified lake deposits. All are capped by younger lava cascades that spilled over the rim of the Esplanade. A conceptual diagram showing the relationships between these units is shown in Figure 23.4.

The Esplanade Cascades. The Esplanade platform is essentially covered by a series of basalt flows between Vulcan's Throne and Whitmore Wash (mi 179-188; 286-302 km). In this area all of the tributary valleys dissecting the Esplanade are filled with basalt, in some cases with more than 30 individual flows. The present cascades over the Esplanade represent only the most recent of a series of flows that spilled over the Esplanade into the inner gorge.

Cinder Cones and Dikes. At Toroweap and

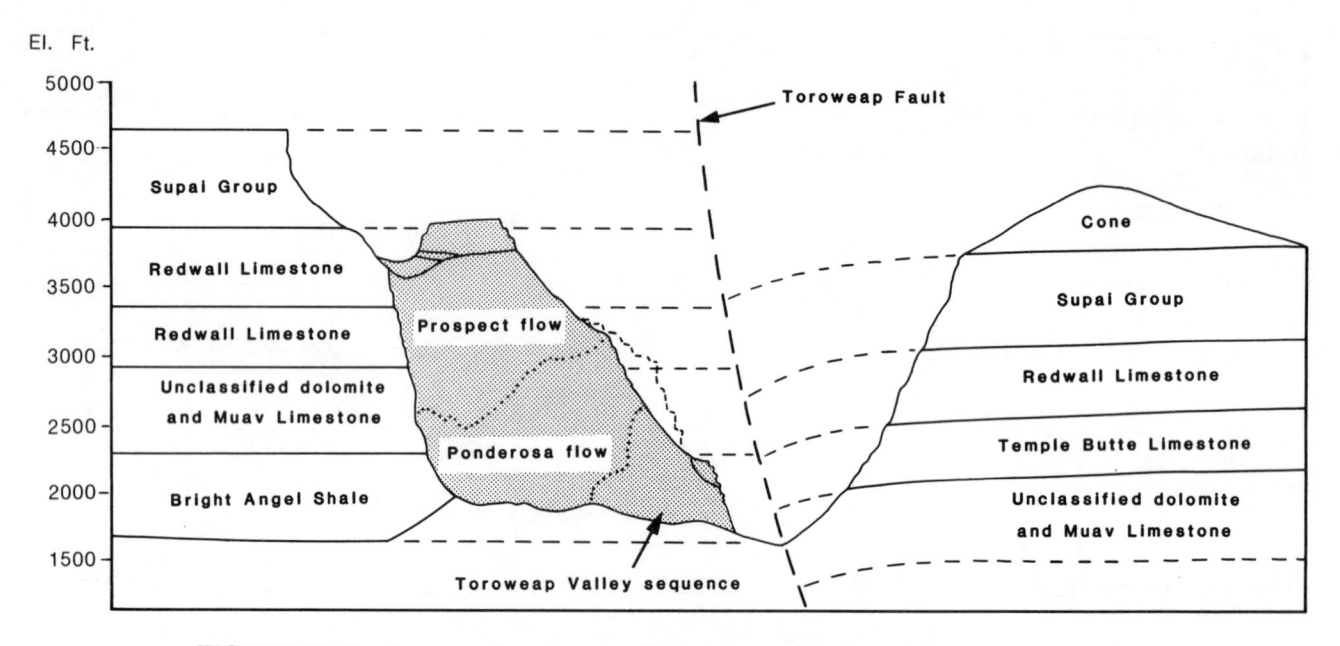

FIGURE 23.3b Cross-section showing flow units at the mouth of Prospect Canyon.

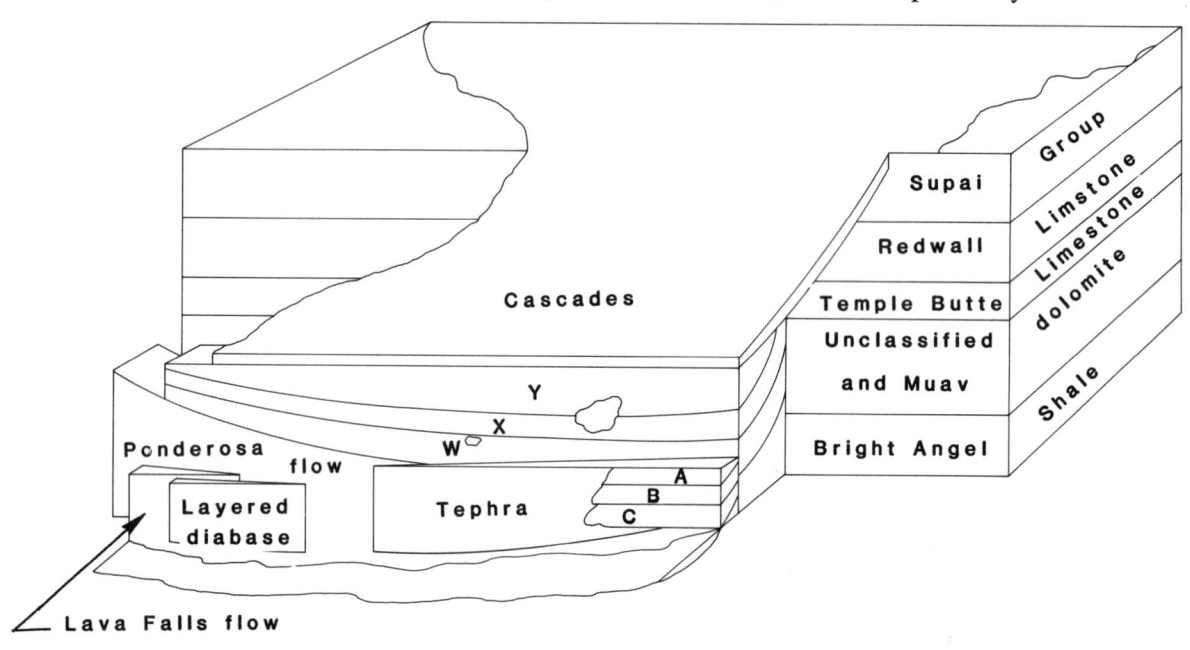

FIGURE 23.4 Generalized diagram showing the relationships of the major flow units of the Esplanade platform.

Prospect Canyons, prominent volcanoes are perched so close to the rim of the inner gorge that parts of the cones have sloughed off into the canyon below. Probably everyone who has seen the area has asked, "why are the cones developed at the very rim of the canyon?" and "why were there not more extrusions within the canyon itself?" Our study (Hamblin, in press) reveals that there were a number of volcanic eruptions deep within the canyon and even on the steep canyon walls of the inner gorge. These, of course, were rapidly modified by erosion and mass movement, even as they formed, so that now only relatively small remnants of stratified ash still cling to the steep canyon walls of the inner gorge marking sites of the former eruptive centers in the canyon. The largest remnant of cinder cones in the inner gorge is located at mile 181.2 (291.6 km).

Remnants on the South Side of the River. The large and complex sequence of basalt remnants preserved on the north wall of the canyon is in stark contrast to the south wall, which appears to be scrubbed clean of most basalt that once filled this part of the inner gorge. In all probability, the flows preserved on the north side originated on the north side, and as they entered the inner gorge displaced the Colorado River, causing it to impinge upon the south wall where maximum erosion subsequently occurred. Repeated cascades from the north continued to protect the north wall from erosion and progressively shifted the Colorado River toward the south. However, two relatively small remnants of basalt are still preserved at the mouth of small tributary canyons on the south wall, providing important evidence for the level of lava dams in this area. The largest is at mile 183 (294.5 km) where six units are exposed.

The Buried Canyon At Mile 183

One of the most remarkable exposures of basalt in the Grand Canyon is on the north side of the river at mile 183 (294 km) half way between Toroweap Valley and Whitmore Wash. Here a segment of the Grand Canyon was filled with a sequence of older basaltic flows, preserved now as a buried canyon (Figure 23.5). The basalts are not remnants clinging to the wall of the canyon as is true at most places. Rather, the basalts fill the entire lower part of the inner gorge, just as they did when they originally clogged the canyon and formed a dam. The floor and walls of the original canyon are exposed and preserve a segment of one of the lava dams in perfect cross-section.

The buried canyon is filled with 9 major flow units that total 650 feet (200 m) in thickness; most of the flows can be seen in Figure 23.5. The top of the highest flow is at an elevation of 2,480 feet (756 m), or 860 feet (260 m) above present river level. The flows are overlain by a deposit of river gravel 160 feet (50 m) thick, which in turn is capped by a young volcanic cone. Two significant erosional surfaces found in the sequence of flows indicate that the ancient canyon was not filled during a single volcanic eruption, but rather by several episodes of extrusion and the formation of a series of small lava dams, followed by erosion and then deposition of river gravels.

Younger Flow Units

After this segment of the Grand Canyon was filled with the sequence of basalts just described, the river channel migrated to the south, forming a meander at the margin of the dam. The river then began to excavate a new channel south of the original canyon. This new channel was cut through Paleozoic strata down to the present level of the river before being filled with still younger intracanyon flows (fig. 23.5).

Isolated High Level Remnants, Mile 184-185

Numerous isolated remnants of basalts are preserved from mile 183.5 to 184.5 (295.3 - 296.9 km), and when seen from river level they present a seemingly bewildering complex sequence of events. Aerial photographs show that some of the remnants are small cascades that spilled over the Esplanade and flowed down the small steep tributaries toward the Colorado River. Most, however, are clearly remnants of high level lava dams.

Whitmore Wash

The Whitmore Wash area (mi 188; 302 km) is important in the study of lava dams in the canyon because it provides evidence of a series of younger lava dams, most of which are poorly preserved upstream. As is the case in Toroweap Valley and Prospect Canyon, Whitmore Wash is structurally controlled by a major fault (Hurricane fault), and has been filled with a series of basalt flows that filled the canyon.

The flows that fill Whitmore Wash consist of numerous thin flows, most of which range from 6 to 20

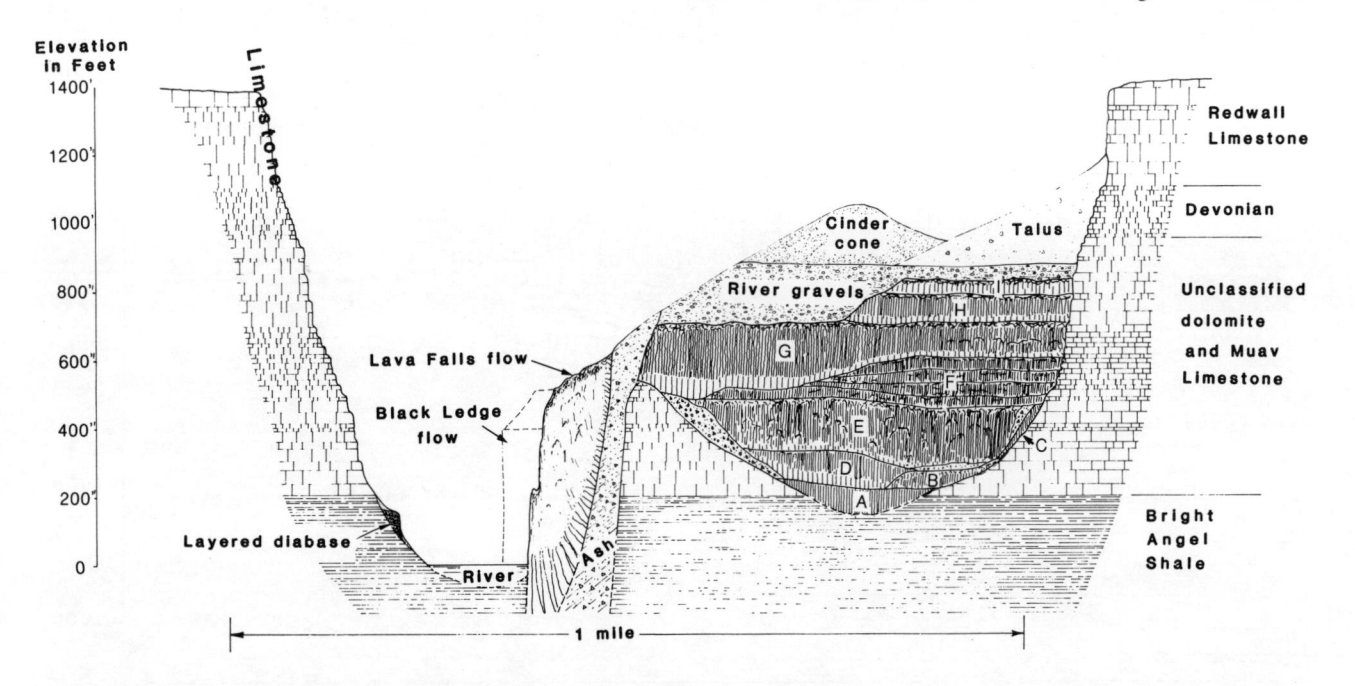

FIGURE 23.5 Cross-section showing stratigraphic and geomorphic relations of the older and younger lava dams at mile 181 (291 km). Lava flows that formed the Buried Canyon dam are labeled A-I. Remnants of younger dams are preserved on the walls of the present inner gorge.

feet (2 to 6 m) in thickness (fig. 23.6). Each flow is separated by a thin zone of ash and cinders. No significant deposits of alluvium occur between the flows indicating that they were extruded in rapid succession. The youngest flows of the Whitmore series had their source in cascades that spilled into Whitmore Wash from the Esplanade platform about 4 miles (6 km) up the canyon.

A large remnant of a thick, intracanyon flow, believed to be a remnant of the Ponderosa(?) dam, is juxtaposed against the sequence of thin basalts that fill the eastern valley (fig. 23.6), implying that the flows in Whitmore Wash are older than the youngest of the flows in Prospect Canyon(?). A small remnant of the Black Ledge(?) (fig. 23.6) flow occurs near the western part of the exposure.

Remnants Of Younger Dams Downstream From Whitmore Wash

Remnants of a sequence of younger basalt flows are preserved from mile 179 (288 km), below Toroweap Valley, downstream to mile 263 (423 km). They are preserved in the areas of least vigorous erosion. Some remnants are more than a mile (1.6 km) in length, and their geomorphic and stratigraphic relations are for the most part quite clear.

Lava Falls Flow. Remnants of the Lava Falls flow are found only in the area between Toroweap Valley and Whitmore Wash. It consists of a single flow unit, the top of which is 600 feet (180 m) above the Colorado River. The best exposures are on the north bank above Lava Falls Rapids, at mile 179.5 (288.8 km), at mile 184 (296.1 km), and on the south bank at mile 185.5 (298.5 km).

Black Ledge Flow. The Black ledge flow consists of a single flow unit, the top of which is 370 feet (113 m) above river level in the vicinity of Whitmore Wash. From there the thickness of the flow gradually decreases downstream. The Black ledge flow is a dense black aphanitic rock that contains a few scattered olivine phenocrysts less than 1.8" (4.6 cm) in diameter. It is the most extensive flow in the Grand Canyon, and can be traced from the foot of Toroweap Valley at mile 179 (km 288) downstream to mile 254 (409 km), a distance of 75 miles (121 km).

Layered Diabase Flow. The Layered diabase flow consists of as many as 15 individual flows that are characterized by crude, blocky columnar jointing,

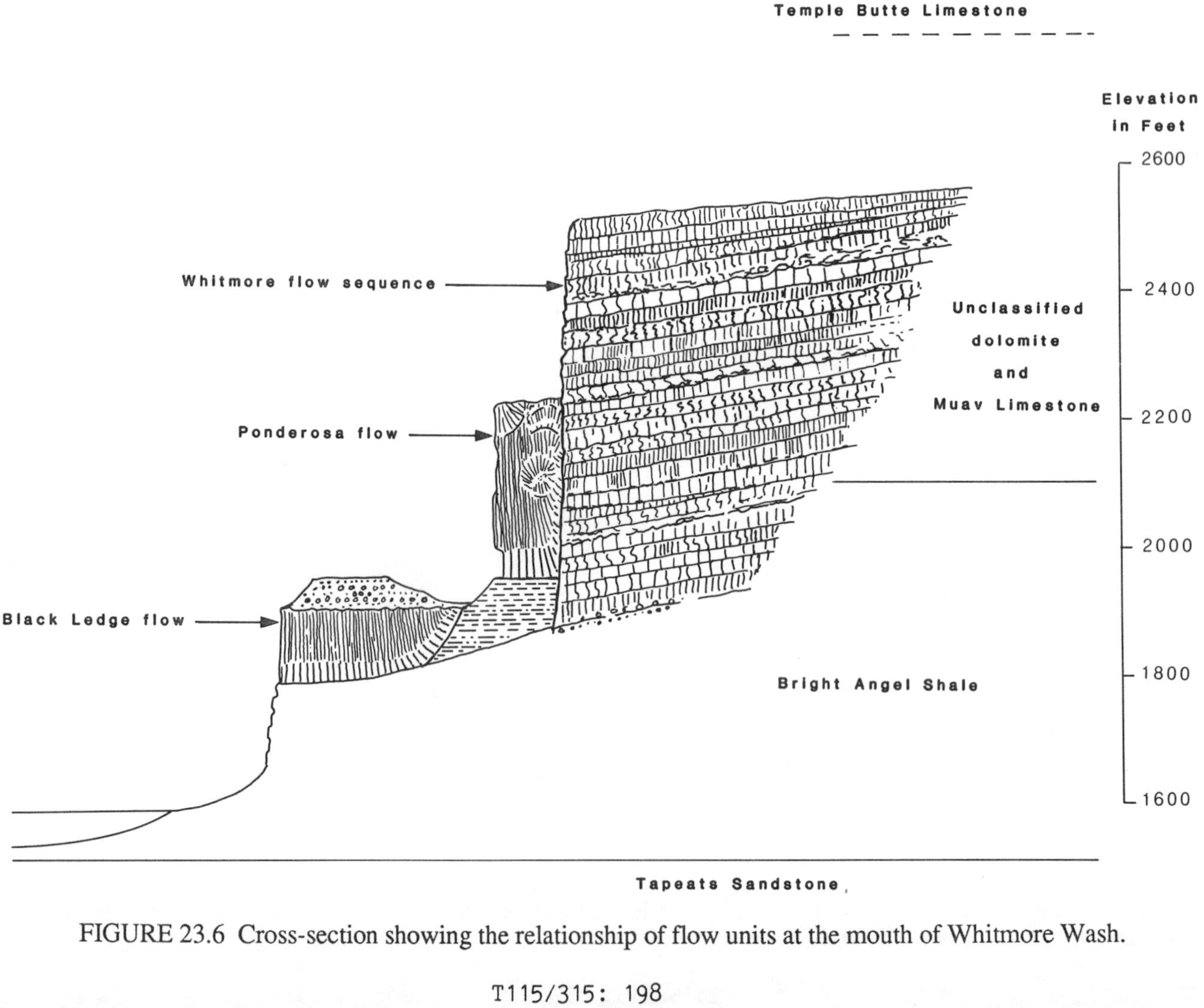

FIGURE 23.6 Cross-section showing the relationship of flow units at the mouth of Whitmore Wash.

without evidence of the common three-part columnar structure. The rock is coarse-grained, highly vesicular, and exhibits diabasic texture. The top of the flow is 298 feet (90 m) above the Colorado River at mile 182.2 (293 km), but decreases in elevation downstream.

Gray Ledge Flow. Remnants of the Gray ledge flow are widespread between mile 179 (288 km) and 195 (314 km), and in most exposures are juxtaposed against the Black ledge flow. The Gray ledge flow is characterized by a unique internal structure of columnar joints that consist of abnormally thick basal columns having straight and well-defined smooth boundary surfaces. Many individual columns are more than 8 feet (2.5 m) in diameter and 20 feet (6 m) high. This coarse jointing stands out in contrast to the thin, sinuous columns typical of most other flow units. The entablature jointing, in contrast, is sinuous and discontinuous and erodes to form a very ragged surface. The upper surface is largely ash and cinders. Most remnants of the Gray ledge flow are capped by thick, stratified river gravels that locally are 150 feet (45 m) thick. Most of the gravel consists of well-rounded, moderately well-sorted basalt boulders.

AGE OF FLOWS

Relative Ages

The relative age of most flows within the Grand Canyon is clearly expressed by juxtaposition. The age of the small isolated remnants, however, remains somewhat uncertain but can be correlated with larger remnants on the basis of rock type, elevation, and internal structures. The relative age of the major dams has thus been firmly established by geologic mapping.

Isotopic Ages

A number of isotopic dates have been obtained for some of the flows and provide a time framework into which most of the major volcanic events can be placed. The K-Ar age of the base of the Toroweap Valley sequence is 1.16±0.18 Ma (McKee and others, 1968). Other flows were sampled during 1972 and dated by Brent G. Dalrymple of the U.S. Geological Survey. Many had excess argon, but reliable dates for four flows in the Buried canyon sequence, the "D" flow, the layered diabase and the massive diabase, were obtained (table 23.1; Brent. G. Dalrymple, personal communication, 1988). Additional samples of the Whitmore Dam and the Black ledge flow were dated for this study by Paul Damon (personal communication 1988; table 23.1). Unfortunately the oldest dams have not been dated because samples could not be obtained from the inaccessible outcrops in the vertical canyon walls. The exact date of the onset of late Cenozoic volcanic activity therefore remains uncertain.

Polarity of flows

In addition to sampling for age dates in 1972, many of the flows younger than the Toroweap Valley sequence were sampled by C. Sherman Grommé (personal communication, 1988) of the U.S. Geological Survey and their magnetic polarity determined. All samples had normal polarity implying that, except for the older dams, most of the eruptions probably occurred since 720,000 years ago and are of Brunhes age.

DEVELOPMENT AND DESTRUCTION OF LAVA DAMS

Rates of Formation of Dams.

Although the formation of a lava dam in the Grand Canyon was a significant event, dramatically changing the canyon morphology, the time necessary for the construction of a lava dam was remarkably short. Observations of basaltic eruptions in historic times indicate that most eruptions occur in a matter of days or weeks. The major flows in the Grand Canyon, most of which were 100 to 200 feet (30 - 60 m) thick, probably moved tens of miles (km) down the Colorado River in a matter of days.

Rates of Reservoir Fill

The hydrologic data from the Bureau of Land Management 1963-1964 Lake Mead Survey (Lara and Sanders, 1970) provide the basic information from which we are able to calculate the rates at which the lakes formed behind the various lava dams were filled with water and subsequently filled with sediment. These data also provide some indication on the time necessary for a lava dam to be completely eroded away (table 23.1). We have calculated that a lake formed behind a lava dam 100 feet (30 m) high would fill and overflow in 2.3 days. Higher lava dams (300 to 400 ft; 91 to 122 m), would overflow in 17 days. Dams 1,000 feet (304 m) high would overflow in 250 days. A dam 2,000 feet (610 m) high would overflow in 23 years.

The volume of sediment carried by the Colorado River has been measured for many years by the Bureau of Reclamation. These data indicate that reservoirs behind the lava dams became "silted-up" in only a few hundred years at most; many of the smaller reservoirs became silted up in a few months (table 23.1). Thus, the sediment load of the Colorado River was soon transported over the dam providing the tools for abrasion of the stream and normal downcutting of the river channel soon after the dam was formed.

Erosion and Destruction of Dams

Normal downcutting of the dam by abrasion was undoubtedly a significant process of erosion and began as soon as water overflowed the dam. It reached maximum efficiency when the lake silted up and a normal sediment load was transported over the dam. Another important process by which the dams were eroded was undoubtedly the migration of rapids and waterfalls that initially formed on the downstream end of the flow

Rates of waterfall migration are well documented in some rivers and provide insight into possible rates at which the lava dams were destroyed. Niagara Falls, for example, has migrated headward at the rate of 3 feet (1 m) per year, and in the last 8,000 years, has migrated a distance of more than 11 miles (18 km). If this figure is typical for the migration of a waterfall on a large river, the lava dams in the Grand Canyon, which were generally less than 20 miles (32 km) long would require a maximum of 20,000 years to be completely destroyed by headward migration of waterfalls alone. From these data we can tentatively conclude that every lava dam was destroyed in less than 20,000 years after it formed.

From the foregoing observations, the 12 lava dams identified in the western Grand Canyon occupied the canyon for only a relatively brief period of time. From the beginning of volcanic activity to the present, lava dams existed probably no more than a total of 240,000 years of perhaps a 1.2 million year interval of time.

VOLCANIC HISTORY

Prospect and Ponderosa Dams

The oldest dams in the Grand Canyon were formed by thick flows that almost filled the inner gorge to the rim of the Esplanade platform (fig. 23.7b). These are preserved near the mouth of Prospect Canyon and are referred to as the Prospect Dam (fig. 23.3a). The top of the dam was at an elevation of 4,000 feet (1,220 m), or 2,330 feet (710 m) above present river level. From the elevation of the remnant of the Prospect Dam we are able

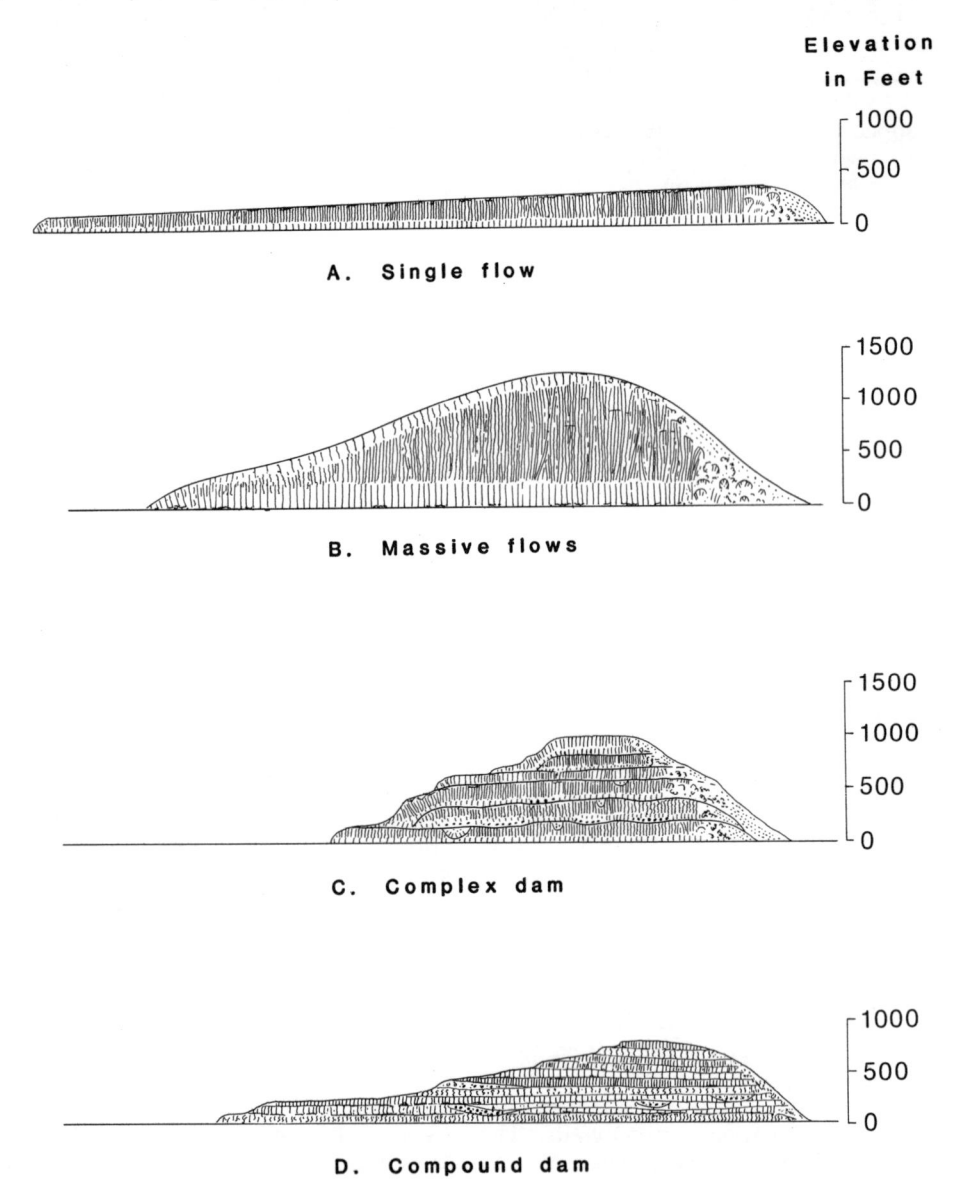

FIGURE 23.7 Types of lava dams in the Grand Canyon. A) A simple dam formed by a single flow 150 feet to 600 feet (45-180 m) thick; B) High dam formed by massive flows more than 800 feet (244 m) thick; C) Complex dam built from multiple flow units 50 to 200 feet (50-61 m) thick; (D) Compound dam composed of numerous flows 10 to 30 feet (3-9 m) thick.

to reconstruct the shoreline of the lake that formed behind it. Prospect Lake was the largest and deepest lake to form in the Grand Canyon. It must have extended all the way up through the Grand Canyon into Lake Powell near the region of Bullfrog, Utah, a distance of 324 miles (151 km) (fig. 23.8a). The rather extensive terrace deposits of gravel, sand, and silt in the Bullfrog area of Lake Powell occur at elevations of approximately 4,000 feet (1,120 m) and may be remnants of the delta of the Colorado River formed in the Prospect Lake.

The Ponderosa Dam was the second major dam to form in the Grand Canyon. It was similar to but somewhat smaller than the Prospect Dam, and was constructed by a single flow at least 1,000 feet (300 m) thick. The top of the barrier was at an elevation of 2,800 feet (850 m), about 1,130 feet (345 m) above river level. The best exposures are in the alcove east of Prospect Canyon, and at the west end of the Esplanade cascades on the north side of the river at mile 181 (291 km). The lake that formed behind Ponderosa Dam was 126 miles (203 km) long and extended through the Park Headquarters area of the Grand Canyon upstream approximately to Nankoweap Rapids.

Toroweap, Esplanade, and Buried Canyon Dams

After destruction of the massive Prospect and Ponderosa dams, a sequence of three smaller, complex dams were built from multiple flow units. They are referred to as the Toroweap, Esplanade, and Buried Canyon dams from large exposures preserved on the north side of the inner gorge at the mouth of Toroweap Canyon at mile 179 (288 km), below the Esplanade Cascades at mile 181 (291 km), and at the Buried Canyon at mile 183 (294 km).

The lakes formed by these dams were among the larger lakes to form in the Grand Canyon. They extended all the way through the present National Park Visitors area and upstream into the vicinity of Lees Ferry ranging from approximately 108 to 177 miles (174-285 km) long (fig. 23.8b). Throughout much of the Park region, the lakes were approximately 750 feet (225 m) deep. In the region of the Visitors Center the lakes essentially flooded all of the Granite Gorge with the shoreline being very close to the vertical wall of the Tapeats Sandstone.

The gravel, sand, and silt that forms the terrace deposits at Lees Ferry (approximate elevation 3,600 feet; 1,100 m) may have been deposited as a delta built by the Paria and Colorado Rivers where they emptied into the Toroweap Lake. Likewise, the silt deposits that form the main floor of Havasu Canyon (elevation 3,195; 974 m) represent a major remnant of Toroweap Lake deposits.

Whitmore Dam

One of the most obvious expressions of lava flows damming the Colorado River is in the vicinity of Whitmore Wash from mile 187 to 190 (300-305 km). Viewed from the air above mile 190 (305 km), large remnants of a sequence of relatively thin lava flows are seen filling Whitmore Wash and adjacent canyons. Downstream this sequence is preserved as prominent terraces that outline large segments of the original dam on both sides of the canyon.

The Whitmore Dam was distinctive in that it was composed of numerous thin flow units, most of which range from 10 to 20 feet (3-6 m) in thickness. More than 40 individual flow units fill the valley of Whitmore Wash. The lake behind the Whitmore Dam was close to the same elevation as the Esplanade Dam so that its configuration and shoreline would have been similar.

Younger Dams

Five younger dams were formed in the Grand Canyon ranging in height from 200 to 600 feet (60-180 m). Most of these dams were constructed from a single flow that extended downstream several tens of miles. The lakes that formed behind the younger dams were barely more than 660 feet (200 m) wide and 40 to 77 miles (64-124 km) long, hardly occupying more than the present river channel.

RATES OF EROSION

The series of lava flows within the Grand Canyon provide an unusually well documented record of rates of erosion, slope retreat, and the adjustment of streams to equilibrium. This is possible because the relative ages of the 12 lava dams have been determined by juxtaposition, and radiometric dates provide a series of important time bench marks. It is clear that the Colorado River was able to erode through the series of lava dams at an astounding rate. The cumulative thickness of basalt in the 12 lava dams was 11,300 feet (3,444 m). Thus, the Colorado River has actually cut through a vertical thickness of more than two miles of rock in slightly more than a million years. In all probability the actual downcutting through the basalts took place in significantly less time than a million years because there were large time intervals between the destruction of dams when the Colorado River was not influenced whatsoever by the presence of lava. The best estimates based on the time necessary for the erosion and destruction of the 12 dams, having a cumulative thickness of 11,300 feet (3,444 m), is approximately 200,000 years. During the intervening time the Colorado River flowed at its normal gradient with neither large scale deposition nor erosion taking place.

One of the most significant conclusions of our studies (Hamblin, in press) with respect to processes of erosion and canyon cutting is that after each lava dam was formed, the Colorado River rapidly eroded through the dam down to its original level, but no farther. This process of re-excavating the canyon took place at least 12 times, during the last ~1.2 million years or so.

If these interpretations of rates at which the dams were destroyed are correct, then it is clear that the Colorado River had the capacity to erode with remarkable speed through practically anything placed in its way. These data strongly support the concept that the regional

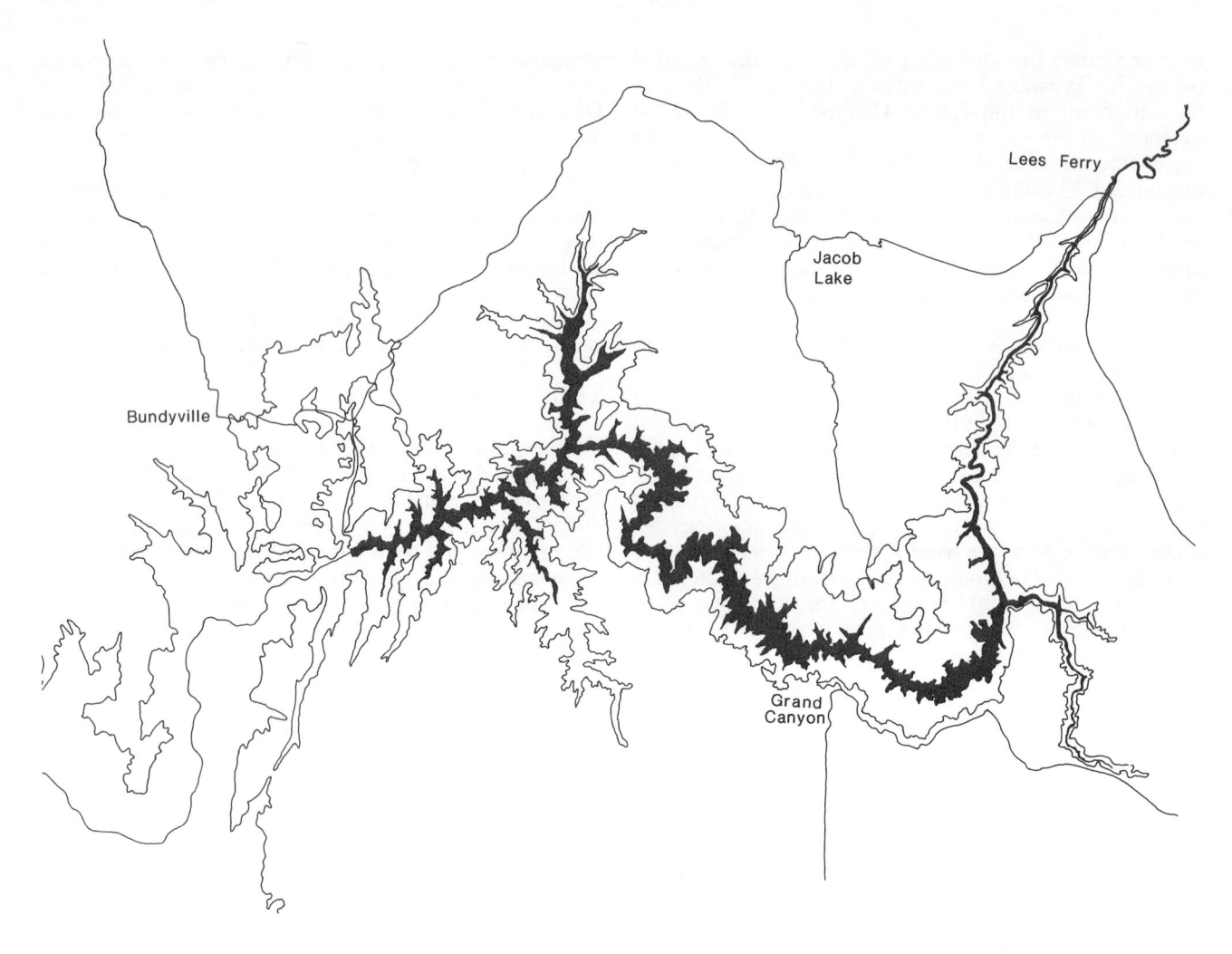

FIGURE 23.8a Map showing the extent of Prospect lake. The Prospect dam was the highest lava dam across the Colorado River. The lake formed behind the dam, extended all the way through the Grand Canyon and up into Utah. It formed more than 1.2 million years ago.

profiles of large river systems such as the Colorado River are in a state of equilibrium, and that when the equilibrium is disturbed by any means, such as the formation of a lava dam, uplift, or movement along faults, erosion to arrive at the profile of equilibrium takes place almost instantaneously in a geologic time frame.

Rates of Slope Retreat

In addition to providing an insight into the capacity of a river to downcut and to maintain a profile of equilibrium, the remnants of the lava dams provide important documentation concerning rates of slope retreat. Throughout the western Grand Canyon where remnants of lava dams still remain, one fact is completely clear; the remnants of these dams commonly are preserved only in the more protected parts of the canyon, i.e., on the insides of meander bends, beneath young cascades, in protected alcoves, and in many areas only as hanging valleys in the mouth of minor tributaries. Those remnants are preserved as relatively thin slivers adhering to the canyon wall.

Actual downcutting through the dam by abrasion along the river channel would produce a vertical gorge, only 200 to 250 feet (60-75 m) wide (the average width of the Colorado River channel). Slope retreat has been responsible for removing the rest of the basalt in the inner gorge. For example, at the present time the inner gorge at the level of the Temple Butte Limestone is roughly 2,000 feet (610 m) wide. Six dams were built up to this elevation or higher. This means that at that level approximately 900 feet (275 m) of slope retreat occurred on each side of the canyon in order to remove the basalt of each of the six dams. This is a total of 5,400 feet (1,645 m) of slope retreat.

This fact clearly indicates that with the destruction of each lava dam and the rapid re-establishment of the river gradient to its original profile there also was a rapid and contemporaneous retreat of the slopes back to their original profiles. The process occurred so quickly that the original profile of the canyon was re-established before the formation of the next lava dam. In each case after a lava dam was eroded, the basalt retreated to within a few feet of the original canyon wall. Then the process of slope retreat essentially stopped. In many places, the process of slope retreat completely removed the basaltic

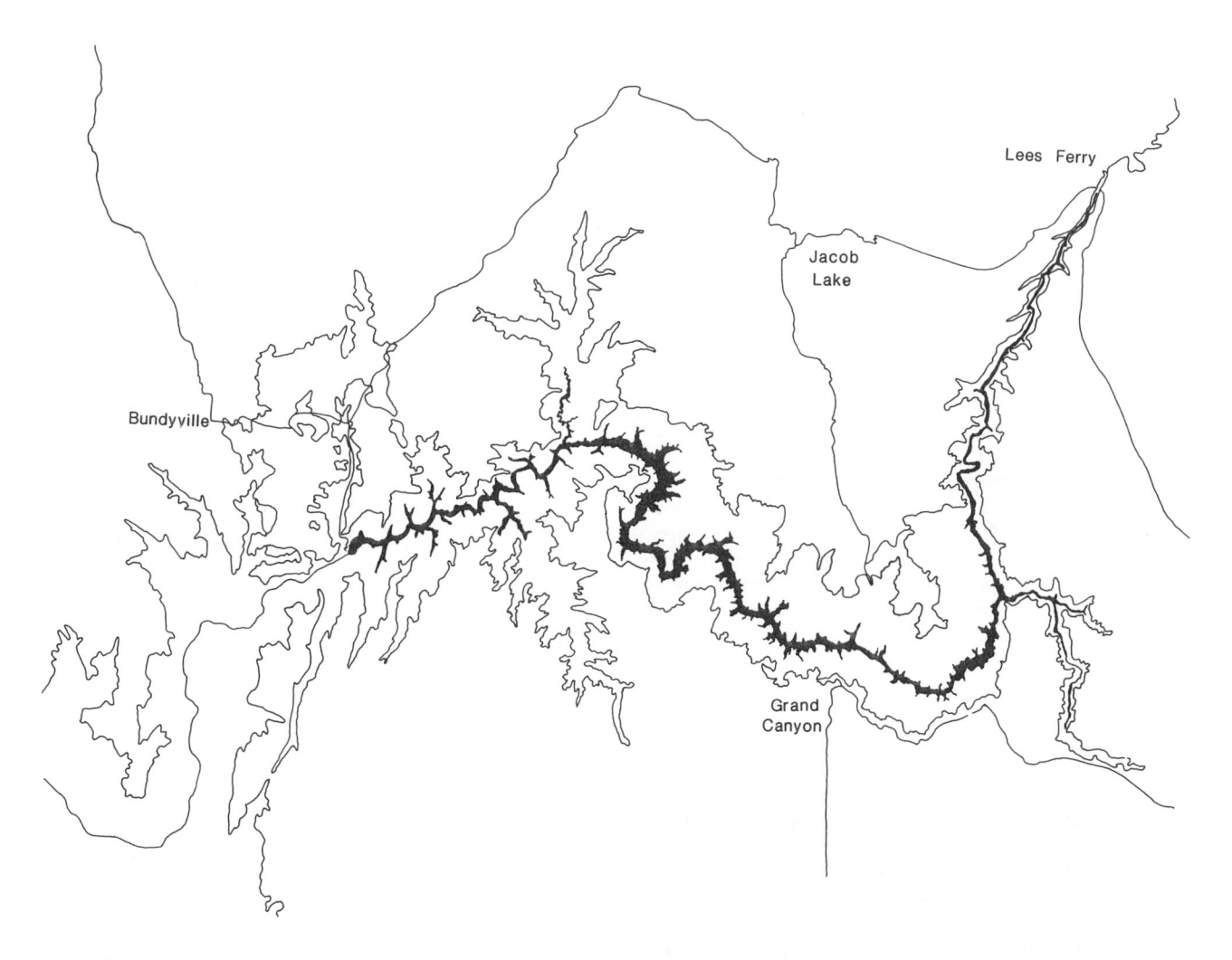

FIGURE 23.8b Map showing the extent of Toroweap lake. The Toroweap lake extended upstream through the Grand Canyon to the area of Lees Ferry. The shoreline throughout much of the canyon was near the cliff of the Tapeats Sandstone.

flows, but the process of slope retreat did not enlarge the canyon or proceed beyond the profile of the original canyon walls.

Several important conclusions can be derived from these facts: 1) The energy system that causes slope retreat in the inner gorge of the Grand Canyon has the capacity to cause the canyon walls to recede at the rate of an average of one mile per million years. 2) The slopes recede rapidly to a profile of quasi-equilibrium. They then recede at a very slow rate. (3) Slope retreat is delicately balanced with the stream gradient. Renewed downcutting of the stream channel is accompanied by renewed slope retreat. (4) The profile of the Grand Canyon is apparently in a state of quasi-equilibrium. (5) Periods of major rapid slope retreat are initiated by tectonic uplift.

The conclusion that can be drawn is that slope retreat is also extremely rapid until a profile of equilibrium is reached. Slope retreat then continues at an extremely slow rate.

LAKE DEPOSITS

The lakes that formed in the Grand Canyon behind the lava dams were unusual from the standpoint of their geomorphic setting, origin, and history. They were rare lakes in a deep canyon, similar in most every respect to the present man-made reservoirs such as Lake Mead and Lake Powell. Studies concerning sedimentation in these reservoirs provide an important insight into sedimentation in the lava dam lakes. The Colorado River carries a tremendous load of gravel, sand, and silt and was by far the major source of sediment input. In the Grand Canyon the only major tributary is the Little Colorado River. Havasu Creek and Kanab Creek are other large tributaries, but both are less than 60 miles (96 km) long and carry small sediment loads. Thus, the Colorado River flows through the Grand Canyon without appreciable tributary input.

When the river was blocked by a lava dam, coarse grained river sand and silt would be deposited as deltaic sediments at the point where the Colorado River entered the lake and would be deposited in large foreset layers beyond a basal gravel and prograded farther out into the lake. Finer grained sediments would be carried in suspension out into the deeper part of the reservoir. The deltaic sand and silt deposited at the head of the lake would ultimately prograde over the finer sediment.

The lakes formed in the Grand Canyon were surrounded by steep canyon walls. Beaches would rarely develop along the shore, and without significant input from the tributaries the major process operating along the shores of the lakes would be mass movement. Slope wash, rock falls, and general downslope migration of colluvium operating along the canyon walls would continue as subaqueous processes in the lake. A thin mantle of coarse subaqueous slope debris would be deposited close to the canyon walls contemporaneous with the deposition of lake silts. Each tributary would continue to transport and deposit detritus as small deltaic deposits at the heads of tributary bays in the lakes.

A general model of the lake sediments is as follows. Colluvium would accumulate along all slopes and form a subaqueous apron of talus up to several tens of feet thick. Silt and fine sand would fill the interstices of the talus. Beyond the zone affected by slope processes, the major type of sediment deposited would be mud and silt. Tributary stream gravels and coarse-grained deltaic sediments would be a significant facies in all tributary channels as a result of flash floods; some of the larger tributaries near the upstream reaches of the lake might construct deltas of sand and gravel. Deltaic deposits of the Colorado River would prograde downstream over the silts if the lake is maintained for an appreciable interval of geologic time.

Upon destruction of the dam, most of the wet unconsolidated lake sediment would easily be flushed out of the steep canyon leaving only minor remnants of sediment close to the valley walls. This material would be dominantly slope wash and colluvium interlayered with silt. It could be locally cemented with travertine deposits and preserved as remnants clinging to the canyon walls.

The largest and most complete sequence of sediments that may have accumulated in lakes behind lava dams is found in Havasu Canyon. Silt and fine sand fill the canyon to an elevation of 3,195 feet (974 m) and form a flat surface extending from the vicinity of Havasu Springs to Supai Falls, a distance of approximately 2 miles (3 km). Other large deposits that possibly are lake sediments include those at Elve's Chasm, the terrace deposits at Unkar Rapids and Lees Ferry, and the material at Red Slide (mile 175; 281 km).

Data on the elevation of these and other smaller sedimentary deposits, though fragmentary, are tantalizing. When plotted on the longitudinal profile of the present river they reveal that the elevations of the height of the deposits show a strong clustering at levels corresponding to the height of the major lava dams. The clustering between 2,350 and 2,480 feet (716-755 m) corresponds well to the general height of the Lava Falls D, Whitmore, Buried Canyon, and Esplanade dams, ranging in elevation from 2,259 feet to 2,500 feet (688-762 m). Clustering between 3,220 and 3,480 feet (981-1,060 m) is at the general level of the Massive Prospect dam (elevation 3,640 ft; 1,110 m). Other levels (2,800 to 2,800 ft; 853-866 m) correspond with the Ponderosa (2,800 ft; 853 m), and the 2,960 to 3,000 feet (902-914 m) corresponds to the Toroweap level (3,091 ft; 942 m). Even the 2,050-foot (625-m) level corresponds with the Black ledge (2,033 ft; 620 m). This correlation, apparently more than fortuitous, lends support to the conclusion that many of the later Cenozoic sedimentary deposits in the Grand Canyon may indeed be related to the lava dam lakes.

CHAPTER 24: QUATERNARY TERRACES IN MARBLE CANYON AND EASTERN GRAND CANYON, ARIZONA

Michael N. Machette and John N. Rosholt,
U.S. Geological Survey, Federal Center, Denver, Colorado

STRATIGRAPHY AND CHRONOLOGY OF TERRACE DEPOSITS

The Colorado River system drains most of the Central Rocky Mountain and the Colorado Plateau provinces. Within this region, the Colorado and Green Rivers are the major arteries of the system. These two rivers join about 31 miles (50 km) southwest of Moab, in east-central Utah.

From this point southward to the Grand Wash Cliffs east of Lake Mead, the Colorado River is incised in a series of deep canyons cut through resistant limestone, sandstone, and basement rock, and, to a much lesser extent, flows through narrow basins excavated in nonresistant shale and unconsolidated sediment. Because of the Colorado's deep entrenchment in narrow canyons across the Colorado Plateau province, sediments that record the Quaternary history of the river are rarely preserved. However, fluvial deposits preserved in a few basins along the river's course provide brief glimpses of its history, and one area within Grand Canyon National Park provides an opportunity to study Colorado River sediments deposited during the past 0.75 Ma.

The Colorado River in the Grand Canyon--a general overview

The late Tertiary erosional history of the Colorado River has been a topic of much discussion and heated debate for the past century (see Hunt, 1969), but the advent of K-Ar dating and recent study of Miocene fluvial-transport directions in northern Arizona (Lucchitta, 1984) have provided a fairly coherent, although general picture of the pre-Quaternary evolution of the Colorado River system. The Colorado River flowed south and west across the southern Colorado Plateau province (southern Utah and northernmost Arizona) during the late Miocene, whereas tributaries heading to the south in northern Arizona flowed north and northwest across the present lower Grand Canyon. Basalt flows as young as 5.5 Ma fill some of these north-trending stream channels and thus establish a maximum early Pliocene age (<5 Ma) for incision of the lower Grand Canyon (downstream from the Kaibab Plateau; fig. 24.1). Conversely, in this same area basalt flowed over the rim of the Grand Canyon and to near present river level about 1 Ma. Thus, it appears that the lower part of the Grand Canyon was well entrenched by 1 Ma, or perhaps earlier as discussed in Chapter 18. Upstream, dating of Quaternary fluvial deposits (Machette and others, 1986) indicates that the river cut or recut about 450 feet (150 m) in the past million years in response to headward entrenchment of the river system. The age estimates and heights presented here suggest that deposition of major fluvial terraces along the Colorado River is controlled by dramatic changes in climate (i.e., glacial to interglacial), and in turn changes in vegetative cover and ultimately sediment yield for the river system. However, the dating control is not accurate enough to prove the climatic hypothesis.

Unfortunately, a similar understanding of Quaternary entrenchment elsewhere along the trunk streams of the river system is yet to unfold, due mainly to a lack of age control of local depositional sequences. The results of dating within the Grand Canyon, coupled with studies of soil development and terrace heights along the main arteries, could provide the basic framework for a better understanding of the Quaternary history of Colorado River system.

Number and ages of terraces preserved in the Grand Canyon

Within the Grand Canyon, the best-preserved sequence of Colorado River deposits are found between the tributaries of Basalt Canyon and Tanner Canyon (fig. 24.2), where the nonresistant Precambrian Dox Sandstone has been eroded to form a 2-2.5 mile- (3-4 km-) wide valley (between river mile 66 and 72; fig. 24.2). This valley is typical of the basins that form in easily eroded sediments along the Colorado River. In this basin there are at least seven major levels of piedmont-slope alluvium graded to and overlapping main-stem alluvium of the Colorado River (fig. 24.3). The contact between the piedmont-slope (reddish brown) and main-stem (light gray) facies is conspicuous, as is the marked difference in degree of rounding of gravel clasts in each facies. The piedmont-slope alluvium forms a thin cap on the main-stem alluvium, although some high-level alluvial deposits have been eroded to a lag of bouldery gravel on rounded hilltops (for example, level 7, fig. 24.3).

The main-stem alluvium that forms terraces along the Colorado River is sandy, pebble- to cobble-size gravel derived primarily from resistant rocks exposed in the Rocky Mountains. The alluvium typically fills channels 32 to 66 feet (10-20 m) deep in the Dox Sandstone, and the base of successively younger alluvial deposits is commonly at progressively lower heights above the river. The heights used for terrace levels throughout this discussion are from the constructional surface of the overlapping piedmont-slope alluvium which, in general, is only 6 to 13 feet (2-4 m) thick near the Colorado

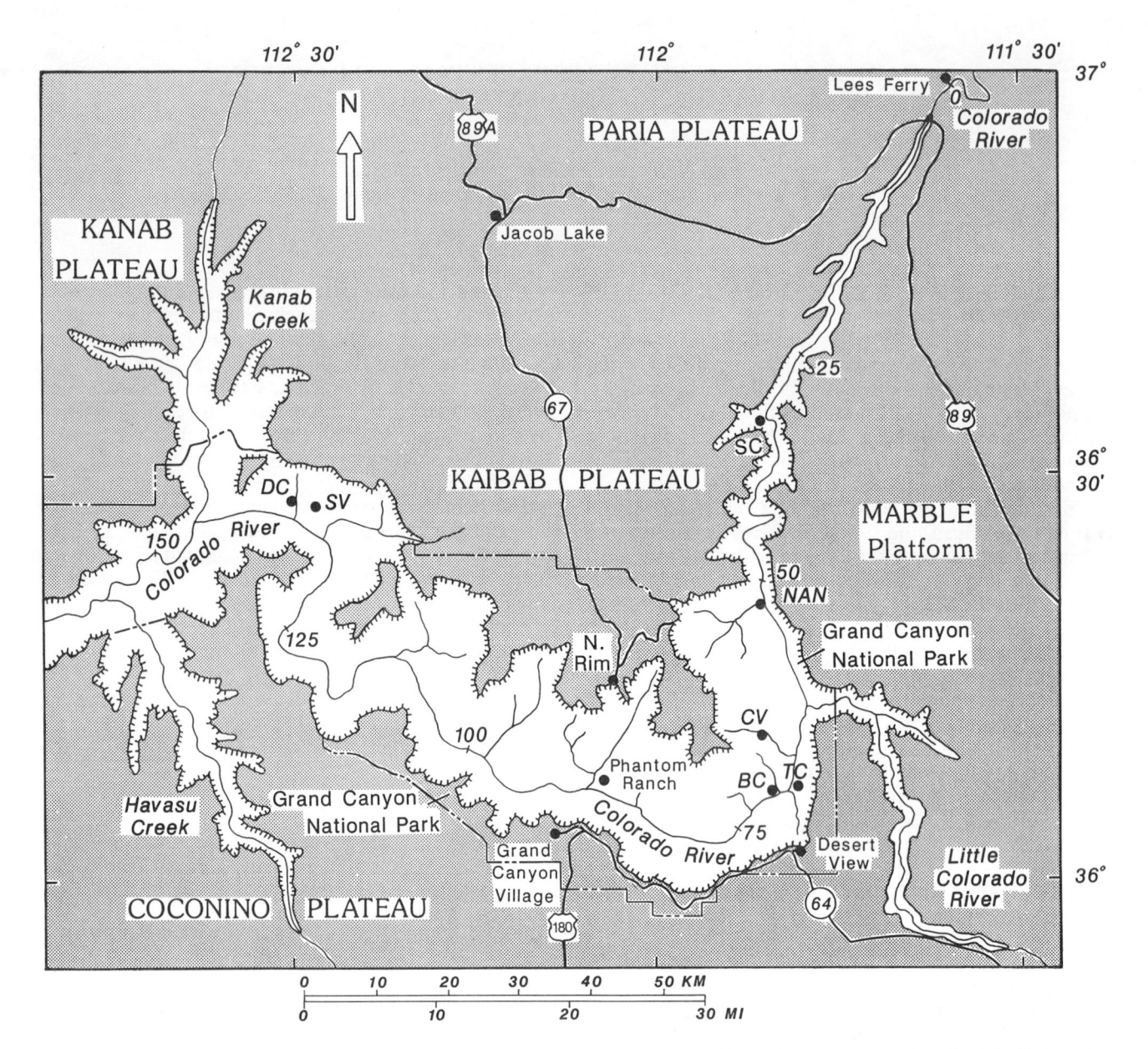

FIGURE 24.1. Index map showing physiographic, geologic, and cultural features associated with Colorado River in Grand Canyon National Park, northern Arizona. Letters indicate study areas mentioned in the text: SC--Stanton's Cave, NAN--Nankoweap Creek, TC--Tanner Creek, and BC--Basalt Creek. Other letters indicate areas of sampling by the authors, but not reported here. Numbers along river indicate mileage downstream from Lees Ferry, Arizona.

River; thus, the true elevation for each terrace level is accordingly less.

The youngest deposits along the Colorado River (levels 1 and 2) are restricted to the valley floor and adjacent floodplain (Table 24.1). Level 1 is formed by flood-plain deposits that lie as much as 6 to 10 feet (2-3 m) above the river (fig. 24.3) and have weakly developed soils (ochric A horizon). Remnants of uppermost Pleistocene alluvium (level 2) are preserved 16-23 feet (5-7 m) above the river behind the Holocene floodplain, but its piedmont-slope facies generally remains protected from the river in major side streams such as Comanche Creek. Level 2 deposits commonly have a zonal soil profile consisting of an ochric A, cambic B, and calcic C horizon (stage I to weak stage II morphology; see Machette, 1986 for discussion of calcic soils). Terrace levels 3 and 4 are well preserved east of the river between Basalt and Tanner Canyons, whereas terrace levels 5 and 6 are best preserved west of the river and south of Basalt Canyon. The soils on terrace levels 3-5 are characterized by thin (eroded) A horizons, thin light-reddish-brown argillic B horizons, and stage II to III calcic horizons that are best developed on the higher terrace levels. Terrace level 7, the highest in the area, has been found only on a ridge locally known as Hilltop Ruins (see Stevens, 1983, p. 73) above Unkar Rapids (mi 72.5). Here, the soil has been eroded and all that remains is a thick layer of oxidized parent material.

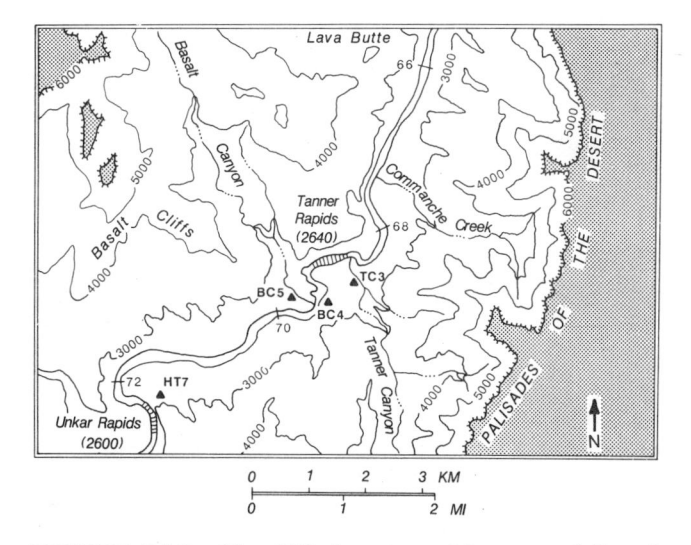

FIGURE 24.2. Simplified topographic map of Grand Canyon near Tanner Canyon and Basalt Canyon (Colorado River miles 66-72; km 106-116) showing sample localities mentioned in text (i.e., BC5 refers to terrace level 5 near Basalt Canyon). Topographic contours are in feet.

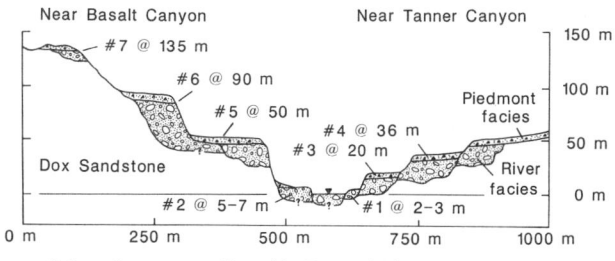

FIGURE 24.3. Diagrammatic cross section of Quaternary deposits (river and piedmont-slope facies) in Basalt Canyon-Tanner Canyon area (see fig. 24.2) showing ancient river levels (numbered young to old) and heights (in m) above present Colorado River (solid triangle shows present river level).

Terrace level	Height above Colorado River (m)	U-trend age (ka)	[1] Cumulative rate of downcutting (m/ka) Mean	Range
1	2-3	[2] 5±5	0.50	(0.2+)
2	5-7	[3] 15±5	0.40	(0.25-0.70)
3	25	40±24	0.60	(0.39-1.56)
4	36	75±15	0.48	(0.40-0.60)
5	50	150±30	0.33	(0.28-0.42)
Nankoweap Rockfall	<60	210±25	<0.29	(0.26-0.32)
7	135	>700	<0.19	

[1] Rates based on height and U-trend age.
[2] Estimate based on correlation with Holocene.
[3] Estimate based on correlation with Pinedale outwash.

TABLE 24.1. Summary of river levels and uranium-trend age estimates used to calculate cumulative rates of downcutting of Colorado River in the upper Grand Canyon. [n.d., not determined]

In this area, ages for the alluvium that forms terrace levels 3, 4, 5, and 7 were determined using the uranium-trend method (Rosholt, 1985; Szabo and Rosholt, in press). This method has been used elsewhere in the western U.S. to date fluvial, glacial, and eolian deposits that are less than 700 ka. The age determinations from Colorado River deposits are based on analysis of 6-8 samples collected from each soil profile. The soil on terrace level 3 (TC3, fig. 24.2) is weakly developed because it is in coarse, well-drained, main-stem alluvium (or perhaps it has been eroded); most of the other soils analyzed are from the fine grained piedmont-slope facies. The age determined from terrace level 3 (TC3) is 60±60 ka; the large error limit (100 percent) is caused by little isotopic variation with depth, which typically occurs in well-drained and (or) quartz-rich parent materials and at sites where the soil has been eroded. Samples of the intact soil from terrace level 3 near Nankoweap Creek (fig. 24.1) yielded an age of 40±24 ka (fig. 24.4A). Although both age estimates have large error limits, the latter date may indicate a major depositional event (during isotope stage 3?) that predates the maximum extent of Pinedale glaciation (isotope stage 2) in the Rocky Mountains (15-20 ka; Pierce, 1979).

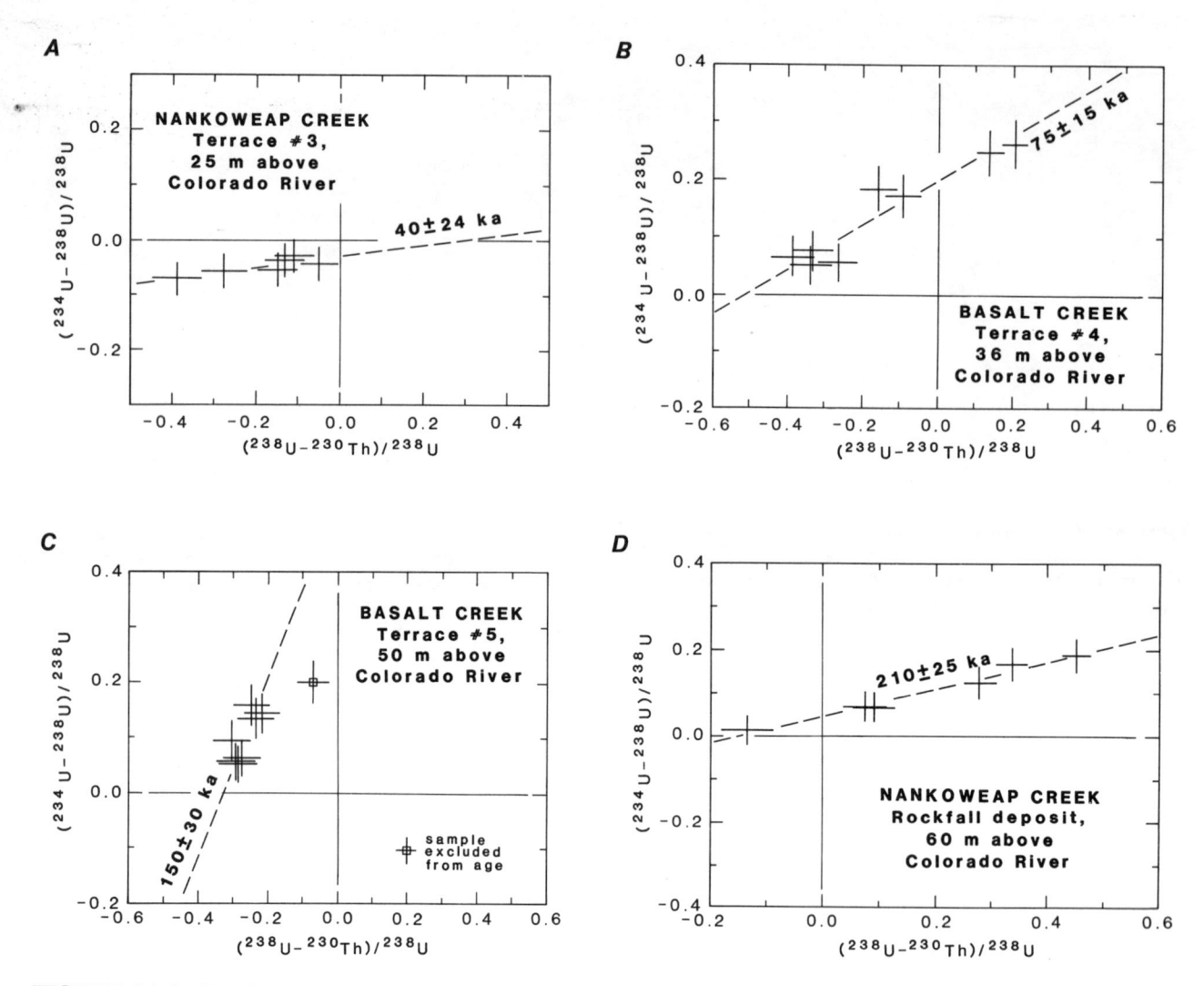

FIGURE 24.4. Uranium-trend age determinations for four deposits along Colorado River: A) terrace level 3, 40±24 ka; B) terrace level 4, 75±15 ka; C) terrace level 5, 150±30 ka; and D) rockfall debris (pre-level 5), 210±25 ka.

Soil from terrace level 4 (sample BC4, fig. 24.2), which is about 52 feet (16 m) above level 3 between Tanner and Basalt Canyons, yielded an age of 75±15 ka (fig. 24.4B). The small error limit and near doubling in elevation from level 3 to 4 suggest that level 4 is a major depositional event distinct from that of level 3. Level 4 may correlate with an interglacial event (isotope stage 4; 60–75 ka) or perhaps the preceding glacial event (isotope stage 5a; 80 ka). Sample BC5 from terrace level 5, which is 164 feet (50 m) above the Colorado River in this area, yielded an age estimate of 150±30 ka (fig. 24.4C). This age suggests deposition during the Bull Lake glaciation (isotope stage 6), which culminated about 150 ka in the Rocky Mountains (Pierce, 1979), or during the succeeding interglaciation (isotope stage 5e, 120–130 ka).

Deposits in the Nankoweap area at river mile 52.5 (km 84.5) (fig. 24.1), record a major rockfall that was probably first eroded by the river when it stood at level 5 (this reasoning is based on height above stream). The rockfall debris (Qrf, fig. 24.5) is preserved on both the north and south banks of Nankoweap Creek, which flows into the Colorado from the west. The rockfall fills a deep valley cut into old alluvial-fan and river gravel (labeled pre-Qrf on fig. 24.5) of unknown age. The level of the river channel at the time of the rockfalls emplacement is problematic; perhaps Elston's (chapter 21) suggestion of an earlier, deep incision stage is evidenced here. Although the surface of the rockfall has been scoured by the river, there are thin remnants of alluvial fans and terraces corresponding to levels 3 and 4 on the south side of Nankoweap Creek. Soil from the highest intact part of the rockfall, 197 feet (60 m) above the Colorado River (see Hereford, 1984b, fig. 9-2) yielded an age of 210±25 ka. This age is consistent with the degree of development of the soil on this erosional level of the rockfall (strong stage III carbonate morphology, B horizon probably eroded). Fan alluvium deposited on the rockfall and graded to terrace level 4 has a thin argillic B horizon and stage II calcic horizon. As previously mentioned, soil from the 82-foot- (25-m-) high terrace (level 3) at Nankoweap yielded an age of 40±24 ka (fig. 24.4A).

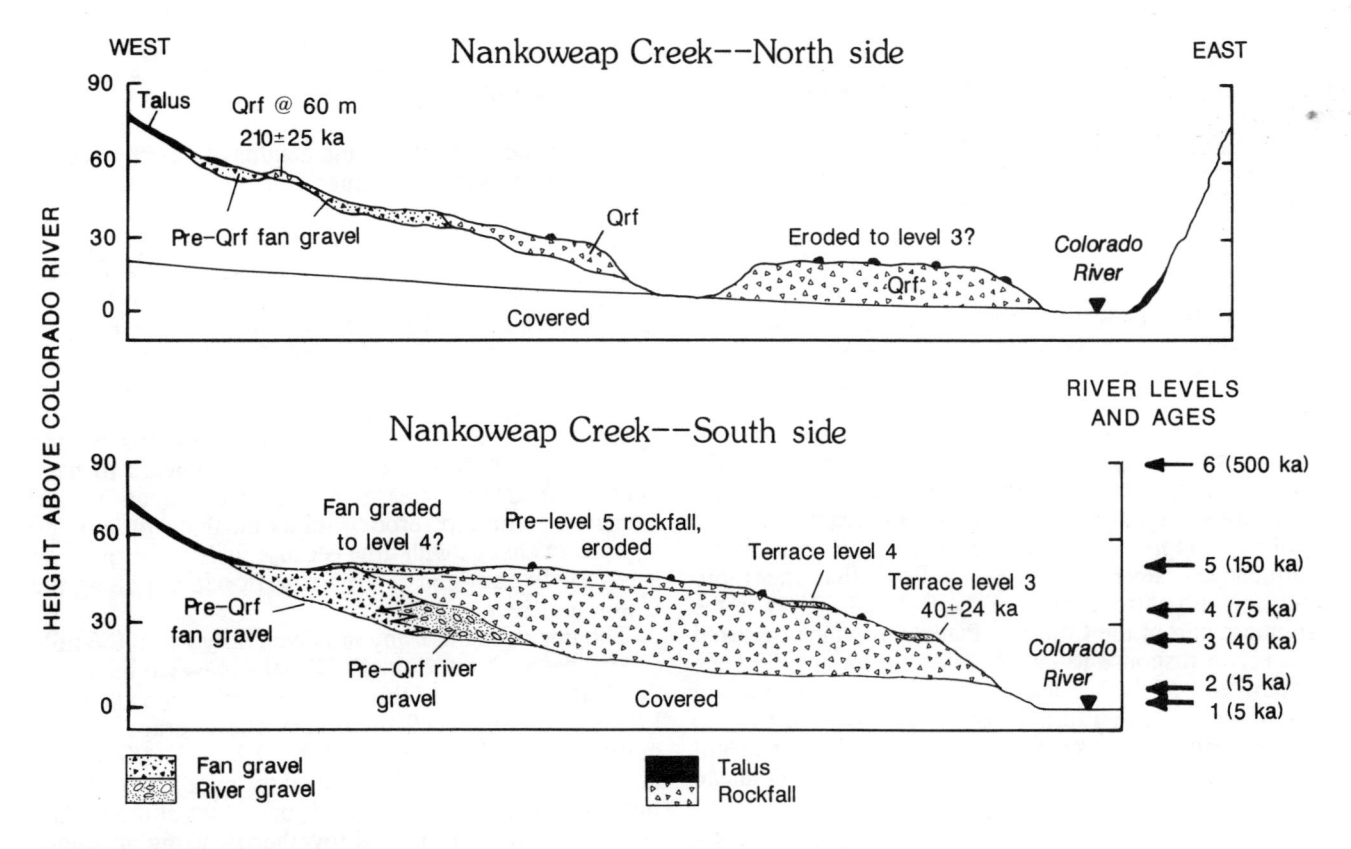

FIGURE 24.5. Diagrammatic cross sections of talus, rockfall, river, and alluvial-fan deposits north and south of Nankoweap Creek at its junction with Colorado River (river mile 52; km 83.6). Stratigraphic relations show rockfall probably predates river level 5 (150 ka). Solid triangles show present river level.

Hereford (1984b) suggested that the Nankoweap rockfall was responsible for damming the Colorado to a level high enough to inundate Stanton's Cave (fig. 24.1), 144 feet (44 m) above the river 20.5 miles (33 km) upstream. This hypothesis seems reasonable because the eroded top of the rockfall is generally 148-164 feet (45-50 m) above stream level (fig. 21.5), and it is as much as 197 feet (60 m) above the river on the north side of Nankoweap Creek; however, the rockfall may well predate the Quaternary terraces which have been etched across it (chapter 21). Age determinations from materials found in the cave provided a minimum time limit for its flooding. Driftwood in the cave has been radiocarbon dated at 43,700±700 yrs B.P. (Hereford, 1984b, footnote 2); a larger specimen of driftwood was reanalyzed (USGS Lab. No. W-5720) and yielded a radiocarbon age of >40,000 yrs B.P. (Meyer Rubin, written communication, 1986). Overlying the wood is a thick bed of anhydrite which has fallen from the ceiling of the cave; the anhydrite has downward-increasing closed-system uranium-series ages of 30 to 59 ka (Rosholt and others, 1987). Additionally, beneath the driftwood there is river silt which was later infiltrated by gypsum; this deposit yields a uranium-trend age of 55±25 ka for the start of gypsum infiltration and a minimum age for the river silt. Thus, the results from three different types of radiometric age determinations point to a major damming of the Colorado River at or before 60 ka. The one remaining consideration is the need to maintain a rockfall dam at Nankoweap from about 210 ka until 60 ka. Such a long life for the dam seems unlikely, considering the vigorous action of the Colorado River. This leads us to speculate that the age determinations from the wood and silt are absolute minima (i.e., 43.7 and 55 ka versus 210 ka) and that the age estimates from the anhydrite reflect the period when gypsum was seeping into the cave and dehydrating to form an anhydrite crust on the cave ceiling. Using these arguments, we place the time of cave flooding closer to 210 ka than to 60 ka. For comparison, the alluvial chronology at Basalt Creek (about 60 km downstream) suggests deposition 164 feet (50 m) above present level of the Colorado at 150 ka and deposition 118 feet (36 m) above present level about 75 ka.

The two highest river terrace levels in the Basalt Canyon area (fig. 24.2) are poorly documented. In their reconnaissance study, Machette and others (1986) probably lumped several discrete depositional units within level 6. Terrace level 6 remains undated, but strong development of its calcic soil and its 290-foot (90-m) height suggest an age of about 500 ka (fig. 24.6). Mapping of deposits at this level would probably reveal multiple terraces and a corresponding range of ages (perhaps 300-600 ka). The highest river terrace in the study area is designated level 7, and its deposits are preserved on a ridge informally named Hilltop Ruins about 443 feet (135 m) above the river at Unkar Rapids (HT7 on fig. 24.2). Oxidized sandy to clayey piedmont-slope alluvium that overlies river gravel yielded an uranium-trend age of more than 700 ka.

Rates of downcutting inferred from uranium-trend dating and degree of soil development

The age determinations mentioned here suggest a history of cyclic incision and aggradation as the Colorado River downcut throughout the Quaternary. Although the river has remained at nearly the same level for the past 1 Ma or more in the lower Grand Canyon, in the upper Grand Canyon it has deepened its channel about 493 feet (150 m) since early Quaternary time (about 750 ka). The river systems history in early Quaternary and Pliocene time is largely problematic at this point. Lucchitta's (1984) reconstruction from downstream suggests that most of the earlier downcutting in the upper Grand Canyon and all of the downcutting in the lower Grand Canyon was accomplished between 5 and 1 Ma, whereas Elston and Young (chapter 18) call upon a 3-stage, cut-fill-cut sequence. Nevertheless, we believe that the rapid incision and entrenchment of the Colorado River in the southern part of the Colorado Plateau province occurred largely in response to late(?) Miocene to late Pliocene epeirogeny of the western U.S. and block uplift along the Grand Wash fault. Continued downcutting to present time suggests that the river system has not reached equilibrium (grade) within the Colorado Plateau province.

Rates of downcutting for the past 750 ka -- upper Grand Canyon

The age determinations and heights of river terraces in the areas of Basalt and Tanner Canyons and Nankoweap Creek (Table 24.1) provide evidence for accelerated downcutting by the Colorado River in late Quaternary time. Figure 24.6 is a plot of terrace height as a function of age and indicates a gradual increase in downcutting rates from about 0.3 ft (0.1 m)/1,000 yrs (from 300 ka to 700 ka) to as much as 2.3 ft (0.7 m)/1,000 yrs (from present to 50 ka). The apparent seven-fold increase in rate (fig. 24.6) is probably influenced by short-term cycles of cutting-and-filling and a narrow observation window (50,000 yrs), but there still appears to have been a bonafide increase in rate of downcutting. In terms of volume of eroded material, there may be no significant change with time because incision below level 3 is largely confined to a narrow floodplain along the Colorado River.

The upper Grand Canyon is generally about 0.6 mile (1 km) deep. If the Colorado River excavated the upper 2,790 feet (850 m) of the Canyon in Pliocene and early Pleistocene time (5.0-0.75 Ma), the average rate of cutting would have been 0.6 ft (0.2 m)/1,000 yrs, or about two times the rate suspected for middle Quaternary time (750-130 ka). However, if one looks at the middle and late Quaternary record together by using an age of

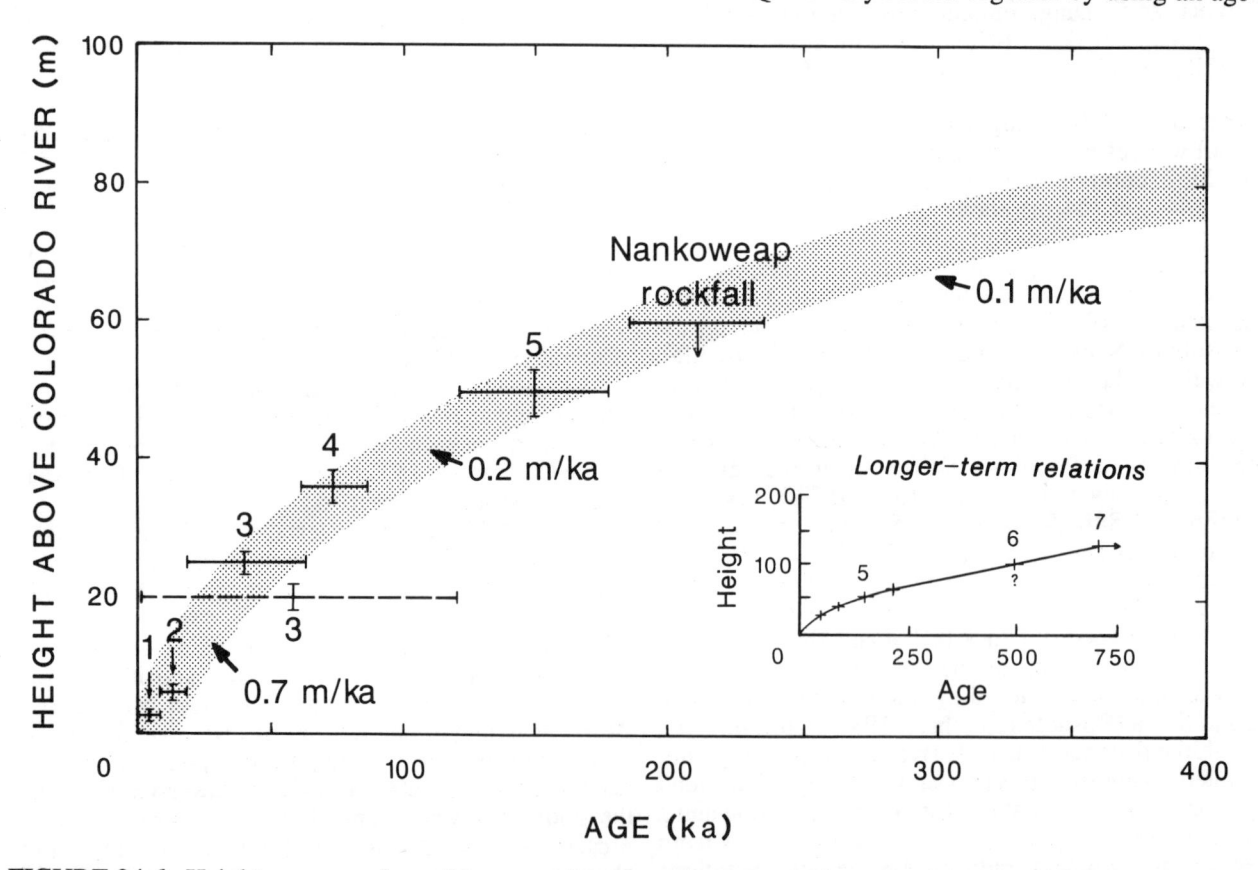

FIGURE 24.6. Heights, ages, and possible rates of downcutting inferred from fluvial and rockfall deposits along Colorado River near Nankoweap Creek and in Basalt Canyon-Tanner Canyon area. Smooth broad band includes most of the variation in heights and ages and shows general trend for past 200-400 ka. Inset graph shows middle Quaternary (750-130 ka) trend inferred from river levels 5-7.

700 ka for level 7 (443 feet (135 m) above the Colorado River), the average rate of downcutting has been ≤0.6 ft (0.19 m)/1,000 yrs. The rates of downcutting for these two time intervals (present to 0.75 Ma and 0.75-5.0 Ma) are remarkably similar, considering how little we really know about the Pliocene and early Quaternary history of the Colorado River.

Aerial view of Hurricane fault looking north from Three Springs Canyon (mi 216; km 348). Pennsylvanian Watahomigi Formation (A) on the west abuts Proterozoic Vishnu Schist (B) on the east. Down-to-the-west displacement is about 2,400 feet (730 m) at Three Springs Canyon.

CHAPTER 25: BRECCIA PIPES AND ASSOCIATED MINERALIZATION IN THE GRAND CANYON REGION, NORTHERN ARIZONA

Karen J. Wenrich
U.S. Geological Survey, Denver, Colorado

Peter W. Huntoon
Department of Geology, University of Wyoming, Laramie, Wyoming

INTRODUCTION

Thousands of solution-collapse breccia pipes crop out in the canyons and plateaus of northwestern Arizona; some are host to high-grade, uranium-rich mineral deposits (fig. 25.1). The mineralized pipes are enriched in a large suite of elements. Besides U, the metals Ag, Cu, Pb, V, and Zn have been mined from various Grand Canyon breccia pipes; the mineralized rock is also commonly enriched in As, Ba, Cd, Co, Mo, Ni, and Se (Wenrich, 1985).

The breccia pipes formed as sedimentary strata collapsed into dissolution caverns in the underlying Mississippian Redwall Limestone. Upward stoping through the upper Paleozoic and lower Mesozoic strata, involving units as high as the Triassic Chinle Formation, produced vertical, rubble-filled, pipe-like structures (fig. 25.2). A typical pipe is approximately 300 feet (90 m) in diameter and extends upward as much as 3,000 feet. (910 m). Only a few pipes have been observed to exist below the Redwall Limestone; these involve the upper 200 feet (60 m) of Cambrian and Devonian carbonates exposed in the cliffs above the Colorado River near the Grand Wash Cliffs. The stoping process created the extensive brecciation of the rock between the steep walls of the pipe (fig. 25.3). All breccia has moved downward in the pipes, settling into chaotic orientations. The brecciated core of each pipe abuts generally well-stratified, little deformed country rock (fig. 25.4). The contact between the breccia core and the country rock forms a permeable vertical zone, commonly referred to as the ring fracture. Secondary mineralizing fluids circulated along the ring fracture producing gamma radiation anomalies and depositing supergene copper and zinc minerals. Bounding the core of the pipe is "a zone tens of feet or less in width containing outwardly-dipping concentric ring fractures" (Verbeek and others, 1988). The brecciated core of most pipes is not exposed on the plateau surface, yet the underlying pipe is commonly manifested on the surface in the form of a large (sometimes as much as ten times the pipe diameter) collapse basin surrounded by concentric inward-dipping beds; this basin has been referred to as the "collapse cone" by Krewedl and Carisey (1986).

Collapse probably began shortly after deposition of the Redwall Limestone, about 300 m.y. ago, coincident with infilling of karst depressions by the Upper Mississippian and Lower Pennsylvanian(?) Surprise Canyon Formation. Dissolution of the Redwall Limestone and upward stoping of the overlying strata either continued throughout the late Paleozoic and early Mesozoic, or ceased after Mississippian time and was reactivated again during Late Triassic time. No pipes have been observed in strata younger than Triassic. Although such strata have been removed by erosion across most of northwestern Arizona, the pipes are probably no younger than Triassic because no breccia clasts in pipes have been identified that originated from post-Triassic strata.

The breccia pipes tend to occur in clusters. As many as six per square mile have been mapped in some areas of the Grand Canyon, while elsewhere tens of square miles have no surface expressions of any breccia pipes (Billingsley and others, 1986a, 1986b; Wenrich and others, 1986a, 1986b). Mineralized pipes are also clustered. This is especially obvious in the area of Hack Canyon, where the Hack 1, 2, 3, and old Hack Canyon mines (4 separate pipes) occur within one square mile (fig. 25.1).

Breccia pipes that crop out along the cliffs of the Grand Canyon are obvious to the trained observer, however those exposed on the plateau surfaces are very subtle. Pipes are mapped on the plateau surfaces through the recognition of circular features, particularly those with inward-dipping beds (fig. 25.5). Shallow circular depressions and vegetation anomalies are also suggestive of underlying breccia pipes.

Recognition of breccia pipes on the northern Arizona high plateaus is complicated by the occurrence of gypsum and limestone collapses contained solely within the Permian Toroweap Formation and Kaibab Limestone. Although the dissolution of gypsum in both of these formations produces surface collapse features which are similar in morphology to those associated with breccia pipes, they are shallow seated and unmineralized. However, gypsum dissolution can enhance the surface expression of some of the deep-seated breccia pipes. For example at the Pigeon mine (fig. 25.6), the actual pipe is less than 300 ft (90 m) in diameter, but the collapsed surface expression over the pipe on the Kaibab Plateau is about 0.5 mile (0.8 km) in diameter.

MINING HISTORY

Mining of breccia pipes in the Grand Canyon region began during the late 1870's, at which time essentially all

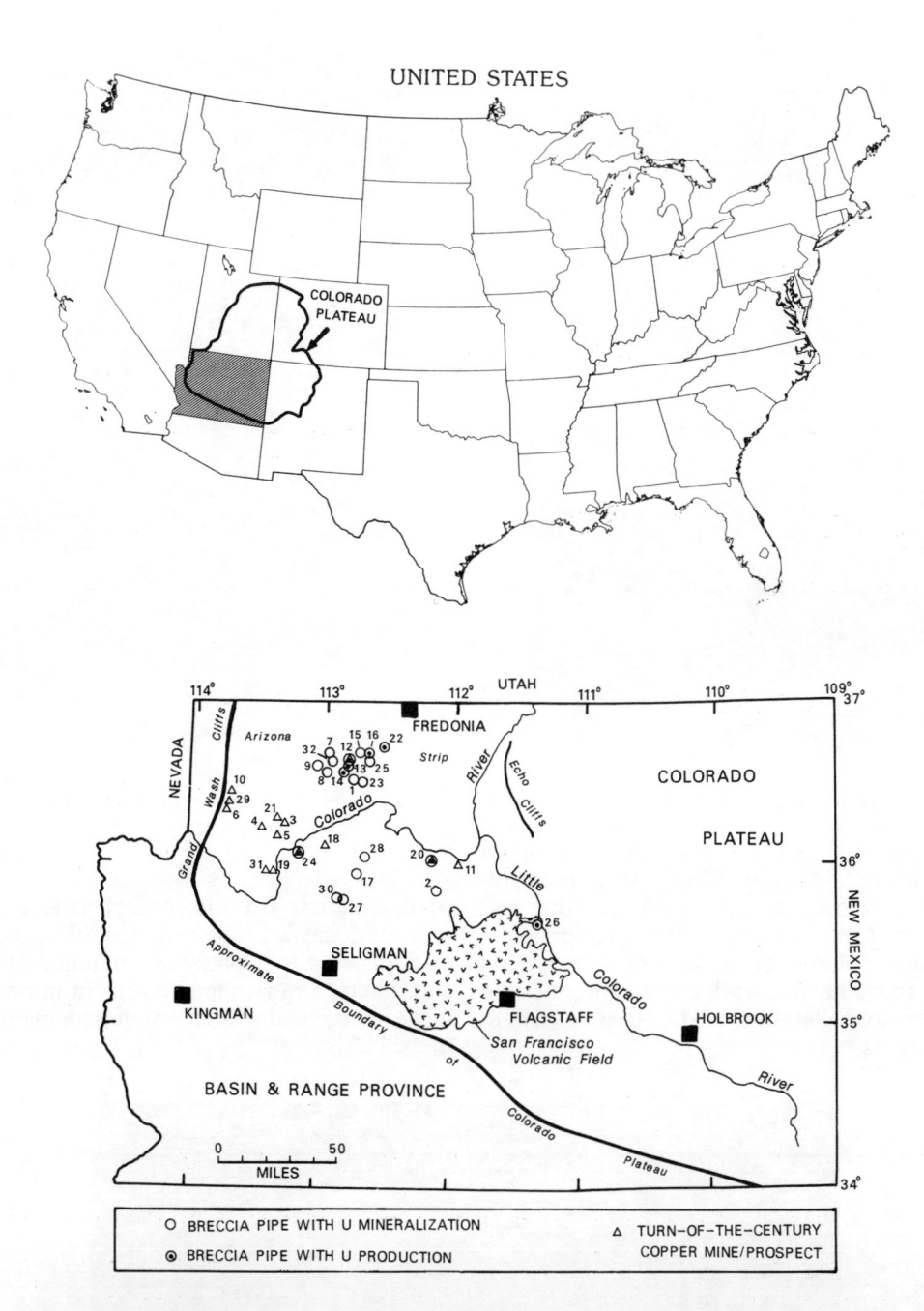

FIGURE 25.1A. Index map of the U.S.A. showing the location of Figure 25.1B (stipled pattern) and the approximate outline of the Colorado Plateau. 25.1B. Index map of northern Arizona, showing the locations of mineralized breccia pipes, and the San Francisco volcanic field that buries terrane with high potential for breccia pipes. Numbers refer to the following mines: (1) Arizona I, (2) Canyon, (3) Chapel, (4) Copper House, (5) Copper Mountain, (6) Cunningham, (7) DB-1*, (8) EZ-1, (9) EZ-2, (10) Grand Gulch, (11) Grandview, (12) Hack 1, (13) Hack 2, (14) Hack 3, (15) Hermit, (16) Kanab north, (17) Lynx*, (18) Mohawk Canyon, (19) Old Bonnie Tunnel, (2) Orphan, (21) Parashant, (22) Pigeon, (23) Pinenut, (24) Ridenour, (25) Rim*, (26) Riverview, (27) Rose, (28) Sage*, (29) Savannic, (30) SBF, (31) Snyder, (32) What*. (Pipes indicated by "*" were taken from Kwarteng and others, 1988, fig. 1).

mining was for copper. In 1951, uranium was discovered in the existing copper mine at the Orphan breccia pipe (fig. 25.7). During the period 1956-1969, the Orphan mine yielded 4.26 million pounds of U_3O_8 with an average grade of 0.42% U_3O_8 (Chenoweth, 1986). In addition to uranium, 6.68 million pounds of

FIGURE 25.2. Upward stoping from the Mississippian Redwall Limestone resulted in thousands of breccia pipes, such as this one in the western Grand Canyon (Wenrich and others, 1987, pipe #338). More than 300 feet (90 m) of downdropped Redwall Limestone and lower Supai Group breccia clasts in a finely-comminuted matrix form this column of rock, which has been downdropped into the Redwall Limestone. Some limonite and hematite alteration can be observed within the breccia.

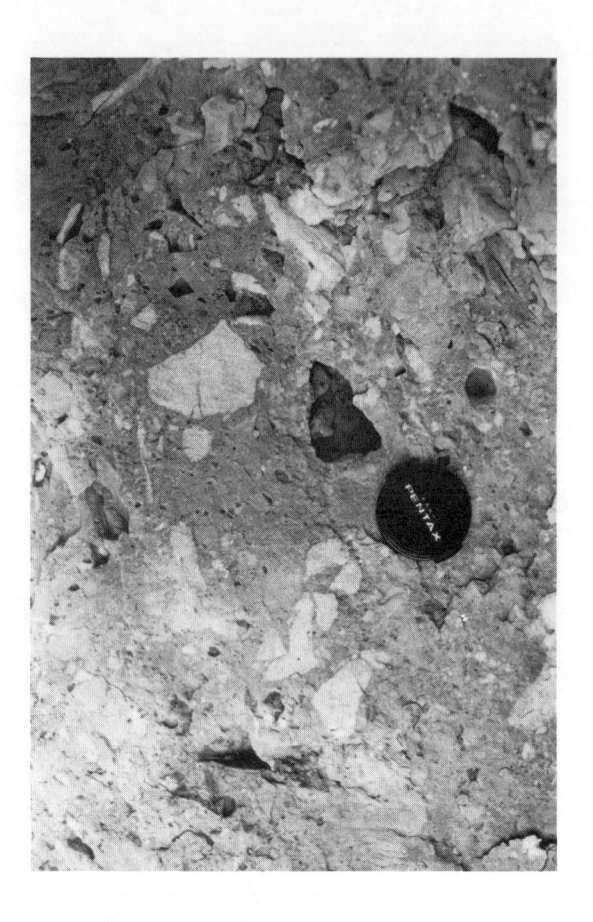

FIGURE 25.3. Brecciated Esplanade Sandstone from a pipe in Mohawk Canyon contains bleached angular clasts of Esplanade in a finely comminuted matrix (Wenrich and others, 1986a, pipe #242). In mineralized breccia pipes ore minerals occur within such matrix material, not in the clasts.

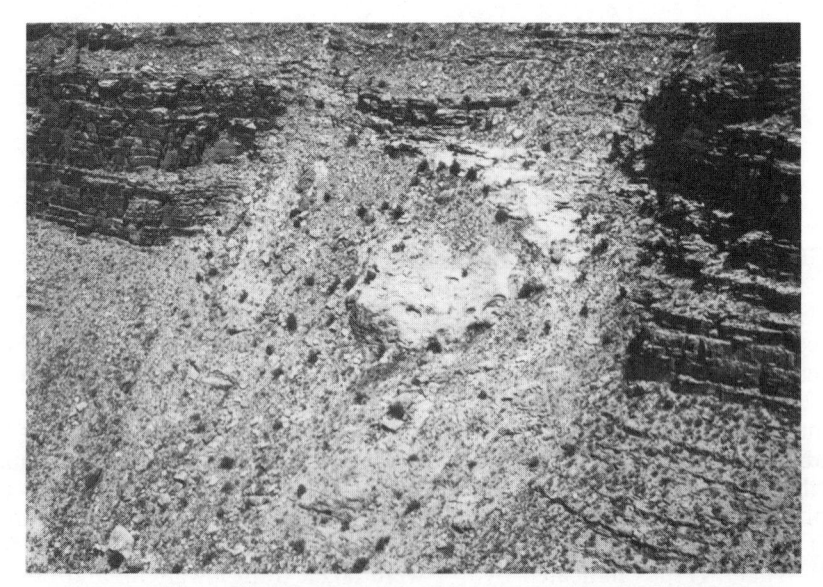

FIGURE 25.4. The brecciated core of each pipe abuts generally well-stratified, little deformed country rock, as can be seen in this pipe located in a tributary canyon to 196-Mile Creek (Wenrich and others, 1986a, pipe #260). Note the bleaching of the breccia against the otherwise red Esplanade Sandstone.

FIGURE 25.5. Shallow structural basins on the high plateaus, such as the East Collapse, shown above (located on the Marble Plateau), are commonly the surface expression of the upper part ("cone") of the breccia pipes. Note the concentric series of inward-dipping beds of Triassic Chinle Formation. This pipe contains oxidized nodules rich in Cu and Ag, and some radioactively-anomalous surface rock with up to 160 ppm U

copper, 107,000 ounces of silver, and 3,400 pounds of V_2O_5 were recovered from the ore (Chenoweth, 1986). Between January 1980 and December 1986, Energy Fuels mined 10 million pounds of uranium from breccia pipe deposits on the North Rim (fig. 25.1) at an average grade of 0.65% U_3O_8 (Mathisen, 1987). These deposits include the Pigeon (fig. 25.6), Hack 1, Hack 2, and Hack 3 pipes.

REDWALL CAVE SYSTEMS IN THE GRAND CANYON REGION

Four aerially extensive dissolution events in the Grand Canyon region have imposed temporally distinct systems of caves involving the Redwall Limestone. These are discussed below in chronological order:

1. An extensive karst developed on the emergent Redwall surface during Late Mississippian time (McKee and Gutschick, 1969). The Grand Canyon breccia pipes are associated with this Mississippian karst; cavities from this karst served as nucleation points for upward stoping. The modern Yucatan karst provides an excellent analogue for Late Mississippian conditions in the Grand Canyon. Active dissolution during this early event coincided with incision of Late Mississippian channels that were subsequently infilled with the Chesterian-Morrowan(?) Surprise Canyon Formation (Billingsley and Beus, 1985). The depth of the karst was greater than, but approximately proportional to, the depth of the channels. In the western Grand Canyon region, some deep channels were incised 400 feet (122 m) and in places the paleokarst in that vicinity involved carbonates below the base of the Redwall Limestone.

2. Laramide uplift and attendant erosional beveling, particularly in the western Grand Canyon region, imposed strong hydraulic gradients and introduced fresh ground waters into newly exposed and near-surface parts of the Redwall section. The result was cave formation in the upper part of the Redwall Limestone on the Hualapai Reservation. Many of the passages in this network remain open and some appear to be reexcavated from the Mississippian paleokarst.

3. Continued Tertiary uplift of the Grand Canyon region imposed ever stronger hydraulic gradients that led to development of caves in the Muav through the Redwall Limestones. Many of these post-Laramide caves are superimposed along fractures associated with Miocene and younger extensional faults. They developed under artesian conditions in buried parts of the Redwall section. Remnants of these caves are best preserved in the eastern Grand and Marble canyons as laterally extensive, vertical slots dissolved along extensional fractures.

4. A modern karst now conducts ground water from the plateaus to the Colorado River. This drainage network also utilizes passageways localized along the late Tertiary extensional fault zones, but in contrast to its predecessors, its caves tend to occupy positions at the base of the Muav Limestone. In locations such as Marble Canyon, where the Muav Limestone is not exposed, passages have developed in whatever carbonate interval lies in close proximity to the floor of the canyon. The passages in the modern system tend to be more widely spaced than predecessor systems, yet they effectively serve as regional drains for the surrounding plateaus (Huntoon, 1974). The modern caves developed

FIGURE 25.6. The Pigeon mine, located in Snake Gulch (a tributary to Kanab Creek) on the North rim, went into production for uranium in 1984. The pipe, around 300 feet (90 m) in diameter, lies within a broad structural basin that is more than 0.5 mile (0.8 km) in diameter. Note the sharply inward-dipping beds located on the right side of the photograph.

coincidentally with erosion of the Grand Canyon, so they are Pliocene(?) or younger in age.

STRUCTURAL LOCALIZATION OF BRECCIA PIPES

Distinct alignments of breccia pipes occur at various locations in northern Arizona. They are particularly notable on the Marble Plateau (Sutphin and Wenrich, 1988; Sutphin, 1986) where one distinct N45W line of 10 pipes extends for 15 miles (25 km), and includes 2 pipes containing Cu-, Ag-, and U-bearing minerals. These alignments probably reflect fracture traces that existed in the Redwall Limestone prior to breccia pipe development. Joints studied by Roller (1987) on the Redwall Limestone-capped Hualapai Plateau revealed that northeast- and northwest-trending fracture sets predate deposition of the overlying Pennsylvanian and Permian Supai Group. Groundwater circulation was localized along these fractures leading to the development of many linear cave passages in the Redwall karst. It appears that especially large cavities dissolved at intersecting cross joints, which later served as nucleation points for upward-stoping breccia pipes.

MINERALOGY OF BRECCIA PIPE OREBODIES

The breccia pipes contain an extensive mineral suite (Wenrich, 1985), which can be roughly divided into 5 stages, some of which may overlap (Wenrich and Sutphin, 1988; Hoyt B. Sutphin, personal communication, 1988): (1) The paragenetic sequence begins with the deposition of calcite, dolomite, barite, siderite, anhydrite, and kaolinite; (2) A second stage of mineralization is characterized by minerals rich in Ni, Co, As, Fe, and S, such as siegenite, bravoite, pyrite, millerite, gersdorffite, rammelsbergite, niccolite, arsenopyrite, and marcasite; (3) The third stage of mineralization was characterized by the formation of Cu-Fe-Zn-Pb sulfides; (4) Uraninite, the only primary U-bearing phase that has been observed within the breccia pipe orebodies, was precipitated within the coarsely crystalline calcite matrix, in minor vugs, and as rims around detrital quartz grains. Extensive silicification is rare within the breccia pipe orebodies, although some quartz overgrowths formed subsequent to uraninite mineralization. A black, glassy pyrobitumen, locally abundant in some breccia pipes, but absent in most, appears to have formed subsequent to the uraninite. (5) In some pipes, late copper sulfide mineralization resulted in the formation of bornite, chalcocite, djurleite, digenite, and covellite. Later supergene alteration formed an assemblage of nonopaque supergene minerals, such as malachite, azurite, and brochantite.

Uranium mineralization occurred in most breccia pipes at roughly around 200 Ma (based on work completed on a large set of U-Pb isotopic analyses by Ludwig and others, 1986). Ore-forming fluids that deposited the sphalerite, calcite, and dolomite had minimum temperatures ranging between 80 °C and 173 °C with salinities always >9 weight percent NaCl equivalent, although most commonly >18 weight. percent NaCl equivalent (Wenrich, 1985). Thus, most of the sulfides and certainly the sphalerite, calcite, and dolomite were deposited from a saline brine similar to that which formed the Mississippi Valley-type deposits. In contrast, the late-stage uraninite was probably deposited from ground water.

FIGURE 25.7. Mining in the Orphan pipe, just west of Grand Canyon Village, began in 1893 when Daniel Hogan staked a claim for copper. In 1951 uranium was discovered on the mine dumps, and production soon began for uranium. The pipe outcrops in the Coconino Sandstone and Hermit Shale. The adits shown here were Hogan's original mines from which copper ore was hauled out of the canyon around the turn of the century on the backs of burros (note the old wheelbarrow in the foreground). The vertical contact of the finely-comminuted Coconino Sandstone (far right of photo) against the normally flat-lying Coconino (hosting the adits, which lie in the ring fracture zone) delineates the boundary of the Orphan pipe located to the right of the photo.

DISCUSSION

The timing of mineralization during the Triassic Period appears to coincide with both (1) the formation of a magmatic arc in southwestern Arizona that extended northwestward into southeastern California and western Nevada (Woodward-Clyde Consultants, 1982), and (2) the uplift of Precambrian and Paleozoic rocks on the back, or northeast side, of the magmatic arc into the Mogollon highlands (Bilodeau, 1986). Uplift of the highlands probably activated the circulation of (1) brines (from which the base metals were precipitated) in the Mississippian, Pennsylvanian, and Permian strata, trapped there since deposition, and (2) groundwater from the highlands carrying uranium from the volcanic arc or the unroofed Precambrian granitic rocks. Existing or developing breccia pipes in the Grand Canyon area provided vertical conduits linking the stacked aquifers in the stratigraphic section. Flow was necessarily upward in the pipes in accord with large-basin hydraulic gradient configurations. Triassic circulation of groundwater at the level of the Redwall Limestone would account for the reactivated dissolution of the Mississippian paleokarst zone and dissolution of breccia clasts in existing or incipient pipes. This provided (1) a mechanism that allowed the pipes to propagate further upward and (2) vertical conduits for fluid flow of the brines trapped in the Pennsylvanian and Permian sediments. These brines precipitated the base metals and produced the first three stages of mineralization. Apparently some brecciation continued subsequent to these mineralizing events because many of these minerals have been fractured (Hoyt B. Sutphin, personal communication, 1987).

This hydrodynamic model moves uranium-rich waters northward from their source in the Triassic Mogollon Highlands through the Redwall carbonate section and upward through the pipes (fig. 8). Implicit for uranium mineralization is the need for upward-circulating mineralizing fluids to encounter either reducing waters higher in the section, or a reducing medium such as the pyrite-rich sulfide deposits previously precipitated from the brines. Although petroleum products, such as the pyrobitumen precipitated during stage 4 subsequent to the uraninite, are abundant in a few pipes, and sparse petroleum products have been found in sphalerite fluid inclusions, they are not present in all pipes. Yet, the upward migrating uranium-bearing fluids and associated H_2S could have encountered petroleum fluids in the overlying aquifers in the Permian section, which would have provided an additional reductant to the sulfides already present. As the Triassic highlands were eroded or tectonically severed, hydraulic gradients progressively diminished, eventually bringing the mineralization event to an end. The uraninite and sulfide cement in the matrix of the breccia probably acted as a sealer to the orebodies, preventing further fluid movement and additional brecciation.

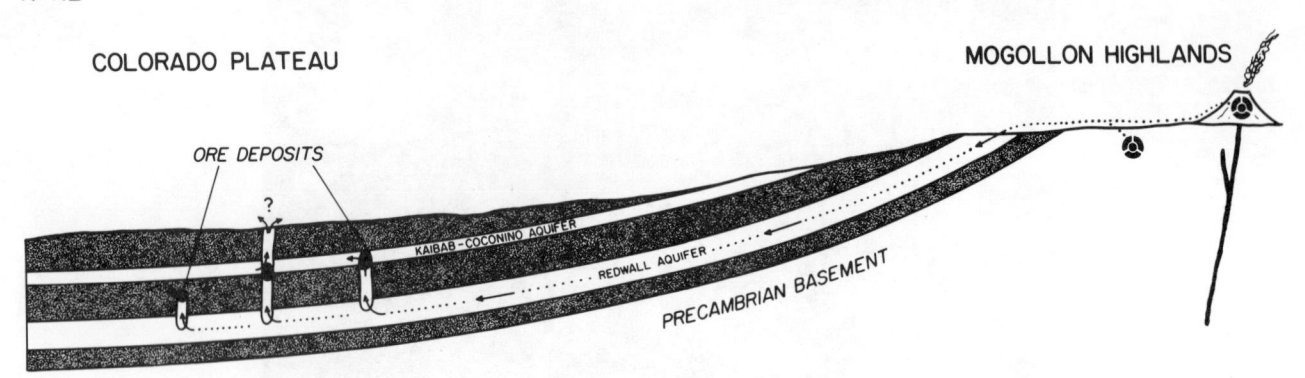

FIGURE 25.8. Generalized circulation system that produced orebodies in the Grand Canyon breccia pipes. Vertical dimension is 2,000 m; horizontal dimension is many hundreds of kilometers.

Aerial view from above the Colorado River near mile 190 (km 306) of 190-mile landslide, western Grand Canyon. Unclassified (Late Cambrian) dolomite beds at A (displaced) match beds at B (in place). The base of the landslide rests on lower half of Rampart Cave Member of Muav Limestone (Middle Cambrian) at C. D - Bright Angel Shale; E - Tapeats Sandstone; F - travertine.

CHAPTER 26: GRAVITY TECTONICS, GRAND CANYON, ARIZONA

Peter W. Huntoon
Department of Geology and Geophysics,
University of Wyoming, Laramie, Wyoming

Several gravity tectonic processes have been identified in the Grand Canyon that ranged from slow to rapid and that involved both ductile and brittle failure. Only gravity tectonic processes producing large scale features are treated here, and these include valley anticlines, high angle gravity faults, and rotational landslides.

VALLEY ANTICLINES[1]

The axis of a valley anticline (fig. 26.1) coincides exactly with the strike of the Colorado River between Fishtail and Havasu Canyons (mi 139-157; km 224-253), and extends discontinuously westward to Parashant Canyon (Billingsley and Huntoon, 1983). Structurally identical river anticlines also occur in the principal tributary canyons along this 60 mile (100 km) reach, such as in Kanab and Tuckup Canyons. The occurrence of the anticlines along the sinuous trends of the canyons reveals a genetic link between the two. Huntoon and Elston (1980) deduced that the anticlines are the product of flowage of the saturated shaly parts of the Cambrian Muav Limestone and underlying Bright Angel Shale toward the canyons. The driving force for this deformation is the huge difference in lithologic load between the heavily loaded rocks under the 2,100-foot (640-m) high canyon walls and the unloaded canyon floors (Sturgul and Grinshpan, 1975). Similar valley anticlines are missing in the easternmost Grand Canyon and Marble Canyon where the same Cambrian units also occur at river level. Their absence is explained by the lack of shaly rock in the eastern Cambrian section that could deform ductily when saturated.

Where best developed, the limbs of the valley anticlines dip away from the river at angles as great as 60°, and folding extends more than 800 feet (250 m) into the canyon walls. The anticlinal bulge is a horizontal shortening structure across the river in which the shortened beds arch upward into the void provided by the canyon. Deformation in the anticlines ranges from ductile to brittle.

Numerous sets of minor low angle conjugate thrust faults that parallel the axes of the anticlines occur in the Muav Limestone. The intersecting faults in these sets respectively dip toward and away from the river. They are more abundant and have larger offsets in locations where the Muav Limestone is steeply folded, although

FIGURE 26.1. Valley anticline developed along the strike of the Colorado River upstream from the mouth of Havasu Canyon, Grand Canyon, Arizona. Note dip of strata away from the river. View is downstream near mile 147 (km 237).

[1]Editors note: These structures, which follow the course of the Colorado River and locally its tributary streams in narrow, steep walled canyons, also are referred to as "river anticlines."

they also occur in beds that dip as little as 2° (Huntoon and Elston, 1980, p. 4). The thrust faults accommodate horizontal shortening across the anticlines by allowing for structural thickening and shortening of the deformed beds.

Steeply dipping kink bands also parallel the anticlines and have formed in response to layer parallel shortening across the fold axis (Reches and Johnson, 1978, p. 282-284). Such bands can be mistaken in outcrop for small displacement reverse faults. Most of the observed kink bands displace the rocks upward toward the river and the bands containing the kinked strata dip toward the river. The kink bands are generally no wider than 6 feet (2 m).

GRAVITY FAULTS

High angle gravity faults in the eastern Grand Canyon were described by Huntoon (1973) as map scale brittle failures unrelated to deep seated tectonic processes. As shown on Figure 26.2, the faults are generally restricted to Paleozoic strata above the base of the Bright Angel Shale, and are very steeply dipping to near vertical. Displacements are greatest at the top of the section and die out downward. The faults are laterally discontinous and rarely can be traced more than 3 miles (5 km) along strike. These faults most commonly cut across and are restricted to narrow ridges, which are flanked by deep canyons that expose the Bright Angel Shale.

Huntoon (1973, p. 122) concluded that the faults represent brittle failure accompanying minor ductile deformation within the Bright Angel Shale. They develop as the overlying column of rock, in some cases 4,000 feet (1,220 m) high, settles into its ductile shale base.

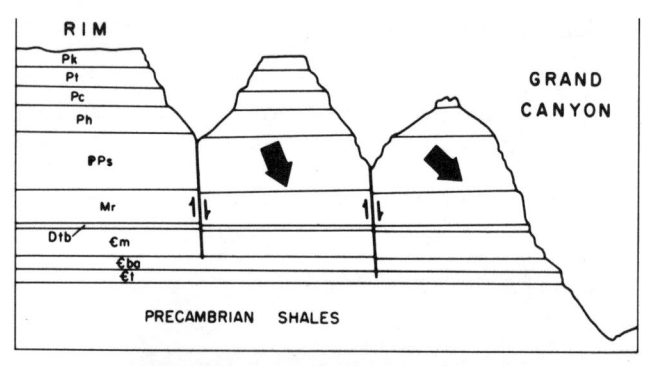

High angle gravity faults are very important in the morphologic evolution of Grand Canyon scenery. Once such faults form, erosion proceeds rapidly along them, thereby segmenting what was formerly a ridge between two canyons into a string of buttes. As erosion progresses, the buttes become isolated from the plateau. Ironically the faults that initiate the butte forming process vanish because they eventually erode. Consequently, faults no longer can be found around some deeply eroded old buttes such as Solomon Temple in the eastern Grand Canyon (Huntoon and others, 1986).

ROTATIONAL SLIDES

The most important gravity tectonic structures in terms of canyon widening are rotational slides. Rotational slides involve massive blocks or rows of blocks (fig. 26.3) that detach from the canyon walls and glide into the canyon along listric normal faults. The listric faults are concave upward thereby causing the descending masses to rotate backward against the glide planes. The displaced blocks are commonly called toreva blocks in local parlance (Reiche, 1937).

Rotational slides most commonly occur in two settings in the Grand Canyon. The largest involve the Redwall cliff, which is inclusive of the Pennsylvanian-Permian Supai Group at the top and Cambrian Muav Limestone at the base. This section is commonly 1,600 or more feet (490 m) thick depending on how much of the underlying Bright Angel Shale has become involved in the slide. Much smaller rotational slides and calving involve the Cambrian Tapeats Sandstone, 250 to 400 feet (75 to 120 m) thick, which have detached above the ductile Late Proterozoic Galeros Formation in the eastern Grand Canyon (Huntoon and others, 1986). Rotational slides can form wherever cliffs develop above ductile strata, so other cliffed sections also spawned slides in the Grand Canyon.

Large rotational slides involving the entire Redwall cliff develop where oversteepened slopes on the underlying Bright Angel Shale can no longer provide sufficient buttressing to carry the lithostatic loads imposed on the shale at the base of the cliff. Huntoon (1975) proposed that attainment of stable Redwall profiles along the Colorado River develop through a series of catastrophic rotational slides shortly after the Bright Angel Shale is exhumed by downcutting of the river. Saturation of the shale undoubtedly facilitates the process.

Incision of the Bright Angel Shale has not occurred uniformly in time along the length of the Grand Canyon owing to structural undulations on the shale surface. Consequently, the inception of rotational sliding varied from location to location. It has yet to occur between 140-Mile and Stairway Canyons because the Bright Angel Shale has not been exposed or is just beginning to be incised along this reach.

Figure 26.3 illustrates the steps involved in attaining a stable profile once the Bright Angel Shale is incised.

Permian:
  Pk - Kaibab Formation
  Pt - Toroweap Foramtion
  Pc - Coconino Sandstone
  Ph - Hermit Shale
Permian and Pennsylvanian:
  PPs - Supai Group

Mississippian:
  Mr - Redwall Limestone
Devonian:
  Dtb - Temple Butte Formation
Cambrian:
  Єm - Muav Limestone (includes unclassified dolomite)
  Єba - Bright Angel Shale
  Єt - Tapeats Sandstone

Heavy arrow shows relative movement of the butte.

FIGURE 26.2. Erosion along high angle gravity faults that develop across ridges allow buttes to separate from the ridges. Faults develop as buttes settle into ductile Cambrian and Proterozoic shales.

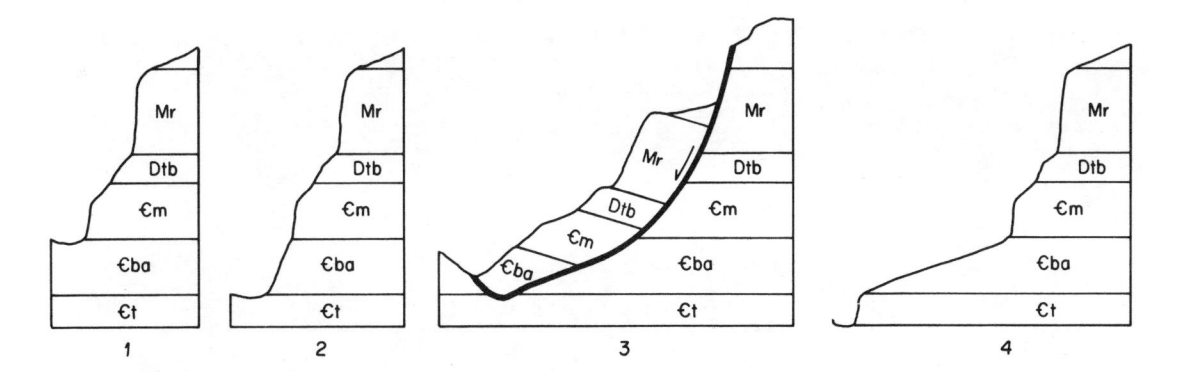

FIGURE 26.3. Stabilization of cliff profiles through rotational landsliding, Grand Canyon, Arizona. 1. Channel is downcut to top of Bright Angel Shale, profile is stable. 2. Channel has cut through Bright Angel Shale, shale slope is too steep to buttress overburden load. 3. Profile fails through rotational sliding. 4. Erosion of debris results in stable cliff profile. Mr - Redwall Limestone, Dtb - Temple Butte Formation, Cm - Muav Limestone, Cba - Bright Angel Shale, Ct - Tapeats Sandstone.

Each step is preserved and well exposed along the Colorado River in the reach between Tapeats Creek and Kanab Canyon (mi 134-144; km 216-232).

The river has just begun to excavate through the Muav Limestone and into the Bright Angel Shale near the mouth of Kanab Canyon. The result is a narrow, sheer-walled canyon less than 0.3 miles (0.5 km) wide but more than 1,600 feet (500 m) high. As downcutting continues, the foot of the limestone cliff will rest on a progressively heightening shale cliff. Without lateral buttressing, the section above the shale will ultimately fail, and toreva blocks will slide into the canyon. Such collapse has occurred north of the river in the not too distant geologic past immediately upstream between Fishtail and Deer Creek Canyons. Exposures farther to the east in Surprise Valley reveal that progressive tiers of blocks calved off the canyon wall until the slope on the Bright Angel Shale became sufficiently wide to support the cliffed section above it. Surprise Valley, north of Cogswell Butte, contains two or more rows of toreva blocks which are shown in Figure 26.4. These fell into and blocked a former tributary to Tapeats Canyon illustrating that the same widening process also operates in tributary canyons as well as along the main stem of the Colorado River.

The tier of huge blocks that calved off the north wall west of Deer Creek Canyon completely filled a 1 mile (1.6 km) reach of the Colorado River channel. The blockage forced the river out of its channel, where it subsequently cut a parallel channel 0.3 mile (0.5 km) to the south (Huntoon, 1975, p. 5). The youthfulness of this slide is revealed by the fact that the floor of the blocked channel west of Deer Creek Falls is the same elevation as the present channel and that through-flowing tributary drainage has not been reestablished through the slide debris. Sediments that were ponded behind the blockage are still preserved in thick sections on the south side of the Colorado River 2 miles (3.2 km) upstream from Deer Creek Canyon

Halfway between Deer Creek and Tapeats Canyons, on the south side of Cogswell Butte, is a second blocked channel shown on Figure 26.5. The floor of the channel is 100 feet (30 m) above present river level, attesting to its antiquity. It may date from Pliocene time. Blockage of the course of the Colorado River by this slide caused the river to be shifted southward where it cut its present channel.

The most unusual toreva block found in the Grand Canyon is Carbon Butte in the eastern Grand Canyon (fig. 26.6), described in detail by Ford and Breed (1970). Carbon Butte is comprised of a Paleozoic section that extends from the Bright Angel Shale through the Redwall Limestone, and it is located between the east and west forks of Carbon Canyon (mi 64.6; km 103.9). The butte detached from the Redwall cliff on a listric slip surface, bottoming between the Cambrian Tapeats Sandstone and the underlying Late Proterozoic Kwagunt Formation. The detached mass slid southward trailing behind it large blocks of Tapeats Sandstone that were torn from its base. The detached block came to rest one mile (1.6 km) south and 1,800 feet (548 m) below its starting point. The track it followed was a valley eroded on resistant Precambrian strata along the south-plunging axis of a Proterozoic syncline.

Large landslides and rotational slides are common in the western Grand Canyon downstream from Whitmore Wash. As in the eastern Grand Canyon, failure of the Bright Angel Shale was responsible for most of the slides, and the overlying strata up through the Esplanade Sandstone typically comprise the detached masses. Numerous examples of blocked stream channels exist in the region, the most notable being a Colorado River blockage by a huge detachment that measures 1.2 miles (1.9 km) along strike across from 205-Mile Canyon. This slide displaced the Colorado River eastward along the toe of the slide. Several large slides also occur in tributary canyons in the western Grand Canyon.

FIGURE 26.4. Row of rotational slide blocks that have fallen from the Redwall cliff into Surprise Valley, Grand Canyon, Arizona. A second, closer row of blocks lies buried by alluvium in the center of the valley. View is toward the northeast.

FIGURE 26.5. Former channel of the Colorado River that was blocked by a Pliocene(?) rotational slide from the south side of Cogswell Butte (right) between Tapeats and Deer Creek Canyons, Grand Canyon, Arizona. Notice that the river excavated a new canyon south of the blockage, and that the floor of the filled channel lies 100 feet (30 m) above the modern channel attesting to the antiquity of the slide. View is downstream to the west.

FIGURE 26.6. Carbon Butte (small left–dipping butte in center) detached from the ridge to the left and glided 1 mile (1.6 km) down a shallow valley eroded along the axis of a syncline in Proterozoic strata trailing in its wake blocks of Cambrian Tapeats Sandstone torn from its base. Notice that the Redwall section was subsequently eroded from the ridge from which Carbon Butte fell. View is toward the northeast; Chuar Butte in background.

CHAPTER 27: MINING ACTIVITY IN THE GRAND CANYON AREA, ARIZONA

George H. Billingsley
U.S. Geological Survey, Flagstaff, Arizona

Man has been interested in mineral resources of the Grand Canyon area for many hundreds of years. Deposits of salt, hematite, copper and silver minerals, lead, asbestos, and uranium have been developed in the area, and efforts have been made to find economic deposits of gold and oil. American Indians carried out the earliest recognized mining activity in the Grand Canyon, at least by the late 1700's and probably centuries earlier. Others have searched for mineral wealth in the Grand Canyon since the 1880's. Efforts to protect and preserve the area of the Grand Canyon began in 1882. Water, or the lack of it, has been a notable factor in development of the area.

Early Mining

Salt was a valued commodity among the Hopi and Navajo Indians, used for ceremonial and domestic purposes. However, domestic use was probably not too well accepted because, except for salt stalactites, the salt has the texture and consistency of salt-flavored plaster. The area of the salt mines (mi 63, km 101; fig. 27.1) is sacred to the Indian people today. Salt may have been gathered elsewhere in the Grand Canyon where the Tapeats Sandstone crops out, but there are no records of such activities. Ground water seeping along fractures in the Bright Angel Shale and bedding planes of the Tapeats Sandstone commonly pick up sufficient sodium chloride and other minerals to form incrustations on Tapeats cliffs and ledges where seepages occur.

Hematite and its weathered products were another valuable commodity to the Northeastern Pai Indians. Red siltstone and claystone were mined from a cave in the north wall of Diamond Creek Canyon. The red claystone, when mixed with deer tallow, formed a red skin pigment valued for not only for its bright color, but also for its protective properties against sunburn and cold. By 1827 or 1830, fur trappers ranged west from the Rio Grande (New Mexico) country and brought in cheap manufactured goods including makeup, which gradually ended demand for the red pigment (Dobyns and Euler, 1976).

Scattered deposits of green and blue copper minerals in the Grand Canyon were utilized by Indians many centuries ago. The pretty stones probably were used for jewelry and trading, and perhaps also for ceremonial purposes; bits of malachite and azurite are found in some of the earlier Indian occupation sites.

Anglo-American prospectors began to roam the Grand Canyon country as early as 1850, but no records are known of their early prospecting and mining activities. Only after passage of the Federal Mining Law of 1872, which set up location and recording procedures, could the pattern of prospecting be revealed in the Grand Canyon. Hundreds of prospectors probably tramped the waterless side canyons in search of gold and other valuable minerals. Most of them may have survived the summer heat of 125 °F on the canyon floor and the winter cold of -20 °F, or more, on the rims. Because of the harsh desert and rugged inverted mountain terrain, a few prospectors certainly must have left their bones, dreams of riches, and burros behind when a lonely death overtook them.

Copper

Copper, silver, lead, and asbestos minerals were found in the Grand Canyon during the late 1800's and early 1900's. Small copper deposits are scattered throughout the Grand Canyon and on the surrounding plateaus. Copper deposits known to the local Indians were traded to early prospectors. For example, the Grand Gulch mine on the Grand Wash Cliffs (fig. 27.1) was bought from the Indians for a horse and some flour (Crampton, 1972). No native copper or silver is known to have come from the canyon area.

Most of the copper in the Grand Canyon is found in breccia pipes (chapter 25) in upper Paleozoic formations ranging from the Mississippian Redwall Limestone to the Permian Kaibab Formation. Smaller amounts of copper are found in faults and shear zones in Early to Late Proterozoic rocks between miles 64 and 78 (km 103-125, chapter 1). Copper minerals also occur as stratiform deposits in intraformational breccia beds within the Kaibab Formation near the Jacob Lake, Laguna Lake, and Anita areas on the Colorado Plateau (fig. 27.1).

Most copper deposits in the Grand Canyon area were too small and low grade to warrant commercial development or make it worth the trouble of getting the ore out with a few burros. A few mines, such as the Grandview, Grand Gulch, Hacks Canyon, and Orphan Lode (fig. 27.1), made it a paying venture to pack hand-sorted, high-grade ore out on burros or in small wagons, as long as each load averaged about 20 percent or higher in copper. These conditions plus the long haulage distance from the Grand Canyon to processing plants in central Arizona or Utah, coupled with a significant drop in copper prices in 1907, forced the abandonment of many copper mining operations within and around the canyon. Some mines were successful if one remembers that "successful" is a relative term, but they were

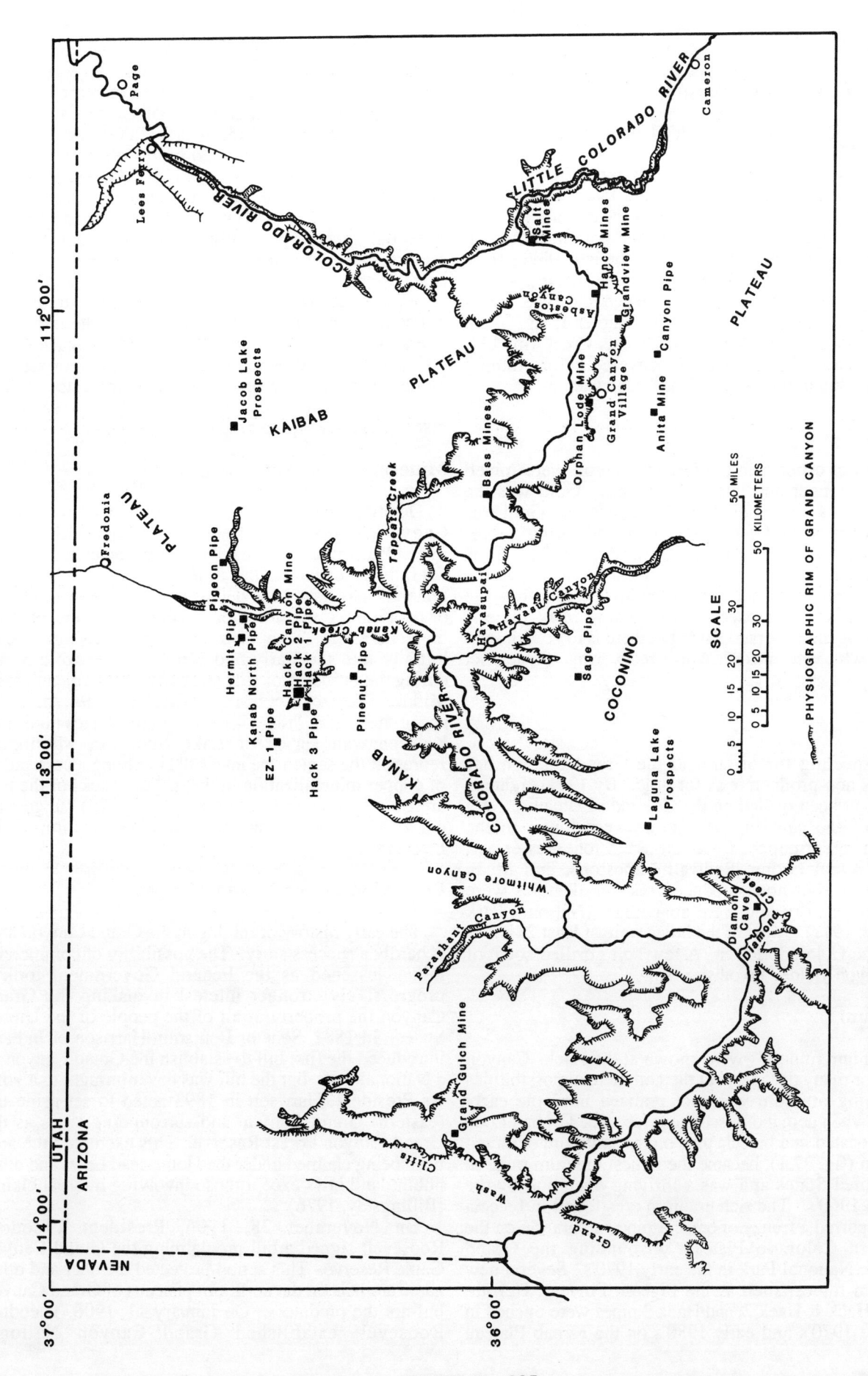

FIGURE 27.1. Map showing the location of some mines in the Grand Canyon area, Arizona.

successful only for a short period of time.

Silver, Lead, and Asbestos

Most of the silver and lead ores were mined in Havasu Canyon (mi 157, km 253, fig. 27.1). They are commonly associated with breccia pipes in Paleozoic rocks. Small amounts of silver, lead, and copper minerals also occur in fractures and shear zones in Early and Middle Proterozoic rocks but not in economic quantities. Asbestos was mined in the late 1800's from Middle Proterozoic rocks where diabase sills intrude the Bass Limestone, such as at the Hance and Bass mines (mi 77 and 111, km 124 and 179, chapter 1; fig. 27.1). Similar asbestos deposits near Tapeats Creek (mi 134, km 216) probably were not discovered before mining was prohibited in this part of the Grand Canyon.

Gold

Many prospects in the Grand Canyon were mined first for copper and later for uranium. Gold was the mineral most often sought with no success. Gold was discovered in the Colorado River sands during a river expedition by John Wesley Powell in 1871. It is disseminated in tiny amounts almost anywhere in the river sand, having been carried far downstream from the source rocks. A prospector would say, "Yes, there is gold along the Colorado River but there is too much sand mixed with it to make it worth recovering" (Waesche, 1934).

Oil

Prospecting for oil around the Grand Canyon has been as non-productive as for gold. By 1961, eighteen holes had been drilled on the Colorado Plateau, mostly north of the canyon. Oil stains were found in the Toroweap Formation, Coconino Sandstone, lower part of the Supai Group, Redwall Limestone, and Muav Limestone, but none produced oil and all holes were classified as dry and were abandoned (National Park Service, 1977). Most of the holes were at least 25 miles from the Colorado River. A few holes drilled south of the canyon yielded no oil shows.

Uranium

Uranium minerals were known at the Hacks Canyon mine for many decades, but the commercial possibilities of mining uranium were not realized until the early 1950's when demand raised the price. The Orphan Lode mine, located in a breccia pipe on the south rim of Grand Canyon (fig. 27.1), became the richest uranium mine in the United States and was a thriving operation by the middle 1960's. The rich uranium ores found in breccia pipes spurred a frenzy of breccia-pipe exploration on the southern Colorado Plateau surrounding the Grand Canyon National Park in the early 1970's. Several new uranium mines, such as the Pigeon, Pinenut, Hermit, EZ-1, Hack 1, Hack 2, and Hack 3 pipes were opened in the late 1970's and early 1980's on the Kanab Plateau north of the canyon (fig. 27.1). Other uranium deposits, such as the Canyon, Kanab North, and Sage pipes, are potential mines when uranium prices increase. Not all breccia pipes contain uranium or other metals. In fact, only 2 or 3 percent of the nearly 2,000 breccia pipes known thus far on the Colorado Plateau contain economic ore deposits of any kind. Many show minor amounts of uranium minerals, but the decrease in uranium prices in the late 1980's has made several potential breccia-pipe mines uneconomical. These factors, plus environmental concerns, have halted most uranium operations at this time. Only the richest breccia pipes are being mined today because their high concentrations of uranium ore make them paying operations from which monies spent in getting the mines into operation can be recovered. Breccia pipes are still being discovered by several mining interests, but the expensive exploratory drilling needed to establish reserves is not keeping pace with discoveries.

Mineralization Ages

Detailed studies of mineral deposits of the Grand Canyon region are needed to understand the mineralization process, and the ages of mineralization are not fully understood at this time. Copper and other metal minerals may have been deposited in the Paleozoic rocks of the Grand Canyon region during several episodes of mineralization. They also may have been introduced initially into the Early and Middle Proterozoic rocks during the intrusion of sills and subsequent volcanism in Middle Proterozoic time. Walcott (1894, p. 519) speculated that "the Grand Canyon series and the Keweenawan series of Lake Superior, Michigan, represent the same time interval", implying a correlation of copper mineralization in the igneous rocks of the two areas. Wenrich and Huntoon (chapter 25) suggest an early Mesozoic age for uranium mineralization of the breccia pipes.

Protection of the Grand Canyon

The early history of mining in the Grand Canyon area is hardly a success story. The possibility of commercial mines lessened as the Federal Government took a progressively stronger interest in making the Grand Canyon the property of all of the people of the United States. In 1882, Senator Benjamin Harrison of Indiana introduced the first bill to establish the Grand Canyon as a National Park, but the bill was never brought to a vote. As President, Harrison in 1893 acted to set aside the (eastern) Grand Canyon and surrounding areas as the Grand Canyon Forest Reserve. This exempted the area from being claimed under the Homestead Laws and other public-land laws except those involving mineral claims (Billingsley, 1976).

On November 28, 1906, President Theodore Roosevelt signed a bill proclaiming the Grand Canyon Game Reserve. This action protected the deer and other game animals on the north rim plateaus of Grand Canyon but not the predators. On January 11, 1908, Theodore Roosevelt established Grand Canyon National

Monument under the Act for the Preservation of American Antiquities. The primary effect of this proclamation was to forbid prospecting and mining on all lands in the (eastern) Grand Canyon that were not already covered by valid mineral claims (Billingsley, 1974). Most of the mines were abandoned by 1908 anyway, and active mines faced increasing haulage costs and lower copper prices, which forced most of the smaller mining operations out of business by the 1920's.

On February 26, 1919, President Woodrow Wilson signed a bill proclaiming Grand Canyon National Monument as a National Park. Congress revised the boundaries of Grand Canyon National Park in 1927. A new Grand Canyon National Monument was proclaimed by President Herbert C. Hoover on December 22, 1932, which included many tributary canyons to the Colorado River as well as parts of the north and south rims of the central Grand Canyon area (from about mile 145 to 184; km 233-296, chapter 1). Throughout this period, efforts were made to acquire small parcels of private or State-owned land within the National Park by gift or exchange; in this way many mining claims were added to the park (Hughes, 1967).

Significant expansion of the Grand Canyon National Park boundaries in 1974 protected virtually all tributary canyons within the physiographic boundaries of Grand Canyon from future mining; exceptions were the Parashant and Whitmore Canyon areas (mi 188-200, km 302-322, chapter 1), where active uranium exploration was being conducted, and parts of the Hualapai, Havasupai, and Navajo Indian Reservations (figs. 4.4 and 4.5). A contract agreement between Cotter Cooperation, owners of the Orphan Lode mine, and the National Park Service allowed mining to continue under park lands until March 1988, at which time the mine would become the property of the National Park. John Hance's old asbestos claim in Asbestos Canyon (mi 78, km 125, chapter 1) is the last mining claim that had not reverted to the National Park by 1988. The 312.42 acre-parcel was purchased in the early 1900's by William Randolph Hearst, and its trade to the National Park System for other property outside the Park is being negotiated as of the time of this writing (John Ray, Park Ranger, Grand Canyon National Park, personal communication, 1988).

Water

The action of water through geologic time has been responsible for creation of the Grand Canyon, Since humans entered the Grand Canyon, however, water has proven to be the most valuable and most often sought-after commodity. Water has played a vital role. Without it, there could have been at best only limited facilities for the visitors who look into and across the canyon, and no river to support human populations downstream or on which to float through the canyon. More money has been spent on water-related problems in the southwestern United States than on any other commodity. In this generally waterless area that has an expanding population, development of Colorado River water for farming, and for domstic and industrial use, has cost more than development of all the metals that have been found in the Grand Canyon region. The limited water resources of the Colorado River have become an ever more valued commodity with time as human population has increased. Money yet to be spent on managing and distributing the water is almost beyond comprehension.

Communities near the Grand Canyon area are few and far apart. Their growth is restricted by difficulties in developing water resources. The lack of water has inadvertently helped to preserve the Grand Canyon and vicinity as a natural wilderness area. Once prospectors dreamed of extracting mineral wealth from the Grand Canyon; now many people dream of seeing and experiencing the Canyon for its varied beauty and wilderness.

CHAPTER 28: BAT CAVE GUANO MINE, WESTERN GRAND CANYON, ARIZONA

Peter W. Huntoon
Department of Geology and Geophysics,
University of Wyoming, Laramie, Wyoming

The Bat Cave occurs in the Muav Limestone on the north side of Lake Mead at river mile 266 (km 428). This natural cave was the site of a fascinating and expensive guano mining operation in which the product was lifted out of the canyon by means of an aerial tramway that terminated on the south canyon rim. The highest of the tramway towers is situated on the Supai Group on the south rim more than 3,600 feet (1,100 m) above the lake.

Following a few previous attempts, U.S. Guano Corporation began developing Bat Cave as a guano mine in 1956 to produce bat guano from a prominent Muav Limestone cave 800 feet (244 m) above the lake on the north wall of the canyon (Billingsley, 1974). The guano deposit was mistakenly estimated to contain 100,000 tons (New York Times, 1957). Actual production reached about 1,000 tons at a cost of approximately $3.5 million (George Billingsley, personal communication, 1988). The deposit, which met or exceeded 6 percent nitrate, extended 2,000 feet (600 m) into the cave and was mined by manually loosening it with hoes, whereupon it was moved through a 10-inch vacuum hose to the loading bin at the entrance (Beatty, 1962). From there it was transported by tram via a cable tramway for a 9,820-foot-long (3417 m), 3,000-foot (925 m) climb to the tower visible on the south rim. Next it was trucked to Kingman, Arizona, for sacking and distribution.

The operation ceased in the early 1960's as a result of depletion of reserves. In 1962, several months after the mine was closed, a jet aircraft from Nellis Air Force Base collided with the tram cable, causing the cable to collapse into the canyon. The pilot returned to his base with 6 to 8 inches of wing tip missing. Local legend has it that a Federal damage payment to the mine operator finally returned the mining venture to profitability.

CHAPTER 29: SMALL METEORITE IMPACT IN THE WESTERN GRAND CANYON, ARIZONA

Peter W. Huntoon
Department of Geology and Geophysics,
University of Wyoming, Laramie, Wyoming

The site of a former village of the Hualapai Indians lies 2,000 feet (610 m) below the canyon rim in Meriwitica Canyon, 3 miles (4.8 km) upstream from the mouth of Spencer Canyon. The Hualapai lived in isolation in adobe type dwellings on a 1/2 mile- (0.8 km-) wide, 2.5 mile- (4 km-) long alluvial terrace developed behind a 200 foot- (60 m-) high travertine dam below Meriwhitica Spring. The spring, which discharges from the Rampart Cave Member of the Muav Limestone, supplied ample drinking and irrigation water. The flat alluviated canyon floor and spring combined to provide a setting similar to the one currently occupied by the Havasu Tribe in Havasu Canyon, 51 miles (82 km) to the northeast.

In 1921 or 1924(?), a meteor streaking through the early evening darkness slammed into the north wall of Milkweed Canyon across from the village (Grant Tapije, personal recollection, 1986). The meteorite impact left a visible but not particularly obvious scar on the Redwall cliff (Young, 1978, p. 288). One photo of the scar is known (Richard Young, personal communication, 1986). The Indians took this terrifying event as a bad omen and fled the canyon the next day, never to return permanently. The Indians occasionally, visit Meriwhitica Canyon, and there was an attempt to establish a small farm in the valley in the 1940's, but that venture was short lived (Richard Young, personal communication, 1986). Time has obliterated the original village.

REFERENCES

Aldrich, L.T., Wethergill, G.W., and Davis, G.L., 1957, Occurrence of 1350 million-year-old granitic rocks in western Grand Canyon: Geological Society of America Bulletin v. 68, p. 655-656.

Allmendinger, R.W., Hauge, T.A., Hauser, E.C., Potter, C.J., Klemperer, S.L., Nelson, K.D., Knuepfer, P., and Oliver, J., 1987, Overview of the COCORP 40N transect, western United States, the fabric of an orogenic belt: Geological Society of America Bulletin, v. 98, p. 308-319.

Anderson, C.A., and Creasey, S.C., 1958, Geology and ore deposits of the Jerome area, Yavapai County, Arizona: U..S Geological Survey Professional Paper 308, 185 p.

Anderson, R.E., and Huntoon, P.W., 1979, Holocene faulting in the western Grand Canyon, Arizona, discussion and reply: Geological Society of America Bulletin, v. 90, p. 221-224.

Armstrong, R.L., Eisbacher, G.H., and Evans, P.D., 1982, Age and stratigraphic-tectonic significance of Proterozoic diabase sheets, Mackenzie Mountains, northwestern Canada: Canadian Journal of Earth Sciences, v. 19, p. 316-323.

Babcock, R.S., in press, The older Precambrian, in Beus, S.S., and Morales, Michael, eds., Grand Canyon Geology: New York, Oxford University Press and Museum of Northern Arizona.

Babcock, R.S., Brown, E.H., and Clark, M.D., 1974, Geology of the Older Precambrian rocks of the Upper Granite Gorge of the Grand Canyon, in Breed, W.S., and Roat, E.C., eds., Geology of the Grand Canyon: Grand Canyon Natural History Association, Grand Canyon, Arizona and Museum of Northern Arizona, Flagstaff, Arizona, p. 1-19.

Babcock, R.S., Brown, E.H., Clark, M.D., and Livingston, D. E., 1979, Geology of the Older Precambrian rocks of the Grand Canyon, Part II, The Zoroaster plutonic complex and related rocks: Precambrian Research, v. 8, p. 243-275.

Baker, V.R., 1984, Flood sedimentation in bedrock fluvial systems, in Koster, E.H., and Steel, R.J., eds., Sedimentology of gravel and conglomerates: Canadian Society of Petroleum Geologists Memoir 10, p. 87-98.

Beatty, W.B., 1962, Geology and mining operations in U.S. Guano Cave, Mohave County, Arizona: Castro Valley, California, Cave Research Associates, Cave Notes, v. 4, p. 40-41.

Bennett, V.C., and DePaolo, D.J., 1987, Proterozoic crustal history of the western United States as determined by Neodymium Isotopic Mapping: Geological Society of America Bulletin, v. 99, p. 674-685.

Berggren, W.A., Kent, D.V., Flynn, J.J., and Van Couvering, J.A., 1985, Cenozoic geochronology: Geological Society of America Bulletin, v. 96, p. 1407-1418.

Best, M.G., and Brimhall, W.H., 1970, Late Cenozoic basalt types in the western Grand Canyon region, in Hamblin, W.K., and Best, M.G., eds., The western Grand Canyon district: Utah Geological Society, Guidebook to the geology of Utah, no. 23, p. 56-74.

Best, M.G., and Brimhall, W.H., 1974, Late Cenozoic alkali basaltic magmas in the western Colorado Plateau and the Basin and Range transition zone, U.S.A., and their bearing on mantle dynamics: Geological Society of America Bulletin, v. 85, p. 1677-1690.

Best, M.B., McKee, E.H., and Damon, P.E., 1980, Space-time-composition patterns of late Cenozoic mafic volcanism, southwestern Utah and adjoining areas: American Journal of Science, v. 280, p. 1035-1050.

Beus, S.S., 1973, Devonian stratigraphy and paleogeography along the western Mogollon Rim, Arizona: Flagstaff, Museum of Northern Arizona Bulletin no. 49, 36 p.

Beus, S.S., Carothers, S.W and Avery, C.C., 1985, Topographic changes in fluvial terrace deposits used as campsite beaches along the Colorado River in Grand Canyon: Arizona-Nevada Academy of Science Journal, v. 20, p. 111-120.

Beus, S.S., Dalton, R.O., Stevenson, G.M., Reed, V.S., and Daneker, T.M., 1974, Preliminary report on the Unkar Group (Precambrian) in Grand Canyon, Arizona, in Karlstrom, T.N.V., Swann, G.A., and Eastwood, R.L., eds., Geology of northern Arizona, with notes on archeology and paleoclimate, Pt. 1, Regional studies: Geological Society of America, Rocky Mountain Section Meeting, Flagstaff, Arizona, p. 34-53.

Bhattacharji, S., 1967, Scale model experiments on flowage differentiation in sills, in Wyllie, P.J., ed., Ultramafic and related Rocks: New York, John Wiley, p. 69-70.

Bhattacharji, S., and Smith, C.H., 1964, Flowage differentiation: Science, v. 145, p. 150-153.

Billingsley, G.H., 1974, Mining in the Grand Canyon, in Breed, W.J., Roat, E.C., eds., Geology of the Grand Canyon: Flagstaff, Museum of Northern Arizona and Grand Canyon Natural History Association, p. 170-178.

Billingsley, G.H., 1976, Prospectors proving ground: Arizona Historical Journal, v. 17, no. 1, p. 69-88.

Billingsley, G.H., 1978, A synopsis of stratigraphy in the western Grand Canyon: Flagstaff, Arizona, Museum of Northern Arizona, Research Paper 16, 27 p.

Billingsley, G.H., and Beus, S.S., 1985, The Surprise Canyon Formation, an Upper Mississippian and Lower Pennsylvanian(?) rock unit in the Grand Canyon, Arizona: U.S. Geological Survey Bulletin 1605A, p. A27-A33.

Billingsley, G.H., and Huntoon, P.W., 1983, Geologic map of Vulcan's Throne and vicinity, western Grand Canyon, Arizona: Grand Canyon Natural History Association, scale 1:48,000.

Billingsley, G.H., and McKee, E.D., 1982, Pre-Supai buried valleys, in McKee, E.D., ed. The Supai Group of Grand Canyon: U.S. Geological Survey Professional Paper 1173, p. 137-147.

Billingsley, G.H., Wenrich, K.J., and Huntoon, P.W., 1986a, Breccia pipe and geologic map of the southeastern Hualapai Indian Reservation and vicinity, Arizona: U.S. Geological Survey Open-File Report, 86-458B, Flagstaff, Arizona, scale 1:48,000, 26 p., [In review]

Billingsley, G.H., Wenrich, K.J., and Huntoon, P.W., 1986b, Breccia pipe and geologic map of the southwestern Hualapai Indian Reservation and vicinity, Arizona: U.S. Geological Survey Open-File Report 86-458D, Flagstaff, Arizona, scale 1:48,000.

Bilodeau, W.L., 1986, The Mesozoic Mogollon highlands, Arizona--an Early Cretaceous rift shoulder: Journal of Geology, v. 94, p. 724-735.

Blackwell, D.D., 1978, Heat flow and energy loss in the western United States, in Smith, R.B., and Eaton, G.P., eds., Cenozoic tectonics and regional geophysics of the western cordilleran: Geological Society of America Memoir 152, p. 175-208.

Bloeser, Bonnie., Schopf, J.W., Horodyski, R.J and Breed, W.J., 1977, Chitinozoans from the Late Precambrian Chuar Group of the Grand Canyon, Arizona: Science, v. 195, p. 676-679.

Boyce, J.M., 1972, The structure and petrology of the Older Precambrian crystalline rocks, Bright Angel Canyon, Grand Canyon Arizona: Flagstaff, Arizona, Northern Arizona University, M.S. Thesis.

Brathovde, J.E., 1986, Stratigraphy of the Grand Wash Dolomite (Upper? Cambrian), western Grand Canyon, Mohave County, Arizona: Flagstaff, Arizona, Northern Arizona University, M.S. Thesis 140 p.

Breed, W.J., and Ford, T.D., 1973, Chapter two-and-a-half of Grand Canyon history, or the Sixty Mile Formation: Plateau, Museum of Northern Arizona, Flagstaff, Arizona, v. 46, no. 1, p. 12-18.

Bressler, S.L., 1981, Preliminary poles and correlation of the Proterozoic Uinta Mountain Group, Utah and Colorado: Earth and Planetary Science Letters, v. 55, p. 53-64.

Bressler, S.L., and Butler, R.F., 1978, Magnetostratigraphy of the late Tertiary Verde Formation, central Arizona: Earth and Planetary Science Letters, v. 38, p. 319-330.

Bressler, S.L., and Elston, D.P., 1980, Declination and inclination errors in experimentally deposited specularite-bearing sand: Earth and Planetary Science Letters, v. 48, p. 227-232.

Brian, N.J., and Thomas, J.R., 1984, 1983 Colorado River beach campsite inventory, Grand Canyon National Park, Arizona: Division of Resources Management, Grand Canyon National Park Report, 56 p.

Brown, E.H., Babcock, R.S., Clark, M.D., and Livingston, D.E., 1979, Geology of the Older Precambrian rocks of the Grand Canyon, Part I, Petrology and structure of the Vishnu Complex: Precambrian Research v. 8, p. 219-241.

Burkham, D.E., 1986, Trends in selected hydraulic variables for the Colorado River at Lees Ferry and near Grand Canyon, Arizona, 1922-84: U.S. Bureau of Reclamation Glen Canyon Environmental Studies Report, 51 p.

Butler, R.F., and Taylor, L.H., 1978, A middle Paleocene paleomagnetic pole from the Nacimiento Formation, San Juan Basin, New Mexico: Geology, v. 6, p. 495-498.

Campbell, Ian., and Maxson, J.H., 1938, Geological studies of the Archean rocks at Grand Canyon, Carnegie Institution of Washington Yearbook 37, p. 359-364.

Chenoweth, W.L., 1974, The karstic ground water basins of the Kaibab Plateau, Arizona: Washington, D.C., American Geophysical Union, Water Resources Research, v. 10, p. 579-590.

Chenoweth, W.L., 1986, The Orphan Lode mine, Grand Canyon, Arizona, a case history of a mineralized, collapse-breccia pipe: U.S. Geological Survey Open-file Report 86-510, 126 p.

Clark, M.D., 1976, The geology and petrochemistry of the Precambrian metamorphic rocks of the Grand Canyon, Arizona, Leicester, England, University of Leicester, Ph.D. Dissertation.

Clark, M.D.,1978, Amphibolitic rocks from the Precambrian of Grand Canyon: mineral chemistry and phase petrology: Mineralogical Magazine, 42, 199-207.

Clark, M.D.,1979, Geology of the Older Precambrian rocks of the Grand Canyon, Part III, Petrology of the mafic schists and amphibolites: Precambrian Research v. 8, p. 277-302.

Cloud, Preston, 1988, Oasis in Space: New York, Norton and Company, 506 p.

Coates, R.R., 1968, Basaltic andesites, in Hess, H.H., and Poldervaart, Arie, eds., Basalts, the Poldervaart treatise on rocks of basaltic composition: New York, John Wiley, p. 689-736.

Conway, C.M., 1986, Field guide to Early Proterozoic strata that host massive sulfide deposits at Bagdad, Arizona, in Nations, J.D., Conway, C.M., and Swann, G.A., eds., Geology of central and northern Arizona: Geological Society of America, Guidebook, Rocky Mountain Section Meeting, Flagstaff, Arizona, p. 140-157.

Cooley, M.E., 1960, Analysis of gravels in Glen-San Juan Canyon Region, Utah and Arizona: Arizona Geological Digest, v. III, p. 19-24.

Cooley, M.E., (comp.), 1967, Arizona highway geologic map: Arizona Geological Society, scale 1:1,000,000.

Cooley, M.E., and Akers, J.P., 1961, Ancient erosional cycles of the Little Colorado River, Arizona and New Mexico: U.S. Geological Survey, Professional Paper 424-C, p. C244-C248.

Cooley, M.E., Aldridge, B.N., and Euler, R.C., 1977, Effects of the catastrophic flood of December 1966, North Rim area, eastern Grand Canyon, Arizona: U.S. Geological Survey Professional Paper 980, 43 p.

Crampton, C.G., 1972, Land of living rock, the Grand Canyon and the high plateaus, Arizona, Utah, and Nevada: Alfred A. Knopf, New York, 268 p.

Dalton, R.O., Jr., 1972, Stratigraphy of the Bass Formation: Flagstaff, Arizona, Northern Arizona University, M.S.Thesis, 140 p.

Dalton, R.O., Jr., and Rawson, R.R., 1974, Stratigraphy of the Bass Limestone, Grand Canyon, Arizona [abs.]: Geological Society of America, Abstracts with Programs, v. 6, p. 437.

Damon, P.E., 1968, Application to the dating of igneous and metamorphic rocks with the Basin Ranges, in Hamilton and Farquhar, F.D., eds., Radiometric dating for geologists: New York, Interscience, p. 36-44.

Damon, P.E., 1979, Continental uplift at convergent boundaries: Tectonophysics, v. 61, p. 307-319.

Damon, P.E., Shafiqullah, Muhammed, and Scarborough, R.B., 1978, Revised chronology for critical stages in the evolution of the lower Colorado River [abs.]: Geological Society of America Abstracts with Programs, v. 10, no. 3, p. 101.

Daneker, T.M., 1974, Sedimentology of the Precambrian Shinumo Quartzite, Grand Canyon, Arizona [abs.], Geological Society of America Abstracts with Programs,v. 6, no. 5, p. 438.

Daneker, T.M., 1975, Sedimentology of the Precambrian Shinumo Sandstone, Grand Canyon, Arizona: Flagstaff, Arizona, Northern Arizona University, M.S. Thesis, 195 p.

Darton, N.H., 1910, A reconnaissance of parts of northwestern New Mexico and northern Arizona: U.S. Geological Survey Bulletin, 485, 88 p.

Davis, G.A., Anderson, J.L., Frost, E.G., and Shackelford, T.J., 1980. Mylonitization and detachment faulting in the Whipple-Buckskin-Rawhide Mountains terrane, southeastern California and western Arizona, in Crittenden, M.D., Coney, P.J., and Davis, G.H., eds., Cordilleran metamorphic core complexes: Geological Society of America Memoir 153, p. 79-129.

Davis, G.H., 1978, Monoclinal fold pattern of the Colorado Plateau in Matthews, Vincent III, ed., Laramide folding associated with basement block faulting in the western United States: Geological Society of America Memoir 151, p. 215-233.

Dickinson, W.R., 1981, Plate tectonic evolution of the southern cordillera: Arizona Geological Society Digest, v. 14, p. 113-135.

Dickinson, W.R., Klute, M.A., Hayes, M.J., Janecke, S.U., Lundin, E.A., McKittrick, M.A., and Olivares, M.D., 1987, Paleogeographic and paleotectonic setting of Laramide sedimentary basins in the central Rocky Mountain region: Geological Society of America Bulletin, v. 100, p. 1120-1130.

Dobyns, H.F., and Euler, R.C., 1976, The Walapai people: in Dobyns, H.F. and Griffin, J.I., eds., Published by Indian Tribal Series, Phoenix, Arizona, Library of Congress Catalog Number 76-19828, 106 p.

Dokka, R.K., Mahaffie, M.J., and Snoke, A.W., 1986, Thermochronologic evidence of major tectonic denudation association with detachment faulting, northern Ruby Mountains-East Humboldt Range, Nevada: Tectonics, v. 7, p. 995-1006.

Dolan, Robert, Howard, A.D., and Gallenson, A., 1974, Man's impact on the Colorado River in the Grand Canyon: American Scientist, v. 62, p. 392-401.

Dolan, Robert, Howard, A.D., and Trimble, D., 1978, Structural control of the rapids and pools of the Colorado River in the Grand Canyon: Science, v. 202, p. 629-631.

Doser, D.I., and Smith, R.B., 1982, Seismic moment rates in the Utah region: Seismological Society of America Bulletin, v. 72, p. 525-551.

Drever, H.I., and Johnston, R., 1967, Picritic minor intrusions, in Wyllie, P.J., ed., Ultramafic and related Rocks: New York, John Wiley, p. 71-82.

Dutton, C.E., 1880, Report on the geology of the High Plateaus of Utah, with atlas: U.S. Geological Survey Rocky Mountain Region, 307 p.

Eddy, Clyde, 1929, Down the world's most dangerous river: New York, Frederic A. Stokes Co., p. 66-71.

Elston, D.P., 1978, Oligocene and Miocene development of mountain region and environs, central Arizona: Evidence for timing of plateau uplift and erosion [abs]: Geological Society of America Abstracts with Programs, v. 10, no. 3, p. 104.

Elston, D.P., 1979, Late Precambrian Sixtymile Formation and orogeny at the top of the Grand Canyon Supergroup, northern Arizona: U.S. Geological Survey Professional Paper 1092, 20 p.

Elston, D.P., 1984a, Magnetostratigraphy of the Belt Supergroup--a synopsis: Missoula, Montana Bureau of Mines and Geology, Special Publication 90, Abstracts with Summaries, Belt Symposium II, p. 88-90.

Elston, D.P., 1984b, Rocks, landforms, and landscape development in central Arizona, in Smiley, T.L., Nations, J.D., Péwé, T.L., and Schafer, J.P., eds., Landscapes of Arizona: New York, University Press of America, p. 151-173.

Elston, D.P., 1986, Magnetostratigraphy of Late Proterozoic Chuar Group and Sixtymile Formation, Grand Canyon Supergroup, northern Arizona: Correlation with other Proterozoic strata of North America [abs]: Geological Society of America Abstracts with Programs, v. 18, no. 5, p. 353.

Elston, D.P., 1988a, Grand Canyon Supergroup, Northern Arizona: stratigraphic summary and preliminary paleomagnetic correlations with parts of other North American Proterozoic successions: in Reynolds, S.J., and Jenney, J.P., eds., Summary of Arizona Geology; Arizona Geological Society Digest, v. 17, [In press].

Elston, D.P., 1988b, Paleomagnetic poles from Cambrian, Devonian, Pennsylvanian, and Permian strata of Arizona and Montana [abs.]: Geological Society of American Abstracts with Programs, v. 20, no. 7, p. A350.

Elston, D.P., and Bressler, S.L., 1977, Paleomagnetic poles and polarity zonation from Cambrian and Devonian strata of Arizona: Earth and Planetary Science Letters, v. 36, p. 423-433.

Elston, D.P., and Bressler, S.L., 1980, Paleomagnetic poles and polarity zonation from the Middle Proterozoic Belt Supergroup, Montana and Idaho: Journal of Geophysical Research, v. 85, p. 339-355.

Elston, D.P., and Bressler, S.L., 1984, Devonian pole from Montana, and refined Paleozoic polar path for North America [Abs]: American Geophysical Union Transactions (EOS), v. 65, no. 45, p. 864-865.

Elston, D.P., and Bressler, S.L., in review, Magnetostratigraphically controlled polar path and polarity zonation from Middle and Late Proterozoic rocks of North America.

Elston, D.P., and Grommé, C.S., 1974, Precambrian polar wandering from Unkar Group and Nankoweap Formation, eastern Grand Canyon, Arizona, in Karlstrom, T.N.V., Swann, G.A., and Eastwood, R.L., eds., Geology of northern Arizona with notes on archaeology and paleoclimate, Pt. 1, Regional studies: Geological Society of America, Rocky Mountain Section Meeting, Flagstaff, Arizona, p. 97-117.

Elston, D.P., and Grommé, C.S., 1979, Paleomagnetic correlation of Middle Proterozoic strata of Arizona and Lake Superior [abs]: American Geophysical Union Transactions (EOS), v. 60, no. 18, p. 236.

Elston, D.P., and Grommé, C.S., 1984, Stratigraphically controlled polar path from Proterozoic sequences of North America [abs]: American Geophysical Union Transactions (EOS), v. 65, no. 45, p. 865.

Elston, D.P., and Grommé, C.S., in review, Magnetostratigraphic pole path and polarity zonation from Middle Proterozoic rocks of Grand Canyon, northern Arizona: Correlations with Middle Proterozoic rocks of central Arizona and Keweenawan rocks of Lake Superior.

Elston, D.P., and McKee, E.H., 1982, Age and correlation of the Late Proterozoic Grand Canyon disturbance, northern Arizona: Geological Society of America Bulletin, v. 93, p. 681-699.

Elston, D.P., and Purucker, M.E., 1979, Detrital magnetization in red beds of the Moenkopi Formation (Triassic), Gray Mountain, Arizona: Journal of Geophysical Research, v. 84,, p. 1653-1665.

Elston, D.P., and Scott, G.R., 1973, Paleomagnetism of some Precambrian basaltic flows and red beds, eastern Grand Canyon, Arizona: Earth and Planetary Science Letters, v. 18, p. 253-265.

Elston, D.P., and Scott, G.R., 1976, Unconformity at the Cardenas-Nankoweap contact (Precambrian), Grand Canyon Supergroup, northern Arizona: Geological Society of America Bulletin, v. 87, p. 1763-1772.

Epis, R.C., and Chapin, C.E., 1975, Geomorphic and tectonic implications of the post-Laramide, late Eocene erosion surface in the southern Rocky Mountains in Cenozoic history of the southern Rocky Mountains: Geological Society of America Memoir 144, p. 45-74.

Evitt, W.R., 1963, A discussion and proposals concerning fossil dinaflagellates, hystrichospheres, and acritarchs, II: National Academy of Sciences, Proceedings, v. 49, p. 298-302.

Feng, J., Perry, E.C., and Horodyski, R.J., 1986, Sulfur isotope geochemistry of sulfate of the Belt Supergroup and the Grand Canyon Supergroup [Abs]: Geological Society of America Abstracts with Programs, v. 18, no. 5, p. 353.

Fenneman, N.M., 1946, Physical divisions of the United States: U.S. Geological Survey Map: Washington, D.C., scale 1:700,000.

Fisher, R.A., Dispersion on a sphere: Proceedings of the Royal Society of London, Series A, v. 217, p. 295-305.

Fisher, D.J., Erdman, C.E., and Reeside, J.B., 1960, Cretaceous and Tertiary formations of the Book Cliffs, Carbon, Emery, and Grand Counties, Utah, and Garfield and Mesa Counties, Colorado: U.S. Geological Survey Professional Paper 332, 80 p.

Fitton, J.G., James, D., Kempton, P.D., Ormerod, D.S., and Leeman, W.P., 1988, The role of lithospheric mantle in the generation of late Cenozoic basic magmas in the western United States, in Menzies, M.A., and Cox, K.G., eds., Continental and oceanic lithosphere--similarities and differences: Journal of Petrology (Special Volume), p. 331-349.

Fleischer, R.L., Price, P.B., and Walker, R.M., 1975, Nuclear tracks in solids, principles and applications: Berkeley, University of California Press, 605 p.

Ford, T.D., and Breed, W.J., 1970, Carbon Butte, an unusual landslide in the Grand Canyon: Plateau, Museum of Northern Arizona, Flagstaff, v. 43, p. 9-15.

Ford, T.D., and Breed, W.J., 1972a, The Chuar Group of the Proterozoic, Grand Canyon, Arizona: International Geological Congress, 1972, Montreal, Canada, 24th, Proceedings, Section 1, Precambrian Geology, p. 3-10.

Ford, T.D., and Breed, W.J., 1972b, The problematical Precambrian fossil Chuaria: International Geological Congress, 1972, Montreal, Canada, 24th, Proceedings, Section 1, Precambrian Geology, p. 11-18.

Ford, T.D., and Breed, W.J., 1973, Late Precambrian Chuar Group, Grand Canyon, Arizona: Geological Society of America Bulletin, v. 84, p. 1243-1260.

Ford, T.D., Breed, W.J., and Mitchell, J.S., 1972, Name and age of the upper Precambrian basalts in the eastern Grand Canyon: Geological Society of America Bulletin, v. 83, p. 223-226.

French, A.N., and Van Der Voo, Rob, 1979, The magnetization of the Rose Hill Formation at the classical site of Graham's fold test: Journal of Geophysical Research, v. 77, p. 7688-7696.

Gilbert, G.K., 1874, Preliminary geological report, in U.S. Army Engineer Dept., Progress report upon geographical and geological explorations and surveys west of the one hundredth meridian, in 1872: Washington, D.C., U.S. Government Printing Office, App. D., p. 48-52.

Gilbert, G.K., 1875, Report on the geology of portions of Nevada, Utah, California, and Arizona, examined in the years 1871 and 1872, in U.S. Army Engineer Dept., Report upon geographical and geological explorations and surveys west of the one hundredth meridian, in charge of First Lieut. Geo. M. Wheeler, vol. 3, Geology: Washington, D.C., U.S. Government Printing Office, p. 17-187.

Giletti, B.J., and Damon, P.E., 1961, Rubidium-strontium ages of some basement rocks from Arizona and northwestern Mexico: Geological Society of America Bulletin v. 72, p. 639-644.

Gleadow, A.J.W., Duddy, I.R., Green, P.F., and Hegarty, K.A., 1986a, Fission track lengths in the apatite annealing zone and the interpretation of mixed ages: Earth and Planetary Science Letters, v. 78, p. 245-254.

Gleadow, A.J.W., Duddy, I.R., Green, P.F., and Lovering, J.F., 1986b, Confined fission track lengths in apatite: a diagnostic tool for thermal history analysis: New York, Springer-Verlag, Contributions to Mineralogy and Petrology, v. 94, p. 405-415.

Gleadow, A.J.W., Duddy, I.R., and Lovering, J.F., 1983, Fission track analysis: a tool for the evaluation of thermal histories and hydrocarbon potential: Australian Petroleum Exploration Association Journal, v. 23, p. 93-102.

Gleadow, A.J.W., Hurford, A.J., and Quaife, R.D., 1976, Fission track dating of zircon: improved etching techniques: Earth and Planetary Science Letters, v. 33, p. 273-276.

Gomez, Ernest, 1979, Geology of the south central part of the New River Mesa Quadrangle, Cave Creek area, Maricopa County, Arizona: U.S. Geological Survey Open-File Report 79-1312, Flagstaff, Arizona, scale 1:12,000, 156 p..

Gomez, Ernest, and Elston, D.P., 1978, Oligocene and Miocene development of the mountain-desert region boundary, Cave Creek Arizona [abs.]: Geological Society of America Abstracts with Programs, v. 10, no. 3, p. 107.

Graf, W.L., 1979, Rapids in canyon rivers: Journal of Geology, v. 87, p. 533-551.

Gray, R.L., 1964, Cenozoic geology of Hindu Canyon, Mohave County, Arizona: Arizona Academy of Science Journal, v. 3, no. 1, p. 39-42.

Gresans, R.L., 1981, Extension of the Telluride erosion surface to Washington State and its regional and tectonic significance: Tectonophysics v. 79, p. 145-164.

Hamblin, W.K., 1965, Origin of reverse drag on the downthrown side of normal faults: Geological Society of America Bulletin, v. 76, p. 1145-1164.

Hamblin, W.K., 1970, Late Cenozoic basalt flows of the western Grand Canyon, in Hamblin, W.K., and Best, M.G., eds., The western Grand Canyon District: Utah Geological Society, Guidebook to the geology of Utah, no. 23, p. 21-38.

Hamblin, W.K., 1989, Late Cenozoic lava dams in the western Grand Canyon: Provo, Utah, Brigham Young University geology studies, special publication 6, 350 p.

Hamblin, W.K., and Rigby, J.K., 1968, Guidebook to the Colorado River, Part 1, Lees Ferry to Phantom Ranch in Grand Canyon National Park: Provo, Utah, Brigham Young University geology studies, v. 15, Pt. 5, 84 p.

Hamilton, W.B., 1982, Structural evolution of the Big Maria Mountains, northeastern Riverside County, southeastern California, in Frost, E.G., and Martin, D.L., eds., Mesozoic-Cenozoic tectonic evolution of the Colorado River region, California, Arizona, and Nevada: San Diego, California, Cordilleran Publishers, p. 1-27.

Harrison, J.., and Peterman, Z.E., 1982, Adoption of geochrono metric units for divisions of Precambrian time: American Association of Petroleum Geologists Bulletin, v. 66, p. 801-804.

Harrison, T.M., Armstrong, R.L., Naeser, C.W., and Harakal, J.E., 1979, Geochronology and thermal history of the Coast Plutonic Complex, near Prince Rupert, British Columbia: Canadian Journal of Earth Sciences, v. 16, p. 400-410.

Hartman, J.H., 1984, Systematics, biostratigraphy, and biogeography of Latest Cretaceous and early Tertiary Viviparidae (Mollusca, Gastropoda) of southern Saskatchewan, western North Dakota, eastern Montana, and northern Wyoming: Minneapolis, Minnesota, University of Minnesota, Ph.D. Dissertation, 919 p.

Haq, B.U., Hardenbol, J., Vail, P.R., 1987, Chronology of fluctuating sea levels since the Triassic: Science, v. 235, p. 1156-1167.

Hayes, P.T., and Drewes, Harold, 1978, Mesozoic depositional history of southeastern Arizona, in Callender, J.F., Wilt, J.C., and Clemons, R.E., eds., Land of Cochise: New Mexico Geological Society, 29th field conference guidebook, p. 201-207.

Hendricks, J.D., and Lucchitta, Ivo, 1974, Upper Precambrian igneous rocks of the Grand Canyon, Arizona, in Karlstrom, T.N.V., Swann, G.A., and Eastwood, R.L., eds., Geology of northern Arizona, with notes on archaeology and paleoclimate, Pt. 1, Regional studies: Geological Society of America, Rocky Mountain Section Meeting, Flagstaff, Arizona, p. 65-86.

Hereford, Richard, 1984a, Climate and ephemeral-stream processes: Twentieth-Century geomorphology and alluvial stratigraphy of the Little Colorado River, Arizona: Geological Society of America Bulletin, v. 95, p. 654-668.

Hereford, Richard, 1984b, Driftwood in Stanton's Cave--the case for temporary damming of the Colorado River at Nankoweap Creek in Marble Canyon, Grand Canyon National Park, Arizona, in Euler, R.E., ed., The archeology, geology, and paleobiology of Stanton's Cave, Grand Canyon National Park, Arizona: Grand Canyon Natural History Association Monograph 6, p. 99-106.

Hereford, Richard, 1986, Modern alluvial history of the Paria River drainage basin, Southern Utah: Quaternary Research, v. 25, p. 293-311.

Hintze, L.F., 1975, Geological highway map of Utah: Provo, Utah, Brigham Young University Department of Geology, special publication 3, scale 1:1,000,000.

Hjulström, F., 1935, Studies of the morphological activity of rivers as illustrated by the river Fyris: University of Uppsala (Sweden) Geological Institute Bulletin, v. 25, p. 221-227.

Horodyski, R.J., 1986, Paleontology of the late Precambrian Chuar Group, Grand Canyon, Arizona [abs.]: Geological Society of America Abstracts with Programs, v. 18, no. 5, p. 362.

Horodyski, R.J., and Bloeser, Bonnie., 1983, Possible eukaryotic algal filaments from the Late Proterozoic Chuar Group, Grand Canyon, Arizona: Journal of Paleontology, v. 57, p. 321-326.

Howard, A.D., and Dolan, Robert, 1976, Changes in fluvial deposits of the Colorado River in the Grand Canyon caused by Glen Canyon Dam: First Conference on scientific research in the National Parks, Nov. 9-12, New Orleans, Louisiana, Proceedings, p. 845-851.

Howard, A.D., and Dolan, Robert, 1981, Geomorphology of the Colorado River in the Grand Canyon: Journal of Geology, v. 89, p. 269-298.

Hughes, J.D., 1967, The story of man at Grand Canyon: Grand Canyon Natural History Association Bulletin, 14, p. 57-93.

Hunt, C.B., 1956, Cenozoic geology of the Colorado Plateau: U.S. Geological Survey Professional Paper 291, 74 p.

Hunt, C.B., 1969, Geologic history of the Colorado River, in The Colorado River region and John Wesley Powell: U.S. Geological Survey Professional Paper, 669, p. 59-130.

Hunt, C.B.,1974, Grand Canyon and the Colorado River, their geologic history in Breed, W.J., and Roat, E.C., eds., Geology of the Grand Canyon: Flagstaff, Arizona, Museum of Northern Arizona, p. 129-141.

Huntoon, P.W., 1973, High angle gravity faulting in the eastern Grand Canyon, Arizona: Plateau, Museum of Northern Arizona, Flagstaff, Arizona, v. 45, p. 117-127.

Huntoon, P.W., 1974, The karstic ground water basins of the Kaibab Plateau, Arizona: Washington, D.C., American Geophysical Union, Water Resources Research, v. 10, p. 579-590.

Huntoon, P.W., 1975, The Surprise Valley landslide and widening of the Grand Canyon: Flagstaff, Arizona, .Plateau, Museum of Northern Arizona, , v. 48, p. 1-12.

Huntoon, P.W., 1977a, Cambrian stratigraphic nomenclature and ground-water prospecting failures on the Hualapai Plateau, Arizona: Dublin, Ohio, National Water Well Association, Water Well Journal Publishing Company. Ground Water, v. 15, p. 426-433.

Huntoon, P.W., 1977b, Relationship of tectonic structure to aquifer mechanics in the western Grand Canyon district, Arizona: Water Resources Research Institute, Water Resources Series, no. 66, 51 p.

Huntoon, P.W., 1981, Grand Canyon monoclines, vertical uplift or horizontal compression?: Laramie, Wyoming, University of Wyoming, Contributions to Geology, v. 19, p. 127-134.

Huntoon, P.W., 1988, Early Tertiary reorganization of stresses, southwestern Colorado Plateau, Arizona, observed strain and a hypothesis on cause [abs.]: Geological Society of America Abstracts with Programs, v. 20, no. 3, p. 170.

Huntoon, P.W., Billingsley, G.H., Breed, W.J., Sears, J.W., Ford, T.D., Clark, M.D., Babcock, R.S., and Brown, E.H., 1976, Geologic map of the Grand Canyon National Park, Arizona: Grand Canyon Natural History Association and Museum of Northern Arizona, Flagstaff, Arizona, scale 1:62,500.

Huntoon, P.W., Billingsley, G.H., Breed, W.J., Sears, J.W., Ford, T.D., Clark, M.D., Babcock, R.S., and Brown, E.H., 1986, Geologic map of the eastern part of the Grand Canyon National Park, Arizona: Grand Canyon Natural History Association and Museum of Northern Arizona, Flagstaff, Arizona, scale 1:62,500.

Huntoon, P.W., Billingsley, G.H., and Clark, M.D., 1981, Geologic map of the Hurricane fault zone and vicinity, western Grand Canyon, Arizona: Grand Canyon Natural History Association, Grand Canyon, Arizona, scale 1:48,000.

Huntoon, P.W., Billingsley, G.H., and Clark, M.D., 1982, Geologic map of the Lower Granite Gorge and vicinity, western Grand Canyon, Arizona: Grand Canyon Natural History Association, Grand Canyon, Arizona, scale 1:48,000.

Huntoon, P.W., and Elston, D.P., 1980, Origin of the river anticlines, central Grand Canyon: U.S. Geological Survey Professional Paper 1126A, 9 p.

Huntoon, P.W., and Sears, J.W., 1975, Bright Angel and Eminence faults, eastern Grand Canyon, Arizona: Geological Society of America Bulletin, v. 86, p. 465-472.

Hurford, A.J., 1986, Cooling and uplift patterns in the Lepontine Alps, South Central Switzerland and an age of vertical movement on the Insubric fault line: New York, Springer-Verlag, Contributions to Mineralogy and Petrology, v. 92, p. 413-427.

Iddings, J.P., 1894, Notes on the petrographic character of the lavas, in Walcott, C.D., Precambrian igneous rocks of the Unkar terrain, Grand Canyon of the Colorado, Arizona: U.S. Geological Survey Annual Report 14, pt. 2, p. 497-524.

Johnson, P.W., and Sanderson, R.B., 1968, Spring Flow to the Colorado River, Lees Ferry to Lake Mead: Phoenix, Arizona, Arizona State Land Department, Water Resources Report 34, 26 p.

Karlstrom, K.E., Bowring, S.A., and Conway, C.M., 1987, Tectonic significance of an Early Proterozoic two-province boundary in central Arizona: Geological Society of America Bulletin 99, p. 529-538.

Keller, G.R., Smith, R.B., and Braile, L.W., 1975, Crustal structure along the Great Basin-Colorado Plateau transition from seismic reflection studies: Journal of Geophysical Research, v. 80, p. 1093-1098.

Keller, G.R., Smith, R.B., Braile, L.W., Heaney, R., and Shurbet, D.H., 1976, Upper crustal structure of the eastern Basin and Range, northern Colorado Plateau, and middle Rocky Mountains from rayleigh wave dispersion: Seismological Society of America Bulletin, v. 66, p. 869-876.

Kelley, V.C., 1955, Monoclines of the Colorado Plateau: Geological Society of America Bulletin, v. 66, p. 789-804.

Kent, D.V., and Opdyke, N.D., 1980, Paleomagnetism of Siluro-Devonian rocks from eastern Maine: Canadian Journal of Earth Sciences, v. 17, p. 1653-1665.

Keroher, G.C., 1970, Lexicon of geologic names of the United States: U.S. Geological Survey Bulletin 1350, 848 p.

Keyes, Charles, 1938, Basement complex of the Grand Canyon: Pan American Geologist, v. 20, no. 2, p. 91-116.

Kieffer, S.W., 1985, The 1983 hydraulic jump in Crystal Rapid: implications for river-running and geomorphic evolution in the Grand Canyon: Journal of Geology, v. 93, p. 385-406.

Kieffer, S.W., 1987a, The rapids and waves of the Colorado River, Grand Canyon Arizona: U.S. Geological Survey, Open-File Report 87-096, Flagstaff, Arizona, 98 pp.

Kieffer, S.W., 1987b, Hydraulics of the rapids of the Colorado River, Grand Canyon, Arizona, a 20-minute video: U.S. Geological Survey Open-File Report 86-503, Flagstaff, Arizona, 1986.

Kieffer, S.W.,1988, Hydraulic maps of the ten major rapids in the Grand Canyon: U.S. Geological Survey Miscellaneous Investigations Maps A-J.

Koons, E.D., 1948, High level gravels of western Grand Canyon: Science, v. 107, p. 475-476.

Koons, E.D., 1955, Cliff retreat in the southwestern United States: American Journal of Science, v. 253, p. 44-52.

Krewedl, D.A., and Carisey, J.C., 1986, Contributions to the geology of uranium mineralized breccia pipes in northern Arizona: Arizona Geological Society Digest, v. 16, p. 179-186.

Kuno, H., 1968, Differentiation of basalt magmas, in Hess, H.H., and Poldervaart, Arie, eds., Basalts, the Poldervaart treatise on rocks of basaltic composition: New York, Wiley and Sons, p. 689-736.

Kwarteng, A.Y., Chavez, P.S., Jr., Wenrich, K.J., and Goodell, P.C., in press, Application of Landsat Thematic Mapper digital data to the exploration for uranium-mineralized breccia pipes in northwestern Arizona: ERIM Meeting, Houston, Texas 14 p.

Lander, E.B., 1977, A review of the Oreodonta (Mammalia,

Artiodactylia), Parts I, II, and III: Berkeley, California, University of California, Ph.D. Dissertation, 474 p.

Lara, J.M., and Sanders, J.I., 1970, 1963-1964 Lake Mead Survey: Denver, Colorado, U.S. Bureau of Reclamation, Res. Rep. Reg. OCE-70-21, 162 p

Laursen, E.M., Ince, Simon, and Pollack, Jack, 1976, On sediment transport through Grand Canyon: Third Federal Interagency Sedimentation Conference Proceedings, p. 4-76 to 4-87.

Leeman, W.P., 1974, Late Cenozoic alkali-rich basalt from the western Grand Canyon area, Utah and Arizona--isotopic composition of strontium: Geological Society of America Bulletin, v. 85, p. 1691-1696.

Leopold L.B., 1969, The rapids and the pools-Grand Canyon in the Colorado River Region and John Wesley Powell: U.S. Geological Survey Professional Paper 669, p. 131-145.

Levings, G.W., 1980, Water resources in the Sedona area, Yavapai and Coconino Counties, Arizona: Arizona Water Commission Bulletin 11, Phoenix, Arizona. Prepared by the U.S. Geological Survey in cooperation the Arizona Water Commission, 3 maps, scale 1:24,000, 37 p.

Lindsay, E.H., and Lundin, R.F., 1972, An Oligocene oreodont (Mammalia, Artiodactylia) from central Arizona: Journal of Paleontology, v. 46, p. 115-119.

Lingley, W.S., 1973, Geology of the Older Precambrian rocks in the vicinity of Clear Creek and Zoroaster Canyon, Grand Canyon, Arizona: Bellingham, Washington, Western Washington State, M.S. Thesis.

Lingley, W.S., 1976, Some structures in Older Precambrian rocks of the Clear Creek-Cremation Creek area, Grand Canyon National Park, Arizona: Arizona Geological Society Digest, X, p. 27-35.

Livingston, D.E., Brown, E.H., Babcock, R.S., and Clark, M.D., 1974, Rb-Sr whole rock isochron ages for "Older" Precambrian plutonic and metamorphic rocks of the Grand Canyon, Arizona [abs.]: Geological Society of America Abstracts with Programs, v. 6, no. 6, p. 848.

Longwell, C.R., 1936, Geology of the Boulder Reservoir floor, Arizona-Nevada: Geological Society of America Bulletin, v. 47, p. 1393-1476.

Loughlin, W.D., and Huntoon, P.W., 1983, Compilation of available ground water quality data for sources within the Grand Canyon of Arizona: Laramie, Wyoming, University of Wyoming Department of Geology and Geophysics, 9 p., and appendices.

Lucchitta, Ivo, 1972, Early history of the Colorado River in the Basin and Range province: Geological Society of America Bulletin, v. 83, p. 1933-1948.

Lucchitta, Ivo, 1979, Late Cenozoic uplift of the southwestern Colorado Plateau and adjacent lower Colorado River region: Tectonophysics, v. 61, p. 63-95.

Lucchitta, Ivo, 1984, Development of the landscape in northern Arizona--the country of plateaus and canyons, in Smiley, T.L., Nations, J.D., Péwé, T.L., and Schafer, J.P., eds., Landscapes of Arizona: New York, University Press of America, p. 269-302.

Lucchitta, Ivo, 1987, The mouth of the Grand Canyon and edge of the Colorado Plateau in the upper Lake Mead area, Arizona, in Beus, S.S., ed., Rocky Mountain Section Centennial Field Guide: Boulder, Colorado, Geological Society of America, v. 2., p. 365-370.

Lucchitta, Ivo, 1988, Canyon maker - a geological history of the Colorado River: Flagstaff, Arizona, Museum of Northern Arizona, Plateau, v. 39, no. 2, 32 p.

Lucchitta, Ivo, and Hendricks, J.D., 1972, Precambrian lava flows at Basalt Canyon, eastern Grand Canyon National Park, Arizona [abs.]: Geological Society of America Abstracts with Programs, v. 4, no. 5, p. 388-389.

Lucchitta, Ivo, and Hendricks, J.D., 1974, Spilitic alteration of the Precambrian Cardenas Lavas, Grand Canyon, Arizona [Abs.] Geological Society of America Abstracts with Programs, v. 6, no. 5, p. 454-455.

Lucchitta, Ivo, and Hendricks, J.D., 1983, Characteristics, depositional environment, and tectonic interpretations of the Proterozoic Cardenas lavas, eastern Grand Canyon, Arizona: Geology, v. 11, p. 177-181.

Lucchitta, Ivo, Hendricks, J.D., and Hereford, Richard, 1983, Revision to the Middle Proterozoic paleomagnetic path for the eastern Grand Canyon, Arizona, on the basis of reassignment of two key stratigraphic units [abs.]: Geological Society of America Abstracts with Programs, v. 14, no. 5, p. 422.

Lucchitta, Ivo, and Suneson, Neil, 1982, Relationship of core complexes, Cenozoic extensional tectonism and the stable craton near the Colorado Plateau margin in western Arizona [abs.]: Geological Society of America Abstracts with Programs, v. 14, no. 4, p. 182.

Ludwig, K.R., Rasmussen, J.D., and Simmons, K.R., 1986, Age of uranium ores in collapse-breccia pipes in the Grand Canyon area, northern Arizona [abs.]: Geological Society of America Abstracts with Programs, v. 18, no. 5, p. 392.

McCabe, Chad, and Van der Voo, Rob, 1983, Paleomagnetic results from upper Keweenawan Chequamegon Sandstone: Implications for red bed diagenesis and Lake Precambrian apparent polar wander of North America: Canadian Journal of Earth Sciences, v. 20, p. 105-112.

McCabe, Chad, Van der Voo, Rob, Wilkinson, Bruce, and Devaney, Kathleen, 1985. A Middle/Late Silurian paleomagnetic pole from limestone reefs of the Wabash Formation, Indiana, U.S.A.: Journal of Geophysical Research, v. 90, p. 2959-2965.

McKay, E.J., 1972, Geologic map of the Show Low quadrangle, Navajo County, Arizona: U.S. Geological Survey geologic quadrangle map CQ-973.

McKee, E.D., 1933, The Coconino Sandstone--its history and origin, in Papers Concerning the Paleontology of California, Arizona, and Idaho: Carnegie Institution of Washington, Publication 440, p. 77-115.

McKee, E.D., 1938a, Original structures in Colorado River flood deposits of Grand Canyon: Journal of Sedimentary Petrology, v. 8, p. 77-83.

McKee, E.D., 1938b, The environment and history of the Toroweap and Kaibab Formations of northern Arizona and southern Utah: Carnegie Institution of Washington, Publication 492, 268 p.

McKee, E.D., 1963, Nomenclature for lithologic subdivisions of the Mississippian Redwall Limestone, Arizona: U.S. Geological Survey Professional Paper 475-C, p. 21-22.

McKee, E.D., 1975, The Supai Group--subdivision and nomenclature: U.S.Geological Survey Bulletin, 1395-J, p. 1-11.

McKee, E.D., 1982, The Supai Group of Grand Canyon: U.S. Geological Survey Professional Paper 1173, 504 p.

McKee, E.D., and Gutschick, R.C., 1969, History of the Redwall Limestone of northern Arizona: Geological Society of America Memoir 114, 726 p.

McKee, E.D., Hamblin, W.K., and Damon, P.E., 1968, K-Ar age of lava dam in Grand Canyon: Geological Society of America Bulletin, v. 79, p. 133-136.

McKee, E.D., and McKee, E.H., 1972, Pliocene uplift of the Grand Canyon region-time of drainage adjustment: Geological Society of America Bulletin, v. 83, p. 1023-1931.

McKee, E.D., and Resser, C.E., 1945, Cambrian history of ;the Grand Canyon region: Carnegie Institution of Washington Publication 563, 168 p.

McKee, E.D., and Schenk, E.T., 1942, The lower canyon lavas and related features at Toroweap in Grand Canyon: Journal of

Geomorphology, v. 5, p. 245-273.

McKee, E.D., and Weir, G.W., 1953, Terminology for stratification and cross-stratification in sedimentary rocks: Geological Society of America Bulletin, v. 64, p. 381-390.

McKee, E.D., Wilson, R.F., Breed, W.J., and Breed, C.S., eds., 1967, Evolution of the Colorado River in Arizona: Flagstaff, Arizona, Museum of Northern Arizona Bulletin 44, 67 p.

McKee, E.H., and Noble, D.C., 1974, Rb-Sr age of the Cardenas Lavas, Grand Canyon, Arizona, in Karlstrom, T.N.V., Swann, G.A., Eastwood, R.L., eds., Geology of northern Arizona with notes on archaeology and paleoclimate, Pt. 1, Regional studies: Geological Society of America, Rocky Mountain Section Meeting, Flagstaff, Arizona, p. 87-96.

McKee, E.H., and Noble, D.C., 1976, Age of the Cardenas Lavas, Grand Canyon, Arizona: Geological Society of America Bulletin, v. 87, p. 1188-1190.

McKenzie, D., and Bickle, M.J., 1988, The volume and composition of melt generated by extension of the lithosphere: Journal of Petrology, v. 29, p. 625-679.

McNair, A.H., 1951, Paleozoic stratigraphy of part of northwestern Arizona: American Association of Petroleum Geologists Bulletin, v. 35, p, 503-541.

Machette, M.N., 1986, Calcic soils of the southwestern United States, in Weide, D.L., ed., Soils and Quaternary geology of the southwestern United States: Geological Society of America Special Paper 203, p. 1-21

Machette, M.N., Rosholt, J.N., and Bush, C.A., 1986, Uranium-trend ages of Quaternary deposits along the Colorado River, Grand Canyon National Park, Arizona [abs.]: Geological Society of America Abstracts with Programs, v. 18, no. 5, p. 393.

Marley, M.C.,1987, The influence of hydraulics on boat trajectories and boating accidents on the Colorado River, Grand Canyon, Arizona: U.S. Geological Survey Open-File Report, 87-576.

Mathisen, I.W., Jr., 1987, Arizona Strip breccia pipe program: Exploration, development, and production [abs.]: American Association of Petroleum Geologists Bulletin, v. 71, no. 5, p. 590.

Matthes, G.H., 1947, Macroturbulence in natural streamflow: American Geophysical Union, Transactions (EOS), v. 28, no. 2, p. 255-262.

Maxson, J.H., 1961, Geologic map of the Bright Angel quadrangle, Grand Canyon National Park, Arizona: Grand Canyon Natural History Association, scale 1:48,000

Maxson, J.H., 1967, Preliminary geologic map of the Grand Canyon and vicinity, Arizona, eastern section: Grand Canyon Natural History Association, scale 1:62,500.

Maxson, J.H., and Campbell, Ian, 1935, Stream fluting and stream erosion: Journal of Geology, v. 43, p. 729-744.

Mayer, L., 1979, Evolution of the Mogollon Rim in central Arizona: Tectonophysics, v. 61 p. 49-62.

Metcalf, L.A, 1982, Tephrostratigraphy and potassium-argon age determinations of seven volcanic ash layers in the Muddy Creek Formation of southern Nevada: Reno, Nevada, Water Resources Center, University of Nevada System, 187 p.

Molenaar, C.M., 1983, Major depositional cycles and regional correlations of upper Cretaceous rocks, southern Colorado Plateau and adjacent areas, in Reynolds, M.W., and Dolly, E.D., Mesozoic paleogeography of the west-central United States: Rocky Mountain Paleogeography Symposium 2, Rocky Mountain Section, Society of Economic Paleontologists and Mineralogists, p. 201-224.

Moore, R.T., 1968, Mineral deposits of the Fort Apache Indian Reservation, Arizona: Arizona Bureau of Mines Bulletin 177, p. 1-34.

Morris, W.A., 1977, Paleolatitude of glaciogenic upper Precambrian Rapitan Group and the use of tillites as chronostratigraphic marker horizons: Geology, v. 5, p. 85-88.

Naeser, C.W., 1979a Fission-track dating and geologic annealing of fission tracks, in Jager, E., and Hunziker, J.E., eds., Lectures in Isotope Geology: New York, Springer-Verlag, p. 154-169.

Naeser, C.W., 1979b, Thermal history of sedimentary basins: Fission-track dating of subsurface rocks: Society of Economic Paleontologists and Mineralogists Special Publication 26, p. 109-112.

Naeser, C.W., 1981, The fading of fission tracks in the geologic environment--data from deep drill holes: New York, Pergamon Press, Nuclear Tracks, v. 5, p. 248-250.

Naeser, C.W., Bryant, B., Crittenden, M.D., Jr., and Sorensen, M.L., 1983, Fission-track ages of apatite in the Wasatch Mountains, Utah: An uplift study, in Miller, D.M., Todd, V.R., and Howard, K.A., eds., Tectonic and stratigraphic studies in the eastern Great Basin: Geological Society of America Memoir 157, p. 29-37.

Naeser, N.D., and Naeser, C.W., 1984, Fission-track dating, in Mahaney, W.C., ed., Quaternary dating methods: Amsterdam, Elsevier, p. 87-100.

National Park Service, 1977, Environmental assessment, management option for examination of uranium leases, Lake Mead National Recreational Area, Arizona-Nevada: National Park Service, Denver Service Center Report D-1190, 123 p.

National Research Council, 1987, River and Dam Management: A Review of the Bureau of Reclamation's Glen Canyon Environmental Studies: Washington, D.C., National Academy Press, 203 p.

Nations, J.D., Wilt, J.C., Hevly, R.H., 1985, Cenozoic paleogeography of Arizona, in Flores, R.M., and Kaplan, S.S., eds, Cenozoic Paleogeography of the West-Central United States, Rocky Mountain Paleogeography Symposium 3: Rocky Mountain Section, Society of Economic Paleontologists and Mineralogists, p. 335-355.

New York Times, 1957, March 27, Helicopter builds an aerial span to guano cave in Grand Canyon, p. 43, 47.

Noble, L.F., 1914, The Shinumo quadrangle, Grand Canyon district, Arizona: U.S. Geological Survey Bulletin 549, 100 p.

Noble, L.F., 1922, A section of the Paleozoic formations of the Grand Canyon at the Bass Trail: U.S. Geological Survey Professional Paper 131 p. 23-73.

Noble, L.F., and Hunter, J.F., 1916, A reconnaissance of the Archean complex of the Granite Gorge, Grand Canyon, Arizona: U.S. Geological Survey Professional Paper 98-I, p. 95-113.

Ormerod, D.S., Hawkesworth, C.J., Rogers, N.W., Leeman, W.P., and Menzies, M.A., 1988, Tectonic and magmatic transitions in the Western Great Basin, U.S.A.: Nature, v. 333, p. 349-353.

Osborn, E .F., 1969, Role of oxygen pressure in the crystallization and differentiation of basaltic magma: American Journal of Science, v. 257, p. 609-627.

Palmer, H.C., and Hayatsu, A., 1975, Paleomagnetism and K-Ar dating of some Franklin lavas and diabases, Victoria Island: Canadian Journal of Earth Sciences, v. 12, p. 1439-1447.

Park, J.K.,1981a, Analysis of the multicomponent magnetization of the Little Dal Group, Mackenzie Mountains, Northwest Territories, Canada: Journal of Geophysical Research, v. 86 p. 5134-5146.

Park, J.K., 1981b, Paleomagnetism of the Late Proterozoic sills in the Tsezotene Formation, Mackenzie Mountains, Northwest Territories, Canada: Canadian Journal of Earth Sciences, v. 18, p. 1572-1580.

Park, J.K., and Aiken, J.D., 1986, Paleomagnetism of the Late Proterozoic Tsezotene Formation of northwestern Canada: Journal of Geophysical Research, v. 91, p. 4955- 4970.

Pasteels, P., and Silver, L.T., 1966, Geochronologic investigations in the crystalline rocks of the Grand Canyon

Arizona [abs.]: Geological Society of America Special Paper 87, p. 124.

Peirce, H.W., 1984, The Mogollon Escarpment: Arizona Bureau of Geology and Mineral Technology Fieldnotes, v. 14, no. 2, p. 8-11.

Peirce, H.W., 1985, Arizona's backbone: the transition zone, Phoenix, Arizona, Arizona Bureau of Geology and Mineral Technology, Fieldnotes, v. 15, no. 3, p. 1-6.

Peirce, H.W., Damon, P.E., and Shafiqullah, M.uhammed, 1975, An Oligocene(?) Colorado Plateau edge in Arizona: Tectonophysics, v. 61, p. 1-24.

Perry, F.V., Baldridge, W.S., and DePaolo, D.J., 1988, Chemical and isotopic evidence for lithosphere thinning beneath the Rio Grande rift: Nature, v. 332, p. 432-434.

Peterson, Fred, 1984, Fluvial sedimentation on a quivering craton, influence of slight crustal movements on fluvial processes, Upper Jurassic Morrison Formation, western Colorado Plateau: Sedimentary Geology, v. 38, p. 21-49.

Peterson, Fred, 1986, Jurassic paleotectonics in the west-central part of the Colorado Plateau, Utah, and Arizona, in Peterson, J.A., ed., Paleotectonics and sedimentation in the Rocky Mountain region, United States: American Association of Petroleum Geologists Memoir 41, p. 563-596.

Peterson, J.A., and Smith, D.L., 1986, Rocky Mountain paleogeography through geologic time, in Peterson, J.A., ed., Paleotectonics and sedimentation in the Rocky Mountain region, United States: American Association of Petroleum Geologists Memoir, p. 3-19.

Phoenix, D.A., 1963, Geology of the Lees Ferry area, Coconino County, Arizona: U.S. Geological Survey Bulletin 1147, 86 p.

Pierce, K.L., 1979, History and dynamics of glaciation in the northern Yellowstone National Park area: U.S. Geological Survey Professional Paper 729-F, 90 p.

Potochnik, André., 1987, Fluvial response to the Laramide-Mid-Tertiary tectonic transition along the southeastern margin of the Colorado Plateau [abs]: Geological Society of America Abstracts with Programs, v. 19, no. 7, p. 809.

Powell, J.W., 1874, Report of exploration in 1873 of the Colorado of the West and its tributaries: Smithsonian Institution Annual Report, p. 18-21.

Powell, J.W., 1876, Report on the geology of the eastern portion of the Uinta Mountains and a region of country adjacent thereto: Washington, D.C., U S Geological and Geographical Survey of the Territories (Powell), 218 p.

Price, W.E., Jr., 1950, Cenozoic gravels on the rim of Sycamore Canyon, Arizona: Geological Society of America Bulletin, v .62, p. 501-508.

Purucker, M.E., Elston, D.P.,and Shoemaker, E.M., 1980, Early acquisition of characteristic magnetization of red beds of the Moenkopi Formation (Triassic), Gray Mountain, Arizona: Journal of Geophysical Research, v. 85, p. 997-1127.

Racey, J.S., 1974, Conodont biostratigraphy of the Redwall Limestone at east-central Arizona: Tempe, Arizona, Arizona State University, M.S. Thesis, 199 p.

Ragan, D.M., and Sheridan, M.F., 1970, The Archean rocks of the Grand Canyon, Arizona [abs.]: Geological Society of America Abstracts with Programs, v. 2, no. 2, p. 132-133.

Randle, T.J., and Pemberton, E.L., 1987, Results and analysis of STARS modeling efforts of the Colorado River in Grand Canyon: U.S. Bureau of Reclamation, Glen Canyon Environmental Studies Report, no. 11, 41 p.

Rapp, S.D., MacFadden, B.J., and Schiebout, J.A., 1983, Magnetic polarity stratigraphy of the early Tertiary Black Peaks Formation, Big Bend National Park, Texas: Journal of Geology, v. 91, p. 555-572.

Rawson, R.R., and Turner, C.E., 1974, The Toroweap Formation, a new look, in Swann, G.A., Karlstrom, T.N.V., and Eastwood, R.L., eds., Guidebook to the geology of northern Arizona, with notes on archeology and paleoclimate, Pt. 1, Regional studies: Geological Society of America, Rocky Mountain Section Meeting, Flagstaff, Arizona, p. 155-190.

Rawson, R.R., and Turner-Peterson, C.E., 1979, Marine-carbonate, sabkha and eolian facies transitions within the Permian Toroweap Formation, northern Arizona, in Baars, D.L., ed., Permianland: Durango, Colorado, Four Corners Geological Society Guidebook, 9th Field Conference, p. 87-99.

Reches, Ze'ev, 1978, Development of monoclines, part I, structure of the Palisades Creek branch of the east Kaibab monocline, Grand Canyon, Arizona: in Matthews, Vincent III, ed., Laramide folding associated with basement block faulting in the western United States, Geological Society of America Memoir 151, p. 235-271.

Reches, Ze'ev, and Johnson, A.M., 1978, Development of monocline, part II, theoretical analysis of monoclines: in Matthews, Vincent III, ed., Laramide folding associated with basement block faulting in the western United States, Geological Society of America Memoir 151, p. 235-271.

Reed, V.S., 1974, Stratigraphy of the Hakatai Shale, Grand Canyon, Arizona [abs,]: Geological Society of America Abstracts with Programs, v. 6, no. 5, p, 469.

Reiche, P., 1937, The toreva block, a distinctive landslide type: Journal of Geology, v. 45, p. 538-548.

Reynolds, M.W., and Elston, D.P., 1986, Stratigraphy and sedimentation of part of the Proterozoic Chuar Group, Grand Canyon, Arizona [abs]: Geological Society of America Abstracts with Programs, v. 18, no. 5, p. 405.

Reynolds, M.W., Palacas, J.G., and Elston, D.P., 1988, Potential petroleum source rocks in the Late Proterozoic Chuar Group, Grand Canyon, Arizona, in U.S. Geological Survey Circular 1025, p. 49-50.

Reynolds, S.J., Florence, F.P., Welty, J.W., Roddy, M.S., Currier, D.A., Anderson, A.V., and Keith, S.B., 1986, Compilation of radiometric age determinations in Arizona: Arizona Bureau of Geology and Mineral Technology Bulletin 197, 258 p.

Ritter, S.M., 1983, Conodont biostratigraphy of Devonian-Pennsylvanian rocks, Iceberg Ridge, Mojave Country, northwest Arizona: Provo, Utah, Brigham Young University, M.S. Thesis, 54 p.

Robinson, H.H., 1913, The San Franciscan volcanic field: U.S. Geological Survey Professional Paper 76, 213 p.

Roller, J.A., 1987, Fracture history of the Redwall Limestone and lower Supai Group, western Hualapai Indian Reservation, northwestern Arizona: U.S. Geological Survey Open-File Report 87-359, Flagstaff, Arizona, 33 p.

Rosholt, J.N., 1985, Uranium-trend systematics for dating Quaternary sediments: U.S. Geological Survey Open-File Report 85-298, Denver, Colorado, 48 p.

Rosholt, J.N., Emslie, Steve, and Stevens, Larry, 1987, Paleoclimatic and paleohydrologic significance of uranium-series ages of anhydrite and gypsum in caves, Grand Canyon National Park, Arizona [abs.]: Geological Society of America Abstracts with Programs, v. 19, no. 5, p. 330.

Ross, R.J., Jr., Naeser, C.W., Izett, G.A., Obradovich, J.D., Bassett, M.G., Hughes, C.P., Cocks, L.R.M., Dean, W.T., Inghan, J.K., Jenkins, C.J., Richards, R.R., Sheldon, P.R., Toghill, P., Whittington, H.B., and Zalasiewicz, J., 1982, Fission-track dating of British Ordovician and Silurian stratotypes: Great Britain Cambridge University, Geological Magazine, v. 119, p. 135-153.

Rowley, P.D., Anderson, J.J., Williams, P.L., and Fleck, R.J., 1978, Ages of structural differentiation between the Colorado Plateau and Basin and Range provinces in southwestern Utah: Geology, v. 6, p. 51-55.

Roy, J.L., 1977, La position stratigraphique déterminee paléomagnétiquement de sédiments carbonifères de Minudie

Point, Nouvelle Ecosse: à propos de l'horizon repère magnétique du carbonifère: Canadian Journal of Earth Sciences, v. 14, p. 1116-1127.

Roy, J.L., and Park, J..K.,1969, Paleomagnetism of the Hopewell Group, New Brunswick: Journal of Geophysical Research, v. 74, p. 594-604.

Roy, J.L., and Park, J.K., 1974, The magnetization process of certain red-beds: vector analysis of chemical and thermal results: Canadian Journal of Earth Sciences, v. 11, p. 437-471.

Roy, J.L., and Robertson, W.A., 1968, Evidence for diagenetic remanent magnetization in the Mariongouin Formation: Canadian Journal of Earth Sciences, v. 5, p. 275-285.

Roy, J.L., and Robertson, W.A., 1978, Paleomagnetism of the Jacobsville Formation and the apparent polar path for the interval 1100-1670 m.y. for North America: Journal of Geophysical Research, v. 83, p. 1239-1304.

Rusho, W.L and Crampton, C.G., 1981, Desert river crossing: Salt Lake City, Utah, Peregrine Smith, Inc, 126 p.

Sbar, M.L., Barazang, M., Dorman, J., Scholz, C.H., and Smith, R.B., 1972, Tectonics of the intermountain seismic belt, western United States, microearthquake seismicity and composite fault plane solutions: Geological Society of America Bulletin, v. 83, p. 13-28.

Schmidt, J.C., 1986, Changes in alluvial deposits, upper Grand Canyon, Fourth Federal Interagency Sedimentation Conference, v. 2, Las Vegas, Nevada, March 24-27, 1986, Proceedings, p. 2-48 to 2-57.

Schmidt, J.C., 1987, Geomorphology of alluvial sand deposits, Colorado River, Grand Canyon National Park, Arizona: Baltimore, Maryland, The Johns Hopkins University, Ph.D. Dissertation, 199 p.

Schmidt, J.C., and Graf, J.B., 1988, Aggradation and degradation of alluvial sand deposits, 1965-1986, Colorado River, Grand Canyon National Park, Arizona: U.S. Geological Survey Open-File Report 87-555, 120 p.

Schuster, R.L., 1986, Landslide dams: processes, risk, and mitigation: American Society of Civil Engineers, Conference Proceedings, Seattle, Washington, April 7, 1986, Geotechnical Special Publication no. 3.

Sears, J.W., 1973, Structural geology of the Precambrian Grand Canyon Series, Arizona: Laramie Wyoming, University of Wyoming, M.S.Thesis, 100 p.

Shafiqullah, Muhammed, Damon, P.E., Lynch, D.J., Reynolds, S.J., Rehrig, W.A., and Raymond, R.H., 1980, K-Ar geochronology and geologic history of southwestern Arizona and adjacent areas: Arizona Geological Society Digest, V. XII, p. 201-260.

Sharp, R.F., 1940, A Cambrian slide breccia, Grand Canyon, Arizona: American Journal of Science, 238, p. 658-672.

Shoemaker, E.M., Squires, R.L., and Abrams, M.J., 1975, The Bright Angel, Mesa Butte, and related fault systems of northern Arizona: California Institute of Technology, Jet Propulsion Laboratory Technical Report 32-1597, p. 23-41

Shoemaker, E.M., and Stephens, H.G., 1969, The Green and Colorado River canyons observed from the footsteps of Beaman and Hillers 97 years after Powell [abs.]: Geological Society of America Abstracts with Programs, part 5, Rocky Mountain Section, p. 73.

Shuey, R.T., Schellinger, D.K., Johnson, E.G., Allen, L.B., 1973, Aeromagnetics and transition between the Colorado Plateau and Basin Range provinces: Geology, v. 1, p. 107-110.

Silver, L.T., Bickford, M.E., Van Schmus, W.R., Anderson, J.L., Anderson, T.H., and Medaris, L.G., 1977, The 1.4-1.5 b.y. transcontinental anorogenic plutonic perforation of North America [abs]: Geological Society of America Abstracts with Programs, v. 9, no. 7, p. 1176-1177.

Sinno, Y.A., Daggett, P.H Keller, G.R., Morgan, P., and Harder, S.H., 1986, Crustal structure of the southern Rio Grande rift determined from seismic refraction profiling: Journal of Geophysical Research, v. 91, p. 6143-6156.

Skipp, B.A.L., 1969, Foraminifera in McKee, E.D., and Gutschick, R.C., eds., History of the Redwall limestone of northern Arizona: Geological Society of America Memoir 114, p. 173-255.

Smiley, T.L., 1984, Climatic change during landform development, in Smiley, T.L., Nations, J.D., Péwé, T.L., and Schafer, J.P., eds., Landscapes of Arizona: New York, University Press of America, p. 55-77.

Smith, R.B., 1978, Seismicity, crustal structure, and intraplate tectonics of the interior of the western cordillera, in Smith, R.B., and Eaton, G.P., eds., Cenozoic tectonics and regional geophysics of the western Cordillera: Geological Society of America Memoir 152, p. 111-144.

Smith, R B., and Sbar, M.L., 1974, Contemporary tectonics and seismicity of the western United States with emphasis on the intermountain seismic belt: Geological Society of America Bulletin, v. 85, p. 1205-1218.

Spamer, E.E., 1983, Geology of the Grand Canyon: An annotated bibliography, 1857-1982 with an annotated catalogue of Grand Canyon type fossils: Geological Society of America, Microform Publication no. 13, 326 p.

Steiner, M.B., 1988, Paleomagnetism of the late Pennsylvanian and Permian: a test of the rotation of the Colorado Plateau: Journal of Geophysical REsearch, v. 93, p. 2201-2215.

Stephens, H.G., and Shoemaker, E.M., 1987, In the footsteps of John Wesley Powell: Boulder, Colorado, Johnson Books, 286 p.

Stevens, Larry, 1983, The Colorado River in Grand Canyon--a comprehensive guide to its natural and human history: Flagstaff, Arizona, Red Lake Books, 107 p.

Stevenson, G.M., 1973, Stratigraphy of the Dox Formation, Precambrian, Grand Canyon, Arizona: Flagstaff,Arizona, Northern Arizona University, M.S. Thesis, 225 p.

Stevenson, G.M., and Beus, S.S., 1982, Stratigraphy and depositional setting of the upper Precambrian Dox Formation in Grand Canyon: Geological Society of America Bulletin, v. 93, p. 163-179.

Stewart, J.H., Anderson, T.H., Haxel, G.B., Silver, L.T., and Wright, J.E., 1986, Late Triassic paleogeography of the southern Cordillera, the problem of a source for voluminous volcanic detritus in the Chinle Formation of the Colorado Plateau region: Geology, v. 14, p. 567-570.

Stewart, J.H., Poole, F.G., and Wilson, R.F., 1972, Stratigraphy and origin of the Chinle Formation and related Upper Triassic strata in the Colorado Plateau region: U.S. Geological Survey Professional Paper 690, 336 p.

Strand, R.I., 1986, Water related sediment problems: Denver, Colorado, U.S. Bureau of Reclamation, Water Systems Management Workshop 1986, Session 4-1, p. 1-30.

Sturgul, J.R., and Grinshpan, Z., 1975, Finite element model for possible isostatic rebound in the Grand Canyon: Geology, v. 3, p. 169-171.

Sutphin, H.B., 1986, Occurrence and structural control of collapse features on the southern Marble Plateau, Coconino County, Arizona: Flagstaff, Arizona, Northern Arizona University, M.S. Thesis, 139 p.

Sutphin, H.B., and Wenrich, K.J., 1988, Map showing structural control of breccia pipes on the southern Marble Plateau, Arizona: U.S. Geological Survey Miscellaneous Investigations Series Map I-1778, 6 p., 2 plates, scale 1:50,000.

Sutton, R.L., 1974, The geology of Hopi Buttes, in Karlstrom, T.N.V., Swann, G.A., and Eastwood, R.L., eds., Geology of northern Arizona with notes on archaeology and paleoclimate, Part ll, area studies and field guides: Geological Society of America, Rocky Mountain Section Meeting, Flagstaff, Arizona,

p. 647-672.

Szabo B.J., and Rosholt, J.N., in press, Dating methods applicable to the Quaternary, in Morrison, R.B., ed., Quaternary non-glacial geology-Conterminous United States: Geological Society of America, Decade of North American Geology Volume K-2

Turner, R.M., and Karpiscak, M.M., 1980, Recent vegetation changes along the Colorado River between Glen Canyon Dam and Lake Mead, Arizona: U.S. Geological Survey Professional Paper 1132, 125 p.

Twenter, F.R., 1962, Geology and promising areas for ground water development in the Hualapai Indian Reservation, Arizona: U.S. Geological Survey Water Supply Paper 1576A, 38 p.

U.S. Bureau of Reclamation, 1988, Glen Canyon environmental studies final report: [Salt Lake City, Utah, Bureau of Reclamation], 1 volume.

Van Gundy, C.E., 1934, Some observations of the Unkar Group of the Grand Canyon Algonkian: Grand Canyon Nature Notes, v. 9, p. 338-349.

Van Gundy, C.E., 1951, Nankoweap group of the Grand Canyon Algonkian of Arizona: Geological Society of America Bulletin, v. 62, p. 953-959.

Vidal, Gonzalo, 1986, Acritarch-based biostratigraphic correlations and the upper Proterozoic in Scandinavia, Greenland, and North America [abs]: Geological Society of America Abstracts with Programs, v. 18, no. 5, p. 420.

Vidal, Gonzalo, and Ford, T.D., 1985, Microbiotas from the late Proterozoic Chuar Group (northern Arizona) and Uinta Mountain Group (Utah) and their chronostratigraphic implications: Precambrian Research, v. 28, p. 349-389.

Vidal, Gonzalo, and Knoll, A. H., 1983, Proterozoic plankton: in Medaris, L.G., Jr., Byers, C.W., Mickelson, D.M., and Shanks, W.C., eds, International Proterozoic symposium (1981: University of Wisconsin--Madison) Proterozoic geology: Selected papers from an international Proterozoic symposium: Geological Society of America Memoir 161, p. 265-277.

Waesche, H.H., 1934, The Grandview Mine Copper Project, history, geology, and mining methods: Grand Canyon Nature Notes, Grand Canyon Natural History Association, v. 8, no. 12, p. 250-258.

Walcott, C.D., 1883, Pre-Carboniferous strata in the Grand Canyon of the Colorado, Arizona: American Journal of Science, 26, p. 437.

Walcott, C.D., 1890, Study of a line of displacement in the Grand Cañon of the Colorado, in northern Arizona: Geological Society of America Bulletin, v. 1, p. 49-64.

Walcott, C.D., 1894, Pre-Cambrian igneous rocks of the Unkar terrane, Grand Canyon of the Colorado, Arizona, with notes on the petrographic character of the lavas by J.P. Iddings: U.S. Geological Survey Annual Report 14, part 2, p. 497-524.

Walcott, C.D., 1895, Algonkian rocks of the Grand Canyon of the Colorado: Journal of Geology, v. 3, p. 312-330.

Walcott, C.D., 1899, Precambrian fossiliferous formations: Geological Society of America Bulletin, v. 10, p. 199-244.

Walen, M.B., 1973, Petrogenesis of the granitic rocks of part of the Upper Granite Gorge, Grand Canyon, Arizona: Bellingham, Washington, Western Washington University, M.S. Thesis.

Webb, R.H., 1987, Occurrence and geomorphic effects of streamflow and debris flow floods in northern Arizona and southern Utah, in Mayer, L., and Nash, D., eds., Catastrophic flooding: Boston, Allen and Unwin, p. 247-265.

Webb, R.H., Pringle, P.T., Reneau, S.L., and Rink, G.R., 1988, Monument Creek debris flow, 1984: Implications for formation of rapids on the Colorado River in Grand Canyon National Park: Geology, v. 16, p. 50-54.

Webb, R.H., Pringle, P.T., and Rink, G.R., 1987, Debris flows in tributaries of the Colorado River, Grand Canyon National Park, Arizona: U.S. Geological Survey Open-File Report 87-118, 64 p.

Wegmann, C.E., 1932, Note sur le boudinage: Bulletin de Societie Geologique de France, series 5, p. 477-489.

Wenrich, K.J., 1985, Mineralization of breccia pipes in northern Arizona: Economic Geology, v. 80, p. 1722-1735.

Wenrich, K.J., Billingsley, G.H., and Huntoon, P.W., 1986a, Breccia pipe and geologic map of the northeastern Hualapai Indian Reservation and vicinity, Arizona: U.S. Geological Survey Open-File Report 86-458a, Flagstaff, scale 1:48,000, 29 p.

Wenrich, K.J., Billingsley, G.H., and Huntoon, P.W., 1986b, Breccia pipe and geologic map of the northwestern Hualapai Indian Reservation and vicinity, Arizona: U.S. Geological Survey Open-File Report 86-458C, 32 p., scale 1:48,000.

Wenrich, K.J., and Sutphin, H.B., 1988, Lithotectonic controls necessary for formation of a uranium-rich, solution-collapse breccia-pipe province, Grand Canyon region, Arizona: International Atomic Energy Agency, technical committee meeting on netallogenesis of uranium deposits, Vienna, Austria, March 1987.

Wernicke, B., 1981, Low angle normal faults in the Basin and Range Province nappe tectonics in an extending orogen: Nature, v. 291, p. 645-648.

Wheeler, R.B., and Kerr, A.K., 1936, Preliminary report on the Tonto Group of the Grand Canyon, Arizona: Grand Canyon Natural History Association Bulletin 5, p. 1-16.

White, David, 1929, Flora of the Hermit Shale, Grand Canyon, Arizona: Carnegie Institution of Washington, Publication 405, 221 p.

Wilson, E.D., 1962, A résumé of the geology of Arizona: Phoenix, Arizona, Arizona Bureau of Mines Bulletin 171.

Wilson, E.D., and Moore, R.T., 1959, Structure of Basin and Range Province in Arizona, in Heindl, L.A., ed., Southern Arizona guidebook II: Arizona Geological Society, p. 89-105.

Wilson, R.P., 1986, Sonar patterns of Colorado riverbed, Grand Canyon: Fourth Federal Interagency Sedimentation Conference, v. 2, Las Vegas, Nevada, 1986, Proceedings, p. 5-133 to 5-142.

Woodward-Clyde Consultants, 1982, Geologic characterization report for the Paradox Basin study region, Utah study areas, volume 1, regional overview: Battelle Memorial Institute Office of Nuclear Waste Isolation, ONW1-290.

Yochelson, E.L., 1979, Charles D. Walcott--America's pioneer in Precambrian paleontology and stratigraphy, in Kupsch, W.O., and Sarjeant, W.A.S., eds., History of Concepts in Precambrian Geology: Geological Association of Canada Special Paper 19, p. 261-292.

Young, R.A., 1966, Cenozoic geology along the edge of the Colorado Plateau in northwestern Arizona: St. Louis, Missouri, Washington University, Ph.D. Dissertation, 167 p.

Young, R.A. 1970, Geomorphological implications of pre-Colorado and Colorado tributary drainage in the western Grand Canyon region: Plateau, Museum of Northern Arizona, Flagstaff, Arizona, v. 42, p. 107-117.

Young, R.A., 1978, Discussion of Cambrian stratigraphy nomenclature and ground water prospecting failures on the Hualapai Plateau, Arizona: Ground Water, v. 16, p. 287-289.

Young, R.A., 1979, Laramide deformation, erosion and plutonism along the southwestern margin of the Colorado Plateau: Tectonophysics, v. 61, p. 25-47.

Young, R.A., 1982, Paleogeomorphic evidence for the structural history of the Colorado Plateau margin in Arizona, in Frost, E.G., and Martin, D.L., eds., Mesozoic-Cenozoic tectonic evolution of the Colorado River region, California, Arizona, and Nevada: San Diego, California Cordilleran Publishers, p. 29-39.

Young, R.A., 1985, Geomorphic evolution of the Colorado Plateau margin in west-central Arizona, a tectonic model to distinguish between the causes of rapid, symmetrical scarp retreat and scarp dissection, *in* Hack, J.T. and Morisawa, M., eds., Tectonic geomorphology, Binghampton Geomorphology Symposium in Geomorphology International Series 15: London, Allen and Unwin, p. 261-278.

Young, R.A., 1987, Landscape development during the Tertiary, in Colorado Plateau, *in* Graf, W.L., ed., Geomorphic Systems of North America: Geological Society of America Centennial Special Volume 2, p. 265-276.

Young, R.A., 1988, A chronology of episodic Laramide tectonism in northwestern Arizona [abs.]: Geological Society of America Abstracts with Programs, v. 20, no. 3, p. 244.

Young, R.A., and Brennan, W.J., 1974, Peach Springs Tuff, its bearing on structural evolution of the Colorado Plateau and development of Cenozoic drainage in Mohave County, Arizona: Geological Society of America Bulletin, v. 85, p. 83-90.

Young, R.A., Elston, D.P., and McKee, E.H., 1987, Laramide deformation and sedimentation in northern Arizona: Post Turonian through Ypresian events coincident with Cordilleran Fold and Thrust Belt Chronology [Abs.]: Geological Society of America Abstracts with Programs, v. 19, no. 7, p. 903.

Young, R.A., and Hartman, J.H., 1984, Early Eocene fluviolacustrine sediments near Grand Canyon, Arizona: Evidence for Laramide drainage across northern Arizona into Southern Utah [abs.]: Geological Society of America Abstracts with Programs, v. 16, no. 6, p. 703.

Young, R.A., and Huntoon, P.W., 1987, Laramide canyon remnants along the Hurricane fault zone: tectonic disruption by monoclinal folding [abs.]: Geological Society of America Abstracts with Programs, v. 19, no. 7, p. 903.

Young, R.A., and McKee, E.H., 1978, Early and middle Cenozoic drainage and erosion in west-central Arizona: Geological Society of America Bulletin, v. 89, p. 1745-1750.

Zeitler, P.K., 1985, Cooling history of the NW Himalaya, Pakistan: Tectonics, v. 4, p. 127-151.